# A TEXTBOOK
# OF
# ENVIRONMENTAL CHEMISTRY AND POLLUTION CONTROL

## (with Energy, Ecology, Ethics and Society)

# A TEXTBOOK OF ENVIRONMENTAL CHEMISTRY AND POLLUTION CONTROL

## (with Energy, Ecology, Ethics and Society)

**As per the syllabi of RGPV Bhopal, S.V. Technical University Bhilai, UPTU Lucknow, PTU Chandigarh, Anna University Tamil Nadu, JNTU Hyderabad, Anantpur & Kakinada, VTU Belgaum (Karnataka), BPTU Bhubaneshwar, GTUAhmedabad, RTU Jaipur, WBUT Kolkata, Nagpur University, Mumbai University, Pune University, Marathwada University Aurangabad, North Maharashtra University Jalgaon, Amravati University, Cochin University and all other Indian Universities**

**Dr. S.S. Dara**

M.Sc., Ph.D.

*Former Professor and Head*

Department of Applied Chemistry

Visvesvaraya National Institute of Technology, Nagpur

**Dr. D.D. Mishra**

M.Sc., Ph.D., F.I.C.S., M.N.A.Sc.

*Principal*

Technocrats Institute of Technology (C), Bhopal

**S Chand And Company Limited**

**(ISO 9001 Certified Company)**

**S Chand And Company Limited**

**(ISO 9001 Certified Company)**

***Head Office:*** D-92, Sector–2, Noida – 201301, U.P. (India), Ph. 91-120-4682700

***Registered Office:*** A-27, 2nd Floor, Mohan Co-operative Industrial Estate, New Delhi – 110 044, Phone: 011-49731800

www.**schandpublishing.com**; e-mail: **info@schandpublishing.com**

**Marketing Offices:**

Chennai : Ph: 23632120; chennai@schandpublishing.com
Guwahati : Ph: 2738811, 2735640; guwahati@schandpublishing.com
Hyderabad : Ph: 40186018; hyderabad@schandpublishing.com
Jalandhar : Ph: 4645630; jalandhar@schandpublishing.com
Kolkata : Ph: 23357458, 23353914; kolkata@schandpublishing.com
Lucknow : Ph: 4003633; lucknow@schandpublishing.com
Mumbai : Ph: 25000297; mumbai@schandpublishing.com
Patna : Ph: 2260011; patna@schandpublishing.com



*First Edition 1993*
*Subsequent Editions and Reprints 1995, 97, 98, 2000, 2001, 2002, 2004 (Twice), 2005, 2006, 2007, 2008, 2009 (Twice), 2010,*
*Ninth Revised and Enlarged Edition 2011*
*Reprints 2012, 2015, 2018, 2020 (Twice), 2021, 2023 (Twice)*

***Reprint 2024***

**ISBN** : 978-81-219-0883-2 **Product Code:** H3ECP68ENVS10ENAI11R

PRINTED IN INDIA

By Vikas Publishing House Private Limited, Plot 20/4, Site-IV, Industrial Area Sahibabad, Ghaziabad – 201 010 and Published by S Chand And Company Limited, A-27, 2nd Floor, Mohan Co-operative Industrial Estate, New Delhi – 110 044.

# PREFACE TO THE NINTH REVISED AND ENLARGED EDITION

As we know human resource is not only the backbone of a country but also a source of its development. For that, it is essential to acquainted with the new inventions, researches and necessities with time. Books are the main source for that. Hence it is necessary to revise and modify the syllabus from time to time in this age of science. Probably, keeping this fact in view, the UGC and AICTE are always doing this job. They have revised the syllabi enormously. In light of this, it was necessary to revise the famous book "A Textbook of Environmental Chemistry and Pollution Control" written by Late Dr. S.S. Dara, because it was not fulfilling the requirements of the students as per the latest curricula.

In this ninth revised and enlarged edition of the book, all essential changes have been made as per the revised curricula of all Indian technical and non-technical universities. Keeping in view of the academic requirements of the engineering colleges, several new topics like **ecosystem, social issues and the environment; marine, thermal and nuclear pollutions; natural resources; human population and the environment; and ethics and moral values are added in this edition**. These new topics and additional data, diagrams are of immense use. The new topics added in this edition are written in a simple and an easy-to understand language keeping in view the mental level of the students so that the students may acquire the complete knowledge of the subject. The text of this edition is more justifiable. In the book, case studies and examples are also given at appropriate places. In this way, now the book is complete, clear, attractive and organized as per the latest syllabi of different universities.

While preparing the manuscript, I have consulted i.e., books, journals, websites, etc. a large number of works, and so much originality cannot be claimed in the contents and, at best, it is a compiled work with a novel presentation. I take this opportunity to place on record my indebtedness to all those books, journals, websites which I have consulted during the preparation of the manuscript.

I offer my thanks to the which management and editorial team of S. Chand & Company Pvt. Ltd., for their contribution in converting the book into revised edition.

**Dr. D.D. MISHRA**

# ENVIRONMENTAL CHEMISTRY AND POLLUTION CONTROL

## CONTENTS

# CONTENTS

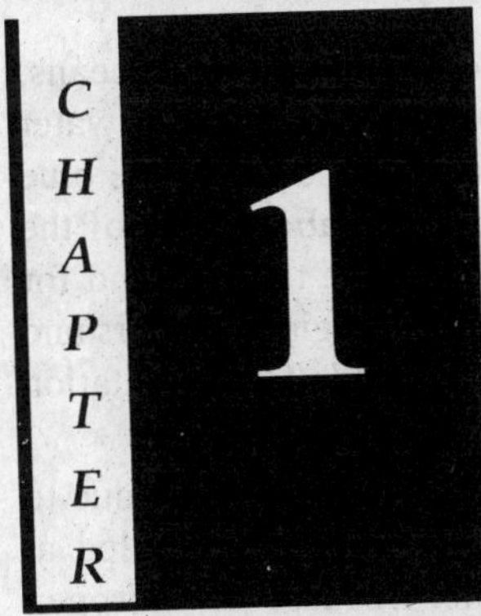

# Environmental Chemistry

## ENVIRONMENTAL CHEMISTRY

Environmental Chemistry deals with the study of the various chemical phenomena taking place in the environment. In a broader perspective, it comprises of the study of the chemical species existing in the various segments of the environment, their sources, pathways, reactions and their consequences on the activities of human beings and other life-forms. Thus, environmental chemistry may be considered as a multi-disciplinary study, involving physical and life-sciences, meteorology, agriculture, public health, engineering etc. The basic concepts of environmental chemistry is interesting not only to the scientists engaged in various scientific and engineering activities but also to the personnel involved in resource planning and material management. It is now universally realised that any future developmental activity has to be viewed in the light of its ultimate environmental impact. The tremendous increase in industrial activity during the last few decades and the release of obnoxious industrial wastes into the environment, have been of considerable concern in recent years from the point of view of environmental pollution. Environmental pollution on one hand and deforestation and population explosion on the other, are threatening the very existence of life on the earth. This situation can improve only if people from all walks of life realize the importance of environmental protection. Hence, environmental education (which includes basic concepts of environmental chemistry) at all levels of non-formal and formal education, is of paramount importance.

## ENVIRONMENTAL SEGMENTS

The environment may broadly be considered to comprise of the following four segments.

(1) Lithosphere (2) Hydrosphere

(3) Atmosphere (4) Biosphere.

**(1) Lithosphere.** The mantle of rocks constituting the earth's crust is called lithosphere. For all practical purposes, the soil covering the rocks (which results from physical, chemical and biological processes during weathering) is also considered to be an important part of the lithosphere. The soil mainly consists of complex mixture of inorganic and organic matter and water. The inorganic mineral constituents include complex mixture of silicates of Na, K, Ca, Al and Fe; oxides of Fe, Mn and Ti; and carbonates of Ca and Mg. The organic matter, which constitutes not more than about 5% of the soil, mainly determines the productivity of the soil. It consists of biologically active components such as polysaccharides, nucleotides, organo-phosphorous and organo-sulphur compounds, sugars and humic materials. The clay minerals and humus present in the soil possess a very high cation exchange capacity and thus help in supplying essential trace metals to the plants as nutrients (The cations in the soil are exchanged for $H^+$ from the carbonic acid present in the soil and thus supply the trace metals to the plants through the roots).

(2) **Hydrosphere.** This includes all the surface and ground water resources *viz*., oceans, seas, rivers, streams, lakes, reservoirs, glaciers, polar ice caps, ground-water and the water locked in rock-crevices and minerals lying deep below the earth's crust. Earth is called the blue planet, because about 80% of its surface is covered by water. But, however, about 97% of the earth's water resources is locked-up in the oceans and seas, which is too saline to drink and for the direct use for agricultural and industrial purposes. About 2.4% is trapped in giant glaciers and polar ice-caps. Thus not even 1% of the total world's water resources is available for exploitation by man for domestic, agricultural and industrial purposes.

Water exhibits unusual properties, as compared to the hydrides of VI Group elements, due to the presence of hydrogen bonding. Without hydrogen bonding, water would have boiled at – 100°C and in that case life on the earth would have been impossible. Because of hydrogen bonding, water is a liquid at room temperature, with melting point of 0°C and boiling point of 100°C at atmospheric pressure*.

(3) **Biosphere.** This is the region of the earth where life exists and includes a global girdle extending from about 10,000 m below sea-level to 6,000 m above sea-level. Thus, the biosphere covers the entire realm of living organisms and their interactions with the other segments of the environment, namely the lithosphere, the hydrosphere and the atmosphere.

(4) **Atmosphere.** The atmosphere comprises of a mixture of gases (*e.g.*, $N_2$, $O_2$, $CO_2$, Ar etc) and it extends upto about 500 kms above the surface of the earth. A constant exchange of matter takes place between the atmosphere, biosphere and hydrosphere. Their relative weights are of the following order :

| Biosphere | : | Atmosphere | : | Hydrosphere |
|---|---|---|---|---|
| 1 | : | 300 | : | 69,100 |

The weight of the atmosphere is approximately (4.5 to 5) $\times 10^{15}$ metric tons. The atmospheric temperature, pressure and density vary considerably with altitude. Thus, the atmospheric temperature varies from – 100°C to + 1200°C, depending upon the altitude. The atmospheric pressure at sea-level is 1 atmosphere, whereas at 100 km above sea-level, it drops to $3 \times 10^{-7}$ atmosphere. The atmospheric density at the surface of the earth is about 0.0013 g/cu.m. which decreases sharply with increasing altitude and gradually thins out into space. At about 600 km and above, the atoms and molecules describe free elliptical orbits in the earth's gravitational field.

The atmosphere surrounding the earth acts as a gaseous blanket, protecting the earth from dangerous cosmic radiations from outer space and helps in sustaining life on the earth. The atmosphere screens the dangerous u.v. radiations from the sun (< 300 nm) and transmits only the radiations in the range 300 nm to 2500 nm, comprising of near ultra-violet, visible and near infra-red radiations and ratio-waves (0.01 to $4 \times 10^5$ nm).

The atmosphere plays a vital role in maintaining the heat balance on the earth by absorbing the i. r. radiation received from the sun and re-emitted by the earth. In fact, it is this phenomenon, called "the Greenhouse effect", which keeps the earth warm enough to sustain life on the earth. Apart from this, the important gaseous constituents of the earth, *viz.*, $O_2$, $N_2$ and $CO_2$ play important roles in sustaining life on earth. Oxygen supports life on earth, Nitrogen is an essential macro-nutrient for plants (via nitrogen fixation and fertilizer manufacture) and Carbon dioxide is essential for photosynthetic activity of plants. Moreover, atmosphere is a carrier of water from oceans to land, which is so vital for the hydrologic cycle.

*It has density of 1.0 gm/$cm^3$; surface tension of 73 dynes per cm at 20°C; viscosity of 0.01 poise at 20°C; dielectric constant of 80; and specific heat of 1 cal per gram per °C.

Any major disturbance in the atmospheric composition, either by extraordinary or anthropogenic activities, may lead to disastrous consequences and may even endanger the very survival of life on the earth.

## COMPOSITION OF THE ATMOSPHERE

The composition of clean, dry air, near sea-level in given in Table 1.1.

**Table 1.1. Composition of clean, dry air, near sea-level**

| *Components* | *Content* | |
|---|---|---|
| | *per cent by volume* | *ppm* |
| **(a) Major components** | | |
| Nitrogen ($N_2$) | 78.09 | 7,80,900 |
| Oxygen ($O_2$) | 20.94 | 2,09,400 |
| Water Vapour ($H_2O$) | 0.1-5 | 1,000-50,000 |
| **(b) Minor Components** | | |
| Argon (Ar) | 0.934 | 9,340 |
| Carbon dioxide ($CO_2$) | 0.032 | 320 |
| **(c) Trace Components** | | |
| Neon (Ne) | 0.00182 | 18.2 |
| Helium (He) | 0.000524 | 5.24 |
| Methane ($CH_4$) | 0.00018 | 1.8 |
| Krypton (Kr) | 0.00011 | 1.1 |
| Nitrous oxide ($N_2O$) | 0.000025 | 0.25 |
| Hydrogen ($H_2$) | 0.00005 | 0.5 |
| Xenon (Xe) | 0.0000087 | 0.087 |
| Sulphur dioxide ($SO_2$) | 0.0000002 | 0.002 |
| Nitrogen dioxide ($NO_2$) | 0.0000001 | 0.001 |
| Ammonia ($NH_3$) | 0.000001 | 0.01 |
| Carbon monoxide (CO) | 0.000012 | 0.12 |
| Ozone ($O_3$) | 0.000002 | 0.02 |
| Iodine ($I_2$) | Traces | Traces |

The Proportions of the various components of the atmosphere more or less remain constant upto a height of about 16 km from the earth's surface. Above this height, gravitational separation begins, although this becomes significant above 130 km. The weight of the atmosphere is about 4.5 to 5 × $10^{15}$ metric tonnes, which is about one millionth of the total weight of the earth.

## ATMOSPHERIC STRUCTURE

The atmosphere, which extends upto about 500 km above the earth's surface, can be broadly divided into four major regions, with widely varying temperatures, even within each region. Some important characteristics of the four major atmospheric region *viz*., Troposphere, Stratosphere, Mesophere and Thermosphere, are summarised in Table 1.2 and Fig. 1.

**Table 1.2. Characteristics of the major regions of the atmosphere**

| Name of the region | Height above the earth's surface, km. | Temperature range °C | Major chemical species present |
|---|---|---|---|
| Troposphere | 0 – 11 | 15 to – 56 | $O_2$, $N_2$, $CO_2$, $H_2O$ |
| Stratosphere | 11 – 50 | – 56 to – 2 | $O_3$ |
| Mesosphere | 50 – 85 | – 2 to – 92 | $O^+_2$, $NO^+$ |
| Thermosphere | 85 – 500 | –92 to 1200 | $O^+_2$, $O^+$, $NO^+$ |

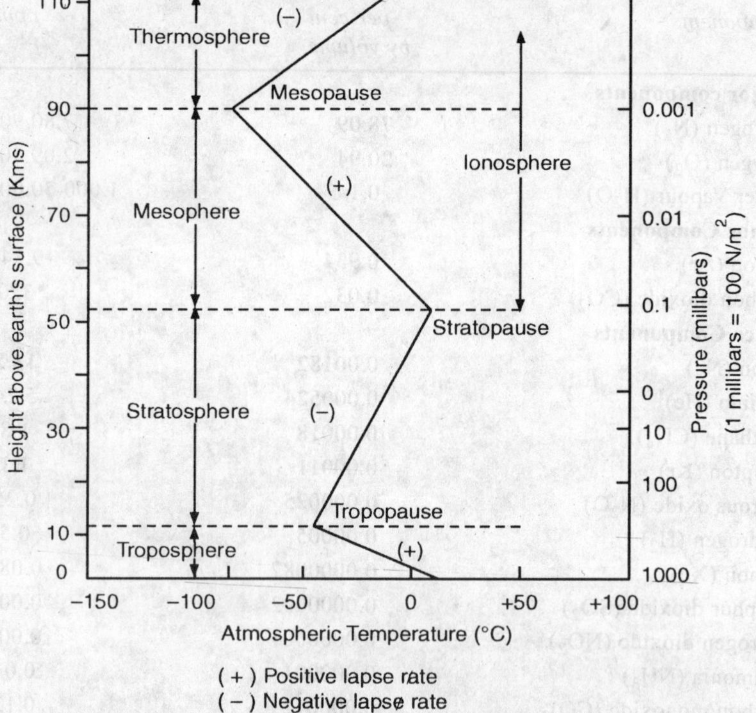


**Fig. 1.** Major atmospheric regions with temperature and pressure profile.

**(1) Troposphere.** This is the region nearest to the earth's surface and extends upto an altitude of 11 km. The upper limit may vary by a few kilometers, depending upon temperature, nature of the terrestrial surface and some other factors. The troposphere acounts for over 70% of the atmospheric mass. The composition of air in this region remains more or less constant in the absence of any significant air pollution. This is mostly due to the turbulence and constant circulation of air masses, as a result of convection currents arising from differential heating and cooling rates between the equator and the poles. The density of air, in this region, decreases exponentially with increasing altitude. The troposphere contains most of the water, cloud and particulate matter of the atmosphere

The temperature of air in the troposphere decreases fairly steadily with increasing altitude from the ground temperature to a temperature of about –56°C. It can be seen from Fig. 1 that the temperature-altitude curve then changes its slope (*i.e.*, the temperature starts increasing with increasing altitude) rather suddenly in a narrow transitional layer at the top of the troposphere, known as the "Tropopause", which is usually at an altitude of 10 km to 20 km. The temperature of the tropopause is the least at the equator.

The change of temperature with height is called the "lapse rate". The decrease of temperature with increasing altitude in the troposphere is called positive lapse rate. The transition from positive lapse rate to negative lapse rate at the tropopause marks what is called the "temperature inversion".

**(2) Stratosphere.** The region above the tropopause is called the stratosphere. In this region, the temperature-altitude curve shows a warming trend with increasing altitude *i.e.*, it exhibits a negative lapse-rate. The temperature in this stratospheric region continues to increase with height, until 50 kms, where the temperature attains a maximum of –2°C. This warming up tendency in the stratosphere is due to the absorption of solar *u.v.* radiation by ozone, whose concentration in this region is in the range of 1 to 5 ppm by volume and this is responsible for the negative lapse-rate. The air in this region is very dry and the clouds and convection currents from the troposphere normally do not penetrate into it.

The presence of ozone in the stratosphere serves as a shield to protect life on the earth from the harmful effects of the solar *u.v.* radiations. Moreover, it serves as a source of heat for separating the quiescent stratosphere from the turbulent troposphere.

Because of the quiescent nature of the stratosphere, the molecules and particles in the region have long residence times. This is significant from the point of view of atmospheric pollution because any pollutant reaching this region may spell long term global hazard, as compared to their impact in the troposphere, which is much denser and more turbulent.

The region immediately above the stratosphere (above 50 km height) is called "Stratopause" which is the second transitional layer that is relatively warm. This is not much coller than the earth's surface. It reflects sound waves from earth back to the surface.

**(3) Mesosphere.** This is the region above the "Stratopause" and extends upto 85 km height. In this region, the temperature again decreases with height *i.e*, it exhibits a positive lapse-rate. This is due to relatively low levels of ozone and other species that can absorb *u.v.* radiations from the sun. The temperature, at the top of the Mesosphere, reaches about –92°C. Immediately above the Mesophere is another transitional layer, called "Mesopause" which is the region of minimum or coldest temperature in the atmosphere (*i.e*, about –100°C)

**(4) Thermosphere.** This is the region immediately above the mesopause, where the temperature rises very rapidly with increasing altitude, exhibiting a negative lapse-rate. The maximum temperature that is attained in this region is about 1200°C. This region is characterised by low pressures and low densities. The atmospheric gases present in this region (e.g., oxygen and nitric oxide) absorb the solar radiations in the far ultra-violet region and undergo ionisation.

The region above the stratosphere, in the altitude range of 50 km to 100 km, is called "Ionosphere". In this region, positive ions e.g., $O_2^+$, $O^+$, $NO^+$ etc and electrons exists at significant levels. These charged species persist for long periods of time, without mutual neutralization, due to the rarefied conditions existing in the region.

The troposphere, the stratosphere and the mesosphere are fairly uniform in composition. From air pollution point of view, troposphere is of particular significance.

## RADIATION BALANCE OF THE EARTH

The earth continuously receives energy from the sun, a part of which is absorbed, while the remaining is emitted back into space. The earth maintains its heat balance within narrow limits, due to complex mechanisms of energy-transfer. Hence, optimum climatic conditions are maintained for supporting life.

The sun radiates energy, like a black body at 6000°K. Out of the solar radiations reaching the earth, 92% consists of radiations in the range 315 to 1400 nm. 45% of this is in the visible region (400 to 700 nm). The earth absorbs radiations mainly in the visible region and emits radiation in the infra-red region (2–40 μ, with maximum at 10 μ).

The quantity of solar energy per unit time passing through unit area at right angles to the direction of the solar beam measured just outside the earth's atmosphere, is called the "solar-constant" (S). The value of the solar flux reaching the earth's upper atmosphere is estimated to be about 1400 watts $m^{-2}$ $min^{-1}$. The heat equivalent of the solar radiation reaching the earth is estimated to be $2.68 \times 10^{24}$ joules per year. Out of this energy, 50% is reflected before it reaches the earth's surface, 15% is reflected by the earth's surface, 5.3% is absorbed by the soil, 27.8% is utilized for heating and evaporation of water and the remaining 1.9% is absorbed by the marine and land vegetation. The energy transport plays a vital role in maintaining the earth's radiation balance and this proceeds through various mechanisms, such as the following:

(1) Conduction of energy through the interaction of atoms, molecules and other species.

(2) Convection of energy via massive air circulation.

(3) Radiation of energy from the earth in the infrared region (2 to 40 μ).

(4) Re-absorption of most of the outgoing infra-red radiation (2 to 40 μ) by water vapour (4-8 μ), $CO_2$ (12 to 16.3 μ) and other gases such as $CH_4$ and re-emitting a part of this radiation to the earth's surface. This process, called the "green-house effect" (discussed later) is mostly responsible for maintaining the climatic features and temperature of the earth.

Further, the particulate matter in the atmosphere, which is released by natural forces (*e.g.*, wind, sea sprays and volcanoes) or by anthropogenic activities (*e.g.*, agricultural and industrial activities which release dust, fumes, smoke) also exert cooling or heating effect, depending upon the nature of the particles which may reflect, scatter or absorb radiations.

The fraction of the incident solar radiation that is reflected and scattered back into the space is called "albedo". Deforestation and the consequent soil erosion, industrial operations and other such activities also influence the earth's radiation balance by altering the 'albedo'.

## CHEMICAL SPECIES AND PARTICULATES PRESENT IN THE ATMOSPHERE

The various chemical species present in atmosphere have been summarised in Table–1.1 and Table –1.2.

**Ions.** The region of the atmosphere which is at an altitude of 50 km to 100 km, is called "Ionosphere" because appreciable levels of electrons and positive ions, such as $O_2^+$, $O^+$, $NO^+$, etc exist in this region with considerable residence times. The *u.v.* radiations from the sun are mostly responsible for the formation of the ionic species in ionosphere. During night times, when the *u.v.* radiations are not present, the positive ions tend to recombine with the free electrons, yielding the neutral species from which they are originated initially. This process occurs more rapidly in the lower regions of the ionosphere where the concentration of these species is relatively high.

**Radicals.** Apart from the ionic species, the atmosphere also consists of highly reactive free radicals, generated by photo-chemical reactions *e.g.*, HO•, $CH_3$•, $SO_2$•, ROO•, $NO_2$•, and HCO•. The free radicals, which may be organic or inorganic, comprise of atoms or groups of atoms with impaired electrons having short half-lives. Owing to their high reactivity, they interact with other chemical species available, to propagate chain reactions until chain termination takes place due to destruction of one of the free radicals in the chain, by any of the various possible mechanisms. Free radicals play significant role in the formation of photo-chemical smog.

**Particles.** Several types of particles exist in the atmosphere in varying sizes from 0.1 μ to 10 μ. Even pure air may contain several hundreds of particles per c.c., whereas hightly polluted air may contain about a lakh of particles per c.c. whereas highly polluted air may contain about a lakh of particles per c.c. Particles of colloidal dimensions are called "aerosol>". The word "particulates" is generally used to describe the particles present in aerosols. Aerosols of natural origin having diameter <0.2 μ are called "Aitkin" particles. The various particles existing in nature include dust, smoke, fog, pollen grains, volcanic ash and bacteria. Dusts, mist, smoke, smog and fumes may also result from anthropogenic activities. Inorganic particulate, such as iron oxide, calcium oxide etc., may result from combustion of coal and metallurgical operations. Particulate lead results from automobile exhausts. Aerosol mists are generated from oxidations of $SO_2$ to $SO_3$ which in presence of water vapour forms droplets of $H_2SO_4$.

Organic particulate mater arises from automobile exhausts, combustion of fuels and evaporation of organic matter from vegetation. Certain organic particulates contain polycyclic aromatic hydrocarbons (PAH) which are carcinogenic. They are generated from the pyrolysis of higher paraffins present in fuels and in some vegetable matter.

Atmospheric particulates may enter the respiratory tract and cause serious health hazards. Removal of obnoxious particulates from gaseous emissions is an important step in controlling air pollution.

The size and chemical characteristics of the particles in atmosphere are more significant than their concentration. The particles in the range of 0.1 μ to 1 μ manifest in several important effects in the atmosphere, such as the following:

(1) They play a vital role in maintaining the radiation balance and heat balance of the earth.

(2) They provide nuclei for condensation of water vapour.

(3) They are responsible for fog and cloud formation.

(4) They may bring about heterogeneous gas-phase reactions. The particles may absorb gases and catalyze some reactions. For example, the decomposition of ozone is catalyzed by a solid surface.

(5) Particles absorb and destroy free radicals, thereby reducing the rate of free radical chain reactions.

(6) The absorption of gases on some solid surfaces may change the absorption spectrum of the gas for sun light. For instance, oxygen absorbed on carbon particles might absorb sunlight more strongly than what free oxygen would do.

(7) Particles would help in several types of reactions in the atmosphere such as

(*a*) Catalysing the oxidation of $SO_2$ by $O_2$ or $O_3$ in aqueous droplets.

(*b*) Neutralization reactions between sulphuric acid droplets and lime stone dust.

(*c*) Reaction between $O_3$ or $NO_2$ with salt particles in water droplets.

(*d*) Photochemical reactions involving dusts and aerosols.

## REACTIONS IN THE ATMOSPHERE

The various chemical and photochemical reactions taking place in the atmosphere, mostly depend upon the temperature, composition, humidity and intensity of sun light. Thus the ultimate fate of a chemical species in the atmosphere depends upon these parameters. Photochemical reactions take place in the atmosphere by the absorption of solar radiations in the u.v. region. Absorption of photons by chemical species gives rise to electronically excited molecules, which can bring about certain reactions, which are not possible under normal laboratory conditions.

excepting at higher temperatures and in prèsence of chemical catalysts. The electronically excited molecules produced by the absorption of a photon may undergo any of the following changes:

(*a*) Reaction with other molecules on collision

(*b*) Polymerisation

(*c*) Internal rearrangement

(*d*) Dissociation

(*e*) De-excitation by fluorescence or de-activation by return to the original state.

Any of the first four changes stated above may serve as an initiating chemical step or a primary process. The three steps involved in an overall photochemical reaction are : (1) Absorption of radiation, (2) Primary reactions, and (3) Secondary reactions.

The various chemical species that can undergo photo-chemical reactions in the atmosphere include $NO_2$, $SO_2$, $HNO_3$, $N_2$, Ketones, $H_2O_2$, organic peroxides and several other organic compounds and aerosols such as metal oxides.

**Reactions involving oxides of nitrogen.** The oxides of nitrogen present in the atmosphere are $N_2O$, NO and $NO_2$. Nitrous oxide mainly originates from various microbiological processes whereas NO and $NO_2$ mainly originate from the combustion of fossil fuels and other anthropogenic activities. The NO and $NO_2$ together are referred to as $NO_x$, which is an important constituent of polluted air.

In the stratosphere, $N_2O$ undergoes photochemical decomposition to NO, which in turn, depletes the protective ozone layer:

$$N_2O + h\nu \longrightarrow N_2 + O$$

$$N_2O + O \longrightarrow NO + NO$$

$$NO + O_3 \longrightarrow NO_2 + O_2$$

The most important primary photochemical reaction is the dissociation of $NO_2$.

$$N_2O + h\nu \longrightarrow NO + O$$

The NO so formed may be oxidised by $O_3$ or more slowly by $O_2$, thus leading to a cyclic chain reaction. The chain may be broken only when the $NO_2$ is completely converted into $HNO_3$ by hydration and catalytic oxidation in presence of aerosols, fog or photochemical smog. The reactions may take place in the following sequence:

$$2\,NO + O_2 \longrightarrow 2\,NO_2$$

$$NO + O_3 \longrightarrow NO_2 + O_2$$

$$2NO_2 + O_3 \longrightarrow N_2O_5 + O_2$$

$$4NO_2 + O_2 + 2H_2O \longrightarrow 4HNO_3$$

$$N_2O_5 + H_2O \longrightarrow 2HNO_2$$

$HNO_3$ and $HNO_2$ may also undergo photochemical dissociation as follow

$$HNO_2 + h\nu \longrightarrow NO_2 + H\bullet$$

$$HNO_2 + h\nu \longrightarrow NO + HO\bullet$$

$$HNO_3 + h\nu \longrightarrow NO_2 + HO\bullet$$

In the stratosphere, $NO_2$ may react with the HO• free radical, forming $HNO_3$:

$$NO_2 + HO\bullet \longrightarrow HNO_3$$

The $HNO_3$ so formed is removed as acid rain or is converted into particulate nitrates due to neutralisation by $NH_3$ or particulate lime.

**Oxidation of Sulphur dioxide.** Sulphur dioxide, present in the atmosphere, absorbs solar radiation in the range of 300–400 nm and produces electronically excited states of $SO_2$. This undergoes oxidation to $SO_3$ and in presence of water vapour, this is conveted to $H_2SO_4$.

$$SO_2 + h\nu \longrightarrow SO_2\bullet$$

$$SO_2\bullet + O_2 \longrightarrow SO_4\bullet$$

$$SO_4\bullet + O_2 \longrightarrow SO_3 + O_3$$

$$SO_3 + H_2O \longrightarrow H_2SO_4$$

The overall reaction in presence of sunlight and relative humidity 30-90% may be represented as

$$SO_2 + \tfrac{1}{2} O_2 + H_2O \longrightarrow H_2SO_4$$

This photochemical oxidation of $SO_2$ leading to the formation of $H_2SO_4$ aerosol is greatly accelerated in the presence of olefinic hydrocarbons and oxides of nitrogen which are usually present in "photochemical smog".

**"Photochemical Smog"** is initiated by the photochemical dissociation of $NO_2$ and the consequent secondary reactions involving unsaturated hydrocarbons, other organic compounds and free radicals, leading to the formation of organic peroxides and ozone. This phenomenon takes place during sunny days with low winds and low level inversion. The photochemical smog and the consequent formation of aerosols reduce the visibility, cause irritation to eyes and damage plants and rubber goods.

The oxidation of $SO_2$ can also take place by interaction with the free radical HO• present in photochemical smog

$$SO_2 + HO\bullet \longrightarrow HOSO_2\bullet$$

$$HOSO_2\bullet + O_2 \longrightarrow HOSO_2\,O_2\bullet$$

$$HOSO_2\,O_2\bullet \xrightarrow{NO_2} HOSO_2\,O\bullet + NO_3\bullet$$

(sulphate)

Chemical oxidation of $SO_2$ may also take place in water droplets present in aerosols. This reaction is accelerated in presence of $NH_3$ and catalysts *e.g.*, oxides of Mn, Fe, Cu, Ni.

Solid particles, such as soot, bring about catalytic oxidation of $SO_2$ by providing a heterogeneous phase for contact. Soot is formed during combustion of solid and liquid fuels in domestic and industrial operations and automobile emissions.

Sulphur dioxide is a pollutant responsible for smog formation, acid rains, and corrosion of metals and alloys.

**Oxidation of Organic Compounds.** Organic compounds such as hydrocarbons, aldehydes and ketones absorb solar radiation and undergo various photochemical and chemical reactions involving free radicals. Some of these reactions are catalysed by particulate matter such as soot and metal oxides. Some of the intermediates and final products formed contribute to photochemical smog formation.

$$\underset{\text{Methane}}{CH_4} + O \longrightarrow H_3C\bullet + HO\bullet$$

$$CH_4 + HO\bullet \longrightarrow H_3C\bullet + H_2O$$

$$H_3C + O_2 + M\ (\text{Third body}) \longrightarrow H_3\,COO\bullet + M$$

$$\underset{\text{Propylene}}{CH_3CH = CH_2} + HO\bullet \longrightarrow CH_3 - CH\bullet - CH_2OH$$

$$\underset{\text{Aldehyde}}{RCHO} + h\nu \longrightarrow R\bullet + HCO$$

$$\underset{\text{Ketone}}{R_2CO} + h\nu \longrightarrow R\bullet + R\bullet\ CO$$

where R is an alkyl or aryl radical or even a hydrogen atom. The alkyl or aryl radicals may react with oxygen to form a peroxyl radical, which in turn may react with another oxygen atom to give $O_3$. The peroxyl radical may react with $NO_2$ to given peroxyacyl nitrate (PAN), formaldehyde (which is an irritant) and various polymeric compounds which reduce visibility.

$$R\overset{O}{\overset{\|}{C}}OO\bullet + NO_2 \longrightarrow \underset{\text{(PAN)}}{R\overset{O}{\overset{\|}{C}}OO\,NO_2}$$

## GREENHOUSE EFFECT

The earth is heated by sunlight and some of the heat that is absorbed by the earth is radiated back into space. However, some of the gases in the lower atmosphere, acting like glass in a greenhouse, allow the solar radiations (in the range 300 to 2500 nm, i.e., near u.v., visible and near I.R. region, while filtering the dangerous u.v. radiations i.e. $< 300$ nm) but do not allow the earth to re-radiate the heat into space. In other words, these gases in the atmosphere are transparent to the sun-light coming in but they strongly absorb the infra-red radiation, which the earth sends back as heat. A part of the heat so trapped in these atmospheric gases is reemitted to the earth's surface. The net result is the heating of the earth's surface by this phenomenon, called the "Greenhouse effect". The gases that are responsible for this Greenhouse effect are $CO_2$, water vapour, $CH_4$ and man-made chloroflourocarbons (CFC's). Water vapour strongly absorbs I.R.

radiations in the range 4000 to 8000 nm and $CO_2$ in the range 12,000 to 16,300 nm. The radiations in the range 8000 to 12,000 nm escape unabsorbed and this is known as the region of atmospheric window.

Carbon dioxide is released by volcanoes, oceans, decaying plants as well as human activities, such as deforestation and combustion of fossil fuels. Automobile exhausts accounts for 30% of $CO_2$ emissions in developed countries.

Methane is released from coal mines, decomposition of organic matter in swamps, rice paddy cultivation, guts of termites in forest debris and stomachs of ruminants.

Chloroflourocarbons (CFC's) are used as coolants in refrigerators, propellants in aerosol sprays, plastic foam materials like "Thermocoles" or "Styrofoam" and in automobile air-conditioners.

In fact, the "Greenhouse gases" (particularly $CO_2$ and water vapour) are responsible for keeping our planet warm and thus sustaining life on the earth. If the Greenhouse gases are very less or totally absent in the earth's atmosphere, then the average temperature on the earth would have been at sub-zero levels. But, however, if the concentration of Greenhouse gases is larger, they may trap too much of heat, which may threaten the very existence of life on the earth. For instance, the $CO_2$ present in the atmosphere of the planet Venus, is about 60,000 times more than that in the earth's atmosphere. Hence the average temperature of Venus is about 425°C, thus making the existence of life impossible there.

Oceans and bio-mass are the major sinks for the atmospheric $CO_2$. Oceans convert $CO_2$ into soluble bicarbonates. The photosynthetic activity in the green plants increases with increase in $CO_2$ level in the atmosphere. Forests are the places where lot of photo-synthetic activity occurs. They also act as vast reservoirs of fixed but readily oxidisable carbon in the form of vegetation, wood and humus. Hence, forests maintain a balance in the atmospheric $CO_2$ level. Therefore, deforestation definitely upsets this balance and increases the atmospheric $CO_2$ level.

It is estimated that the atmospheric $CO_2$ content has increased by 25% during the last two centuries. This is mostly attributed to the industrial revolution over these two centuries. This is one of the reasons for the slight increase in the global temperature (about 0.5°C). Since the concentrations of the Greenhouse gases have been continuously increasing because of deforestation, industrialization, increased burning of fossil fuels, mining, exhausts from increasing number of automobiles and other anthropogenic activities, there is an increasing concern about the possible "global warming". Some scientists fear that if proper precautions are not taken, the concentration of the green-house gases in the atmosphere may double within the next 50–100 years. If this happens, the average global temperature may increase by 4 to 5°C. This will increase the evaporation of surface waters, which may influence climatic changes depending upon the pattern of cloud formation. For instance, low-level dense clouds may exert cooling effect whereas high level thin cloud formation may exert heating effect due to increased greenhouse effect.

The projections from computer modelling regarding the climatic changes that could be triggered off due to "global warming" reveal alarming scenarios. Even 1.5°C raise in surface temperature can adversely affect the food production in the world. Thus, the wheat growing zones in the northern latitude may be shifted from the USSR and Canada to the polar regions *i.e.*, from fertile soils to poor soils near the north pole. The biological productivity of the ocean would also decrease due to warming of the earth's surface layer, which in turn, may reduce the transport of the nutrients from deeper layers to the surface by vertical circulation. The computer modelling also indicate the following effects due to "global warming": melting of the polar ice caps; dry areas becoming drier; humid areas like the Amazon suffering more intense tropical storms; drastic drop in food production, particularly in lands within 35 degrees north and south of the

Equator; increased breeding of pests and diseases due to more humid conditions; shorter, wetter and warmer winters; and longer, hotter and drier summers, particularly in mid-continental areas. Global warming may also trigger increased thermal expansion of oceans and melting of glaciers, which results in lifting up of the sea-level by 20 cm to 1.5 meters by the later part of the 21st century. Thus, cities like Bombay, Miami, London, Venice, Bangkok and Leningrad may become extremely vulnerable. Defences against the raising sea-levels and expanding oceans are very difficult and expensive, which many nations of the world cannot afford. Further, a global temperature raise, even about 1.5°C, is likely to cause floods, hurricanes, tornadoes, apart from raising of the sea-level due to melting of the polar ice caps and inundating coastal cities like Madras, Sydney, New York and Boston.

There are differences of opinion among experts regarding the dynamics and effects of "global warming" due to the complexity of natural phenomena that might be operating simultaneously. More accurate future climatic projections will be possible with better super-computer models, based on greater understanding of the complex natural climatic forces involved. But until that time, the possible devastating effect due to "global warming" by the "greenhouse effect" cannot be underestimated. Some of the steps suggested to minimise the "Greenhouse effect" include reduction in the use of fossil fuels, encouraging the use of alternative sources of energy (*e.g.* solar, geothermal, wind, bio-gas, etc). Conservation of forests, extensive afforestation, encouraging community forestry, reduction in the use of automobiles, research in development of more efficient automobiles engines, ban on CFC's and nuclear explosions, development of environmentally compatible technologies with the help of intensive inter-disciplinary research, effective check on the growth of population and imparting of non-formal and formal environmental education.

## FORMATION AND DEPLETION OF OZONE IN THE STRATOSPHERE

Ozone is an important chemical species present in the stratosphere. At an altitude of about 30 km, its concentration is about 10 ppm. The ozone layer present in the stratosphere acts as a protective shield for the life on earth. It strongly absorbs ultra-violet radiations from the sun in the region 220-330 nm and thereby protects the life on earth from severe radiation damage, such as DNA mutation and skin cancer. Thus only a small fraction of u.v. radiation reaches the lower atmosphere and the earth's surface.

Ozone is formed in the stratosphere by photochemical reaction :

$$O_2 + h\nu\ (242\ \text{nm}) \longrightarrow O + O$$

$$O + O_2 + M\ (\text{third body, such as } N_2 \text{ or } O_2) \longrightarrow O_3 + M$$

The third body absorbs the excess energy liberated by the above reaction and thereby the ozone molecule is stabilized. Thus, ozone is constantly formed in the stratosphere. However, it is also destroyed by chlorine, released due to volcanic activity and also by reaction with (a) Nitric oxide, (b) Atomic oxygen and (c) Reactive hydroxyl radical, which are also present in the atmosphere*, by the following reactions:

(a) $O_3 + NO \longrightarrow NO_2 + O_2$

(b) $O_3 + O \longrightarrow O_2 + O_2$

(c) $O_3 + HO^{\bullet} \longrightarrow HO_2 + HOO^{\bullet}$

$HOO^{\bullet} + O \longrightarrow HO^{\bullet} + O_2$

*In atmos phere, NO comes from chemical and photo-chemical reactions in the atmosphere, supersonic jets, nuclear explosions etc., $Cl_2$ comes from CFC's and volcanoes; and OH comes from bio-mass burning and from natural water systems.

Ozone, in the stratosphere, is also found to be destroyed by man-made chloroflourocarbons (CFC's), which are used as coolants in refrigerators, air-conditioners, propellants in aerosol sprays and in plastic foams, such as "Thermocole" or "Styrofoam". The CFC molecules, escaping into the atmosphere, decompose to release chlorine in the ozone layer (by photo-dissociation) and each atom of chlorine, thus liberated, in capable of attacking several ozone molecules.

$$Cl + O_3 \longrightarrow ClO + O_2$$

The reaction is followed by

$$ClO + O \longrightarrow Cl + O_2$$

which regenerates Cl atoms, so that a long chain process is involved, which conserves Cl atoms. The environmental hazards of CFC's were recognised as early as 1970. In fact, temporary thinning in the stratospheric ozone layer, leading to the formation of "Ozone hole" was actually detected over the Antarctica during September to November in 1985. Reported increase in cases of skin cancer in South Australia are also attributed to u.v. radiations reaching the earth, due to depletion of ozone layer, over that part of the world temporarily.

The detection of the "Ozone hole" over Antarctica in 1985 attracted the attention of scientific community in the world. The U.S. immediately banned the use of CFC's in spray cans. Further, in the year 1987, twenty four nations of the world signed the Montreal Protocol, which aims at 35% reduction in the global production of the CFC's by the year 1999. Simultaneously, efforts to produce chlorine-free substitutes have also started. In fact, synthesis of a product called HFC-134a has already been reported as an effective substitute for CFC. The use of hydrofluorocarbons (HFC's), hydrochlorofluorocarbons (HCFC's), and methyl cyclohexane (MCH) as substitutes for CFC's is envisaged for several applications.

## LAPSE RATE & TEMPERATURE INVERSION

The change of atmospheric air temperature with altitude is called the "lapse rate". As we go high in the atmosphere, the pressure decreases. A gas, that is free to expand, has a tendency to move from high pressure region to low pressure region. While doing so, the gas expands and if there is no opportunity to exchange its heat with the surroundings, the gas undergoes adiabatic cooling. This adiabatic lapse rate can be calculated with the help of pressure-volume-temperature relationship, the adiabatic expansion and the change of pressure with the altitude. The value for adiabatic lapse rate has been calculated to be about—10°C/km for dry air and about—6°C to —7°C/km for humid air, the higher values occurring at greater altitudes where the water content is less.

Under conditions of adiabatic lapse rate, a smoke plume rises upwards into the atmosphere by virtue of its low density (because of its higher temperature) until it reaches air of similar temperature and similar density. However, in many practical situations, the lapse rate is found to be either greater or lesser than that of the adiabatic lapse rate due to external heating and cooling effects. From air-pollution point of view, the following two situations are of great importance :

(*i*) "Super adiabatic lapse" where the lapse rate is considerably higher than that of adiabatic. On a clear summer day, the air near the earth's surface gets heated rapidly to reach a condition where the lapse rate is super adiabatic. In such a situation, the atmosphere is said to be in an unstable equilibrium. It is under such conditions when the atmospheric pollutants are rapidly dispersed due to considerable vertical mixing of air.

(*ii*) "The negative lapse rate" or "temperature inversion" when the temperature increases with increasing altitude. Inversion occurs in the atmosphere mainly due to the following two reasons:

(a) Radiations inversion:- Such a situation occurs when the air near the earth's surface gets cooled because of the loss of heat by the earth at night by emitting longwave radiations. Under the conditions of temperature inversion, the air near the ground is denser than the air above and hence very little mixing or turbulence takes place.

(b) Subsidence inversion:- Such a situation occurs when the upper layer of air "subsides" (i.e. descends) during a developing anticyclone (*i.e.* high pressure centre). This subsiding air warms as it contracts. But, the extent of warming will be greater at the upper part than at the lower part near the ground.

Under the conditions of inversion, the denser cold air at the bottom cannot rise up due to lack of any driving force and the atmosphere is said to be stable. Such a situation in dangerous from the atmospheric pollution point of view, because the pollutants in the air cannot be dispersed. Development of inversion conditions generally occur on still and clear nights, when the ground is strongly cooled. Hence, the air near the ground level is also cooled, while the air above is relatively warm. In such a condition, if a pollutant is discharged at or near the ground level, the pollutant tends to concentrate rather than getting dispersed in the air. This increasing intensity of air pollution may lead to disastrous consequences. Inversion is generally destroyed by the next morning due to the solar heating of the ground and the buildings which enable upward current of the warmed air.

Simultaneous occurrence of mist or fog may prolong the duration of inversion by reducing the sunlight reaching the ground.

The probelm of air pollution assumes serious dimensions under conditions of inversion and other meteorological conditions which are adverse to the effective dispersion of pollutants. Inversion is responsible for many air-pollution episodes in the world. The notorious London smog of 1952 the lasted for five days killing four thousand people is an example.

In towns and villages which are situated in valleys, inversion conditions are particularly disastrous, because the dense air charged with pollutants descends down to the valleys and forms a sort of "lake of polluted air".

Fortunately for India, the climatic conditions in many parts are favourable for dispersion for most of the time. But still, every effort should be made to prevent improper and indiscriminate release of pollutants into the atmosphere.

## RADIONUCLIDES IN THE ENVIRONMENT

The various radionuclides present in the environment originate from the following two sources:

**(i) The natural radionuclides:** These arise from the all-pervading primordial radioelements such as uranium and thorium along with their daughter nuclides and potassium.

**(ii) The technology related radionuclides:** These arise from the human activities such as nuclear and thermal power generation, uranium mining and milling, use the radiations and radioisotopes for medical applications, etc. Watching of colour television, air travel and other modern amenities also give rise to additional environmental radiation exposure.

**Natural radiation environment :** An elaborate survey of the environmental radiation using indigenously made sensitive radiation dosimeters by the Bhabha Atomic Research Center, Bombay revealed that population-weighted all India average environmental external radiation exposure projected to the 1991 census figures works out to 734 μ SV/y per person. The average environmental external radiation exposure to the population varies from place to place within our country. For instance, the average exposure value in Andhra Pradesh is 10165 μ SV/y (highest), whereas the corresponding figure in Laccadive is 285 μ SV/y (lowest). However, in certain isolated parts of the beaches containing monazite deposits, the average radiation exposure value was estimated at 5760 μ SV/y per person due to the presence of $^{232}Th$ and its daughter nuclides. [1 SV (sievert) = 100 rem = 1 Joule/kg.].

A significant part of the human exposure to environmental radiation arises from inhalation of radon, thoron and their short lived α – emitting daughter nuclides *e.g.*, $^{212}Bi$, $^{212}Po$, $^{214}Po$ $^{216}Po$ and $^{218}Po$, which are extremely hazardous to the lung tissues. A recent study on the randon inhalation exposures in different parts of our country indicates an average individual exposure of 1440 μ SV/y. The indoor randon levels in western countries are quite high and have been identified as a risk factor for lung cancer similar to smoking cigarattes. However, in our country, thanks to the tropical climate and the consequently high ventilation in our houses, in-door randon exposure is not reckoned to be a serious radiological hazard.

Radiation exposures due to thoron and its short-lived daughter nuclides in air are not usually considered seriously because their concentrations above 1 metre from the ground level are not appreciable. However, in Indian context, these exposures cannot be ignored because a considerable number of people habitually sit and sleep on the Floor. The global average exposure value from this source is estimated at about 260μ SV/y as per UNSCEAR 1988.

The following Table 1.3 shows the per capita radiation dose received by the people in India from radionuclides in the environment from natural and man-made sources:

**Table 1.3: Per capital radiation dose received by the people of India**

| Sr. No. | Source and type of exposure | Contributing radionuclides | Per capita dose rate, μSV/Y |
|---|---|---|---|
| 1. | **Natural sources** | | |
| | (i) External irradiation | $^{238}U$, $^{232}Th$ and their daughters, cosmic radiations and $^{40}K$ | 734 |
| | (ii) Ingestion irradiation | $^{210}Po$, $^{210}Pb$, $^{226}Ra$, $^{228}Ra$, $^{232}Th$, $^{40}K$, $^{14}C$, $^{7}Be$ and $^{87}Rb$ | 330 |
| | (iii) Inhalation irradiation | $^{232}Rn$, $^{218}Po$, $^{214}Po$, $^{220}Rn$, $^{216}Po$, $^{212}Bi$, $^{212}Po$ | 1600 |
| 2. | **Man-made sources** | | |
| | (i) Medical applications | $^{131}I$, $^{99(m)}T$-C, X-ray | 48 |
| | (ii) Stack effluents, liquid effluents and fall out | $^{14}C$, $^{3}H$, $^{41}Ar$, $^{90}Sr$, $^{131}I$, $^{133}Xe$, $^{239}Pu$, $^{137}Cs$, etc. | 7 |
| | (iii) Watching television, air-travel etc. | Soft x-rays, cosmic radiations | 4 |

It is clear from the data given in the above table that the population exposures due to environmental radiation from man-made or technological activities including thermal and nuclear power production, mining and milling operations and from air travel and TV viewing etc. together, is quite small as compared to the natural background radiation constantly received by us. Further, comprehensive surveillance programmes recently conducted by BARC, Bombay clearly indicate that there is absolutely no reason for concern regarding the radionuclides released into the environment from the routine operation of the nuclear installations in our country.

## EI NINO PHENOMENON AND ITS EFFECTS

Oceans and seas play a vital role in global climatic variations. These variations are largely driven by fluctuations in the air-sea momentum and heat exchange. This effect was dramatically expressed in the 1982-1983 "*El Nino*" event, which sparked wide-ranging aberration in weather conditions such as flooding in South California and droughts in Australia, and is believed to be of oceanic origin.

"*El Nino*" event is generally characterized by warming of the coastal waters off the coasts of peru and Ecuador. Under normal conditions, these waters are held relatively cool by the upwelling activity typical of this region of the eastern pacific. The atmospheric circulation of this area is characterized by what is commonly known as the Southern oscillation, i.e, the seasonal shift in atmospheric surface pressure between the Australian Indian ocean region and the South-eastern pacific. This Southern Oscillation has been shown to influence surface pressure, temperature and rainfall variations over many parts of the earth. The disastrous effect of "*El Nino*" events (such as abnormal rainfall variations on a global scale) are postulated to be due to the alternation of the Southern Oscillation by the near-shore surface-warming along peru and Ecuador.

"*El Nino*" effect is generally related to an anomalous weather condition which gives rise to major changes in ocean circulation and biological productivity along the coast of peru. Under normal conditions, upwelling along the coast brings up nutrient-rich deeper water which results in high biological productivity. During "*El Nino*" event, wind patterns along the peruvia coast are changed, the upwelling is interrupted, and the coast is invaded by warm water. This is believed to be due to the reduction in trade-wind intensity following the periods of extremely strong winds which could cause warm tropical waters to pile up along the coast of South America and then further spread Southwards. Absence of upwelling results in a massive mortality of marine organisms and the collapse of the important peruvian anchoy fishery. Such a phenomenon generally occur around Christmas; hence this effect is called "*El Nino*" which is spanish, means "*the child*" (implying, the "christ's child").

"*El Nino*" phenomenon which is believed to occur in cycles, results in reversal of wind flows. "*El Nino*" causes drastic change from very heavy rains and floods to drought, poor crop yields and starvation. Initially, the warm Southward wind currents that appeared every December along the Coastal Ecuador and peru, used to be described as "*El Nino*", but now the word is used for recurrences that are exceptionally intense and persistant, causing climatic disturbances of varying severity around the globe, for more than a year.

Under normal conditions, the trade winds flow from the East to the West, pushing sun-warmed surfaces sea-waters west-wards and exposing cold water to the surface in the East. However, during the "*El Nino*" event, these normal trade winds collapse for even reverse. The weakening of the wind causes a modest change in Sea-surface temperature but the change in wind and pressure becomes drastic. The warm water of the pacific flows back Eastward and Sea-Surface temperature increases significantly on the western coast of South America. Due to this, the wet weather conditions normally present in the Western pacific move to the East and the arid conditions common in the East appear in the West, thereby resulting in heavy rains in South America and possible drought in South-east Asia, India and Southern Africa. Thus, instead of bringing much needed rains to Asia, it takes away the moist air and dumps unwanted rains across the Pacific causing floods. India Asia, it leaves large tract like Indonesia without rains and creates droughts in some South Pacific Islands or even Australia.

The year 1982-83 is supposed to be the worst year under "*El Nino*" for America causing losses to the tune of 13 billion dollars. The year 1997 turned out to be the worst under "*El Nino*" for Asia, particularly Indonesia, which had far below average rainfall. Papua New Guinea too suffered severe drought. The effect of the "*El Nino*" Phenomenon is varied, the worst damage could be the disruption to fish migrations and damage to coral reefs. Those who are dependent on marine ecosystems for their livelyhood would be affected. It has also been documented that the number of births of weddell seals declined every four to six years, possibly coinciding with "*El Nino*" events.

## CLIMATE CHANGE

Though climate is an average weather of an area or environmental factors of an area. These include quantity of light, temperature, humidity, wind, gases, water etc which average for about 30 yrs. Thus the changes in environmental conditions of an area over long period of time is called **climate change**. These changes effect the agriculture, migration of animals, hydrological cycle, thermal gradient between the poles and equator, wind pattern, distribution of rainfall etc.

The scientific and technological revolution has given multiple facilities to mankind, but at the same time man-made (Anthropogenic) activities are responsible for depletion of resources and upsetting the delicate balance between the various components of the environment. They are, excessive use of fossils fuels, deforestation, desertification, loss of fertility of soil, rapid industrialization, increase of automobiles. Changes in the atmosphere conditions resulting in to serious problems like green house effect, depletion of ozone layer and rise of world temperature etc.

The global change in temperature will not be uniform every where and will fluctuate in different regions. The places at higher latitudes will be warmed up more during late autumn and winter than the places in tropics. Poles may experience 2 to 3 times more warming than the global average, while warming in tropics may be only 50 to 100° C on an average. The increased warming at poles will reduce the thermal gradient between the equator and high latitude regions, decreasing the energy available to the heat engine that drives the global weather machine. This will disturb the global pattern of winds and ocean currents as well as the timing and distribution of rainfall. Shifting of ocean currents may change the climate of iceland and Britain, it may result in cooling at a time when rest of the world warms.

## GLOBAL WARMING

The average global temperature is 15°C. The lower most layer of atmosphere i.e., troposphere, traps the heat by a natural process due to the presence of certain gases called **Green house gases**. They are carbon dioxide, ozone, methane Nitrous oxide, Chlorofluorocarbons (CFCs) and water vapours. In presence of these gases the temperature (15°C) would have been –18°C. Thus warming of the earth's climate owing to the increased concentration of green house gases is called **Green house effect**. Therefore, this effect contributes a temperature rise to the tune of 33°C. These gases act like the glass in the botanical green house trapping the reradiated heat near the earth's surface and warming the planet. These gases along with water vapour and clouds absorbs the infrared radiation, trapping heat near the earth's surface. The two predominant green house gases (1) the water vapours whose level in the troposphere has relatively remained constant is controlled by hydrological cycle while (2) $CO_2$ whose level has increased is controlled by the global carbon cycle. Other gases whose levels have increased due to human activities are methane, NO and CFCs. Deforestation has also elevated levels of $CO_2$.

Change the temperature by more than 2°C is disastrous for various ecosystems on the earth including man. Some areas will become inhabitable because of drought or floods following a rise in average sea level.

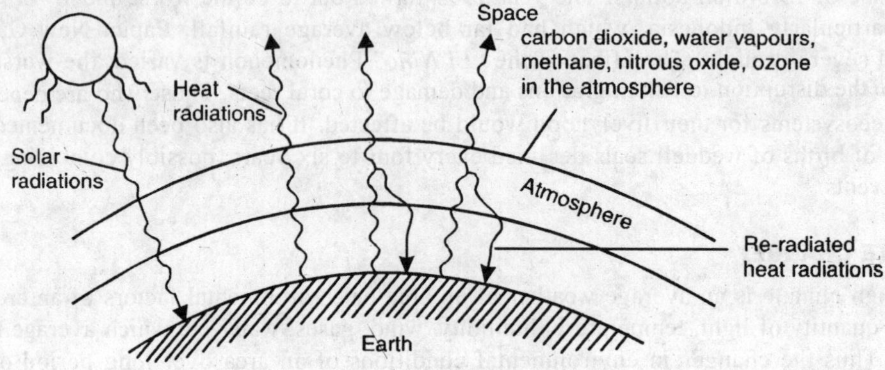


**Fig. 2.** The green house effect.

## GREEN HOUSE GASES

The green house gases present in the troposphere and resulting in an increase in the temperature of air and the earth are discussed here.

**CARBON DIOXIDE ($CO_2$) :** The $CO_2$ is considered as the most dominant factor responsible for the green house effect. The troposphere contains only 00.0375% $CO_2$ (by volume) and its amount is controlled by carbon cycle. The four major pools or reservoirs of carbon are fossil fuels, the atmosphere, the biosphere and the oceans. Its concentration in atmosphere has increased from 290 PPM (1860) to 350 PPM (1990) and is expected to be 700 PPM in 2010. It was studied that in 1976 the world consumption of about 5000 million tons of fossil fuels per year was contributing the equivalent of 2.3 PPM of $CO_2$ to the atmosphere. The net annual increase of $CO_2$ in the air is 0.77 ppm which indicates that 1.60 ppm of fuel generated $CO_2$ is being absorbed elsewhere. In last couple of decades the fossil fuels may to have contributed more $CO_2$ to the atmosphere than the terrestrial biosphere. Clearance of forests is another factor for the increase of $CO_2$.

**CHLORO FLUORO CARBONS (CFCs) :** The main source of CFCs include leaking air conditioners and refrigerators, evaporation of industrial solvents, production of plastic foams, aerosols, propellants (CFC-11) etc. The concentration of CFCs is rising nearly 5% per year. CFCs trap heat 20,000 times more efficiently than $CO_2$ and also destroy ozone layer, thus posing a serious twofold environmental problem. CFCs are responsible for 14 – 24% of global warming. It is said (Dickson and Cicerone 1986) that by the end of year 2050 they alone could contribute more than $CO_2$ to global warming.

**METHANE ($CH_4$)** : It is produced in a number of ways including the action of anaerobic bacteria on vegetation, decomposition of organic matter, incomplete combustion of vegetation, natural gas pipeline leaks, burning of biomass during production and uses of oil and natural gas and petroleum oil etc. It is rising approx. 2% every year. It absorbs 20 – 25% times more heat than $CO_2$.

**NITROUS OXIDE ($N_2O$) :** It is released from nylon products, from burning of biomass and fuels (specially coal). From breakdown of fertilizers in soil, livestock wastes and nitrated contaminated ground water, nylon products etc. It is responsible for about 6% of global warming. Besides trapping heat in the troposphere it also depletes ozone in the stratosphere. It absorbs about 250 times more heat than $CO_2$. The $N_2O$ concentration in atmosphere is 0.3 ppm and is increasing 0.2% annually.

**OZONE :** It comes mostly from hydrocarbons and nitrogen oxides. It causes irritation to eyes and respiratory organs. It decreases the resistance power to infections and aggrevates illness.

## IMPACT OF GLOBAL WARMING

**(i) Climatic change (Increase in global temperature) :** Increase the level of Green house gases causes the global warming have affected the global climate. This effect will increase in future. According to IPCC (1996) the world climate has warmed from 0.3 to 0.6°C during the last century. Over the next century GHGs will increase further by 1°C – 3.5°C December 03 report has predicted the global warming would be 5.5°C higher by 2100 AD. It seems likely that many species will be unable to adjust quickly to global warming and associated climatic change. As a result, biological communities may suffer profoundly. More than 10% of the plant species will not be able to survive. The climatic change also effects rainfall, species composition, plant reproduction cycles and biogeochemical cycles.

**(ii) Effect on Sea levels :** Rising temperatures will cause glaciers to melt and the polar ice caps to shrink. As a result sea level may rise by 0.2 – 1.5 m over the next 50 – 100 yrs. It is proved that sea level have already risen by 10 – 25 cm. If it continues, many low lying areas may be submerged in near future, and it is possible to destroy 20% - 80% of the coastal wetland. Rising sea levels are detrimental to coral reef species, which grow at a precise depth with optimum temperature and water movement. Abnormally high water temperatures in the Pacific Ocean during 1982 – 83 caused the death of symbiotic algae that live inside the coral.

**(iii) Reduction of Biodiversity :** As we have discussed, increased temperatures, inundation of some coastal biological communities and changes in the pattern of distribution of many species over a long period of time are likely to cause reduction in biodiversity in aquatic and terrestrial ecosystems.

**(iv) Effect on Agriculture :** There are different views regarding the effect of global warming on agriculture. It may be positive or negative. However, the effects of this change will vary for $C_3$ (i.e. wheat, rice and beans) and $C_4$ (e.g. maize, millet and sugarcane) plants. As temperature increase with rising $CO_2$ level, some crops may no longer be grown in certain regions. With rise in temperature soil moisture will decrease and evapo-transpiration and pest growth will increase. This will effect certain crops. With increased $CO_2$ concentration some plants will show increased photosynthesis, greater root production, increased nitrogen fixation in root nodules which may increase the growth of plants by 30%.

**(v) Effect on human health :** The global warming will lead to changes in the rainfall pattern in many areas, thereby effecting the distribution of vector borne diseases like malaria filariasis, elephantiasis etc. Warmer temperature and more water stagnation would favour the breeding of mosquitoes, snails and some insects, which are vectors of such diseases. Higher temperature and humidity will increase respiratory and skin diseases. Keeping in view of ill effects of global warming UNEP (United Nations Environmental Programme) is celebrating $5^{th}$ June as **"World Environmental Day"** every year since 1989.

**(vi) Effect on Arctic ecosystems :** Global climate change will have profound effects on arctic ecosystems. Tundra is more sensitive to global climate change than most other ecosystems on earth. According to Shaver *et al.*, 1992 warmer temperature may increase primary production, thereby increasing Carbon input and soil respiration hence increasing carbon output.

**(vii) Ecological disturbance :** Global warming increases the desert. It increases temperature in North America, South Africa, Mexico, India and other countries. Changes of hurricanes, cyclones and floods will be more which will damage the lagoons, estuaries and coral reefs. Global warming may cause extinction of more than one million species of animals and plants by 2050 A.D.

## MEASURES TO CHECK GLOBAL WARMING

To check the global warming following steps are necessary –

(1) Plant more trees (Afforestation)

(2) Control population growth.

(3) Cut down the current rate of CFCs and fossil fuel.

(4) Use of non-conventional source of energy.

(5) Shift from coal to Natural gas.

(6) To trap and use methane as a fuel.

(7) Reduce beef production.

(8) Efficiently remove $CO_2$ from smoke

(9) Use photosynthetic algae to remove atmosphere $CO_2$.

(10) Adopt sustainable agriculture.

(11) Use energy more efficiently.

## ACID RAIN

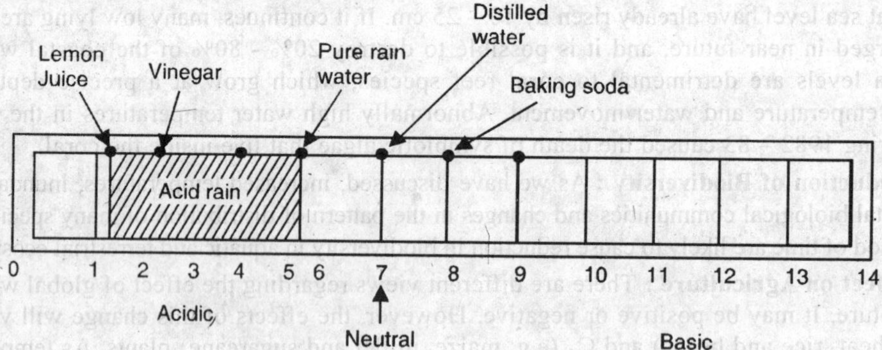


**Fig. 3.** The pH scale of common substances.

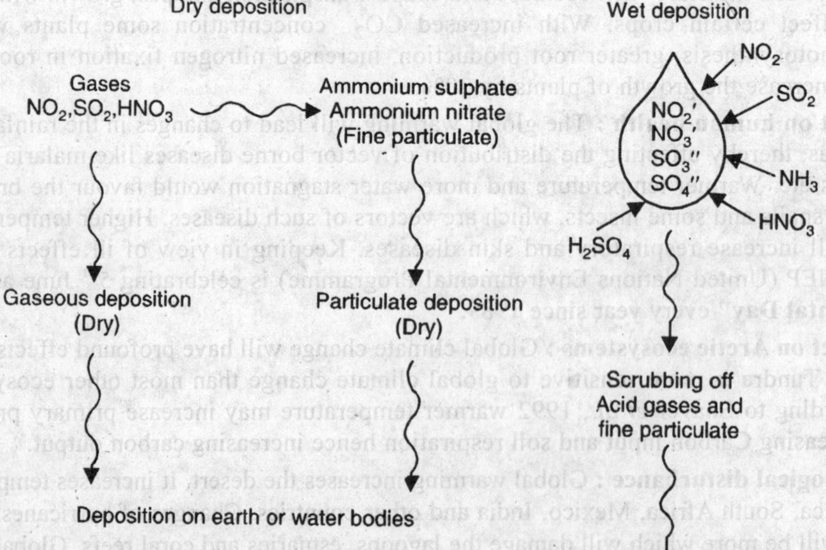


**Fig. 4.** Acid deposition (dry deposition and wet deposition)

Normal rain water is always acidic because of the fact that $CO_2$ present in the atmosphere gets dissolved in it forming carbonic ($H_2CO_3$) acid. Because, the presence of $SO_2$ (Sulphur Dioxide) and $NO_2$ (Nitrogen Oxide) gases as pollutants in the atmosphere, the pH of the rain water is further lowered (as low as 2.4). This is known as **Acid rain**. (Shown in fig. 3 & 4).

"Literally acid rain means the presence of excessive acids in rain waters". Acid rain is a mixture of acids mainly $H_2SO_4$ and $HNO_3$. $H_2SO_4$ are the major contributor (60 – 70%) $HNO_3$ ranks second (30 – 40%) and HCl third.

Acid rain is a man-made and an environmental problem that knows no boundaries. Increasing acidity in natural waters and soils is becoming a problem all over the world. Acid deposition has degraded lakes, swamps and wet lands to such an extent that organisms can not thrive.

## WHERE DOES ACID COME FROM

Acidification of environment is a man-made phenomenon. No doubt that most acid come from human activities i.e. cars, houses, factories, power stations etc. There has always been some acid in rain, coming from volcanoes, swamps and plankton in the oceans, but scientists know that it has increased very sharply over the past 200 years. The acidity is mainly associated with the transport and subsequent deposition of oxides of sulphur, nitrogen and these oxidative products. These oxides are produced by combustion of fossil, fuels, swelters, power plants, automobile exhausts, domestic fires etc. Burning of fossil fuels for power generation contributes to 60 – 70% of the total $SO_2$ emitted globally. Emission of $NO_x$ from anthropogenic sources ranges between 20 – 90 million tones annually over the globe. Acid rain problem has drastically increased due to industrialization.

Researches show that in Swiss, water is not fit for drinking. Pesticides sprayed on crops evaporate and in stratosphere they react with water vapour. Water contaminated with these toxic substances fall as rain. Due to this pH of Delhi water decreased from 7.0 to 6.1 in 1984. The term acid rain was first coined by Robert Angus Swith in 1952. In 1968, scientists pointed out that rain water was becoming increasingly acidic with each passing year.

## HOW ACID RAIN IS FORMED

In high temperature combustion processes most of the nitric oxide originates from atmosphere and some Nitric oxide also released by burning of wood and as a result of microbial nitrification in the soil. Lightning is another source of Nitric oxide.

In day time Nitric oxide is oxidized by oxygen, ozone

$$NO + O_3 \longrightarrow NO_2 + O_2$$

$$H_2O \longrightarrow H^+ + OH$$

$$NO_2 + OH^- \longrightarrow HNO_3$$

Similarly, formulation of $H_2SO_4$ acid in the atmosphere can take place with a wide range of reduced as well as partially oxidised Sulphur compounds, $H_2S$, $CS_2$ etc. These compounds are released from oceans and soil under reducing conditions. The production of $H_2SO_4$ from $SO_2$ may take place homogeneously in the gas phase as :-

$$SO_2 + OH^- + X \longrightarrow HSO_3^- + X$$

Where X = $O_2$ or $N_2$ in atmosphere.

The $HSO_3^-$ so formed can undergo a number of reactions, some of which produce sulphuric acid.

$$HSO_3 + O_2 + X \longrightarrow HOO^- + SO_3^{--} + X$$

$$SO_3^{--} + H_2O \longrightarrow H_2SO_4$$

The hydroperoxy radical $HOO^-$ can also react to give $HNO_3$

$$NO + HOO^- \longrightarrow NO_2 + OH^-$$

$$NO_2 + OH^- + X \longrightarrow HNO_3 + X$$

Thus a number of reactions are taken place, forming different acids.

The chemical composition of rain is highly variable and depends on the geographic location and the influence of natural, anthropogenic chemical processes on the atmosphere of that region.

## EFFECTS OF ACID RAIN

Acid rain exerts both direct and indirect effects on the organisms and materials it comes in contact with. The dry deposition attacks building material, steel and other metals. When deposited in gaseous form it causes direct damage to plants and trees. Wet deposition has direct and indirect effects. It increases the acidity of lakes and rivers and effects aquatic as well as terrestrial ecosystems. Some of the effects may be described as :-

1. A significant reduction in fish population accompanied by decrease in the variety of species in food chains have been observed.
2. Adirondack ponds having high acidity levels, were among the first to lose fish population.
3. Different species reacts differently to acidified lakes. Adult fish can survive in more acidic water having high concentration aluminium than dry fish.
4. Many bacteria and blue green algae are killed due to acidification, disrupting the whole ecological balance.
5. In 1958 at Europe pH of rain water was 5.0 and in Netherland (1962) was 4.5. It damaged the leaves of plants and trees.
6. Forests of West Germany, Switzerland, Czechoslovakia, Swedish were severely effected by acid rain.
7. In North America and Europe, acid rain destroyed crops and forests, reducing agricultural productivity.
8. Acid rain has retarded the growth of pea, beams, radish, potato, spinach, carrots etc.
9. Modern researches show that acid rain leaches Potassium, Calcium, Magnesium etc essential elements from the top soil.
10. Acid lakes have low levels of phytoplankton.
11. The activity of bacteria and other microscopic animals is reduced in acidic water.
12. Broad leafed pond weeds do not grow in acid water.
13. Acid rain causes extensive damage to buildings and structural materials of marble, lime stone, mortar etc.

    Lime stone attacked as :-

    $$CaCO_3 + H_2SO_4 \longrightarrow CaSO_4 + H_2O + CO_2$$

    The attack on marble is termed as **Stone-leprosy.**
14. The Taj Mahal in agra is suffering from $SO_2$, $H_2SO_4$ and other fumes, pollutants released from Mathura refinery.
15. Acid rain corrodes houses, monuments, statues, bridges, fences, railways etc.
16. Acidification can play havoc with human nervous, respiratory and digestive systems by making the person an easy pray to neurological diseases.

## QUESTIONS

(1) Describe the characteristics of different atmospheric segments.

(2) Discus the characteristics of the major regions of the atmosphere.

(3) Write information notes on

(a) Radiation balance of the earth

(b) Chemical species and particulates present in atmosphere

(4) Discuss the various mechanisms of energy transfer which help in maintaining the earth's radiation balance.

(5) Discuss with the help of examples the various chemical and photochemical reactions taking place in the atmosphere and their consequences.

(6) What is Greenhouse effect? What are its consequences? What steps do you suggest to control it?

(7) Discuss the various mechanisms involved in the formation and depletion of ozone in the atmospheres. What are the consequnces of depletion of ozone in the stratosphere? What steps are needed to protect the stratospheric ozone layer?

(8) Discus the origin of various inorganic and particulates in the atmosphere.

(9) Write an information note on the various radionuclides present in the environment with special reference to their origin.

(10) What is "El Nino" effect and what is its significance?

(11) Discuss Global warming and explain how climate change is occurring.

(12) Write brief note on the followings:

(a) Acid rain

(b) Formation of acid in acid rain

(c) Green house gases

(d) Ozone hole

# Environmental Pollution

As we have seen, environmental science is a multidiscilpinary branch of science, involving - chemistry, physics, life sciences, agriculture, public health & medical sciences etc. The word environment is derived from French word "Enviromer" means to "encircle" or 'surround". Hence environment can be defined as one's surroundings which include every thing around the environment. Abiotic environment consists of soil, water, air etc. while the biotic environment includes all other organisms with which the organisms comes into regular contact.

Pollution can be defined as the introduction deliberately or unadvertantly of substances or energy into the environment by man/any other resulting in a deleterious effect. Pollution involves contaminations but contaminations need not constitute pollution. It may be local, regional, transboundary or global. Effects may be direct, indirect or cumulative felt intermittently or constantly, immediate or after. The effects of pollution may be short term or longer term, pose a hazard or a nuisance, be toxic or non-toxic take the form of a chemical, biological, radiation, heat light, noise, dust or odour problem. Environment can be defined in a number of ways. For example,

(1) Environment is the sum of all social, economical, biological, physical or chemical factors which constitute the surroundings of man, who is both creator and moulder of his environment.

(2) Environment refers to the sum total of conditions which surround man at a given point in space and time.

(3) Environment is the representative of physical components of the earth where in man is the important factor influencing his environment.

(4) Environment is a holistic view of the world as it functions at any time, with a multitude of special elemental and socio-economic systems distinguished by quality and attributes of space and mode of behaviour of biotic and abiotic forms.

Huge industrial installations, faster mode of transport urbanisation are the main outcomes of the modern civilization. These and large numbers of others are contributing to what is called, "Environmental Pollution". Pollution of the environment is one of the most horrible ecological crisis to which we are subjected today. The word pollution is derived from Latin word pollution means to make dirty. Thus pollution can also be defined as, "The addition of the constituents to water, air and land, which adversely alter the natural quality of the environment." Pollution is also defined as.

(i) Pollution is the unfavourable alternation of the environment largely because of human activities.

(ii) Pollution is an undesirable change in the physical, chemical or biological characteristics of our air, land and water causing harmful effect on our life or that of other desirable species and cultural assets.

(iii) Pollution means the addition of any foreign material like inorganic, organic, biological or radiological or any physical change occuring in nature, which may harm or effect living organisms directly or indirectly, immediately or after a long time.

We know sometimes in the past the basic amenities (air, land & water) were pure, virgin, undisturbed, uncontainminated and basically most hospitable for living organisms. But the situation is just the reverse today. Thus "Environmental Pollution" is the result of urban industrial technological revolution and speedy exploitation of every bit of natural resources." The craze of progress in agriculture industry, transportation and technology is taken as the general criteria of development of any nation. Such activities of man has created adverse effects on all living organisms in the biosphere. The rapid industrilization has left with us polluted rivers, contaiminated soil, depleted wild life and exhausted ; natural resources. Now the environment has become foul, contaminated, undesirable and therefore harmful for the health of living organisms, including man.

A rapid progress in atomic and in nuclear energy has added a huge amount of radioactive substances in the atmosphere. A number of industrial effluents and emissions, especially toxic gases are spewed in to the air dialy. Thus the environment is deteriorated to such an extent that it has crossed the critical limit and has become lethal to all organisms including man.

## TYPES OF POLLUTION

Pollutants need not be material substances. Noise can be a pollutant and even electromagnetic waves can be pollutants. However, today man is the principal source of pollution, because there is no way for nature to decompose many man made materials and return their elements to the cycle of nature. These substances will just remain and will cause whatever harmful effects they can, until they are somehow dispersed or diluted so that their action is no longer harmful. Thus pollution may be natural (which originates from natural processes such as forest fires, natural organic and inorganic decays, volcanic eruptions, earthquake etc.) or artificial which originates due to activities of man. There are following types of pollution -

I. Air pollution 2. Water pollution
3. Soil or Land pollution. 4. Noise pollution
5. Thermal pollution
6. Radioactive pollution or nuclear pollution.
7. Marine pollution 8. Industrial pollution
9. Biocide pollution 10. Estuarine or oceanic pollution
11. Acid rain pollution 12. Oil pollution
13. Chemical pollution
14. Drug pollution
15. Toxicity pollution 16. Sewage pollution.

Out of these some are important. They will be discussed here.

# AIR POLLUTION

## INTRODUCTION

According to U.S. Public Health Service, "Air pollution may be defined as the presence in the outdoor atmosphere of one or more contaminants or combination thereof in such quantities and of such duration as may be, or may tend to be injurious to human, plant or animal life, or property, or which unreasonably interfere with the comfortable enjoyment of life, or property, or the conduct of business "

The atmospheric structure and composition have been already discussed in the earlier sections of chapter-1. The atmosphere is a dynamic system, which steadily absorbs various pollutants from natural as well as man-made sources, thus acting as a ntural sink. Gases such as CO, $CO_2$, $H_2S$, $SO_2$ and $NO_x$ as well as particulate matter, such as sand and dust, are continually released into the atmosphere through natural activities such as forest fires, volcanic eruptions, decay of vegetation, winds and sand or dust storms. Man-made pollutant e.g. $CO_2$, $NO_x$, $SO_2$, $CO_2$, hydrocarbons, particulates etc are also released into the atmosphere. These have surpassed the pollutants contributed by nature thousand-fold. The magnitude of the problem of air-pollution has increased alarmingly due to population explosion, industrialization, urbanisation, automobiles and other human proclivities for greater comfort. The pollutants travel through the air, disperse and may interact with other substances in the atmosphere before they reach a sink, such as an ocean or a human receptor. If the pollutants enter the atmosphere at a faster rate than are absorbed by the natural sinks, then they gradually accumulate in the air. Such a disturbance in the dynamic equilibrium in the atmosphere by the air pollutants released by anthrapogenic activities resulting in considerable accumulation in the atmosphere may affect the very life on earth and its environment. Further, the dilution and dispersion of the gaseous pollutants in the atmosphere depend upon the meteorological conditions prevailing at a given time.

## CLASSIFICATION OF AIR-POLLUTANTS

The air pollutants may be classified in different ways as follows:

(*a*) **According to origin:**

(*i*) Primary pollutants which are directly emitted into the atmosphere and are found as such, *e.g.*, CO, $NO_2$, $SO_2$ and hydrocarbons.

(*ii*) Secondary pollutants which are derived from the primary pollutants due to chemical or photo-chemical reactions in the atmosphere, *e.g.*, Ozone, Peroxy-acyl nitrate (PAN), Photo-chemical smog, etc.

(*b*) **According to chemical composition :**

(*i*) Organic pollutants, *e.g.*,- Hydrocarbons, aldehydes, ketones, amines and alcohols

(*ii*) Inorganic pollutants

Carbon compounds (*e.g.*, CO and carbonates)

Nitrogen compounds (*e.g.*, $NO_x$ and $NH_3$)

Sulphur compounds (*e.g.*, $H_2S$, $SO_2$, $SO_3$ and $H_2SO_4$)

Halogen compounds (*e.g.*, HF, HCl and metallic fluorides)

Oxidising agents (*e.g.*, $O_3$)

Inorganic particles (*e.g.*, fly ash, silica, asbestos and dusts from transport, mining, metallurgical and other industrial activities).

(*c*) **According to state of matter :**

(*i*) Gaseous pollutants which get mixed with the air and do not normally settle out, *e.g.*, CO, $NO_x$ and $SO_2$.

(*ii*) Particulate pollutants which comprise of finely divided solids or liquids and often exist in colloidal state as aerosols, *e.g.*, - smoke, fumes, dust, mist, fog, smog and sprays.

## AIR POLLUTANTS AND THEIR EFFECTS

The most common air pollutants are : (1) Carbon monoxide (2) Oxides of nitrogen (3) Sulphur dioxide (4) Hydrocarbons and (5) Particulates.

**(1) Carbon monoxide (CO).** The atmosphere air contains 0.1 to 0.12 ppm of carbon monoxide. The natural processes which contribute to carbon monoxide in the atmosphere are volcanic activity, natural gas and marsh gas emissions, electrical discharges in the atmosphere during storms, seed germination etc. However, most of the CO in the atmosphere is due to human activities such as (1) Automobile exhausts (which accounts for 60% of CO in the atmosphere), (2) Forest fires and agricultural burning (*i.e.*, burning of forest debris, crop residue, bushes, weeds and vegetation, which contributes to about 17% of the CO in the atmosphere), and (3) Industrial operations such as electric and blast furnaces in iron and steel industry, petroleum refining, paper industry, gas manufacture and coal mining (which constitutes about 9.6% of CO in the atmosphere). The concentration of CO in city air is about 55 ppm. Soil micro-organisms act as a major sink for CO from ambient atmosphere. It has been demonstrated that 2.8 kg of a soil sample removed 120 ppm of CO from ambient air in 3 hours. Although soil sinks can take care of atmospheric CO, still it exists in significant concentrations in the atmosphere, because the largest CO-producing areas have the least amount of soil sink available around them.

Since transportation sector accounts for about 74% of the entire global CO-emissions, out of which major contribution comes from gasoline-fed internal combustion engines, lot of attention and efforts have been directed towards controlling CO-emission from the sector. Considerable amount of research work is going on along the following directions:

**(a) Modification of engine design.** Designing of more efficient internal combustion engines to minimise the emission of pollutants formed during combustion of gasoline.

The automobile emissions consist of CO, $NO_x$, hydrocarbons and particulates. If some control measures are applied to reduce one of these pollutants, the amount of other pollutants are also affected simultaneously. For instance, a low air-fuel ratio reduces $NO_x$ emissions considerably but increases CO and hydrocarbon emissions. The stoichiometric air-fuel ratio is the ratio at which the right proportion of oxygen is maintained for complete oxidation of C and $H_2$ to $CO_2$ and $H_2O$ respectively. However, automobile carburettors are set to the fuel rich side of the stoichiometric ratio (*i.e.*, greater % of fuel in the air-fuel mixture) because the engine tends to stall on the lean-fuel side (*i.e.*, lessor % of fuel in the air-fuel mixture).

Using leaner air-fuel ratios which results in improved fuel distribution and retarded ignition timing and the use of lower compression ratios are some of the modifications in engine design that have proved successful in checking pollution but with some sacrifice in fuel and power economy.

**(b) Fuel modification and development of substitute fuels.** Efforts are continuing to develop fuel substitutes for gasoline which, on combustion may release lower concentrations of pollutants. Natural gas, methane, steam reformed gasoline, blends of light hydrocarbons etc., have been recommended. With such fuels, emissions due to incomplete combustion are eliminated thereby reducing noxious pollutants in the exhaust gases. However, further technological progress is necessary to enable their wider use. Alcohols and blends of gasoline and alcohol have also been used as fuels. Some of their combustion products such as aldehydes are eye irritants.

Some fuels additive such as barium salts have been also reported to produce excellent results and further studies are in progress in the field of fuel additives too. Extensive toxicological studies are also needed before introducing any new fuel additives.

Tetraethyl lead (TEL), which is widely used as anti knocking additive to gasoline, is the major source of lead pollution in the environment. Lead-free gasoline is being marketed in USA and some parts of Europe in an attempt to curb lead pollution. These lead-free gasoline compositions are more expensive at present but thorough studies (including toxicological) are needed before they are released and recommended for general use.

**(c) Treatment of exhaust gases.** Treatment of exhaust gases with the help of thermal or catalytic reactors in the exhaust pipe has been a topic for extensive research. In two stage catalytic converters, $NO_x$ are reduced to $N_2$ and $NH_3$ in the first converter at elevated temperatures in presence of catalysts such as Pt, Pd and Ruthenium or some base metal alloys in the presence of reducing gases *e.g.*, CO and hydrocarbons. For oxidising CO and hydrocarbons in the second converter, oxidising catalysts of noble metals (e.g., Pt, Ru and Ir) or oxides of CO, Cr, Cu, Ni, Fe and Mn supported on ceramic materials are used in conjunction with suction of additional air to ensure complete oxidation. However, in case of noble metal catalysts, lead-free gasoline will have to be used to prevent poisoning of the catalysts. Further, catalysts suitable for use in automobiles should be cheap, harmless, long-lasting and widely available. As an alternative to catalytic converters, "stratified charge engines" are being developed in Japan. These engines provide for an additional combustion chamber, where a fuel-rich mixture is introduced and ignited with a spark. This sets in combustion at a relatively low temperature, due to which formation of $NO_x$ is minimised. Then the burning mixture enters into the larger main chamber where it gets mixed with a lean-fuel mixture (*i.e.*, higher % of air and lesser % of fuel in the air-fuel mixture). This ensures complete combustion of CO and hydrocarbons without stalling the engine and, at the same time, build up of $NO_x$ is limited because of the lower temperature maintained in the chamber. However extensive research has to be carried out before introducing such engines on a large scale in automobiles.

**(d) Exhaust gas recirculation.** It is another approach that is being studied for the purpose of automobile pollution abatement.

**(2) Oxides of Nitrogen ($NO_x$).** Out of the eight possible oxides of nitrogen, only $N_2O$, NO and $NO_2$ are the important constituents of the atmosphere. The atmospheric background concentrations of $N_2O$, NO and $NO_2$ are 0.25 ppm ; 0.1 to 2 ppm ; and 0.5 to 4 ppm respectively. Although the concentration of $N_2O$ is more in the atmosphere, NO and $NO_2$ are more significant from air pollution point of view and they are usually represented together as $NO_x$. The formation of NO from $N_2$ and $O_2$ is favoured at high temperatures (~1210 to 1765°C) which are usually attained in the combustion processes involving air. Rapid cooling of the combustion products (quenching) prevents the dissociation of NO. The oxidation of NO to $NO_2$ is also favoured at high temperature (~1100 °C) but the amount of $NO_2$ formed is usually not more than 0.5% of the total $NO_x$ present. $NO_2$ is also formed by photolytic reactions in the atmosphere, as discussed earlier. Some of the reactions involved are as follows:

$$N_2 + O_2 \xrightarrow{[1210\text{-}1765\ °C]} 2NO$$

$$2NO + O_2 \longrightarrow 2NO_2$$

$$NO_2 + h\nu \xrightarrow[398\ nm]{} NO + O$$

$$O_2 + h\nu \xrightarrow[242\ nm]{} O + O$$

$$O + O_2 + M \text{ (third body)} \longrightarrow O_3 + M$$

$$O_3 + NO \longrightarrow NO_2 + O_2$$

$$2NO_2 + O_3 \longrightarrow N_2O_5 + O_2$$

$$4NO_2 + 2H_2O + O_2 \longrightarrow 4HNO_3$$

Alternative mechanism for Nitric Acid formation:

(a) $O_3 + NO_2 \longrightarrow NO_3 + O_2$

(b) $NO_3 + NO_2 \longrightarrow N_2O_5$

(c) $N_2O_5 + H_2O \longrightarrow 2HNO_3$

Oxides of nitrogen may be formed either by the natural or artificial fixation of nitrogen from the atmosphere or from nitrogen compounds present in organic matter. The annual global release of $NO_x$ from man-made sources is about $5 \times 10^7$ tonnes, which is only slightly less than that discharged by natural bacterial activity. Oxides of nitrogen are produced by the combustion of coal, oil, natural gas and other organic matter. Thus, $NO_x$ is introduced into the atmosphere from automobile exhausts, incinerators, furnace stacks, coal based power plants and other similar sources. When the fuels are burnt in air, some of the $N_2$ in the air is oxidised to NO. The amount of NO formed depends on the temperature of the flame and the rate of cooling or quenching of the combustion products. Higher flame temperatures and rapid cooling of the combustion products favour the formation of NO.

The average residence times of NO and $NO_2$ in the atmosphere are 4 days and 3 days respectively. They undergo various photochemical and chemical reactions in the atmosphere, leading to the formation of $HNO_3$ which gets precipitated as nitrates rainfall or as dust.

The $NO_x$ from man-made sources may be 10 to 100 times more in urban areas, as compared to rural areas. Even in urban areas, the ambient $NO_x$ levels vary with sunlight and traffic density at any given point of time.

Chemical and photo-chemical reactions involving $NO_2$ and hydrocarbons induced by sunlight are responsible for the formation of photochemical smog.

As discussed in the earlier section, use of two stage catalytic converters can minimize the $NO_x$ from automobile emissions. Similarly, $NO_x$ from power plant emissions can be reduced by 90% by using a two-stage combustion process. The fuel can be first fired at a relatively high temperature using only about 90% of the stoichiometric air required so that only a minimum quantity of NO is formed under these conditions. Then the combustion of fuel may be completed at a relatively low temperature, in the excess air. NO is not formed under these conditions.

(3) **Sulphur dioxide ($SO_2$).** Combustion of any sulphur-bearing materials produces $SO_2$ accompanied by a small quantity of $SO_3$. This mixture is usually denoted as $SO_x$. Nearly 67% of the global $SO_x$ pollution is due to volcanic activity and other natural sources, over which we have no control. The remaining 33% $SO_2$ emission is because of human activities such as combustion of fuels, coal-fired power stations, transportation, refineries, metallurgical operations such as

smelting of suphide ores and chemical plants *e.g.*, manufacture of sulphuric acid. Most of the man-made $SO_x$ pollution is concentrated in urban and industrial areas.

Almost all the sulphur present in liquid and gaseous fuels and about 80% of sulphur present in the solid fuels appears as $SO_x$ in the flue gases. Depending on the sulphur content of the fuel burnt and the conditions of combustion (*e.g.*, % of excess air used), the concentration of $SO_4$ in flue gases varies from 0.05 to 0.4%. However, in metallurgical operations such as smelting of sulphide ores, the $SO_2$ concentration in stack gases may be 5 to 10%.

$SO_2$ is oxidised to $SO_3$ in atmospheric air by photolytic and catalytic processes involving ozone, $NO_x$ and hydrocarbons, giving rise to the formation of photochemical smog. Oxidation of $SO_2$ can take place in presence of catalysts such as $NO_x$, metal oxides, soot and dust. Under ormal humid conditions of the atmosphere, $SO_3$ reacts with water vapour to produce droplets of $H_2SO_4$ aerosol which give rise to the so-called "acid rain" discussed later. The sulphuric acid and sulphate aerosols present in the urban air are smaller than 2μ and hence can easily reach the pulmonary region of lungs, causing serious respiratory problems, particularly in old people.

$$SO_2 + O_3 \longrightarrow SO_3 + O_2$$

$$SO_3 + H_2O \longrightarrow H_2SO_4 \longrightarrow (H_2SO_4)_n$$

$$SO_2 + \tfrac{1}{2}O_2 + H_2O \xrightarrow[\text{metal oxide, soot etc.}]{\text{Catalyst such as}} H_2SO_4 \xrightarrow[\text{aerosol}]{\text{aerosol}} (H_2SO_4)_n$$

Control of $SO_x$ emissions from the anthropogenic activities is contemplated on the following lines:

(1) Removing $SO_x$ from flue gases before letting them out into the atmosphere: Chemical scrubbers such as (a) Lime stone (b) Citric acid are suggested to absorb $SO_2$ from the flue gases.

(*a*) $2CaCO_3 + 2SO_2 + O_2 \longrightarrow 2CaSO_4 + CO_2$

(*b*) $SO_2 + H_2O \longrightarrow HSO_3^- + H^+$

$HSO_3^- + H_2\,cit^- \longrightarrow (HSO_3 \cdot H_2\,cit)^{-2}$

(2) Removing sulphur from the fuels used for combustion: Pyritic sulphur in coal can be removed by grinding and washing in coal washeries. However organically bound sulphur cannot be easily removed from coals. Research is in progress to synthesis special type of micro-organisms using bio-technology, which are capable of converting organically bound sulpur into soluble form.

(3) Utilizing low-sulphur fuels.

(4) Generation of power by alternative energy sources and discouraging fossil-fuel based thermal power-plants.

## ACID RAIN (As a result of above gases)

Rain has always been valued by mankind, because good crops and abundant water supplies are possible only by timely and plentitul rainfall. Summer rains refresh people. Spring rains recharge the aquifers and cleanse the groundwaters. Autumn rains and winter snow help cleansing the air. Rain in general, brings with it a sense of hope, vitality and a promise for the future.

However, over the last few decades, simple rainfall has taken on a threatening complexity in some parts of the world. In these localities, the rain must pass through an atmosphere polluted with oxides of sulphur ($SO_x$) and oxides of nitrogen ($NO_x$). The falling rain and snow react with these oxide pollutants to produce often a mixture of sulphuric acid, nitric acid and water. This is known as acid precipitation or acid rain.

Rain tends to be naturally acidic with a pH of 5.6 to 5.7 due to the reaction of atmospheric $CO_2$ with water to produce carbonic acid. This small amount of acidity is, however, sufficient to dissolve minerals in the earth's crust and make them available to plant and animal life; yet not acidic enough to inflict any damage. Other atmospheric substances from volcanic eruptions, forest fires and other similar natural phenomena also contribute to the natural sources of acidity in rain. Still, even with the enormous amounts of acids created by nature annually, normal rainfall is able to assimilate them to the point where they cause little, if any, known damage. But, it is the contributions of $SO_x$, $NO_x$, etc from anthropogenic activities that disturb this acid balance and convert natural and mildly acidic rain into precipitation with far-reaching environmental consequences.

It has been stated in the earlier sections that $SO_x$ and $NO_x$ emissions into the atmosphere are eventually converted into $H_2SO_4$ and $HNO_3$ droplets due to a series of photochemical reactions and chemical reactions catalysed by other species present in the atmosphere. These acid droplets are partly neutralised with bases *e.g.*, particulate lime, $NH_3$ etc. These salts and the remaining $H_2SO_4$ and $HNO_3$ droplets along with HCl released into the atmosphere by man-made and natural HCl emissions give rise to acidic precipitation, which is popularly known as "Acid Rain." Analysis of acid rain samples showed the following species: $H^+$, $NH_4^+$, $Na^+$, $K^+$, $Ca^{2+}$, $Mg^{2+}$, $SO^{4-}$, $NO^{3-}$ and $Cl^-$. The contributions from the three acids in the "acid rain" are mostly in the order. $H_2SO_4 > HNO_3 > HCl$.

Acid rainfall may occur at a place far away, may be even 500 km to 1000 km from the sources of the pollutants. Some events of acid rainfall in Sweden and Canada have been traced to be due to large scale $SO_x$ emissions from densely populated industrial areas of UK and USA.

Acid rain represents one of the major consequences of air pollution, because of large $SO_x$ and $NO_x$ emissions from big industrial areas into the atmosphere. The longer the $SO_x$ and $NO_x$ remain in the atmosphere, the greater the chances of their oxidation to $H_2SO_4$ and $HNO_3$ by the various photochemical and catalytic chemical reactions. Acid rains may cause extensive damage to materials and terrestrial ecosystems such as water, fish, vegetation, stone, steel, paint, soil and mankind as follows:

(1) Damage to building and structrual materials as well as valuable ancient sculptures, carved from marble, limestone, sand-stone etc., because of pitting and mechanical weakening due to attack by the acidic components. Deformation and degeneration of stone statues in countries like Greece and Italy have been reported to be due to damage by the acid rains. Architectural monuments like Taj Mahal in our country will face a similar threat if proper measures are not taken.

(2) Acidification of soils with the consequent effects on microbial and soil fauna and fixation of nitrogen. Also, indirect effects due to alteration in soil chemistry because of acidification of soils leading to reduced forest productivity.

(3) Foliar damage to crops and forests, leachig of nutrients from leaves and alteration of seed germination characteristics. Damage to young growing plant tissues and the process of photosynthesis, thus hindering the development of plants and threatening their very survival.

(4) Potential effects on aquatic systems, such as acidification, decreased alkalinity and mobilization of metals like aluminium.

(5) Other biological effects on aquatic biota, such as altered species composition among plankton, vegetation and invertebrates, decline in productivity of fish and amphibian, skeletal deformities and increased fish mortality.

(6) Corrosive damage to steel, zinc, oil-based paints and automobile coatings.

(7) Possible effects on humans. Lungs, skin, hair may be affected.

The heavy metals released by acid rain also may cause potential threat to human health. Acidification of drinking water reservoirs and concurrent increases in heavy metal concentrations may exceed public health limits and may cause injurious effects.

The extent of damage due to acid rain depends upon factors such as climate, topography, geology, biota and human activity.

The phenomenon of acid rains is a highly interactive problem and remedial measures to control it are very expensive The potential probability of acid rains is one of the major arguments against installation of coal-based thermal power stations. Keeping in view the potential ill-effects of acid rains, it is highly desirable to control the pollutants (causing acid rains) at source, rather than treating the undesirable consequences of acid rains.

The only practical approach to counter to problem of acid rain is to reduce $SO_x$ and $NO_x$ emissions. The following three general options are considered for this purpose:

**(1) Energy conservation** resulting in reduced fuel consumption and hence slower emissions of $SO_x$ and $NO_x$. Conservation via more efficient fuel use and through improved thermal insulation is also being studied.

**(2) Desulphurization and denitrification** of fuels of stack gases and increased use of fuels naturally low in sulphur content of use to technologies that reduce the $SO_x$ and $NO_x$ emissions. Desulphurization and use of low $NO_x$ producing technologies are the only viable control options today and will perhaps continue to be so for some more time.

**(3) Substitutions** for fossil fuels by other alternative energy forms may offer future solutions to this problem.

Reduction of $SO_x$ emissions can be accomplished by (*a*) removing the sulphur content before the fuel is burnt with the help of techniques such as coal cleaning, coal gasification and desulphurization of liquid fuels (*b*) removing the sulphur content during combustion, as in fluidized-bed combustion, and (*c*) removal of sulphur emissions after combustion, as in stack or flue gas desulphurization systems or scrubbers. The future of $SO_x$ control from traditional fuel sources lies in the perfection of these techniques.

Reduction of $NO_x$ emissions from stationary combustion sources can be achieved by modification of furnance and burner design and/or modification of operating conditions. The combustion modification techniques available now include using 2-stage combustion, precisely controlling air, injecting water during combustion, recirculating flue gases, and/or by altering design of firing chambers. $NO_x$ emissions from mobile combustion sources may be achieved by lowering the combustion temperatures in the engine and catalytic removal of $NO_x$ from exhaust gases using devices such as a 3-way system that reduces carbon monoxide, hydrocarbons and $NO_x$ simultaneously.

Some techniques used for controlling $SO_x$ and $NO_x$ emissions have already been described in earlier sections. More extensive research efforts are necessary to develop economical and effective technologies to control $SO_x$ emissions at the source, to discern the mechanism of long-distance dispersion of $SO_x$ emissions and to monitor and study the various effects of acid rains on plants, animals and humans in greater details.

It is interesting to note that some scientists feel that $SO_x$, an air-pollutant resulting from burning fossil fuel and a major contributor to "acid rain", may also counteract the dangerous phenomenon of "global warming due to the Greenhouse effect", by modifying the cloud formation in such a way as to bounce back the sun-light into space and cool the planet.

**(4) Hydrocarbons**

Hydrocarbons are emitted into the atmosphere by natural biological activity as well as anthropogenic sources such as automobile exhausts; burning of coal, oil, wood and refuse; and solvent evaporation. Anthropogenic sources account for about 15% of the total hydrocarbon emissions into the atmosphere. The annual global hydrocarbon emission to the atmosphere by man-made sources is roughly estimated to be $57 \times 10^7$ tonnes per year. Methane ($CH_4$) is the major hydrocarbon emitted into the atmosphere by natural activities such as anaerobic decomposition of organic matter in water, soil and sediments by micro-organisms. About twenty other hydrocarbons have been identified in ambient atmosphere in areas of heavy vehicular traffic which include ethane, n-butane, n-pentane, isopentane, isobutane, m-xylene, propane, ethylene, acetylene and toluene, The Los Angeles basin (USA) has been extensively studied for hydrocarbon pollution because of its heavy automobile traffic and also because of its topography and climatic features which result in heavy accumulation of hydrocarbons. These hydrocarbons being thermodynamically unstable tend to get oxidised in the atmosphere by a series of chemical and photochemical reactions. This gives rise to the formation of various end-products such as $CO_2$, solid organic particulates (which settle down) and water-soluble acids and aldehydes which are washed down by rain. It is some of these products from photochemicals reactions of hydrocarbons that have harmful effects on human beings rather than the hydrocarbon themselves. The hydrocarbons have a definite role to play in the formation of "photochemical smog" which is characterised by reduced visibility, eye irritation, peculiar odour and damage to vegetation and accelerated cracking of rubber products. (Fig. 2)

## PHOTOCHEMICAL SMOG

When the atmosphere is loaded with large quantities of automobile exhausts during warm sunny days with gentle winds and low level inversion, then the exhaust gases are trapped by the inversion layers with stagnant air masses and simultaneously exposed to intense sunlight. Then, a number of photochemical reactions involving $NO_2$, hydrocarbons and other organic compounds are free redicals take place leading to the formation of ozone, peroxides and other photochemical oxidants in the atmosphere (Figure). This gives rise to the phenomenon of "photochemical smog" which is characterised by the formation of aerosols that reduce visibility, generation of brown hazy fumes that irritate the eyes and lungs, and which cause extensive damge to vegetation and rubber goods. Photochemical smog was observed in some parts of Los Angeles and Denver in USA and that is why it is sometimes referred to as "Los Angeles Smog". Photochemical smog is an oxidising smog and it should be clearly distinguished from the usual reducing "smog" which forms due to the combination of smoke and fog.

For controlling the formation of noxious photochemical pollutants in the atmosphere, the emission of primary pollutants *viz*. hydrocarbons and $NO_x$ should be controlled.

**(5) Particulates**

Some details about particulates have already been presented in a preceding section. Particulates (solid or liquid) are important constituents of the atmosphere. About 2000 million tonnes of particulate matter per year are released from natural agencies such as volcanic eruptions, wind and dust storms, salt spray etc. Man-made activities such as burning of wood, coal, oil and gaseous fuels, industrial processes, smelting and mining operations, fly-ash emissions from power plants, forest fires, burning of coal refuse and agricultural refuse etc., release about 450 million tonnes of particulates per year. The diameter of particulates may range

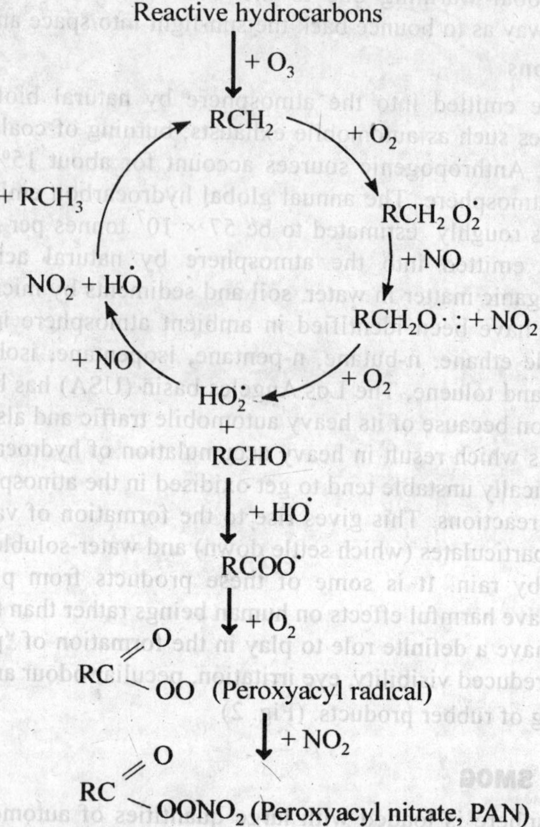


**Fig. 2.** Typical free-radical reactions, involved in Photo-Chemical Smog formation

from $2 \times 10^{-4}$ μ to $5 \times 10^{2}$ μ with varying life times depending upon the size and density of the particles and turbulence of air which control their settling rate. Their number per cubic centimeter may vary from several hundreds to several thousands. In industrial and urban areas, the number of particulates may be more than a lakh per cubic centimeter with mass varying from 60 μg to 2 mg/m$^3$. However, the size of the particulates and their chemical nature are more vital than their number. Owing to their large surface areas, particulates provide excellent sites for absorption of various organic and inorganic species which encourage heterogeneous phase reactions in the atmosphere. The particles are also capable of scattering light and reducing visibility. Particulates include $Fe_3O_4$, $V_2O_5$, CaO, $PbCl_2$, $PbBr_2$, fly ash, aerosols, soot etc. Polycyclic aromatic hydrocarbons (PAH) are important constituents of several organic particulates which are carcinogenic. Soot is a highly condensed product of PAH compounds and can itself absorb many PAH compounds and toxic trace metals *e.g.*, Be, Cd, V, Cr, Ni and Mn as well as carcinogenic organics such as benzo-α-pyrenes.

Fine particulates having size of < 3μ (such as airborne toxic metals like Be and airborne asbestos) which can penetrate through nose and throat, reach the lungs and cause breathing problems and irritation of the lung capillaries. Similarly, pulmonary fibrosis in asbestos mine workers, black-lung disease in coal miners and emphysema in urban populations are attributed to the particulate pollution. Further, airborne particulates such as dust, mist, fumes and soot can cause damage to various materials. Particulates may accelerate corrosion of metals and cause damage to paints and sculptures. The extent of damage depend upon the physical and chemical properties of the particulates. Particulates present in the atmosphere may influence the climate through formation of clouds and snow. They also may absorb solar radiation and reduce visibility.

Particulate matter can be removed from gaseous effluents from industries with the help of devices such as cyclone collectors, settling chambers, wet scrubbers and electrostatic precipitators.

### Characteristics and biochemical effects of some important air pollutants

**1. Oxides of Sulphur ($SO_x$)**

**Characteristics :** $SO_x$ comprises of $SO_2$ (97 to 99%) and $SO_3$ (1 to 3%). It is colourless, heavy, water-soluble gas with pungent and irritating odour. Rapidly diffusing, acid-forming oxidizing agent. Reacts with water forming $H_2SO_3$ and $H_2SO_4$, both of which can react with organic matter, metals and materials.

**Biochemical effects:** Absorbs quickly and irritates the upper respiratory tract. Reacts with cellular constituent chemicals e.g., enzymes. The $H_2SO_4$ formed lowers pH, impairs enzymatic functions and destroys various functional molecules. Leads to bronchial spasms, breathlessness, impaired pulmonary function via airway resistance, impaired lung clearance and increased susceptibility for infection.

**2. Oxides of Nitrogen ($NO_x$)**

**Characteristics :** $NO_x$ mostly comprises of NO, $NO_2$ and $N_2O$. NO is colourless gas and is slightly soluble in water. $NO_2$ is reddish brown gas, somewhat water-soluble, oxidizing agent, can react with water to form $HNO_3$ which is a powerful oxidizing agent and capable of reacting with almost all metals and many organic compounds. $NO_2$ can travel into the respiratory system. It is also involved in the formation of ozone in the atmosphere.

**Biochemical effects:** Oxidises cellular lipids. Forms bonds with haemoglobin and reduces the efficiency of oxygen transport. Disrupts some cellular enzyme systems. Higher levels and prolonged exposures may cause pulmonary fibrosis, inflammation of lung tissues and may eventually lead to death. Causes nitric acid mediated effects some of which are similar to that of $H_2SO_4$ described above. NO can form addition compound with haemoglobin, if it enters the blood stream.

**3. Carbon monoxide (CO)**

**Characteristics :** Colourless, odourless, toxic gas, slightly water soluble but still is extremely dangerous because it has a greater affinity for haemoglobin than that of $O_2$.

**Biochemical effect:** It competitively inhibits combination of $O_2$ and haemoglobin. It attacks haeomoglobin and displaces $O_2$ to form carboxyhaemoglobin, and thus reducing the oxygen carrying capacity of blood.

(Under normal conditions): $O_2 + Hb \longrightarrow \underset{\text{(oxyhaemoglobin)}}{O_2Hb}$

(In presence of CO) : $O_2Hb + CO \longrightarrow \underset{\text{(carboxyhaemoglobin)}}{CO\,Hb} + O_2$

The immediate response to CO-poisoning is loss of judgement, which is responsible for many automobile accidents. Further exposure to higher levels of CO leads to various metabolic disorders such as asphyxiation and causes death. CO-poisoning can be cured by providing fresh $O_2$' which reverses the above reaction.

**4. Ozone ($O_3$) and other photochemical oxidants such as peroxyacetyl nitrate (PAN) Present in photochemical smog.**

**Characteristics :** Ozone is a pale blue gas, fairly water soluble, unstable, sweetish odour. Very reactive oxidizing agent capable of combining with many organic compounds in cells and tissues as well as with rubber and other materials. $O_3$ and PAN are harmful to plants, animals and humans.

**Biochemical effects:** Oxidize cellular constituents. PAN and ozone toxicity is produced via generation of free radicals. The free radicals produced may damage DNA and thus alter cellular genetic integrity too. The toxic effects of ozone are manifested after inhalation and absorption in the lungs causing accumulation of fluids in the lungs (pulmonary edema), damaging lung capillaries and mortality if continued or high level exposures occur. Both $O_3$ and photochemical smog cause irritation of the eyes and respiratory tract. The free radicals produced by $O_3$ and other photochemical oxidants attack the sulphydryl groups on the enzymes and also inactivate enzymes like isocitric dehydrogenase, malic dehydrogenase and glucose-6-phosphate dehydrogenase, which are so much involved and essential for citric acid cycle and generation of cellular energy. $O_3$ may also inhibit the activity of some enzymes involved in synthesis of cellulose and lipids in plants. Among the sulphur containing amino-acids, cysteine is strongly attacked by PAN.

**5. Hydrocarbons (and other volatile organic compounds)**

**Characteristics :** Very reactive. React with many kinds of compounds yielding many kinds of products. Volatile hydrocarbons and other organic compounds participate in atmosphere reactions generating ozone.

**Biochemical effects:** Some of these compouds can react with the constituents of the cells. Carcinogenic hydrocarbons like benzopyrene can react with DNA causing mutations and cancer.

**6. Particulate Matter**

**Characteristics :** Solid particles or liquid droplets including fumes, smoke, dust and aerosols. Solid particular can adsorb various chemicals.

**Biochemical effects :** Effects vary with the nature of the particles. Carbon particles and other particles cause scarring of lungs via complex walling off and fibrogenic reactions leading to a disease condition known as "Pneumoconiosis'. Particles carrying absorbed mutagens lead to damage of DNA in the lungs and elsewhere.

**7. Heavy Metals**

Characteristics: Characteristics depend upon the nature of the particular metal.

**Biochemical effect :** Impairment of enzymatic activity by attacking sulphydryl and other functional groups. Other biochemical effects depend on the chemical nature of the metals involved (vide Chapter VII).

The effect of some important air-pollutants and their sources are summarised in Table 2.1.

**Table 2.1. Sources of some important air pollutants and their effect on human beings and animals**

| Pollutant | Major Sources | Typical Effects |
|---|---|---|
| Carbon monoxide (CO) | Incomplete combustion of fules, automobile exhaust, jet engine emissions, blast furnaces, mines and tobacco smoking | Toxicity, blood poisoning, increased proneness to accidents,CNS impairment. CO combines with haemoglobin, forming carboxy |

| Pollutant | Major Sources | Typical Effects |
|---|---|---|
| | | haemoglobin, which is useless for respiratory purposes and hence leads to death. |
| Sulphur dioxide ($SO_2$) | Combustion of coal, combustion of petroleum products, burning of refuse, petroleum industry, oil refining, power houses, sulphuric acid plants, metallurgical operations and from domestic buning of fuels. | Increased breathing rate and feeling of air-starvation, suffocation, aggravation of asthma and chronic bronchitis, impairment of pulmonary functions, respiratory irritation, sensory irritation, irritation of throat and eyes. |
| Oxides of Nitrogen ($NO_x$) | Automobile exhausts, coal-fired and gas-fired furnaces, boilers, power stations, explosives industry, fertilizer industry, manufacture of $HNO_3$, combustion of wood and refuse. | Respiratory irritation, headache, bronchitis, pulmonary emphysema, impairment of lung defences, edema of lungs, lachrymatory effect, loss of appetite, corrosion of teeth. |
| Hydrogen sulphide ($H_2S$) | Coke ovens, kraft paper mills, petroleum industry, oil refining, viscose rayon manufacturing plants, manufacture of dyes, tanning industry and sewage treatment plants. | Headaches, conjunctives, sleeplessness, pain in the eyes, irritation of respiratory tract, respiratory paralysis, asphyxiation, malodorous. In high concentrations, it may lead to blockage, of oxygen transfer, poisoning cell enzymes and damaging nerve tissues. |
| Chlorine ($Cl_2$) | Accidental breakage of chlorine cylinders, electrolysis of brine, bleaching of cotton pulp and other process industries using chlorine | Irritation to eyes, nose and throat, toxicity, respiratory irritation, lachrymatory effects. In large doses it may cause edema, pneumonitis, emphysema and bronchitis. |
| Hydrogen fluoride (HF) | Glass fibre manufacture, chemical industry, fertilizer industry, aluminium industry, ceramic industry, phosphate rock processing. | Irritation, respiratory diseases, fluorosis of bones, mottling of teeth. |
| Carbon dioxide ($CO_2$) | Combustion of fuels, automobile exhausts, jet engine emissions. | Toxic in large quantities, Hypoxia. |
| Hydrocarbons (HC) | Organic chemical industries, petroleum refineries, automobile exhausts, rubber manufacture. | Some hydrocarbons have carcinogenic effects, lachrymatory effect. |
| Oxidants (*e.g.*, $O_3$) | Photochemical reactions in atmosphere involving organic materials and $NO_2$ etc. Reactions induced by silent electrical discharge and intense u.v. radiations in the atmosphere. | Irritation of lungs, eyes and respiratory tract. Accumulation of fluids in lungs and damage to lung capillaries. These biochemical effects of $O_3$ mostly arise from generation of free radicals which attack the –SH groups present in the enzymes. |

| Pollutant | Major Sources | Typical Effects |
|---|---|---|
| Dusts | Asbestors factories, mining activities, power stations, metallurgical industries, ceramic industry, factory stacks, glass industry, cement industry, foundaries | Respiratory diseases, toxicity from metallic dusts, silicosis and asbestosis from the specific dusts. Asbestos dust causes pulmonary fibrosis, pleural calcification and lung cancer. |
| Ammonia ($NH_3$) | Chemical industries, coke oven refineries, stocks yards, fuel incineration. | Damage to respiratory tracts and eyes, corrosive to mucous membranes. |
| Formaldehyde (HCHO) | Waste incineration, automobile exhausts, combustion of fuels, photochemical reactions. | Irritation to eyes, skin and respiratory tract. |
| Hydrochloric acid (HCl) | By-product from chlorination of organic compounds, combustion of gasoline in which ethylene dichloride is present, burning of coal, paper and chlorinated plastics. | Inflammation and ulceration of upper respiratory tract, clouding of cornea, coughing and choking while inhalation. |
| Radioactive gases and dusts | Radioactive gases are suspended dusts from natural and artificial radioisotope sources. | Somatic effects such as leukaemia and other types of cancer, cataracts and reduction in life expectancy. Genetic effects such as mutations in human gametes. |
| Arsenic (As) | Arsenic containing fungicides, pesticides and herbicides, metal smelters, by-product of mining activities, chemical wastes. | Inhalation, ingestion or absorption through skin can cause mild bronchitis nasal irritation or dermatitis. Carcinogenic activity also is suspected. Attack SH groups of enzymes, coagulate proteins. |
| Beryllium (Be) | Coal, nuclear power and space industries, production of fluorescent lamps, motor fuels and other industrial use. | Damage to skin and mucous membrane, pulmonary damage, perhaps carcinogenic. |
| Boron (B) | Boron producing units, production and use of petroleum fuel and additives, burning coal and industrial wastes, detergent formulations. | Ingestion or inhalation as dust causes irritation and inflammation. Boron hydrides can damages CNS and may result in death. |
| Cadmium (Cd) | Cardmium producing industries, electroplating, welding. By product from refining of Pb, Zn and Cu, fertilizer industry, pesticide manufacture, cadmium nickel batteries, nuclear fission plants, production of TEL used as additive in petrol. | Inhalation of fumes and vapours causes kidney damage, bronchitis, gastric and intestinal disorders, cancer, disorder of heart, liver and brain. Chronic and acute poisoning may reslut. Renal dysfunction, anaemia, hypertension, bone-marrow disorder and cancer. |

| Pollutant | Major Sources | Typical Effects |
|---|---|---|
| Chromium (Cr) | Metallurgical and chemical industries, process using chromate compounds, cement and asbestos units. | Toxic to body tissues, can cause irritation, dermatitis, ulceration of skin, perforation of nasal septum. Carcinogenic action suspected. |
| Lead (Pb) | Automobile emissions, lead smelters, burning of coal or oil, lead arsenate pesticides, smoking mining and plumbing. | Absorption through gastrointestinal and respiratory tract and deposition in mucous membranes, cause liver and kidney damage, gastro-intestinal damage, mental retardation in children, abnormalities in fertility and pregnancy. |
| Zinc (Zn) | Zinc refineries, galvanizing processes, brass manufacture, metal plating, plumbing. | Zinc fumes have corrosive effects of skin and can cause irritation and damage mucous membranes. |
| Manganese (Mn) | Ferromanganese production, welding rods, incineration of organo-manganese fuel additives, manganese containing substances. | Poisoning of CNS, absorption, ingestion, inhalation, or skin contact may cause manganic pneumonia. Respiratory disorders, dermatitis, |
| Nickel (Ni) | Metallurgical industries using nickel, combustion of fuels containing nickel additives, burning of coal and oil, electroplating units using nickel salts, incineration of nickel containing substances, vanaspati manufacture. | cancer of lungs and sinus. |
| Vanadium (V) | Vanadium refining, production of vanadium containing alloys, power plants, burning of oil rich in vanadium. | Gastro-intestinal and respiratory disorders, inhibition of synthesis of cholesterol, heart disease and cancer in case of chronic exposures. |
| Selenium (Se) | Burning of fuels and residual oils, fumes and gases from refinery wastes, incineration of paper and other wastes, natural sources. | Irritation of gastro-intestinal and respiratory tracts, irritation of eyes, nose and throat, damage to lungs, liver and kidneys. |
| Mercury (Hg) | Mining and refining of mercury, organic mercurials used in pesticides, laboratories using mercury. | Inhalation of mercury vapours may cause toxic effects and protoplasmic poisoning. Organo-mercurials are highly toxic and may cause irreversible damage to nervous system and brain. |

**Effects of Air Pollutants on Man and his Environment**

**(1) Damage to materials.** The materials that may be affected by air pollutants include metals; building materials, rubbers, elastomers, paper, textiles, leather, dyes, glass, enamels and surface coatings. The types of possible damage to these materials by air pollutants include corrosion, abrasion, deposition, direct chemical attack and indirect chemical attack. The intensity of damage depends upon factors such as moisture, temperature, sunlight, air movement and of course of nature and concentration of the pollutant. Table 2.2. Summarizes the possible damage to various materials because of air pollution.

### Table 2.2. Manifestation of Air Pollution of Various Materials

| Meterials | Type of damage | Principal air pollutants | Other environmental factors influencing the rate of attack |
|---|---|---|---|
| Metals | Corrosion, soiling and tarnishing | $SO_x$ other acid gases | Moisture, temperature, air, salt |
| Building Materials | Surface erosion leaching, corrosion discoloration | $SO_x$, other acid gases acid mist, sticky particulate matter. | Moisture, freezing, temperature |
| Paints | Discoloration, softened finish, surface erosion | $SO_2$, $H_2S$, $O_3$, sticky particulates | Moisture, sunlight, fungus and other micro-organisms. |
| Textiles | Soiling, spotting, reduced tensile strength | $SO_x$, $NO_x$, other gases, particulates | Moisture, sunlight, physical ware. |
| Leather | Weakening, powdered surface | $SO_x$, other acid gases | Physical ware |
| Paper | Embrittlement | $SO_x$, other acid gases | Moisture, sunlight physical ware |
| Dyes | Fading, colour change | $NO_x$, ozone, oxidants | Sunlight, Moisture. |
| Ceramics | Change in surface appearance | HF, acid, gases | Moisture |
| Rubber | Cracking, Weakening. | Ozone oxidants | Sunlight. |

**(2) Damage to Vegetation.** Air pollutants, such as sulphur dioxide, HF, particulate fluorides, smog, oxidants like ozone, ethylene (from automobiles), $NO_x$, chlorine and herbicide and weedicide sprays exert toxic effects on vegetation. The damage usually manifests in the form of visual injury such as chlorotic marking, banding, silvering or bronzing of the underside of the leaf. Retardation of plant growth may also occur in some cases. The extent of damage to a plant depends upon the nature and concentration of the pollutant, time of exposure, soil and plant condition, stage of growth, relative humidity and the extent of sunlight.

**(3) Damage to farm animals.** Arsenic, lead and fluorides are the main pollutants which cause damage to livestock. These air-borne contaminants accumulate in vegetation and forage anc poison the animals when they eat the contaminated vegetation.

Arsenic occurs as an impurity in coal and many ores. It is also used in insecticides. Livestock near smelting and other industrial operations suffer arsenic poisoning with symptoms like salivation, thirst, liver necrosis, inflammation and depression of central nervous system.

Lead is emitted from metallurgical smelters coke-ovens and coal combustion operations, lead arsenate sprays and automobile exhausts. Lead poisoning occurs in horses and other animals with symptoms such as depression, lethargy, gastritis, paralysis and breathing troubles.

Cattle and sheep are particularly susceptible to flourine toxicity which may cause fluorosis of teeth and bones.

**(4) Darkening of sky and reduction in visibility.** Sky darkening may be caused by heavy smoke and fog or by dust storms. The reduction in visibility may be due to smoke, fog and industrial fumes which contain particulates in the size range of 0.4 to 0.9 u m that scatter light. The intensity of these effects depends upon the particle size, the angle of the sun, aerosol density, thickness of the affected air mass and meteorological factors such as inversion height, wind speed and humidity.

**(5) Effect on human health and human activities.** The effects of air pollution of humans, animals and vegetation has already been discussed in earlier sections. Air pollution can effect the health or workers within the industrial premises, causing absenteeism, sickness and drop in production. Industrial hygiene measures are being taken by many industrial managements to combat these occupational disease. However, apart from the effects on industrial workers, air pollution also affects larger segments of general population.The notorious London smog of 1952, which lasted for 5 days causing 4000 deaths, is an example. Epidemiological and toxicological studies indicate a link between air pollution and respiratory conditions like chronic bronchitis, bronchial asthma, pulmonary emphysema and lung cancer. The vulnerability to air pollution depends upon age, sex, general health status, nutrition, pre-existing diseases, concurrent exposures, concentration and nature of the pollutants involved, extent of exposure, temperature and humidity at the time of exposure. People who are very young or very old infirm, people of poor health, smokers, people with asthma, bronchitis and coronary heart disease and usually more vulnerable. Irritation of nose, eyes and throat and bad odours due to air pollutants cause annoyance, allergey and health hazards.

Air pollution may cause sickness, absenteeism among workers and general lethargy, which naturally result in decrease in efficiency in all facets of human activity.

**Interdependence of Human Activities, Meteorology and Air-Pollution**

The dispersion of air-pollutants at a particular place are influenced by meteorological factors, such as wind speed, wind direction and turbulence. Precipitation and humidity also influence the pollution potential.

**Wind speed and Wind Direction.** The draft and diffusion of air pollutants near the ground level are profoundly influenced by the wind speed and wind direction. The higher the wind speed near the point of discharge of the air pollutants, the more rapidly the pollutants are dispersed and diluted with air. On the other hand, the pollutants tend to be concentrated near the area of discharge when the wind speeds are low. However, in rough terrain, the wind speed and direction near the source may not fully government the movement of the pollutants For instance, hills may deflect the flow of air contaminated by pollutants either vertically and/or horizontally. The extent of deflection depends on the vertical atmospheric stability. In valleys, the winds carrying the pollutants may flow up or down the valley. These meanderings or channeling effects are more pronounced with increasing depth of the valley.

**Atomospheric stability and temperature inversion.** The influence of atmospheric stability and temperature inversion on the upward dispersion and dilution has been already discussed earlier. The decrease of air temperature with height is called the lapse rate. Under conditions of super-adiabatic lapse rate, the atmosphere is unstable and the pollutants are rapidly dispersed, because of the vertical mixing of the air. On the other hand, under conditions of negative lapse rate or temperature inversion, (i.e., when the temperature increases with height), the pollutants in the air do not disperse vartically, because the air near the ground in denser than the air above and hence the turbulence or mixing is very less (since the atmosphere is stable). Thus, air pollution problem may assume alarming dimensions when the climatic conditions are adverse to effective dispersion of the pollutants. Fortunately, climatic conditions in most parts of our country are favourable to effective dispersion of pollutants, as compared to the cold countries.

**Plume characteristics under different lapse conditions.** Chimneys are used with boilers, furnaces etc., to create a natural draft, to provide the necessary air for combustion and to disperse the gaseous pollutants efficiently. The chimney design, particularly the stack height, should be carried out with due consideration of the meteorological factors, such as wind speed and wind direction, to ensure that the population near the factory is not affected. The diffusion profile of the stack gases mainly depends on the stability of the atmosphere. The various types of plume characteristics, along with the corresponding temperature profile are shown in Fig. 2.1.

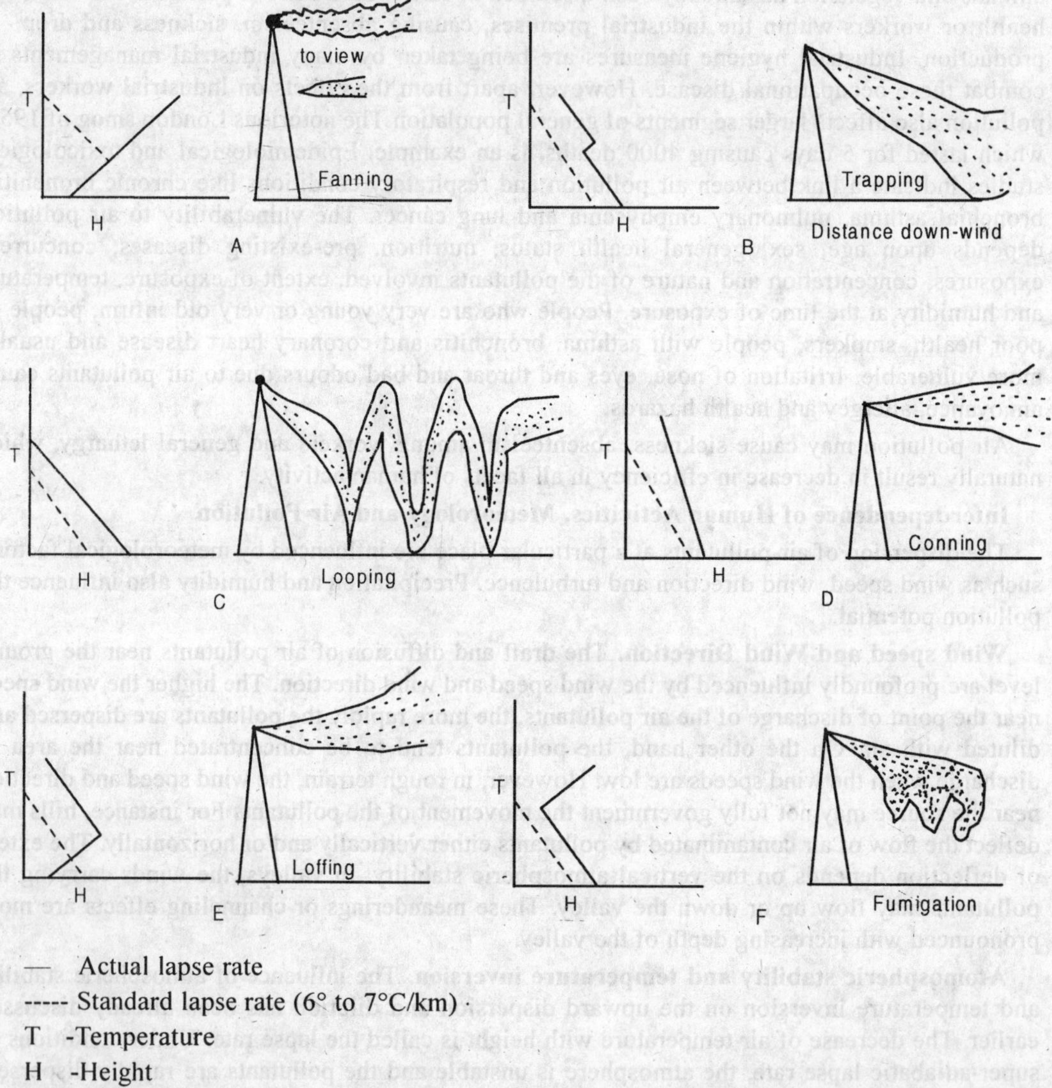


**Fig. 2.1.** Types of smoke plume under different conditions of atmospheric stability

It can be seen from the figure that when the atmosphere is very stable, the gaseous effluent forms a "fanning" plume. Clear sky with light winds during the night are favourable for "fanning" and under these conditions the plume will spread horizontally. When an inversion occurs with light winds aloft and moderate turbulence below, a "trapping" plume may be released. When super-adiabatic lapse rates occurs, "looping" plumes are observed because of the development of large thermal eddies in the unstable air, resulting in bringing high concentrations of the plume gases to the ground for short periods. A clear hot day, accompanied by light winds may produce a "looping" plume. When the atmosphere is cloudy or strongly windly and the turbulence is more mechanical than thermal. "coning' plume is observed. The cone shaped plume tends to reach the ground at greater distances than that in case of "looping". When there is a strong lapse rate above a surface inversion, "lofting" is observed. Under this condition, rapid upward diffusion takes place and emissions will not reach the surface because downward diffusion does not penetrate the inversion layer. When strong lapse rate occurs at the lower layers inversion in upper layers, "fumigation" is observed. Under these conditions, high concentration of the plume gases reach the ground along with full length of the plume due to thermal turbulence. "Fumigation" is usually prevalent with clear skies and light winds during summer.

**Precipitation and Humidity.** The cleaning of pollutants may take place by the three natural processes, namely, fall out, rain out and wash out. Gravitational setting of larger pollutant particles is called "fall out". Pollutants present in cloud particles act as nuclei of condensation and grow to rain drops, which reach the surface by "rain out". When the contaminants in air are swept out by precipitation during its descent to the surface, the process is called "wash out".

The knowledge of local and regional meteorological factors is helpful in location of an industrial plant, the design and use of pollution control equipment and in establishment of air quality criteria.

Just as the extent of air pollution depends on whether conditions, air pollution, in turn, also affects the weather to some extent. For instance, air pollution, may reduce visibility, increase the duration and frequency of fog and reduce the incoming solar radiation. These may affect, to some extent, the human activities. On the other hand, human activities, such as deforestation, massive shifting of surface and ground waters, burning of fossil fuels, automobile emissions etc., may also change tne climatic conditions on the earth.

## AIR QUALITY STANDARDS

Air Quality standards indicate the levels of pollutants that cannot be exceeded during a specified time period in a specified geographic area, with due reference to the method of measurement, units of measurement, concentration and time of exposure. These are derived from air quality criteria, which are in turn derived on the basis of effects of ambient air pollution on human health, vegetation, animals, materials, visibility etc. The "sanitary standard" adopted in USA are worked out with consideration of the local conditions, technical and economic feasibility of controlling the particular pollutant to the desired level etc. The "hygienic standards" adopted in USSR permit only such pollutant concentration which do not cause and harmful or unpleasant effects directly or indirectly on man, which may adversely attect his capacity to work or his physical or mental well being. Hence these standards represent the ideal requirements which every country should strive to attain.

Table 2.3 summarizes some national ambient air quality standards for U.S. proposed in 1971. The primary air-quality standards define the levels judged necessary to protect the public health with an adequate safety margin

**Table 2.3. Ambient Air Quality Standards for U.S.**

| *Pollutant* | *Primary Air Quality Standards* |
|---|---|
| $SO_x$ | (1) 80 μg/m$^3$ (0.03 ppm) annual arithmetic mean |
| | (2) 365 μg/m$^3$ (0.14 ppm) maximum 24 h concentration which should not be exceeded more than once a year. |

| *Pollutant* | | *Primary Air Quality Standards* |
|---|---|---|
| $NO_2$ | | 100 μg/m³ (0.05 ppm) annual arithmetic mean |
| CO | (1) | 10 μg/m³ (9 ppm) maximum 24 h concentration. |
| | (2) | 40 μg/m³ (35 ppm) maximum 1 h concentration. |
| Particulate | (1) | 75 μg/m³ (0.03 ppm) annual geometrical mean. |
| | (2) | 260 μg/m³ (0.1ppm) 24 h maximum concentration which should not be exceeded more than once a year. |
| Oxidants | | 160 μg/m³ (0.08 ppm) max 1 h concentration not to be exceeded more than once a year. |

The standards prescribed for air pollution control in India are as follows:

| | | *Pollutant concentration, μg/m³* | | | |
|---|---|---|---|---|---|
| *S.No.* | *Area category* | *$SO_2$* | *$NO_x$* | *CO* | *Suspended particulate matter* |
| 1. | Industrial and mixed-use areas | 120 | 120 | 5000 | 500 |
| 2. | Residential and rural areas | 80 | 80 | 2000 | 200 |
| 3. | Sensitive areas | 30 | 30 | 1000 | 100 |

Emission standards for industrial environments are also being finalised.

## AIR MONITORING

The harmful effects of air pollution can be assessed only on the basis of adequate data collected at properly selected sampling stations, using well defined sampling procedures and analytical techniques. It is preferable to conduct both physico-chemical and biological monitoring and correlate the results to evolve an integrated approach for air pollution control. The primary objective of the air monitoring programme is to collect basic data on the quality of the pollutants present in air and the influence of topography, population and meteorological factors on air pollution.

## ATMOSPHERIC SAMPLING AND ANALYSIS

The principal objectives of atmospheric sampling are measurement of pollution, collection of base-line information and control data suitable for source detection and trend evaluation. These are helpful to develop air quality criteria, which eventually forms the basis for setting air quality standards. Sampling procedures and developed on the basis of the pollutant being sampled, the collection technique and the device being employed and the sensitivity, specificity and selectivity of the analytical method being used.

In general, a sample of 10 m³ is collected at a rate of 0.003 m³/min to 3.0 m³/min over 1 to 3 hr period. The pollutant concentration is expressed in μg/m³.

The various collecting devices and techniques employed for the particulate and gaseous air pollutants are summarised below:

Collecting techniques employed for Air pollutants

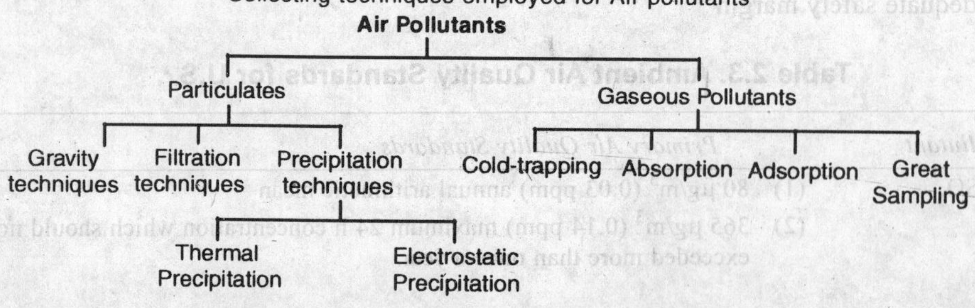

**Gravity techniques** are used to collect settleable particulates (e.g., fly-ash, soot, smoke, etc) in air. Devices used include dust fall sampling instruments, such as dust fall bucket, dust fall jar, etc. Dust fall jar is an open-mouthed glass or polyethylene container with a little water at the bottom. It is exposed to the atmosphere for a month, in a bird-shield on a stand kept 1 m above the sampling floor. The mass of the dust deposits per unit area of jar water surface area is determined after evaporating the water to dryness and weighing. The result is reported in mg/$cm^2$/month or tonnes/$km^2$/month. Sedimentation technique is used for collecting particulates of diameters > 5μ.

**Filtration techniques** are used for collecting suspended particulates that do not settle out early. Devices used include High Volume Sampler, Paper tape sampler, etc. The filter media used are glass fibre (for organic compounds), filter paper (for metals and anions) and silica felts (for trace organic and inorganic species). The pore size of these filters is usually 0.45 μ. The particulate content is usually expressed in terms of μg/$m^3$ of air. High volume air samplers are most commonly used for sampling suspended particles. These are used to pump large volumes of air (upto 2000 $m^3$) at a rate of about 1.7 $m^3$/min and retain particles down to – 0.1μ size on the glass fibre filters.

In a special technique, called 'inertial technique", the polluted air sample is dawn into a sampler provided with obstacles which cause the air-stream to change direction, while the particulates travelling in the initial direction collide with the obstacle. In the dry impingers, the obstacles have an adhesive surface on which the particles are impacted. In the dry impingers of the cascade impactor type, particles can be differentiated with size by using progressively decreasing orifice size. In the wet impingers, the obstacle is immersed in a fluid in which the particulates are collected by impingement.

## PRECIPITATION TECHNIQUES

**Thermal precipitation.** When exposed to a high temperature gradient, suspended particles tend to move to lower temperature region. This principle may be applied for collecting aerosol particles in the range 00.001μ to 10μ. The technique can also be used efficiently for collecting radio-active particulates.

**Electrostatic precipitation.** In this technique electrical charge is used to force radioactive or other particulates in the range of 0.001 to 10 μ to migrate out of the air stream onto a collecting surface.

## COLD TRAPPING OR FREEZE OUT OR CONDENSATION SAMPLING

In this technique, the gaseous pollutants from the air stream are trapped in different collecting chambers, each maintained at progressively decreasing temperature, ranging from 0°C (in an ice bath) to –196°C (in liquid nitrogen bath). This technique is used to collect insoluble or non-reactive vapours, hydrocarbons, radioactive gases, etc., simultaneously using freeze-out sampling trains containing chambers maintained at progressively decreasing temperature.

**Absorption sampling.** In this technique, the desired gaseous contaminant in air is collected by closely contacting with or bubbling through the corresponding absorbent solution. Maintaining suitable bubble size and the requisite residence time are essential for efficient sampling. The absorbents used include water (for absorbing gases *e.g.*, HF), oils (for absorbing hydrocarbons), alkalis (for absorbing acidic gases), and acids (for absorbing alkaline gases). The absorbing devices used include glass scrubbers, packed columns, impingers and counter-current scrubbing systems. A typical fritted glass scrubber used for gaseous air pollutants is shown in Fig. 4.

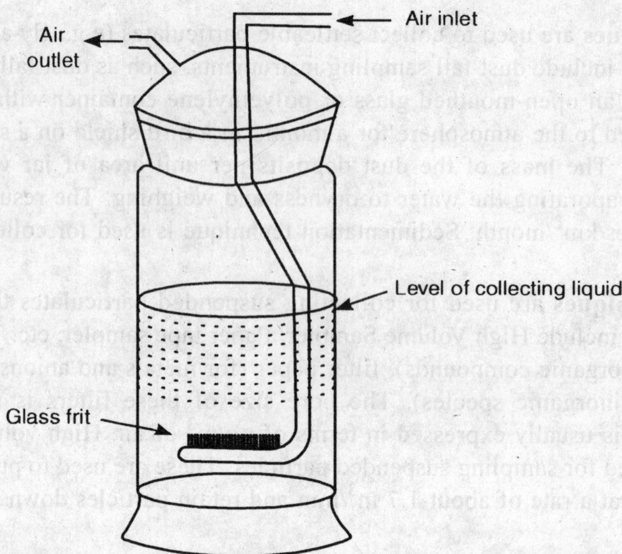


**Fig. 3.** Scrubbers used for sampling gaseous air-pollutants.

**Adsorption sampling.** In this technique, gases and vapours in the polluted air are adsorbed on suitable adsorbents *e.g.*, activated charcoal, activated carbon, activated alumina, silica gel and molecular sieves made from sodium or calcium zeolites. A typical charcoal adsorption tube is shown in Fig. 4. The amount of the pollutant adsorbed is proportional to the surface area of the adsorbent, the physico-chemical characteristics of the adsorbent, the temperature and pressure maintained in the sampling train. When activated charcoal is used as adsorbent, the pollutant adsorbed can be leached with a suitable solvent such as carbon disulphide and the solution is analysed by gas chromatography. In some cases, the sample from the collecting device is directly injected into a gas chromatograph for analysis.

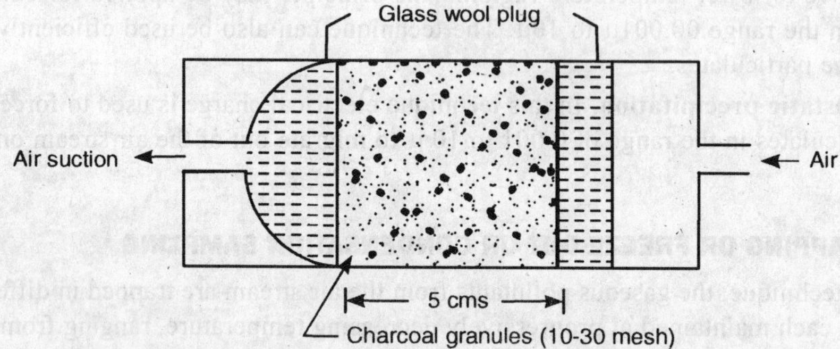


**Fig. 4.** Adsorption tube, packed with charcoal.

**Great sampling:** In this technique, a great sample is taken out between suitable intervals and is allowed to come into contact with a suitable absorbing solution in the sampling apparatus. A liquid- or gas-displacement collector or a deflated plastic bag evacuated flask are commonly employed.

**Analytical and Instrumental Techniques used in the Estimation of Atmospheric Pollutants**

The sampling equipment and analytical methods used for the estimation of some common pollutants are summarized in Table 2.4.

**Table 2.4. Sampling equipment and analytical methods used for the estimation of some pollutants**

| *Sr. No.* | *Name of the pollutant* | *Sampling equipment* | *Method for estimation* |
|---|---|---|---|
| 1. | Dust | Dust fall jar | Gravimetric |
| 2. | Suspended particulates | High Volume Sampler | Gravimetric |
| 3. | Sulphation rate | Lead Candle | Gravimetric |
| 4. | Hydrogen sulphide | Air sampling kit | Methylene Blue Method |
| 5. | Sulphur dioxide | Air sampling kit | West & Gaeke Method, Conductometry; Amperometry |
| 6. | Oxides of Nitrogen | Air sampling kit | Jacob & Hochneisser Method; (Spectrophotometry); Chemiluminiscense |
| 7. | Carbon monoxide | Special sampling | I.R. Spectroscopy; Gas Chromatography. |
| 8. | Hydrocarbon | Special Sampling techniques | Gas chromatography (in conjunction with mass spectrograph for low level of hydrocarbons) |

**1. Dust fall jar.** This is used to determine the settleable particles in air. The dust fall jar comprises of an open-mouthed glass or polythene container with some water in the bottom. The container is usually placed in a bird-shield on a stand 1 m above the sampling floor and is exposed to the atmosphere for a month. The mass of the dust deposited per unit area of jar water surface area is then determined. A dust fall jar and lead candle fixed on a tripod stand are shown in Fig. 5.

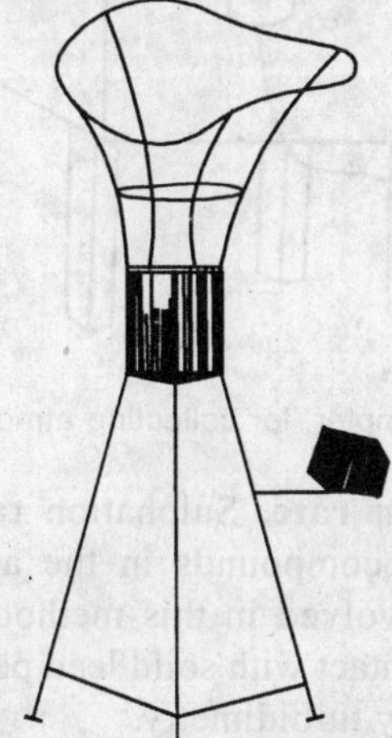

**Fig. 5.** Dust-Fall jar and Lead Candle Fixed on a Tripod stand.

**2. Detemination of suspended particulates with a high volume sampler.** The high volume sampler (Fig. 6) is used to determine the suspended particulates in air that remain for extended periods. This is used for estimating the total levels of particulate matter. This sampler is fixed in a wooden shelter in such a way that air is drawn from the vertically upward flow through a 20 cm × 25 cm glass fibre filter at a flow-rate of about 1.7 $m^3$/min. Particles in the size range of 0.1 to 100 microns diameter are usually collected on the filter. Sampling is usually done for 24 hours.

From the mass of the particulates collected and the volume of the air sampled, the mass concentration of the suspended particulates in the ambient air can be computed as μg/m$^3$ of air. Particulate matter collected on the filter can be incinerated or extracted with a suitable solvent or digested with an oxidizing acid mixture of the subsequent chemical analysis. Charcterization of atmospheric particles is best done with the help of simple of electron microscopy.

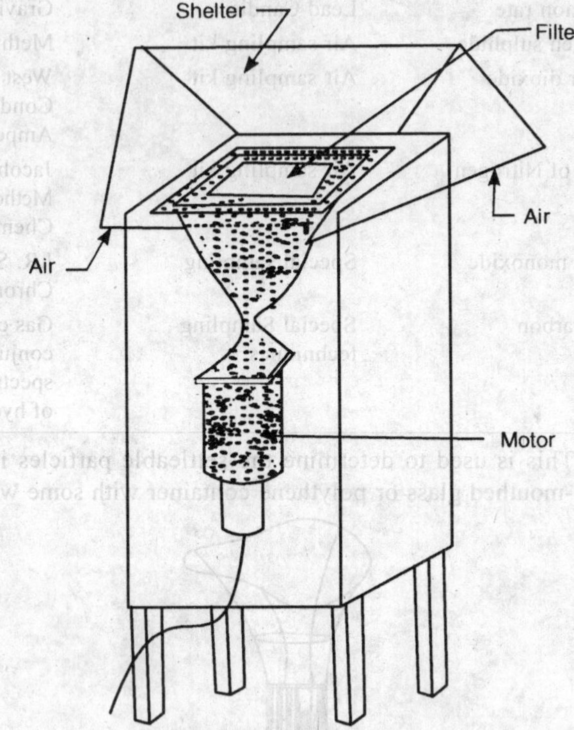


**Fig. 6.** High Volume Sampler, for collecting atmospheric particulate matter.

**3. Determination of sulphation rate.** Sulphation rate is a cumulative index useful for evaluating $SO_2$ and other sulphur compounds in the atmosphere. It is determined by lead peroxide method. The principle involved in this method is that when sulphur compounds in gaseous state come into intimate contact with soild lead peroxide, lead sulphate is formed, which can be estimated by gravimetry or by turbidimetry.

A cylindrical polyethylene or glass tube of area 100 sq. cm. is wrapped with a piece of tapestry cloth or stockinette or gauze bandage (10 cm × 10 cm). A paste containing gum tragcanth mucilage in alcohol and lead peroxide is applied evenly on the fabric, covering the cylinder (only over 100 sq. cm. area) with the help of a spatula and is allowed to dry in a desiccator. The "lead candle" so prepared is installed in a louvered box which keeps the air off but allows the air to come into contact with the fabric surface. After exposure for a month, the $PbSO_4$ formed is estimated in the laboratory by gravimetry or turbidimetry.

Sulphation rate = [Total $BaSO_4$ formed in mg due to sulphation] × 0.343 mg of $SO_3$ per 100 cm$^2$ per day.

(1 mg of $BaSO_4$ = 0.343 mg of $SO_3$).

**4. Estimation of hydrogen sulphide** A measured volume of air at a suitable flow rate is passed through an absorbing solution containing $CdSO_4$ and NaOH, taken in an impinger or an absorption bottle. The solution is then treated with sulphamic acid solution, followed by sulphuric acid, NN-dimethyl P-Phenylene diamine and ferric chloride. The absorbance of the methylene blue formed is determined on a spectrophotometer at 670 nm. The $H_2S$ present can be calculated with the help of a calibration curve prepared with different volumes of standard $Na_2S$ solution. The $H_2S$ concentration is expressed as $\mu g/m^3$ of air passed.

**5. Estimation of sulphur dioxide** $SO_2$ in ambient air in the range 0.005 to 5 ppm can be determined by the modified West & Gaeke method. In this method, the air sample containing $SO_2$ is drawn through a scrubbing solution containing $HgCl_2$ and KCl (which yields $HgCl_4^{2-}$) from potassium tetra-chloromercurate formed). The $SO_2$ present in the air reacts with $HgCl_4^{2-}$ giving $HgCl_2 \cdot SO_3^{2-}$ (dichloro-sulphitomercurate complex). This is allowed to react with formaldehyde, followed by para-rosaniline in acidic medium. The absorbance of the resulting red-violet dye is measured at 548 nm.

The collection efficiency of the scrubbing solution for $SO_2$ is about 95%. If more than 2 ppm of $NO_2$ is present in the air, its interference in the above method of estimating $SO_2$ can be eliminated by the addition of sulphamic acid ($H_2NSO_3H$) which reduces $NO_2$ to $N_2$.

Commercial continuous $SO_2$ analysers are based on conductometry wherein the $SO_2$ is absorbed in $H_2O_2$ solution and the increased conductivity of the solution, due to the formation of $H_2SO_4$, is measured to determine the $SO_2$ content. $SO_2$ monitors, using amperometry, are also available, wherein the changes in electrode potential due to reaction between $SO_2$ and $Br_2$ are mesured to determine the $SO_2$.

**6. Estimation of oxides of Nitrogen** An aqueous solution of $NO_2$ yields equal quantities of nitrate and nitrite ions. The nitrite concentration is determined on the basis of the formation of a reddish purple azo dye produced at pH 2.0 to 2.5 by the coupling of diazotized sulphanilic acid with N-(1-napthyl)-ethylene diamine-dihydrochloride. The colour system obeys Beer's law upto 180 µg of N/I with a 1-cm light path at 543 nm.

$NO_x$ analysis can be best done by chemiluminiscence. Interaction of NO with $O_3$ generates electronically excited $NO_2\cdot$ molecule. This emits radiation in the range 600-3000 nm. This can be measured by a photomultiplier, since the intensity is proportional to concentration.

$$NO + O_3 \longrightarrow NO_2\cdot + O_2$$

$$NO_2\cdot \longrightarrow NO_2 + h\nu$$

The air sample containing NO and $NO_2$ is passed through a thermal converter, which converts $NO_2$ to NO. Analysis of this sample by chemiluminiscence gives total $NO_x$. Analysis of another sample without passing through the thermal converter gives only NO. The difference of the above two analyses corresponds to $NO_2$ in the air sample.

**7. Analysis of carbon monoxide.** Carbon monoxide strongly absorbs I.R. radiation at certain wavelengths. Hence it can be determined by non-dispersive infra-red spectrometry.

Carbon monoxide at levels $< 10$ ppm can be conveniently determined by gas chromatography, using flame ionization detector.

**8. Analysis of hydrocarbons** Hydrocarbons present in air can be collected in an absorption column packed with a suitable absorbent such as porous styrene-divinyl benzene polymer, by sampling about 10 litres of air at a rate of 4 litres/minute. The sample can be desorbed by heating the absorption column for the subsequent gas chromatographic determination, using a flame ionization detector or a mass spectrometer (for low levels).

Apart from the above pollutants, ammonia can be determined by spectro-photometry or potentiometry; volatile pesticides can be determined by gas chromatography; heavy metal ions such as Cu, Cd, Cr, Fe, Mn, Pb, Zn etc., can be determined by Atomic Absorption Spectrometry or Emission Spectroscopy or X-Ray Fluorescence of Neutron Activation Analysis (for Cu, Cr, Fe, Mn, etc). Techniques such as chromatography (for silicates and polycyclic hydrocarbons), potentiometry (for fluorides), Electron spectroscopy (for sulphates) are also used for analysing particulate matter.

**Air-Pollutants from some major industrial and othe sources**

The following table (2.5) summarizes some major air-pollutants from some typical industries and other sources:

**Table 2.5.**

| Source | Typical pollutant emissions |
|---|---|
| Petroleum refineries and petrochemical | $SO_x$, $NO_x$, hydrocarbons, CO, particulate matter, aldehydes industry $NH_3$, odours of mercaptans, smoke. |
| Iron and steel industry | $SO_2$, CO, dust $NO_2$, particulates, acid fumes, oxide fumes, oil and solvent fumes, odours, smoke. |
| Fertilizer industry | $SO_x$, acid mist, $NO_x$, fluorides, $NH_3$, urea dust, fertilizer dust, smoke. |
| Thermal Power Plants | Fly ash, $SO_x$, $NO_x$, particulates such as soot, CO, unburnt coal dust. |
| Chemical Industries | $SO_x$, $NO_x$, $H_2S$ fluorides, hydrocarbons and CO (from organic chemical industries); mercury and $Cl_2$ (from chloralkali plants); and particulates; acid fumes (from acid plants, synthetic fibre plants, etc.) |
| Cement industy | Cement, lime dust, $SO_x$, $NO_x$, fly ash, smoke. |
| Tanneries and leather industry | Mercaptans, sulphides. |
| Paints, pigments and dyes | Nitrobenzene, aniline, thinners, solvents, base materials |
| Paper and paper products | $H_2S$, mercaptans |
| Automobiles | $SO_x$, $NO_x$, HCHO, hydrocarbons, smoke |

## ASSESSMENT OF POLLUTANT EMISSION FROM INDUSTRIES

The assessment of pollutant emissions from various industries can be made on the basis of the following:

(1) Calculation of material balance based on inputs and outputs.

(2) Calculation of emission factor, which is the statistical average of the mass of a pollutant emitted per each source of pollution per unit quantity of material processed or used.

(3) Carrying out stack sampling to actually determine the various pollutants and their quantities emitted from each source. The dust from the gas streams is retained in a suitable filtering media (e.g., filter paper thimble, alundum thimble, membrane filters, fibre glass filters, bubblers, etc.) and the gas passed through is analysed for its constituents.

## AIR POLLUTION FROM AUTOMOBILES

The principal types of automobiles used in our country are: (*i*) Two wheelers and three-wheelers, powered by two-stroke petrol engines (*ii*) passenger cars, powered by four-stroke petrol engines, and (*iii*) trucks and buses, powered by four-stroke diesel engines. The extent of pollution by these automobiles depends upon the engine design, fuel compositon, operating conditions (*e.g.*, idling, cruising, acceleration, deceleration), etc. The emissions from petrol engines are classified as follows:

(*a*) **Exhaust emissions.** These are mostly comprised of CO, $NO_x$, particulates containing lead compounds and unburnt hydrocarbons. The various approaches that are made to control these emissions are : (1) Modification of engine design (2) Modification of operating conditions (3) Treatment of exhaust gases with devices such as absorbers, adsorbers, catalytic converters and after-burners. (4) Modification of alteration of fuels. Some of these have already been discussed earlier.

(*b*) **Crank case emissions.** These include the engine blow-by which leak past the piston, particularly during the compression stroke and the oil vapours produced in the crank case. Obviously, crank case emissions mostly contain hydrocarbons. The quantity of blow-by mostly depends on the design of the engine, condition of the engine and operating conditions. For instance, worn out cylinder and piston rings enhance the blow-by. The crank case emissions can be controlled by installing "positive crank case ventilation systems' which recycle the blow-by gases and crank case ventilation air to the engine intake. Instead of allowing them to escape into the atmosphere.

(*c*) **Evaporative emissions.** It is estimated that an average evaporative emission of hydrocarbons from a passenger car in India is about 20 kg per year. The various methods that are being tried to control the evaporative emissions include (*i*) Storing the fuel vapours in the crank case or absorbing them in a charcoal canister for subsequent recycling to the engine (*ii*) Modifying the fuel by replacing $C_4$ and $C_5$ olefinic hydrocarbons by the corresponding paraffinic hydrocarbons (*iii*) Introducing suitable mechanical devices to combat the problem.

## AIR-POLLUTION CONTROL

The following two basic approaches are used for controlling air pollution:

(*i*) Controlling or confining the pollutants at source. This can be achieved by

(*a*) Modifying the process in such a way that pollutants do not form at all beyond permissible concentrations.

(*b*) Reducing the pollutant concentrations to tolerable levels before they are released to the environment, by use of suitable equipment to destroy, alter or trap the pollutants formed.

(*ii*) Dilution of the pollutants in the atmosphere to permissible levels before they can reach the receptor. This can be achieved by using tall stacks, controlling the process parameters, with due regard to the local meteorological conditions and proper community planning to prevent accumulation of dangerous ground level concentrations within the designated areas.

### Method and equipment used to control gaseous pollutants

**(1) Combustion.** This technique is used when the pollutant contains gases or vapours, which are organic in nature. Flame combustion or catalytic combustion of these pollutants converts them into water vapour and relatively innocuous products, such as $CO_2$. The equipment used for flame combustion include fume incinerators, steam injection or venturi flares and after-burners. Catalytic combustion is resorted to in situations where lower operating temperatures are desirable *e.g.*, burning of waste cracking gases, fumes from paint or enamel baking ovens and coffee roasting processes.

**(2) Absorption.** In this technique, the gaseous effluents are passed through scrubbers or absorbers containing a suitable liquid absorbent to remove or modify one or more of the pollutants present in the gas stream. The efficiency of gas absorption process depends upon: (*a*) The chemical reactivity of the gaseous pollutant in the liquid phase (*b*) The extent of surface contact between the liquid and the gas (*c*) The contact time and (*d*) The concentration of the absorbing medium. The equipment used include plate towers, spray towers, packed towers, bubble-cap plate towers and liquid jet scrubber towers. The gas absorption technique is widely used for removing pollutants like $NO_x$, $H_2S$, $SO_2$, $SO_3$ and fluorides from gaseous effluents. The various absorbing liquids commonly employed are given in Table 2.6

**Table 2.6. Absorbents for some gaseous pollutants.**

| *Pollutant* | | *Absorbent* |
|---|---|---|
| $NO_x$ | – | $H_2O$, aq. $HNO_3$ |
| HF | – | $H_2O$, NaOH |
| $H_2S$ | – | Ethanol amines, NaOH + Phenol (in mole ratio of 3:2), sodium alamine, soda ash, tripotassium phosphate, ammonia liquor from coke ovens, sodium thioarsenate, etc. |
| $SO_2$ | – | Water, alkaline water, suspension of $Ca(OH)_2$, sulphites of Ba or Ca or Na, ethanolamine, dimethyl aniline, 1: 1 mixture of water and xylidine, aluminium sulphate, etc. |

**(3) Adsorption** In this technique, the gaseous effluents are passed through porous solid adsorbents taken in suitable containers. The organic and/or inorganic constituents of the effluent gases are held at the interface of the solid adsorbent by physical adsorption or chemisorption. The efficiency of adsorption depends upon the surface area per unit weight of the adsorbent, other physical and chemical characteristics of the adsorbent and nature and concentration of the gas being adsorbed. The common adsorbents used for various gaseous pollutants are given in Table 2.7.

**Table 2.7. Absorbents for some gaseous pollutants.**

| *Pollutants* | *Absorbent* |
|---|---|
| $NO_x$ | Silica gel, commercial zeolites |
| HF | Lump lime stone, porous pellets of NaF. |
| $H_2S$ | Iron oxide |
| $SO_2$ | Pulverised lime stone or dolomite, alkaline alumina ($Al_2O_3 + Na_2O$) |
| Organic solvent vapours | Activated carbon |
| Petroleum fractions | Bauxite |
| Vapours associated with gases | Alumina, Silica gel, Bauxite |

The preferential adsorption characteristics of some adsorbents render some adsorbents selective for certain applications. For instance, Silica gel-6, activated alumina, and synthetic zeolite or silicate molecular sieves preferentially adsorb water vapour from a mixture of the water vapour and organic pollutants. Desorption of the sorbed gases is usually achieved by increasing the temperature or reducing the pressure.

When the waste gas streams contain higher concentrations of the gases e.g., $NO_x$, $SO_2$, etc., the gases can be recovered economically and used for the manufacture of $HNO_3$, $H_2SO_4$, etc. $SO_x$ from power plants can be removed by injecting pulverised limestone into the boiler furnace. The CaO formed reacts with the $SO_x$ to form calcium sulphite and calcium sulphate. Thus, the $SO_x$ emission into the atmosphere is prevented.

Odorous gases can be controlled by masking, counter-action, absorption in a suitable solvents or adsorption on activated carbon.

**Method and equipment used for controlling particulate emissions**

Particulate materials in ambient air may orginate from stationary as well as mobile sources. The particulate collection devices are based on the size, shape, electrical properties and hygroscopic properties of the particulates concerned. The various devices used may be classified as follows:

***(i)* Mechanical devices.** The devices mostly operate on the basis of the following two mechanisms:

(*a*) Gravity settling in which the velocity of the horizontal carrier gas is reduced adequately so that the particles settle by gravitational force.

(*b*) Sudden change of direction of the gas flow causes the particles to separate out due to their greater momentum.

Settling chambers, buffer chambers and cyclone separators are the commonly used mechanical devices to separate particulates from gases. Settling chambers collect particulate matter by gravity or centrifugal force. These are used in power plants and industries dealing with rock products.

In cyclone collectors, the velocity of the incoming gas stream is transformed into a confined vortex, from which the centrifugal forces drive the suspended particles to the walls of the collector structure. Cyclone collectors are used to remove particulate impurities from rock product industries, iron and steel plants, mining and metallurgical industries.

***(ii)* Filtration Systems.** Dust-laden gases are forced through a porous medium such as woven or filled fabric. The particles are trapped and collected in the filters and the gases devoid of the particles are discharged out. Fibrous or deep-bed filters and cloth bag filters are commonly used. Fabric filter media made from cotton, wool, nylon, dacron, asbestos, silicone coated glass cloth etc., are in use. Cloth and nylon filters are used upto 80-90°C whereas asbestos and silicon covered glass cloth filters can be used upto temperature of 250°C-350°C Wool, orlon and vinyon filters are good for acidic gases whereas cloth, nylon and asbestos filters are good for alkaline. media. Glass fibre filters have superior chemical resistance. Fabric filters are used in industrial dealing with rock products, pigments, etc.

***(iii)* Electrostatic precipitators.** When a gas or an air stream containing aerosols e.g., dust fumes or mist is passed between two electrodes which are electrically insulated from each other and between which appreciable difference in electrical potential exists, then the aerosol particles get precipitated on the electrode that is at a lower potential. An electrostatic precipitator comprise of the following parts:

(*a*) A source of high voltage (*b*) A high voltage discharge electrode (usually negative) of small cross-sectional area, e.g., a wire, and a collecting electrode (usually positive and at ground potential) or large surface area *e.g.*, a tube or plate. (*c*) A device for disposing the collected material and (*d*) An outer housing enclosing the electrode. The following four basic steps are involved in the operation of an electrostatic precipitator: (1) Electrically charging the particles by ionisation (2) Transporting the charged particles by the force exerted upon them in the electric field to the collecting surface (3) Neutralising the electrically charged particles precipitated on the collecting surface and (4) Removing the precipitated particles from the collecting surface by rapping or washing.

In one-stage electrostatic precipitators, ionisation and collection are performed in a single step. In two-stage precipitators, a pre-ionising step is followed by collection. The latter are unsuitable for use with heavy dust concentrations. These are mostly used in air-conditioning plants.

Electrostatic precipitators are the devices of choice when (1) Very large volumes of gases are to be handled (2) Valuable dry material is to be recovered (3) Very high collecting efficiency for the removal of fine particulates is essential and (4) When the gas temperature are very high. Electrostatic precipitators are widely used in power plants, paper and pulp industries, chemical industries such as sulphuric acid plants, iron and steel plants, mining and metallurgical industries, rock products, refineries, carbon black manufacturing industries etc.

**(*vi*) Wet Scrubbers.** Wet scrubbers are used (*a*) When fine particles have to be efficiently removed (*b*) When particulates as well as gaseous contaminants have to be removed (*c*) When the gases to be treated are combustible (*d*) When the volume of the gases being treated is low (*e*) Where cooling is desired and addition of water is not objectionable (*f*) When large variations in process flow-rates have to be accommodated, and (*g*) Where the temperature of the gases to be treated is as high as 300°C or even above.

Wet scrubbers are classified according to the method of particle collection as follows:

(*a*) Liquid carriage type where the gas steam containing the particles is allowed to strike a liquid surface within the collector and the liquid carrying the trapped gas particles moves to a location outside the collector for ultimate disposal.

(*b*) Particle conditioning type where the dust particles in the gas stream is brought into intimate contact with water so that the effective size of the particles is increased due to the formation of heavier water-particles agglomerates. These can be more easily separated from the gas stream by any of the collection mechanisms.

A large variety of wet scrubbers are in use in air pollution control which include ventury scrubber, gravity spray scrubber, wet impinger scrubber, cyclone spray chambers, wet centrifugal scrubber, etc.

Wet scrubbers are used in chemical, mining and metallurgical industries to trap $SO_2$, $NH_3$, metal fumes, etc.

The effective removal of gaseous pollutants and particulates from gaseous effluents from anthropogenic activities is of vital significance. If the air-pollutants are indiscriminately discharged into the atmosphere, the dynamic equilibrium existing in the atmosphere will be disturbed thereby causing adverse effects on man and his environment.

# WATER POLLUTION

## INTRODUCTION

Water is essential for the survial of any form of life. On an average, a human being consumes about 2 litres of water everyday. Water accounts for about 70% of the weight of a human body. About 80% of the earth's surface (*i.e.*, 80% of the total 50,000 million hectares in area) is covered by water. Out of the estimated 1,011 million $km^3$ of the total water present on earth, only 33,400 $m^3$ of water is available for drinking, agriculture, domestic and industrial consumption. The rest of the water is locked up in oceans as salt water, polar ice-caps and glaciers and underground. Owing to increasing industrialization on one hand and exploding population on the other, the demands of water supply have been increasing tremendously. Moreover, considerable part of this limited quantity of water is polluted by sewage, industrial wastes and a wide array of synthetic chemicals. The menace of water-borne diseases and epidemics still threatens the well-being of population, particularly in under-developed and developing countries. Thus, the quality as well as the quantity of clean water supply is of vital significance for the welfare of mankind.

India receives about 1400-1800 mm of rainfall annually. It is estimated that 96% of this water is used for agriculture, 3% for domestic use and 1% for industrial activity. An analysis conducted in 1982 revealed that about 70% of all the available water in our country is polluted. In appreciation of this situation, several steps are being taken to control water pollution.

## CLASSIFICATION OF WATER POLLUTANTS

The various types of water pollutants can be broadly classfied into the following five major categories:

**(1) Organic Pollutants**

The organic pollutants may be further categorized as follows:

**(*a*) Oxygen-demanding wastes.** These include domestic and animal sewage, bio-degradable organic compounds and industrial wastes from food processing plants, meat-packing plants, slaughter-houses, paper and pulp mills, tanneries etc., as well as agricultural run-off. All these wastes undergo degradation and decomposition by bacterial activity in presence of dissolved oxygen (D.O.) This results in rapid depletion of D.O. from the water, which is harmful to aquatic organisms. The optimum D.O. in natural waters is 4-6 ppm, which is essential for supporting aquatic life. Any decrease in this D.O. value is an index of pollution by the above mentioned oxygen demanding wastes. Many aquatic organisms cannot survive at lower D.O. levels in water.

**(*b*) Disease-causing wastes.** These include pathogenic microorganism which may enter the water along with sewage and other wastes and may cause tremendous damage to public health. These microbes, comprising mainly of viruses and bacteria, can cause dangerous water-borne diseases such as cholera, typhoid, dysentry, polio and infectious hepatitis in humans. Hence, disinfection is the primary step in water pollution control.

**(*c*) Synthetic Organic Compounds.** These are the man-made materials such as synthetic pesticides, synthetic detergents (syndets), food additives, pharmaceuticals, insecticides, paints, synthetic fibres, elastomers, solvents, plasticizers, plastics and other industrial chemicals. These chemicals may enter the hydrosphere either by spillage during transport and use or by intentional or accidental release of wastes from their manufacturing establishments. Most of these chemicals are potentially toxic to plants, animals and humans. Some bio-refractory (*i.e.*, resistant to microbial degradation) organics such as aromatic chlorinated hydrocarbons may cause offensive colours, odours and tastes in water, even when present in traces and makes the water (or fish present in it) unacceptable from aesthetic point of view. Non-degradable chemicals, such as alkyl benzene sulphonate from synthetic detergents often lead to persistent foams. Volatile substances, such as alcohols, aldehydes, ethers and gasoline may cause explosion in sewers.

**(*d*) Sewage and agricultural run-off.** Sewage and run-off from agricultural lands supply plant nutrients, which may stimulate the growth of algae and other aquatic weeds in the receiving water body. This unwieldy plant-growth results in the degradation of the value of the water body, intended for recreational and other uses. Further, the water body loses all its D.O in the long run due to the natural biological process of eutrophication and ends up as a dead pool of water.

**(*e*) Oil.** Oil pollution may take place because of oils spills from cargo oil tankers on the seas, losses during off-shore exploration and production of oil, accidental fires in ships and oils tankers, accidental or intentional oil slicks (as in the Gulf War between Iraq and U.S.-led allied forces in the year 1991) and leakage from oil pipe-lines, crossing waterways and reservoirs. Oil pollution results in reduction of light transmission through surface waters, thereby reducing photo-synthesis by marine plants. Further, it reduces the D.O. in water and endangers water birds, coastal plants and animals. Thus, oil pollution leads to unsightly and hazardous conditions which are deleterious to marine-life and sea-food. Oil pollution in seas has been increasing in recent years due to the increase in oil-based technologies, massive oil shipments, accidental oil spillages and international oil slicks during international hostilities.

**(2) Inorganic Pollutants**

Inorganic pollutants comprise of mineral acids, inorganic salts, finely divided metal or metal compounds, trace elements, cyanides, sulphates, nitrates, organometallic compounds and complexes of metals with organics present in natural waters. The metal-organic interactions involve natural organic species, such as fulvic acids and synthetic organic species, such as EDTA. These interactions are influenced by or influence redox equilibria, acid-base reactions, colloid formation and reactions involving micro-organisms in water. Algal growths in water and metal toxicity in aquatic ecosystems are also influenced by these interactions.

Various metals and metallic compounds released from anthropogenic activities add up to their natural background levels in water. Some of these trace metals play essential roles in biological processes, but at higher concentrations, they may be toxic to biota:

The most toxic among the trace elements are the heavy metals, such as Hg, Cd and Pb and metalloids, such as As, Sb and Se. The heavy metals have a great affinity for sulphur and attack the -SH bonds in enzymes, thereby immobilizing the latter. Protein carboxylic acid groups (– COOH) and amino-groups (– $NH_2$) may also be attacked by the heavy metal ions. The heavy metals that may be bound to the cell membranes interfere with the transport phenomena across the cell wall. Heavy metals also tend to precipitate phosphate biocompounds or catalyse their decomposition. Water pollution by heavy metals occurs mostly due to street dust, domestic sewage and industrial effluents.

Polyphosphates from detergents serve as algal nutrients and thus are significant as water pollutants.

**(3) Suspended solids and sediments**

Sediments are mostly contributed by soil erosion by natural processes, agricultural development, strip mining and construction activities. Suspended solids in water mainly comprise of silt, sand and minerals eroded from the land. Soil erosion by water, wind and other natural forces are very significant for tropical countries like India. It is estimated that out of the total land area of 328 million hectares, 175 million hectares are susceptible to degradation by soil erosion. It is also estimated that the continents are losing 5.8 cm of surface soil every 1000 years. About 6000 metric tonnes of soil are washed away into the sea every year, which means that about 5.37 million tonnes of NPK (nitrogen, phosphorous and potassium) fertilizers are washed away into the sea. The erosion leads to qualitative and quantitative degradation of soil in land area. Thus, soil may be getting removed from agricultural land to areas where it is not at all required, such as water reservoirs. Soil particles eroded by running water ultimately find their way into water reservoirs and such a process is called 'siltation'. Reservoirs and dams are filled with soil particles and other solid materials, because of siltation. This reduces the water storage capacity of the dams and reservoirs and thus shortens their life. Such problems are faced with our reservoirs such as Ram Ganga, Hirakud, Nizamsagar, Bhakra and Maithan, and the resultant reduction in live storage capacity of the reservoirs may lead to severe loss of irrigation potential of our country by the end of this century. apart from the filling up of the reservoirs and harbours, the suspended solids present in water bodies may block the sunlight required for photosynthesis by the bottom vegetation. This may also smother sell fish, corals and other bottom life forms. Deposition of solids in quiescent stretches of streams impairs the normal aquatic life in the steams. Further, sludge blankets containing organic solids decompose, leading to anaerobic conditions and formation of obnoxious gases. The tremendous problem of soil erosion can be controlled by proper cultivation practices and efficient soil and forest management techniques.

The organic matter content in sediments is generally higher than that in soils. Sediments and suspended particles exchange cations with the surrounding aquatic medium and act as repositories for trace metals such as Cu, Co, Ni, Mn, Cr and Mo. Suspended solids such as silt and coal may injure the gills of the fish and cause asphyxiation.

**(4) Radioactive Materials**

The radioactive water pollutants may originate from the following anthropogenic activities:

(*a*) Mining and processing of ores, e.g., Uranium tailings.

(*b*) Increasing use of radioactive isotopes in research, agricultral, industrial and medical (diagnostic as well as therapeutic) applications, e.g., $I^{131}$, $P^{32}$, $CO^{60}$, $Ca^{45}$, $S^{35}$, $C^{14}$, $Rb^{86}$, $Ir^{132}$ and $Cs^{137}$

(*c*) Radioactive materials from nuclear power plants and nuclear reactors, e.g., $Sr^{90}$, $Cs^{137}$, $Pu^{248}$, $Am^{241}$.

(*d*) Radioactive materials from testing and use of nuclear weaponry, e.g., $Sr^{90}$, $Cs^{137}$

The radioactive isotopes found in water include $Sr^{90}$, $I^{131}$, $Cs^{137}$, $Cs^{141}$, $Co^{60}$, $Mn^{54}$, $Fe^{55}$, $Pu^{239}$, $Ba^{140}$, $K^{40}$, $Ra^{226}$.

These radioactive isotopes are toxic to life-forms. For instance, $Sr^{90}$, which emanates from testing of nuclear weapons, accumulates in bones and teeth and causes serious disorders in human beings. The maximum permissible level of $Sr^{90}$ in water is 10 pico curies per liter (1 pico curie = $10^{-12}$ curie).

**(5) Heat**

Waste heat is produced in all processes in which heat is converted into mechanical work. Thus, considerable thermal pollution results from thermal power plants, particularly the nuclear-power-based electricity generating plants. In such industries, where the water in used as a coolant, the waste hot water is returned to the original water bodies. Hence the temperature of the water body increases. This rise in temperature decreases the DO content of water, which adversely affects the aquatic life. Moreover, any rise in temperature may increase the susceptibility of aquatic biota to the toxic effects of some chemicals, such as methyl mercury and some polycyclic aromotic hydrocarbons. Reduction of DO in water may alter the spectrum of organisms that can adopt to live the that temperature and that DO levels. Suspended solids in water may also cause bad odours and tastes and also may promote conditions favourable for growth of pathogenic becteria.

If the pollutant concentration in the receiving waters is not within the acceptable limits, adequate steps must be taken to minimize or remove them by suitable treatment technique e.g., sedimentation, filtration, biological oxidation, chemical precipitation of adsorption by activated carbon.

## CHARACTERISATION OF WASTE WATERS

Waste waters are characterised on the basic of various physical, chemical and biological characteristics apart from flow data detail:

**(1) Physical Characteristic.** Colour, Odour, Dissolved Oxygen (D.O.), Insoluble Substances (settleable solids, suspended solids), Corrosive properties, Radio-activity, Temperature range, Foamability, etc.

**(2) Chemical Characteristic.** Chemical oxygen demand (C.O.D.), pH, Acidity of Alkalinity, Hardness, Total Carbon, Total dissolved solids, chlorine demand, known organic and inorganic components such as $Cl^-$, $S^2$, $SO_4^{4-}$, N, P, Pb, Cd, Hg, Cr, As, surfactants, phenols hydrocarbons oils and greases.

**(3) Biochemical characteristics.** Biochemical oxygen demand (B.O.D), presence of pathogenic bactreria etc., and toxicity to man, aquatic organisms, plants and other life forms.

The actual methods used for the treatment of a waste depend upon the characteristics of the particular waste.

**Suspended solids.** The suspeded solids are determined by filtering and aliquot of the sample through a previously weighed sintered crucible or a tared Gooch crucible and drying the crucible at 103°C to 105°C to constant weight. The difference in weight indicated as mg/l gives the suspended solids content of the sample.

**Settleable solids.** The settleable solids content of a sample is obtained by allowing 1 litre of the sample to settle for about 1 hour at 20°C in an Imhoff cone, which is a tapered conical tube. The volume of settleable matter in the cone is recorded as ml/l. The settleable solids may also be expressed in mg/l which can be calculated by the difference between mg/l suspended solids minus mg/l non-settleable matter determined by the procedure described as above.

**Total Solids.** The total solids content of a sample is determined by evaporating a know volume of the sewage of waste water sample, and drying the residue for 24 hours at 103°C to 105°C, followed by weighing. This gives the total solids content of the sample, which includes the dissolved as well as suspended solids.

**Dissolved oxygen (DO).** The measurement of DO gives a ready assessment of purity of water. The determination of dissolved oxygen is the basis of BOD (Biochemical Oxygen Demand) test, which is commonly used to evaluate the pollution strength of waste waters. The determination of DO content is also essential for maintaining aerobic conditions in the receiving waters and also in the aerobic treatment of sewage and industrial waste waters.

The DO content of a water sample is determined iodometrically by the modified Winkler's method. The principle involved in this method is that when manganous sulphate is added to the water sample containing alkaline potassium iodide, manganese hydroxide is formed. This is oxidised to basic manganic oxide by the DO present in the water sample. When sulphuric acid is added, the basic manganic oxide liberates iodine, which is equivalent to the DO originally present in the water sample. The liberated iodine is titrated with a standard hypo solution, using starch as indicator.

**Interference due to nitrite can be eliminated by adding sodium azide to the alkaline potassiun iodide solution used above.**

$$MnSO_4 + 2\,KOH \longrightarrow Mn\,(OH)_2 + K_2SO_4$$

$$2Mn\,(OH)_2 + O_2\,(\text{from DO}) \longrightarrow \underset{\text{Basic manganic oxide}}{2\,MnO\,(OH)_2}$$

$$MnO\,(OH)_2 + 2H_2SO_4 \longrightarrow \underset{\text{Manganic sulphate}}{Mn\,(SO_4)_2} + 3H_2O$$

$$Mn\,(SO_4)_2 + 2KI \longrightarrow MnSO_4 + K_2SO_4 + I_2$$

$$\underset{\text{Sodium thiosulphate}}{2Na_2S_2O_3} + I_2 \longrightarrow \underset{\text{Sodium tetrathionate}}{Na_2S_4O_6} + 2NaI$$

The DO is usually expressed as Mg/l (or ppm)

**Biochemial Oxygen Demand (BOD).** Biochemical oxygen demand represents the quantity of oxygen required by bacteria and other microorganisms during the biochemical degradation and transformation of organic matter present in wastewater under aerobic conditions. BOD test is a very valuable test in the analysis of sewage, industrial effluents and grossly polluted waters. In spite of the inherent limitations, the BOD test is still valued as the best test for assessing the organic pollution. BOD is considered as the major characteristic used in stream pollution control. It gives very valuable information regarding the purification capacity of streams and serves as a guide-line for the Regulatory Authorities to check the quality of effluents discharged into such water bodies.

The BOD test essentially consists of measurement of Dissolved Oxygen Content of the sample, before and after incubation at 20°C for 5 days. If the sample does not contain any oxygen, it is supplied with oxygen and the depletion caused is calculated as the measure of BOD. While carrying out the BOD test, microbial organism (called "seed") may also have to be provided if necessary. The BOD is usually expressed as mg/l. (5 days at 20°C).

**Chemical Oxygen Demand (COD).** The chemical oxygen demand (COD) is a meausre of the oxygen equivalent to that portion of organic matter present in the waste water sample that is susceptible to oxidation by potassium dichromate. This is an important and quickly measured parameter for stream, sewage and industrial waste samples to determine their pollutional strength.

According to the American Society of Testing and Materials (ASTM), COD is defined as the amount of oxygen (expressed in mg/l) consumed under specified conditions in the oxidation of organic and oxidisable inorganic matter, corrected for the influence of chlorides.

The principle involved in the determination of COD is that when the waste water sample is refluxed with a known excess of potassium dichromate in a 50% $H_2SO_4$ solution in presence of $AgSO_4$ (as catalyst) and $HgSO_4$ (to eliminate interference due to chloride), the organic matter of the sample is oxidised to water, $CO_2$ and ammonia. The excess dichromate remaining unreacted in the solution is titrated with a standard solution of ferrous ammonium sulphate. The COD of the sample is calculated as follows:

$$\text{COD in mg/l} = \frac{(V_1 - V_2)\, N \times 8 \times 1000}{X}$$

where $V_1$ and $V_2$ are the volumes of ferrous ammonium sulphate (of normality, N) run down in the blank and test experiments respectively and X is the volume of the sample taken for the test.

Since in the COD test, both the biologically oxidisable and the biologically inert matter are oxidised, the COD value for the sample is always higher the BOD value.

## METHODS AND EQUIPMENT USED IN WASTE WATER TREATMENT

The various methods used in sewage and industrial waste water treatment are as follows:

**(*i*) Preliminary Treatment**

The principal objectives of preliminary treatment are the removal of gross solids (*i.e.*, large floating and suspended solid matter, grit, oil and grease if they are present in considerable quantities.

Large quantities of floating rubbish such as cans, cloth, wood and other larger objects present in waste are usually removed by metal bars, acting like strainers as the waste water moves beneath them in an open channel. The velocity of the water is then reduced in a grit-settling chamber of larger size than the previous channel.

Removal of gross solids is generally accomplished by passing waste water through mixed or moving screens. Different types of these screens are available, which include bar screens (described above), hand raked or mechanical raked screens, drum and wire rope screens.

The modern mechanical screens cum filters include rotary, selfcleaning, gravity type units and circular overhead fed vibratory units. These are costlier, as compared to the conventional bar screens, but are very effective in reducing the suspended solids and BOD. Sometimes, instead of screening, the gross solids in the sewage are cut into small pieces with the help of macerators of comminutors.

Grit (or detritus) is removed in the early stages of treatment in grit channels or tanks to safeguard against any damage to pumps and other equipment by abrasion and also to avoid settling in pipe bends and channels. Grit, being heavier than organic solids, can be separated from organic solids by careful regulation of the flow velocity in the grit tanks. The grit settling chambers are periodically disconnected from the main system to remove the grit manually, for possible use in land-filling, road making and on sludge drying beds. If the waste waters contain appreciable quantities of oil and grease, then it is advisable to remove as much of these as possible, in the preliminary treatment itself to avoid adverse effect on the rest of the plant. This is achieved by passing the waste water through skimming tanks where oil and grease are skimmed off. This process can be rendered more efficient by aeration, chlorination or vacuum flotation.

If the oil and grease are in emulsified condition, as in wool-scouring wastes, ordinary skimming methods are ineffective. In such cases, they may be removed with the help of chemical reagents in primary sedimentation tanks.

**(*ii*) Primary Treatment**

After the removal of gross solids, gritty material and excessive quantities of oil and grease, the next step is to remove the remaining suspended solid as much as possible. The step is aimed at reducing the strength of the waste water and also to facilitate secondary treatment.

**Sedimentation.** The suspended matter can be removed efficiently and economically by **sedimentation**. This process is particularly useful for treatment of wastes containing high percentage of settleable solids or when the waste is subjected to combined treatment with sewage.

The sedimentation tanks are designed to enable smaller and lighter particles to settle under gravity. The most common equipment uised include horizontal flow sedimentation tanks and centre-feed circular clarifiers. The settled sludge is removed from the sedimentation tanks by mechnical scrapping into hoppers and pumping it out subsequently. In a well-designed continuos flow sedimentation tank, about 50% of the suspended solid matter is settled out within two hours of detention time. An efficient sedimentation system is expected to remove about 90% of the suspended solids and 40% of organic matter (thus reducing the BOD).

In waste waters containing larger proportion of industrial wastes, a long detention time helps in mixing and balancing (or equalizing) or the various wastes and safegurds against unduly heavy loads being passed on to the biological purification plants subsequently.

**Sedimentation aids.** Finely divided suspended solids and colloidal particles cannot be efficiently removed by simple sedimentation by gravity. In such cases, mechnical flocculation or chemical coagulation is employed.

In *Mechanical flocculation,* the wastewater is passed through a tank with a detention time of 30 minutes and fitted with paddles rotating at an optimum peripheral speed of 0.43 m/s. Under this gentle stirring, the finely divided suspended solids coalesce into larger particles and settle out. Specialised equipment such as *Clariflocculator* is also available, wherein flocculating chamber is a part of a sedimentation tank.

In *chemical coagulation,* the sewage or other waste water is treated with certain chemicals which form a floc (flocculent precipitate) that absorbs and entrains the suspended and colloidal particles present. The coagulants in comon use are (1) Hydrated lime (2) Alum, $Al_2(SO_4)_3 \cdot 18H_2O$ (3) Copperas, $FeSO_4 \cdot 7H_2O$ (4) Ferric chloride and (5) Chlorinated copperas, $FeSO_4 \cdot Cl$ (mixture of ferric sulphate and chloride). Alum is the most popular coagulant used both in water and waste water treatment. For best results, the chemicals used for coagulation are well mixed with the wastewater in baffled channels followed by mechanical flocculation before sedimentation. Pre-aeration for about 10 minutes before sedimentation is also found to help in th removal of entertained gases like $CO_2$ and $H_2S$ and improved flocculation and separation of oil and grease.

Coagulation is the most effective and economical means to remove impurities.

Sometimes, in addition to the coagulants, other chemicals called "*coagulant aids*" are also used in very small quantities to promote the formation of large and quick settling floc and thereby enhancing coagulation. Activated silica and polyelectrolytes (such as polymers of cyanamide, acrylic or methacrylic acids and their derivatives, and hydrolysed high molecular weight polymers having molecular mass of $10^4$ to $10^6$ of acrylamide or acrylonitrile) are the most commonly used coagulant aids. Polyelectrolytes are generally used in dilute solutions (~ 0.25 ppm). Owing to their selective property, care should be taken in selecting the most suitable polyelectrolytes.

The synthetic coagulant aids work by the following two mechanism:

(*i*) The coagulant aids, having long chain molecular structure, are absorbed on two or more particles, thus drawing them together.

(*ii*) By reducing the charge on the particles and thus reducing the repulsive power of the like charges on the particles.

The sequence of operations in the chemically aided coagulation are (*a*) Addition of lime, if there is no sufficient alkalinity, (*b*) Addition of coagulants, followed by rapid mixing for 4 to 6 minutes, (*c*) Addition of coagulant aids, followed by gentle agitation of slow stirring for about 40 minutes.

**Equalization.** Some industries produce different types of wastes, having different characteristics at different intervals of time. Hence, uniform treatment is not possible. In order to obviate this problem, different streams of effluents are held in big holding tanks for specified periods of time. Each unit volume of waste is mixed thoroughly with other unit volumes of other waste to produce a homogeneous and equalized effluent. Aeration of mechanical agitation with paddles usually give better mixing of the different unit volumes of effluents.

**Neutralization.** Highly acidic or highly alkaline wastes should be properly neutralized before being discharged. Acidic wastes are usually neutralized by teatment with lime stone or lime-slurry or caustic soda, depending upon the type and quantity of the waste. Alkaline wastes may be neutralized by treatment with sulphuric acid or $CO_2$ or waste boiler flue gas.

If both acidic and alkaline wastes are produced in the same plant or at nearby plants, storing them in separate holding tanks and mutual neutralization by mixing them in appropriate proportion is the cheapest method.

**(*iii*) Secondary Treatment**

In secondary treatment, the dissolved and colloidal organic matter present in waste waters is removed by biological processes involving bacteria and other micro-organisms. These processes may be aerobic or anaerobic. In aerobic processes, bacteria and other micro-organisms consumes organic matter as food. They bring about the following sequential changes:

(*i*) Coagulation and flocculation of colloidal matter,

(*ii*) Oxidation of dissolved organic matter to $CO_2$, and

(*iii*) Degradation of nitrogenous organic matter to ammonia, which is then converted into nitrite and eventually to nitrate.

Thus, secondary treatment reduces BOD. It also removes appreciable amounts of oil and phenol. However, commissioning and maintenance of secondary treatment systems are expensive.

The effluent form primary sedimentation tanks is first subjected to aerobic oxidation in systems, such as aerated lagoons, trickling filters, activated sludge units, oxidation ditches or oxidation ponds. Then the sludge obtained in these aerobic processes, together with the obtained in the primary sedimentation tanks, is subjected to anaerobic digestion in the sludge digesters.

Certain micro-organisms, in presence of dissolved oxygen and in proper environmental conditions, utilise organic waste as their food, and convert into simpler compounds such as $CO_2$, $H_2O$, nitrates and sulphates, which are non-pollutants. This process, therefore, can be used to remove organic substances from wastes. Almost all organic substances, with a few exceptions, such as hdyrocarbons and ethers, can be oxidised by aerobic biological treatment. Complex cell tissues and protein materials are also synthesised during this process, which are the agglomerated and removed from the waste by settling. Germicidal and resistant organics, such as cyanides and phenols also can be destroyed by special types of micro-organisms, after prolonged acclimatization periods.

Under anaerobic conditions (i.e., in the absence of dissolved oxygen or gaseous oxygen), certain groups of micro-organisms, e.g., hydrolyte and methane forming organisms, can carry out the digestion of complex organic wastes. The hydrolyte organisms convert complex organic compounds to simple and low-molecular weight organic acids and alcohols. These are then converted by methane bacteria to $CO_2$ and $CH_4$. Anaerobic treatment process can be carried out in depth without the need for large surface area. It can take place in mixed or enriched cultures and can, therefore, be maintained easily on large scale. The process can be applied to most types of substrates excepting a few like lignin and mineral oil. The process is less expensive but the final effluent is less satisfactory, as compared to that from aerobic treatment, because of the dark colour, odour and higher residual BOD.

Anaerobic treatment is mainly employed for the digestion of sludges. However, organic liquid wastes from dairy, slaughter house etc., were treated by this method economically and effectively. The efficiency of this process depends upon pH, temperature, waste loading, absence of oxygen and toxic materials.

Some of the commonly used biological treatment processes are described below:

**(*i*) Aerated Lagoons.** These are large holding tanks or ponds having a depth of 3-5 m and are lined with cement, polythene or rubber. The effluents from primary treatment processes are collected in these tanks and are aerated with mechanical devices, such as floating aerators, for about 2 to 6 days. During this time, a healthy flocculent sludge is formed which brings about oxidation of the dissolved organic matter. BOD removal to the extent of 90% could be achieved with efficient operation. The operation and maintenance are relatively simple. The major disadvantages are the larger space requirements and the bacterial contamination of the lagoon effluent which necessitates further biological purification in maturation pond or by secondary sedimentation and sludge digestion.

**(*ii*) Trickling Filters.** The trickling filters usually consist of circular or rectangular beds, 1m to 3m deep, made of well-graded media (such as broken stones, PVC, coal, coke, synthetic resins, gravel or clinkers) of size 40 mm to 150 mm, over which wastewater is sprinkled uniformly on the entire bed with the help of a slowly rotating distributor (such as a rotary sprinkler) equipped with orifices or nozzels. Thus, the waste water trickles through the media. the filter is arranged in such a fashion that air can enter at the bottom, counter current to the effluent flow and a natural draft is produced. A gelatinous film, comprising of bacteria and aerobic micro-organisms known as "Zooglea", is formed on the surface of the filter medium, which thrive on the nutrients supplied by the sewage or the waste water. The organic impurities in the waste water are adsorbed on the gelatinous film during its passage and then are oxidised by the bacteria and the other micro-organisms present therein. When the thickness of the film on the medium increases, a part of it gets detached and carried away along with the effluent. Hence, the effluent from the trickling filters is allowed to settle in a settling tank to retain the sludge particles and is the discharged. The sludge is then pumped to the sludge digestion unit.

A Schematic representation of a typical Trickling filtration process is given in Fig. 8.

Although trickling filtration is classified as an aerobic process, it is indeed a facultative system. Aerobic bacterial species, *e.g.*, spore forming bacteria and bacillus are mostly present in the upper layer of the filter, whereas anaerobic species, such as Desulfo vibro, are present in the interfaces of the stones. Facultative bacteria, such as Pseudomonas, Alcaligens, Flavo bacterium, Enterobactericeae and Micrococcus are also present in the trickling filters. Algae, fungi ciliates, protozoan, worms, snails, insect, larvae thet feed on micro-organisms are also present.

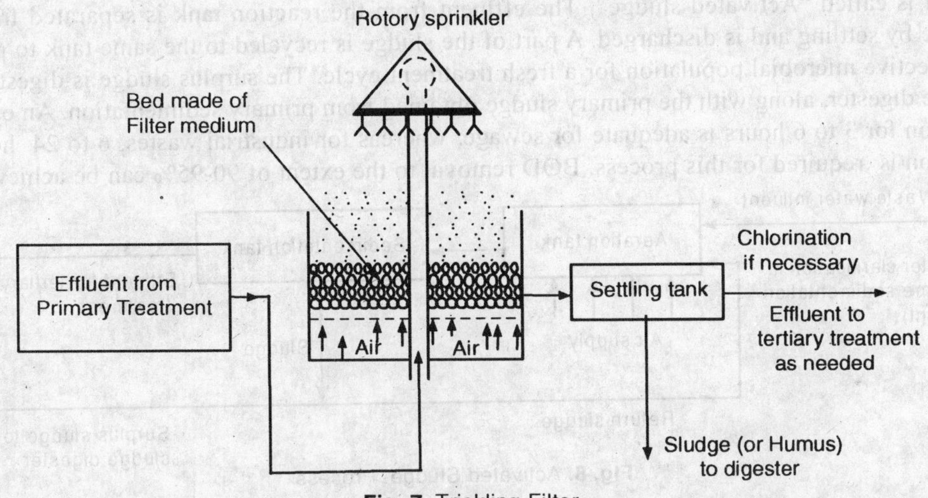


**Fig. 7.** Trickling Filter

By and large, smaller media give better results but they tend to choke easily. Synthetic plastic media are found to be particularly useful for industrial wastes of higher loading. Moreover, they can afford maximum surface area for the formation of microbial film and their light weight helps in greater economy in laying the underdrain of the filter.

The microbial film formed is very sensitive to temperature. The metabolic activity is proportional to the temperature of the waste water passing through the filter. Thus, the efficiency of the filter decreases in winter season.

The efficiency of the filter depends upon the composition of the waste, strength of hydraulic loading, temperature, pH, depth of the filter, the size and uniformity of the filter medium, uniformity of waste water distribution over the filter and proper air supply.

The trickling filter has greater resistance to toxic waste, as compared to the "Activated Sludge Process" and can recuperate more promptly from an overdose of toxic materials. However, shock loads or sudden surges should be avoided lest the efficiency of the filter might be impaired temporarily or even permanently.

Trickling filters are simple to operate and can produce BOD removal to the extent of 65 to 85%, depending upon the rate of filtration. Moreover, constant manual attention is not needed for this process. Trickling filters produce effluents of consistent and better quality.

The disadvantages of the process include the cost of construction and the need for ventilation ducts for the under drain system. The efficiency decreases with increased loading of the waste water. To overcome this problem, the waste water may be diluted with the effluent from the previous treatment.

Trickling filters are effectively used for the treatment of industrial wastes from dairy, distillery, brewery, cannery, food processing, pulp and paper mills, pharmaceuticals, petrochemicals, slaugher house and poulry processing industries.

**(iii) Activated Sludge Process.** This is the most versatile biological oxidation method employed of the treatment of waste water containing dissolved solids, colloids and coarse solid organic matter. In this process, (Fig. 8), the sewage or industrial waste water is aerated in a reaction tank in which some microbial floc is suspended. The aerobic bacterial flora bring about biological degradation of the waste into $CO_2$ and $H_2O$ while consuming some organic matter for synthesising bacteria. The bacterial flora grows and remains suspended in the form of a floc, which is called "Activated sludge". The effluent from the reaction tank is separated from the sludge by settling and is discharged. A part of the sludge is recycled to the same tank to provide an effective microbial population for a fresh treatment cycle. The surplus sludge is digested in a sludge digester, along with the primary sludge obtained from primary sedimentation. An efficient aeration for 3 to 6 hours is adequate for sewage, whereas for industrial wastes, 6 to 24 hours of aeration is required for this process. BOD removal to the extent of 90-95% can be achieved.

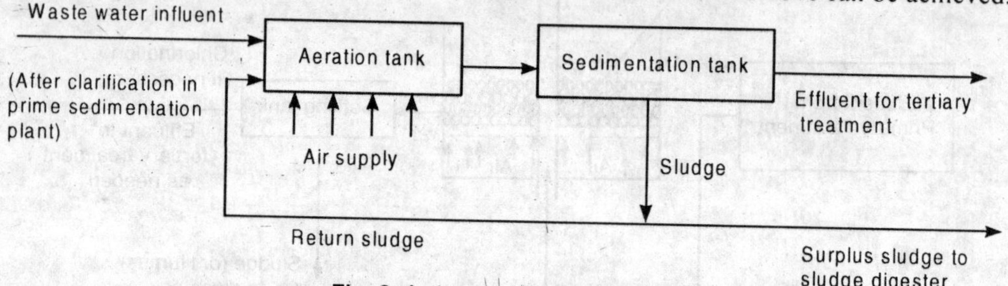


**Fig. 8.** Activated Sludge Process

The microbial flocs formed in this process comprise of Zoogleal masses of living organisms, embedded with their food and slime material and act as active centres for biological oxidation and that is why it is called "Activated Sludge". A young light sludge is preferred to old heavy sludge because the latter would be mineralised and become devoid or oxygen. For this process to be efficient, at least 0.5 ppm oxygen must be present all the time. Oxygen is supplied either by mechanical aeration or by diffused aeration systems.

The micro-organisms should be provided with essential nutrients, such as N and P, which are supplied in the form of urea and mono-orrdi-ammonium hydrogen phosphate. Other nutrients, *e.g.*, K, Mg, Ca, etc are generally present in the waste. Other important factors which determine the efficiency of the activated sludge are pH, temperature and oxidation-reduction potential. The optium pH range for the process is 6.5 to 9.0. Low temperature slows down the rate of metabolism, while high temperature increases the metabolic activity to such an extent that the oxygen is consumed fast, leading to anaerobic conditions.

The presence of synthetic detergents, such as alkyl benzene sulphonates (ABS) and polyethylene glycols, are not susceptible for microbial degradation. They lead to foam formation and make the process difficult and dangerous. Antifoaming compounds are used in such situations.

The performance of the activated sludge process can be assessed with the help of micrbial indicators, as observed by regular microscopic examination of the activated sludges. A good activated sludge contains a relatively high population of free swimming (*e.g.*, stylonichia) and stalked ciliates (*e.g.*, vorticella) apart from a few rotifers. A very high population of dispersed bacteria indicates a poor activated sludge system. The presence of filamentous micro-organisms indicates the deficiency of N and/or P, low pH and low oxygen levels. It may be noted that filamentous bacterial growths' retard floc compaction and settling and result in turbid effluents.

Activated sludge process produces a high quality effluent with relatively small areas. With efficient systems, about 95% BOD removal is possible. The disadvantages include high cost of operation and maintenance, need for careful attention and sensitivity to shock loads of toxic and organic substances.

Activated sludge process was used satisfactorily for the treatment of effluents from food processing, sugar, textile processing, antibiotic manufacturing industries, etc.

Several modifications of the conventional activated sludge process have come into vogue and they mostly differ in the method of air supply. These include: compressed air process, tapered aeration process, extended aeration process, dispersed aeration process, contact stabilization, high rate aerobic treatment process, simplex aeration rotor process and Swedish INKA process.

**(*iv*) Oxidation Ditch.** This can be considered as a modification of the conventional Activated Sludge process. Oxidation ditch (Fig. 9) usually consists of an oval shaped continuous channel, about 1 to 2 m deep and lined with plastic, tar or butyl rubber. Waste water, after screening or comminution in the primary treatment, is allowed into the oxidation ditch. The mixed liquid containing the sludge solids (MLSS) is aerated in the channel with the help of mechanical rotors. Longer retention times are needed. The usual hydraulic retention time is 12 to 24 hrs and for solids, it is 20-30 days. Most of the sludge formed is recycled for the subsequent treatment cycle. The surplus sludge can be dried without odour on sand drying beds.

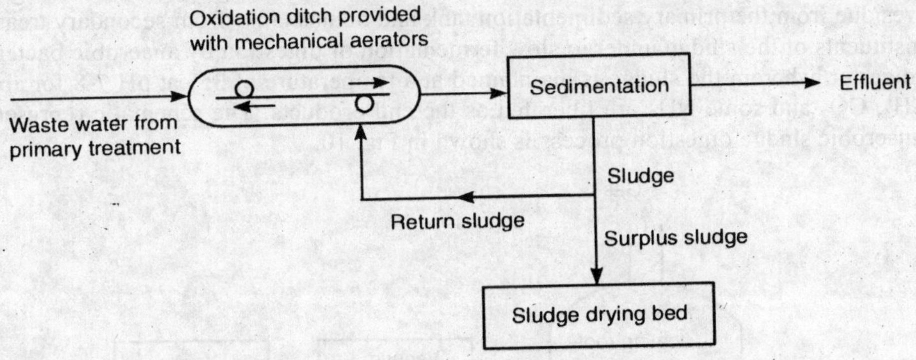


**Fig. 9.** Schematic representation of Oxidation Ditch Process.

The major advantages of the oxidation ditch include simplicity in operation, easy maintenance, low cost of construction, operation and maintenance, overall efficiency and flexibility. This process is generally used for wastes having low BOD.

Oxidation ditch process is used effectively for the treatment of waste water from beet-sugar manufacture, vegetable and fruit canning industry, slaughter house and meat packing industry and edible oil refineries.

**(*v*) Oxidation Pond.** An oxidation pond is a large shallow pond (0.5 m to 1.5 m depth) with arrangements to measure the inflow and outflow. The wastes enter the pond at one end and the effluent is removed at the other end. Stabilization of organic matter in the waste is brought about mostly by bacteria, such as Pseudomonas, Flavo bacterium and Alcaligenes, and to some extent by flagellated protozoa. The oxygen requirement for their metabolism is provided by algae present in the pond. The algae, in turn, utilize the $CO_2$ released by the bacteria for their photo synthesis. Oxidation ponds are also called waste stabilization ponds.

For efficient waste water treatment by this process, an adequate natural aeration, good mixing by wind and proper penetration of sunlight required for photosynthesis of algae are essential. This is the reason why oxidation ponds should be shallow.

Addition of nutrients may be necessary to enhance the growth of algae. This will enhance the amount of oxygen released, which in turn increases the rate of purification of waste water. Any deficiency of oxygen may lead to anaerobic conditions and consequent release of bad odours due to putrefaction of wastes. This is the reason why mechanical aeration is also provided at some places in addition to natural aeration.

Although waste water treatment in oxidation ponds is generally considered as an aerobic process, the purification is performed by a combination of aerobic, facultative and anaerobic processes. The waste water present in the upper part of the pond (which constitutes the major portion of the waste) undergoes aerobic oxidation to $CO_2$ and $H_2O$. Solids present in the waste, which settle as a layer at the bottom, act as anaerobic phase. Here, the organic matter is oxidised by anaerobic bacteria to $CH_4$, $CO_2$, and $NH_3$. The facultative zone exists near the anaerobic phase.

Waste treatment by oxidation pond is cheap and the operation and maintenance are simple. The process can be used for all types of wastes and any degree of purification can be obtained. The process can withstand organic and hydraulic shock loads. The heavy metal ions present in the waste water are precipitated as hydroxides (due to the high pH of the waste water in the oxidation pond) which settle as sludge. However, oxidation ponds require larger space. The effluent from the oxidation ponds may require disinfection or further treatment in a separate maturation pond before final discharge.

**(v) Anaerobic digestion.** This treatment is mainly used for sludge digestion. Sludge is the watery residue from the primary sedimentation tank and humus tank (from secondary treatment). The constituents of the sludge undergo slow fermentation of digestion by anaerobic bacteria in a sludge digester, wherein the sludge is maintained at a temperature of 35° at pH 7-8 for about 30 days. $CH_4$, $CO_2$ and some $NH_3$ are liberated as the end products. The schematic representation of the anaerobic sludge digestion process is shown in Fig. 10.

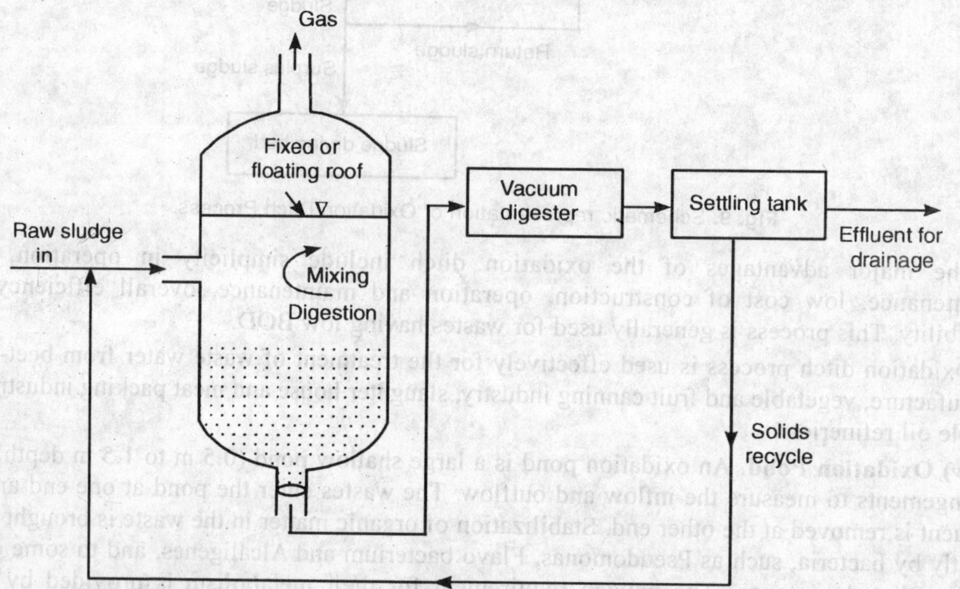


**Fig. 10.** Anaerobic Sludge Digestion Process.

Species of Pseudomonas, Flavo bacterium, Aerobacter, Alcalagenes etc. convert complex organic compounds to low molecular weight organic acids and alcohols. Methano bacterium, Methanosarcina and Methanococcus types of bacteria are responsible for the generation of $CH_4$ in this process. Desulfo Vibrio bacteria reduce sulphates to sulphides and thus, $H_2S$ is released.

The anaerobic digestion process can be accelerated at higher temperatures.

The advantages of anaerobic digestion process are as follows:-

(1) Reduction in volume of the waste by about 65%.

(2) The digested sludge is safer to be used as manure than the undigested sludge.

(3) The digester gas obtained has the following percentage composition by volume: $CH_4$ – 65 to 80%; $CO_2$ – 5 to 30%, $N_2$, $H_2$, $H_2S$ and CO together- about 5%. About 0.6 to 1.25 $m^3$ of the gas is produced per kg of the organic matter destroyed. The calorific value of the gas is about 26 $MJ/m^3$ (*i.e.*, about 700 $BTU/ft^3$). The gas can therefore be used as a fuel to provide the heat required to warm the digestion tanks. In large instalations, it can be used for power generation.

(4) Although anaerobic treatment is a slow process, it is useful for treating small quantities of wastes, containing readily oxidisable dissolved organic solids in liquid form or in finely divided form. the operation and maintenance costs are lesser with this treatment. That is why some liquid wastes containing soluble organics from dairy, slaughter house and paper mill industries have been economically and effectively treated by this process.

## SLUDGE TREATMENT AND DISPOSAL

The sludge from the digester may contain about 90 to 93% water. The sludge is de-watered in drying beds, filter presses or vacuum filters. The de-watered sludge, after chlorination, can be sent for ultimate disposal. The various method used for ultimate disposal include dumping in landfills, incineration, dumping at selected sites in sea, or utilizing as a low grade fertilizer.

## TERTIARY TREATMENT

Tertiary treatment is the final treatment, meant for "polishing" the effluents from the secondary treatment processes, to improve its quality further. The major objectives of tertiary treatment are:

(*a*) Removal of fine suspended solids

(*b*) Removal of bacteria

(*c*) Removal of dissolved inorganic solids

(*d*) Removal of final traces of organic, if it is felt necessary.

Removal of final divided suspended solids can be achieved with the help of micro-strainers and sand filters.

Removal of bacteria, particularly of faecal origin, can be achieved by retaining the effluents from secondary biological treatment plants in maturation ponds or lagoons for specified periods of time. Three or four lagoons arranged in series give an excellent final effluent with very low BOD and low suspended solids. The final effluent is chlorinated if necessary.

Removal of dissolved inorganic solids is a major problem with waste waters from industries such as fertilizers, textile processing, tannery and electroplating. Depending upon the required quality of the final effluent and the cost of treatment that can be afforded in a given situation, any of the following treatment method can be employed:

**Evaporation.** This is an energy intensive and hence an expensive process. It is used only when the recovered solids or the concentrated solutions are reused, *e.g.*, some electroplating wastes. This method is also used when the volume of the waste water to be treated is less. This method is also employed for concentrating radio-active liquid wastes. It permits the recovery of a wide varieties of process chemicals. It is applicable to remove or concentrate chemicals which cannot be accomplished by any other means.

**Ion-exchange.** The use of ion-exchange for de-mineralisation of water is well known. It is widely used for obtaining de-ionised water for use in hig-pressure boilers. This process is now extended to waste water treatment for the removal and recovery of toxic materials from waste water. Ion-exchange process is economical only when the recovered salts are reused in the process, as in electroplating industry. Despite the simplicity of its operation, the method may not be economical if the objective of the treatment is only the removal of dissolved solids from waste water. Special ion-exchangers are available for the retrieval of toxic metal ions from industrial waste water. Applications of ion exhange process in waste water treatment include recovery of Cr. Ni. Phosphate and $H_2SO_4$ from anodising baths; recovery of Cu, Pb and Hg; recovery of Cr from cooling tower below down and removal of cyanides from waste streams.

**Adsorption.** Adsorption by activated carbon is advantageous to remove small quantities of organic contaminants from waste-water. Special adsorbents are commericially available for the removal and retrieval of toxic heavy metal ions from industrial waste water. Activated carbon treatment is particularly useful for the removal of pesticides (*e.g.*, DDT) and carbamate insecticides.

Adsorption processes using activated carbon, peat moss, brown coal and other cellulose materials are finding increasing application in wastewater treatment for the removal of refractory organic, toxic metals, colour etc.

The ALM series or adsorbents developed in Japan incorporate high polymer products containing S- and N-functional groups. The products have very high affinity for heavy metals and can reduce metal levels in mining and industrial wastewaters to less than 1 ppb.

## ELECTRODIALYSIS

This is process in which colloidal/dissolved species are exhanged between two liquids through selective ion-exchange membranes. An electromotive force brings about the separation of the species according to their charge. The semipermeable membranes allow the passage of certain charged species while rejecting the passage of oppositely charged species.

Applications of this process include

(*i*) Concentration of rinse waters to the desired bath strength.

(*ii*) Removal of Cu, Zn, Ni, chromic acid, fluorid and cyanide from process waters.

(*iii*) Recovery of Cr from automobile plating rinse baths.

(*iv*) Recovery of valuable metals and radioactive elements.

(*v*) Control of waters pollution.

(*vi*) In desalination of water.

(*vii*) In purification of plasma protein.

(*viii*) In pharmaceutical, medicinal and photographic industries.

(*ix*) In demineralization of sugars.

## ELECTROLYTIC RECOVERY

In this process electro-chemical reduction of metal ions to elemental metal takes place at the cathode. This process is used to recover Cu, Sn, Ag and other metals from plating, etching or pickling baths. The efficiency of this process has been tremendously increased by innovative designs such as eco-cell involving rotating electrodes (Particularly suitable for concentrated wastes) and extended surface electrolysis (specially suitable for dilute wastes).

**Reverse osmosis:** This is essentially a process in which a semipermeable membrane allows water molecules through and retains the ions. A thin cellophane sheet, supported on a filter cloth can be used as a successful membrane. High pressures are required for useful flow rates and recovery can be achieved by back flushing. High pressure pumps of the order of 500 psi are required.

The various designs of Reverse osmosis equipment include flat plate, tubular, spiral or jelly roll and hollow fiber types.

Applications of reverse osmosis include recovery of valuable components from effluents, recovery of water for reuse, pollution control, recycling of wastewaters and spent chemical from metal plating industry and printed circuit facilities and concentration of waste streams to permit their reuse.

## WASTEWATERS FROM SOME TYPICAL INDUSTRIES : SOURCES, CHARACTERISTICS, EFFECTS AND TREATMENT OPTIONS

## (1) TEXTILE INDUSTRY

### (A) INTRODUCTION

The basic raw materials in a textile industry are cotton, wool or synthetic fibers such as rayon, nylon, acrylic, cellulose acetate or polyester. The textile industry comprises of preparation of the yarn, weaving, knitting and processing.

**(*i*) Cotton textile industry:** The various operations involved in a cotton textile mill are combing, spinning, sizing, weaving and knitting. All these processes except sizing are dry processes. The "grey cloth" thus obtained after the above operations is subjected to the various wet treatment processes such as desizing, scouring, bleaching, mercerizing, dyeing or printing and finishing. All these processes generate considerable volumes of effluents and hence form the major sources of wastwaters.

**(*ii*) Synthetic textile industry :** The fabrics of rayon, cellulose acetate, nylon, acrylic, Dacron, etc. do not require much of dry processing but certain treatments are required for removal of size, antistat and lubricating oil (*e.g.*, removal of PVA, resins, gelatin etc.) used in weaving operations. The wet processes are more or less similar *e.g.*, scouring, dyeing, rinsing, bleaching, finishing, etc. The characteristics of effluents naturally depend upon the type of fiber and the process followed.

The cotton textile industry produces relatively large effluent volumes and higher pollution loads are compared to the synthetic textile industry.

### (B) WASTEWATER SOURCES AND CHARACTERISTICS

**(*i*) Cotton textile industry:** In cotton textile processing, the characteristics of the wastewaters produced in some major operations involved are as follows:

**Sizing:** (Sizing is the process by which the warp thread is sized with starch to give the necessary tensile strength and smoothness required for weaving)

The wastwater is generally coloured and contains starch, polyvinyl alcohol and softeners. It has a high BOD and high content of dissolved and suspended solids.

### DESIZING, SCOURING AND MERCERIZING:

(Desizing is the process used for removing the sizing materials present in the grey cloth with the help of 0.5% $H_2SO_4$ or with the help of enzymes).

(Scouring is the process used to remove natural impurities like greases, waxes and fats by boiling with NaOH, soda ash, sodium silicate, $Na_2O_2$ with small amount of a detergent)

(Mercerization is the process of boiling the cloth with 20% caustic soda solution followed by washing with water. This process gives luster and strength to the fabric)

The wastewater contains starch, acids, alkali silicates and enzymes.

**Bleaching:** (*Bleaching* is used to remove natural colouring materials. It is done using alkaline hypochlorite of chlorine or $Na_2O_2$. Bleaching is usually followed by washing with water and then by scouring treatment with sodium bisulfite to remove traces of alkaili and $Cl_2$)

The effluent contains chlorine, hypochlorites and peroxides.

**Dyeing, printing and finishing:** The effluents contain dyes, alkalies, chromium, phenolics, oil and waxes.

***(ii)* Synthetic textile industry:** The wastewater is generally coloured and contains alkalis, organic solvents, resins, PVA, etc. The effluents has high values of COD and BOD. The origin is waste is more or less same as above.

***(iii)* Wool industry :** The origin of wastewater and the characteristics are given below:

**Scouring:** The effluent is hot, highly coloured and contains greases, soaps, alkalis, detergents and suspended solids. Possess high BOD and COD.

**Oiling :** Olive oil or other vegetable oils

**Sizing :** Starch

**Filling :** Soda ash, soaps and detergents.

**Dyeing :** Dyes, phenolics, chromium salts, alkalis, etc.

### (C) ENVIRONMENTAL EFFECTS OF THE WASTEWATERS

(*i*) The dyes present impart persistent colour to the receiving streams and interfere with photosynthesis of phytoplankton.

(*ii*) The high pH is deleterious to aquatic life.

(*iii*) The colloidal and suspended impurities cause turbidity in the receiving waters.

(*iv*) The oil present interferes with the oxygenation of the receiving water streams.

(*v*) The dissolved minerals increase the salinity of the water, thus rendering it unfit for irrigation purposes.

(*vi*) The toxic chemicals such as Cr, aniline, sulphides, etc. destroy fish and other microbial organisms responsible for self-purification of water streams.

(*vii*) The immediate oxygen demand due to the impurities such as starch, sulphides, nitrites etc. depletes are dissolved oxygen content and adversely affects the aquatic life.

(*viii*) The dissolved solids form incrustations on the surface of the sewers.

(*ix*) The dissolved impurities present cause corrosion in the metallic parts of the sewage treatment plants.

### (D) TREATMENT OPTIONS

Segregation of wastes, screening to remove coarse suspended matter, grease removal (in case of scouring effluents), equalization, neutralization, chemical coagulation to remove colour, suspended and colloidal impurities, aerobic biological treatment (*e.g.*, trickling filtration, activated sludge process, oxidation pond, oxidation ditch or aerated lagoons) and finally tertiary treatment (e.g., reverse osmosis or electrodialysis) to remove dissolved solids, as and when required or affordable.

## (2) PULP AND PAPER INDUSTRY

### (A) INTRODUCTION

The paper manufacturing process mainly consist of the following three steps:

***(i)* Raw material preparation:** In this the cellulosic raw materials **e.g.,** wood, bamboo, cotton liners, bagasse, rags, straw, jute and hemp are slashed and cut into small chips.

(*ii*) **Pulping :** In this, the raw materials are digested with chemicals under high temperature and pressure so as to free the cellulose fibers from the binders *viz.*, lining, resins etc. Several types of pulping processes are available of which the following three are more common.

(*a*) **Sulfate or craft pulping process:** This is a widely used process suitable for any kind of wood. Since the pulp produced is coloured and difficult to bleach and since long and strong fibers are produced in this process, it is used for preparing strong wrapping papers, bag making papers, wall papers, paper boards for cartons, corrugated boards, etc. The process consists of digesting the raw materials with $Na_2S$, NaOH and $Na_2CO_3$ at about 175°C and 120 psi pressure for 2 to 5 hours. During this process, most of the linings are hydrolysed to alcohols and acids. Toxic chemicals such as mercaptans and dimethyl sulphide are also formed. After digestion, the pressure is allowed to drop and the charge (called brown stock) is blown into a pit where the fibers are separated. The remaining waste cooking liquor, known as "black liquor" is sent to another unit for recovering chemicals such as $Na_2S$, Tall oil etc.

(*b*) **Acid sulfide pulping process:** This is used for the manufacture of high grade paper such as bond paper, tissue paper, bread wrap papers etc. In this process, the raw material containing mostly coniferous woods is digested with calcium bisulphite or magnesium bisulfite at 125 to 160°C with steam at a pressure of about 100 psi for about 6 to 12 hours. The pulp produced is dull white in colour and easy to bleach but the fibers are weaker than those obtained is case of sulphate kraft pulping process.

(*c*) **Semi-chemical pulping process :** In this process, lesser amounts of NaOH and $Na_2S$ are employed for pulping and $NaSO_4$ is used as the make-up chemical. This process produces high quality bleaching pulp suitable for manufacturing writing paper, bond paper, tissue, offset, newprint papers and corrugated boards.

After digestion by any of the above processes, washing of the pulp followed by bleaching (with $Cl_2$ or $ClO_2$, NaOH and $CaOCl_2$ in successive stages) must be employed. These steps along with the chemical recovery step contribute to the bulk of the effluents.

(*iii*) **Paper making:** The pulp is disintegrated in a better along with fillers (*e.g.*, clays, $CaCO_3$, $BaSO_4$ or $TiO_2$) rosin or wax emulsion and $Al_2(SO_4)_3 \cdot 18H_2O$ (for providing a gelatinous film on the fiber), dyes and other chemicals are required. Then the pulp is refined. The fibers are cut to the desired size, passed on to stuffing boxes to get the proper consistency and then passed over a moving belt of fine wire. The water passes through the screen while the fibers form a mat on the surface of the screen. This is the sent on the rolls and then to calendering machine for final finishing.

## (B) WASTEWATER SOURCES AND CHARACTERISTICS

All the above steps *viz.*, raw material preparation, pulping, washing, bleaching, chemicals recovery, screening of pulp and paper making contribute to the waste effluents.

The important characteristics of combined effluent of integrated pulp and paper mills with chemical recovery system are dark brown colour, characteristic odour, high content of suspended and dissolved solids (about 1350 mg/1 and 1650 mg/1 respectively), high COD (about 1500 mg/1) and resistant to biological oxidation.

## (C) ENVIRONMENTAL EFFECTS

(*i*) The dark colour of the effluent is due to the lining compounds which are not easily biodegradable and hence it imparts persistent colour to the receiving water streams and inhibits photosynthesis and other natural self-purification process of the water streams.

(*ii*) The immediate oxygen demand of the effluent brings about depletion of oxygen of the receiving stream with the concomitant adverse effects to the aquatic life.

(*iii*) The chemicals present in the effluent, *e.g.*, sulfites, phenols, free chlorine, methyl mercaptan, pentachlorophenol are harmful to fauna and flora of the receiving waters. The settleable materials present may sink to the bottom and interfere with aquatic life.

### (D) TREATMENT OPTIONS

Recovery of by-products, preliminary screening, coagulation, sedimentation, flocculation and flotation to remove suspended matter, chemical treatment to remove colour, activated sludge treatment to remove dissolved organics, lagooning for storage and biodegradation of organic matter.

## (3) ELECTROPLATING INDUSTRY

### (A) INTRODUCTION

In this process, a ferrous or non-ferrous base material is electroplated with Ni, Cr, Cu, Zn, Pb, Cd, Al, Ag, Au, etc., to alter the surface properties of the base metal in order to achieve corrosion resistance, wear resistance, improved or decorative appearance etc. The important steps involved in this process are

**(*i*) Surface Cleaning:** The surface of the base metal is freed from oils and greases by saponification with warm alkali. Greases mineral oil origin are removed by emulsification with alkali mixture containing NaOH, $Na_2CO_3$, $Na_3PO_4$ and $Na_2SiO_3$, etc. or with the help of solvents such as benzene or trichloroethylene. The articles are then washed. The rinse water forms an alkaline effluent.

**(*ii*) Pickling or stripping:** The degreased base metal is then treated with $H_2SO_4$ or HCl to remove scales and rust. The pickled articles are then washed with water. This wash water and the spent pickle liquor (containing ferrous sulphate and unused acids) form the waste effluents.

**(*iii*) Plating :** The pickled articles are electroplated in electrolytic cells containing plating baths as desired. The plating baths may be acidic or occasionally alkaline. Cyanides are generally used in plating baths as they are good oxide solvents and yield brighter and less porous plates. Several types of chemical additives are also used in the plating baths.

Among the plating wastes, cyanide concentrates and spent chromate bath wastes have received particular attention due to their toxicity. These wastes are usually segregated and stored for separate treatment. Apart from them, spent plating bath liquors from Cd, Pb, Zn, Cu, and Ni plating operations, spill overs, floor washes containing the above toxic chemicals and other compounds also form waste effluents from electroplating industry.

### (B) SOURCES AND CHARACTERISTICS OF THE PLATING WASTES

**(*i*) Cleaning :** These operations contribute to alkali wastes containing NaOH, carbonates, silicates, wetting agents and organic emulsifiers. Whenever cleaning is performed with organic solvents, the effluent from this operation consists of the solvents such as trilene, benzene, petrol etc. as well as emulsifiers.

**(*ii*) Stripping or Pickling:** These effluents contain unused acids (e.g., HCl, $H_2SO_4$ and $HNO_3$) and ferrous sulfate.

**(*iii*) Electroplating:** All the constituents of the plating baths contribute to the wastewater steam either through part drag-out, batch dump or floor spill. Electroplating baths may contain Cu, Ni, Ag, Zn, Cd, Cr, Sn, Pb, Fe, ammonia, etc. The anionic components likely to be present include borate, carbonate, fluoborate, cyanide, fluoride, tartrate, phosphate, chloride, sulfide, sulfate, sulfamate, nitrate, etc. Further, many other additives to induce grain refining, deposit brightening, surface levelling, etc. are also added to the plating baths. These include Mo, Se, As, Co, saccharin, aldehydes etc., all of which contribute to the waste streams. Apart from these, contaminants like oil, grease, biodegradable mass suspended solids, etc. may also be present in the wastewaters.

### (C) EFFECTS

(*i*) Plating effluents are highly toxic and corrosive.

(*ii*) Cyanide, chromic acid, chromates, salts of heavy metals, e.g., Cd, Pb, Ni, Zn and Cu present are toxic to aquatic life. Their toxicity to micro-organisms inhibits self-purification property of the streams.

(*iii*) Fe, Sn, etc impact colour to the receiving stream.

(*iv*) Phosphates and nitrates present in the effluent help excessive algal growth which is undesirable.

(*v*) Colloidal and suspended impurities impart unaesthetic appearance to the stream.

(*vi*) Owing to the toxic nature of the effluents, they are not disposed into rivers or waste courses. They are generally discharged into sewers. If cyanide is not completely removed, the HCN gas formed may affect the workers in the sewage treatment plant and sewer system. The organic solvents present may cause explosion in the sewer system. Oils and greases present may interfere with the biological treatment of the sewage. Acidic or alkaline plating effluents may corrode the concrete structures. Suspended impurities present may clog the municipal sewer system.

### (D) TREATMENT OPTIONS

(*i*) Segregation of cyanide waste, chromium wastes and other toxic metal bearing wastes.

(*ii*) **Treatment of cyanide waste by alkaline chlorination :** The pH of the waste is raised to 11 by adding lime or NaOH. Chlorine gas is then passed into the wastewater while agitating it thoroughly for several hours. The cyanide is decomposed under these conditions as per the following chemical reaction:

$$2\,NaCN + 12\,NaOH + 5Cl_2 \longrightarrow 2Na_2CO_3 + 10\,NaCl + 6H_2O + N_2$$

(*iii*) **Treatment of chromium bearing effluents:** The wastewater is to be brought to pH 3 or below by adding a mineral acid if necessary and is then treated with a reducing agent such as $FeSO_4$, $SO_2$ or sodium bisulphite. The Cr (III) formed is then precipitated as the hydroxide by adding lime slurry. The reactions taking place may be represented as follows;

$$H_2Cr_2O_7 + 6\,FeSO_4 + 6\,H_2SO_4 \longrightarrow Cr_2(SO_4)_3 + 3Fe_2\,(SO_4)_3 + 7H_2O$$

$$Cr_2\,(SO_4)_3 + 3\,Ca\,(OH)_2 \longrightarrow 2Cr(OH)_3 + 3\,CaSO_4$$

$$3Fe_2\,(SO_4)_3 + 9\,Ca(OH)_2 \longrightarrow 6Fe(OH)_3 + 9CaSO_4$$

(*iv*) Mixing the effluents obtained after treatement of cyanide and chromium wastes with the other metal-bearing wastes, followed by adjustment of pH to about 9 and settling the precipitates obtained. The sludge may be dried on sand beds and disposed in landfills.

(*v*) Scraping of the floating oils and greases from holding tank and removal of the emulsifed oils by pretreatment with alum and caustic soda before entering into the holding tank.

(*vi*) The rinse waters may be reused after removing the ionic impurities by passing through cationic and anionic exchange resins. Oil is removed from the rinse waters by passing through activated carbon filters before passing through the ion-exchange resins. The effluents from the cation and anion-exchange resins may be treated separately to remove the toxic impurties.

The various treatment technologies available for dealing with electroplating waste problems are given in the following table.

**Table 2.8: Electroplating waste treatment technologies**

| *Waste treatment problem* | *Treatment technology* |
|---|---|
| (1) Removal and/ or recovery of dissolved metals. | Hydroxide precipitation, sulfide precipitation, ion exchange, Membrane filtration, insoluble starch xanthates, peat adsorption, carbon adsorption, electrolytic recovery, high pH precipitation (for complexed metals). |
| (2) Removal of dissolved salts for reuse of water. | Reverse osmosis |
| (3) Recovery of process baths. | Electrodialysis |
| (4) Removal of organics. | Aerobic decomposition, carbon adsorption, resin adsorption. |
| (5) Destruction of cyanides and cyanates. | Electrochemical oxidation, oxidation by chlorine, oxidation by ozone, oxidation by ozone with u.v. radiation, oxidation by $H_2O_2$ (for destruction of cyanides only). |
| (6) Suspended solids removal. | Flotation, centrifugation, ultrafiltration, sedimentation, diatomaceous earth filtration. |
| (7) Sludge dewatering. | Centrifugation, pressure filtration, diatomaceous earth filtration, sludge bed drying, vacuum filtration |
| (8) Reduction of chromium (VI) (from metal finishing and cooling tower blowdowns.) | Electrochemical reduction |
| (9) Concentration and recovery of process chemicals | Evaporation |
| (10) Oil removal. | Centrifugation, coalescing, flotation skimming, ultrafiltration (oil breakdown can be accomplished) by aerobic decomposition.) |

## (4) LEATHER TANNING INDUSTRY

### A. INTRODUCTION

In tanning industry, the animal skins and hides are treated to convert them to non-putrescible and tough leather. The following two processes are generally used:

**(*i*) Beam house processing:** The dry skin is made up of 85% collagen (protein) with small quantities of impurities such as albumin, liquids, globulin and carbohydrates. In this process the following steps are used to remove the impurities and prepare the collagen for tan-yard processing.

**(a) Soaking:** The skins and hides are salted and soaked in water in pits containing wetting agents for a day to remove salts, blood, dirt etc and to restore the moisture content.

**(b) Liming :** The skins and hides are washed with water and again soaked in lime and $Na_2S$ to soften and remove the hairs and other trace protein impurities. The hides are again washed with water.

**(c) Deliming:** The hides are treated with $NH_4Cl$ or $(NH_4)_2$ $SO_4$ and washed again.

**(ii) Tan-yard processing:** This is done by the following two types of tanning processes:

**(a) Vegetable tanning:** This process is used to produce heavy leathers which can be easily tooled and embossed. In this process, the delimed skins and hides are soaked for about 15 days in pits containing vegetable tan liquor (which comprises of extracts of bark, wood, nuts, etc.) and pyrogallol or catechol base. Some types of vegetable tannins and syntans can reduce this tanning time to a few hours only. The tanned materials are then soaked in myrobalan liquor. Then, vegetable oil such as pungam oil is applied to render the hides soft, pliable and tear-resistant. The leather is the dried and dyed.

**(b) Chrome tanning :** This process is used to prepare light and tough leather. The process involves more number of operations but the time required is much less.

**Bating:** The delimed hides are treated with a bating agent (which is a mixture of proteolytic enzymes such as typsin and chymotypsin). This process reduces the pH and swelling, peptizes the fiber, removes the protein degradation products and renders the material smooth, porous and silky.

**Pickling:** The material is treated with $H_2SO_4$ and common salt for a day in a drum.

**Tanning:** The material (leather) in then soaked in chrome tan liquor containing chromium sulphate for 6 hours. Then, $Na_2SO_3$ is added to the leather in the same drum to fix the chrome. It is then taken out of the drum and kept aside for a day to complete the chrome fixation.

**Dyeing:** The tanned leather is washed, neutralised and dyed as required.

**Flat liquoring:** An emulsion of sulphonated oil is applied to the leather to render it soft and pliable.

## B. SOURCES AND CHARACTERISTICS OF TANNERY WASTES

**(i) Soaking:** The effluent contains lot of NaCl, dirt, dung soluble proteins like albumin, organic matter, suspended solids. This putrefies quickly giving bad odour.

**(ii) Liming:** The effluent contains alkali, sulphides, lime, $CaCO_3$, colloidal proteins and their degradation products. It has a high BOD value (4000 to 9000 mg/1).

**(iii) Dehairing and deflushing:** The effluent contains hairs, flesh, sulphides and suspended solids.

**(iv) Deliming:** Effluent is slightly acidic and has BOD values of 1000 to 2000 mg/1.

**(v) Vegetable tanning:** Effluent contains large amount of organic matter, suspended solids, dark colour, bad odour, BOD value is about 12000 mg/1.

**(vi) Bating:** Effluent contains ammoniacal and organic nitrogen, ammonium salts and soluble proteins.

**(vii) Pickling:** Effluent is acidic. Contains excess NaCl.

**(viii) Chrome tanning:** Effluent is acidic, toxic and contains 100 to 200 mg/1 of Cr (III). BOD is about 1000 mg/1

**(ix) Dyeing :** Effluent contains various dyes as used.

**(x) Fat-liquoring:** Effluent contains oils and greases.

### (C) EFFECTS

(*i*) Tannery effluents contain several constituents which are deleterious, irrespective of the fact that where they are discharged viz., into river, stream, sewer, land or sea.

(*ii*) It imparts persistent dull brown colour to the receiving water causing aesthetic and other problems described earlier.

(*iii*) Highly repulsive odour is imparted to the receiving water. The dissolved constituents like proteins are putrefiable.

(*iv*) The acidic or alkaline effluents are corrosive to concrete and metal pipes.

(*v*) Excess NaCl in the effluent is also corrosive and renders the receiving water unsuitable for irrigation.

(*vi*) The effluents may contains pathogenic bacteria.

(*vii*) The dissolved chromium present is toxic to fish and aquatic life and thus affects the natural self-purification property of the stream.

(*viii*) The suspended solids such as hair, flesh, $CaCO_3$, etc. interfere with aeration and photosynthetic activities of the aquatic flora.

(*ix*) If the wastewater is discharged into sewer, the suspended impurities such as $CaCO_3$, hairs etc. may choke the sewerage pipes. The sulphides present in the wastewater cause "crown corrosion" to the concrete structures, etc.

(*x*) The chromium and sulfides present in the wastewater being toxic to microorganisms disrupt the biological treatment operation such as trickling filtration. The suspended lime etc. also interfere with the biological activities in the sewage treatment plants.

(*xi*) The presence of excessive salts and Cr in the wastwaters may deteriorate the quality of the ground water in the affected areas.

### (D) TREATMENT OPTIONS

Primary treatment includes screening to remove hairs, fleshings etc. and then sedimentation. The secondary treatment includes processes such as chemical coagulation and biological treatment. Removal of chromium (by precipitation with lime) followed by activated sludge process gives and effluent of better quality. Oxidation pond and anaerobic lagoon are recommended for small and isolated tanneries. Since excessive quantities of NaCl still remain in solution after any of the above treatments, disposal of the treated effluent still poses a problem.

## (5) FERTILIZERS INDUSTRY

**Fertilizer Raw materials:** P, K, S.

**Fertilizer intermediates:** $H_2SO_4$, $H_3PO_4$, HCl and $HNO_3$

**Fertilizer products:** Urea, ammonium sulphate, ammonium nitrate, normal super phosphates, triple super phosphate, monoammonium and diammonium phosphate, ammonia and other liquid or slurry formulations.

**Sources of waste effluents:** Spill overs from manufacture of acids used as raw materials: spill over of the final fertilizer products; boiler blow-down, cooling waters, etc.

**Characteristics and effects:** Effluent from ammonia production is highly alkaline. Contains excess $NH_3$ from gas scrubbing and gas cleaning operation. Effluent from phosphoric acid manufacture is acidic and contains high amount of phosphates and suspended solids. When discharged into inland surface waters, fertilizer effluents cause eutrophication due to algal bloom. High amounts of fluoride (over 1000 mg/1) present in phosphatic fertilizer effluent enrich the fluoride content of the receiving waters causing dental and skeletal fluorosis to human, abnormal calcification of bones in animals and adverse effects on plants. Presence of Cr, cyanide, ammonia are harmful to aquatic life.

**Treatment Options :** Sedimentation, neutralisation, recycling segregation of wastes, physico-chemical or biological methods to remove $NH_3$, removal or urea by thermal or enzymatic hydrolysis, removal of fluoride (present as fluosilicic acid) and phosphate by precipitating with chalk, lime or double lime treatment, removal of hexavalent chromium by reduction with $SO_2$ in acid medium followed by precipitation of chromium hydroxide by lime treatment, destruction of cyanides by alkaline chlorination, removal or arsenic from segregated $CO_2$- scrubber liquid for isolation or recycling, removal of oils and grease by mechanical oil separation followed by setting and filtration through a bed of activated carbon or coke.

### (6) DAIRY

**Sources of waste effluents:** Spillage, washings of cans, equipments and floor waste from butter and cheese producing units.

**Waste Characteristic:** High BOD, high dissolved organics content, easily putrescibe. Foul odour. Amenable for biological treatment.

**Treatment options:** Equalization, skimming, oil and grease removal by grease traps, biological treatment, *e.g.*, aeration lagoon, oxidation pond or stablization pond. Oxidation ditches with aeration and sludge return arrangements are recommended. Anaerobic lagoons many also be used. The treated effluent may be disposed for spray irrigation.

### (7) RUBBER INDUSTRY

**Source of effluents:** Coagulation of rubber and latex, washings, etc.

**Characteristics :** High BOD, high suspended solids, characteristic odour, variable pH.

**Treatment methods:** Aeration, activated carbon adsorption treatment or chlorination to remove odour and phenolic substances, trickling filtration or activated sludge process.

### (8) SOAP INDUSTRY

**Source of effluents :** Processing wastes, spill overs of chemicals and floor washings.

**Waste characteristics:** Alkali, unsaponified oils, suspended solids. High BOD and COD.

**Treatment methods :** Skimming of floating fatty acids with help of trap tanks, Recovery of glycerine; Primary settlement, neutralization, Secondary treatment before discharge.

### (9) DETERGENT INDUSTRY

**Sources of effluents** : Spill-overs of constituents chemicals, floor washings.

**Waste characteristics:** Usually alkaline pH, contains alkyl benzene sulphonate or other surface active reagents, builders such as phosphates, borax, etc. Not amenable for biological oxidation. Affect aeration and reoxygenation of the stream, causes frothing, toxic to flora and fauna.

**Treatment methods:** Flotation, skimming, precipitation with $CaCl_2$, equalization, pH adjustment, chemical coagulation, settling, activated sludge treatment.

### (10) CANE SUGAR INDUSTRY

**Source of effluents:** Spillage from sugarcane juice extraction and clarification, wash waters from filter press, spray pond overflow, floor washing, spillage from pan boiling.

**Waste characteristics:** Variable pH, High BOD, high dissolved carbonaceous matter, obnoxious odour, suspended solids.

**Treatment methods:** Neutralisation, aerobic oxidation, lagooning or oxidation ponds.

### (11) EDIBLE OIL REFINERY

**Sources of effluents :** Neutralisation of excess fatty acids, spill-overs and filter washings.

**Waste characteristics:** Contains free and saponified oils, High COD, High BOD. High content of dissolved and suspended solids.

**Treatment methods:** Equalization neutralization, coagulation, settling or clarification in a flocculator, aerobic or anaerobic digestion. Reverse osmosis or Electrodialysis.

### (12) OILFIELDS AND OIL REFINERY

**Source of wastes:** Oil pumping, desalting, demulsification and other refining operations

**Waste characteristics:** Contains saline water, oil and drilling mud, odour, contains phenolic and sulphur compounds, High BOD.

**Treatment methods:** Removal of oils, phenols etc., discharge into sea or disposal into subterranean geological formations. If treatment is given the following steps are followed-oil removal, secondary treatment by chemical and biological methods and tertiary treatment-ozonation and activated carbon filtration, final disposal into sea or river.

## WATER FOR MUNICIPAL PURPOSES

Municipal water is mainly used for drinking purposes and for cleaning, washing and other domestic purposes. The water that is fit for drinking purposes is called potable water.

## CHARACTERISTICS OF POTABLE WATER

1. It should be colourless, odourless and tasteless.
2. It should be free from turbidity and other suspended impurities.
3. It should be free from germs, bacteria and other pathogenic organisms.
4. It should not contain toxic dissolved impurities, such as heavy metals, pesticides, etc.
5. It should have a pH in the range 7-8.5.
6. It should be moderately soft, having hardness preferably in the range 50-100 PPm. Its hardness should not be above 150 PPm.
7. It should be aesthetically pleasant.
8. It should not be corrosive to the pipelines and should not cause any incrustations in the pipes.
9. It should not stain clothes.

Table 2.9 gives the WHO (World Health Organization) standards for drinking (Potable) water. Table 2.10 gives the water quality standard in India.

**Table 2.9 Standard (maximum permissible limits) for drinking water as recommended by World Health Organisation (WHO).**

| Parameters | Level WHO standard |
|---|---|
| pH | 6.5 – 9.2 |
| BOD | 6 |
| COD | 10 |
| Arsenic | 0.05 PPm |
| Calcium | 100 PPm |
| Cadmium | 0.01 PPm |

| Parameters | Level WHO standard |
|---|---|
| Chromium | 0.05 PPm |
| Ammonia | 0.5 PPm |
| Copper | 1.5 PPm |
| Iron | 1.0 PPm |
| Lead | 0.1 PPm |
| Mercury | 0.001 PPm |
| Magnesium | 150 PPm |
| Manganese | 0.5 PPm |
| Chloride | 250 PPm |
| Cyanide | 0.05 PPm |
| Nitrate + Nitrite | 45 PPm |
| Polyaromatic hydrocarbons (PAH) | 0.2 PPm |
| Selenium | 0.01 PPm |

**Table 2.10. Water Quality Standard in India**

| Parameters | Standard (Revised 1975) |
|---|---|
| pH | 6.3 – 9.2 |
| Total hardness | 600 PPm |
| Turbidity | 25 PPm |
| Chlorides (as Cl) | 1000 PPm |
| Cyanide (as CN) | 0.05 PPm |
| PAH | 0.2 PPm |
| Fluoride (as F) | 1.5 PPm |
| Nitrate (as $NO_3$) | 45 PPm |
| Phenols | 2 PPm |
| Sulphate (as $SO_4$) | 400 PPm |
| Manganese | 0.5 PPm |
| Mercury | 0.001 PPm |
| Iron | 1 PPm |
| Copper | 1.5 PPm |
| Cadmium | 0.01 PPm |
| Selenium | 0.01 PPm |
| Chromium | 0.05 PPm |
| Lead | 0.1 PPm |
| Arsenic | 0.05 PPm |
| Zinc | 15 PPm |
| Magnesium | 150 PPm |

## TREATMENT OF WATER FOR MUNICIPAL PURPOSES

The municipal water supplied for drinking and other domestic uses should be colourless, odourless, free from suspended impurities, free from germs, bacteria and other pathogenic organisms and should not contain harmful dissolved impurities. Therefore, the raw or impure water obtained by municipalities from sources such as rivers, lakes, wells tube wells, etc. has to be properly treated before supplying for the domestic purpose. The various steps involved in the treatment are as follows:

1. **Aeration:** The raw water is first aerated by bubbling compressed air. This removes bad odours $CO_2$, etc. and also removes iron and manganese by precipitating them as their respective hydroxides.
2. **Settling :** The water is then allowed to stand in large settling tanks. At this stage, some of the heavier impurities present in water settle down by gravity. Also, the bacteria present are partially eliminated due to the UV radiation from sun light.
3. **Coagulation:** The suspended impurities present are then removed by coagulation using lime, soda ash and aluminium sulphate (or ferric alum) as the case may be. The suspended impurities are trapped by the resulting precipitate of $Al(OH_3)$ and settle down at the bottom, thereby bringing about partial clarification of the water. Also, the negatively charged colloidal impurities are neutralized by the trivalent aluminium cation, followed by agglomeration and settling down by gravity.
4. **Filtration:** The partially clarified water is then passed through sand gravity filters. These comprise of rectangular tanks which contain (*a*) a top layer (about 1 meter thick) of fine sand (*b*) a middle layer (0.3 – 0.5 meter thick) of coarse sand, and (*c*) a bottom layer (0.3 – 0.5 metter thick) of graded gravel. A series of porous drains are provided at the bottom of thc gravel layer through which filtered water is collected. The slimy surface layer comprising of finely divided clay, algae, bacteria, etc. formed on the filter bed acts as an effective filtering medium which filters the finely divided residue, suspeded matter and bacteria. The filters are backwashed periodically to remove the precipitated matter from the surface, so as to ensure efficient filtration. Activated carbon may be used for filtration if the water contains odours.
5. **Chlorination:** The filtered water is sterilized by chlorination (by adding chlorine or bleaching powder) to destroy to pathogenic micro-organisms. The water is now pumped to over-head tanks for susbsequent domestic distribution.

## SEWAGE TREATMENT

The wastewater from bathrooms, kitchens, lavatories, etc. is called Domestic Sewage. The waste disposed from factories, laundries, laboratories, business houses, schools, hospitals, etc. also results in Sewage. The spent water from the community as a whole is called *Sanitary Sewage.*

Sewage contains

(*a*) Organic impurities (e.g. Urea (from urine) proteinaceous matters, detergents, biodegradable faeces, animals wastes, fats, carbohydrates, etc.)

(*b*) Inorganic impurities (e.g. nitrates, phosphates, detergents, surfactants, trace metals, other anions and cations)

(*c*) Saprophytic bacteria which are harmless and feed upon organic matter.

(*d*) Pathogenic bacteria such as

(*i*) Vibrio cholerae (which causes cholera)

(*ii*) Shigella dysenteria (which causes bacillary desentry)

(*iii*) Salmonella typhi (which causes typhoid)

(*e*) Industrial wastes, wherever applicable

From the point of view of public health, sewage has to be properly treated.

## OBJECTIVES OF SEWAGE TREATMENT

(*i*) **Stabilization:** This is the process which involves breaking down of organic matter with the help of bacteria into simple substances that to not decompose further. Stabilization can be accomplished with the help of aerobic of anaerobic bacteria.

(*ii*) To render the sewage in-offensive and devoid of its nuisance value.

(*iii*) To prevent contamination of water supplies, thereby protecting aquatic life.

## SEWAGE TREATMENT METHODS

The extent of sewage treatment required mostly depends on the following two characteristics:

(1) The content of suspended solids

(2) The biological oxygen demand (BOD) of the sewage.

The following major treatment method are generally employed:

(1) **Preliminary treatment:** In this treatment, gross solids (eg., large floating and suspended solid matter, grit, oil and grease) are removed by passing through screens, skimming tanks and grit chambers.

(2) **Primary treatment:-** The step is meant to remove the remaining suspended settleable solids, reduce the strength of the waste and to facilitate subsequent secondary treatment. The processes employed include sedimentation, mechanical flocculation and chemical coagulation. After this treatment, about 60% of the suspended solids, 30% COD, 35% BOD, 10% Phosphorous and 20% total nitrogen, are generally reduced.

(3) **Secondary treatment:-** In this treatment step, the dissolved and colloidal organic matter present in the sewage in removed by biological processes involving bacteria and other micro-organism, These processes may be aerobic or anaerobic. They bring about the following sequential changes:

(*a*) Coagulation and flocculation of colloidal matter,

(*b*) Oxidation of dissolved organic matter to $CO_2$,

(*c*) Degradation of nitrogenous organic matter to ammonia, which is then converted into nitrite and eventually to nitrate.

(*d*) Reduction of BOD.

The effluent form primary sedimentation tanks is first subjected to aerobic oxidation in systems such as aerated lagoons, trickling filters, activated sludge unit, oxidation ditches or oxidation ponds. Then the sludge obtained in these aerobic processes, together with that obtained in the primary sedimentation tanks, is subjected to anaerobic digestion in the sludge digesters.

The sludge from the digester which contain 90 to 93% water, is de-watered in drying beds, filter presses or vacuum filters. The de-watered sludge, after chlorination, can be sent for ultimate disposal. The options available for ultimate disposal include dumping in land-fills, incineration, dumping at selected sites in sea, or utilizing as a low-grade fertilizers after composting depending upon the local conditions.

After secondary treatment, about 80% reduction in COD, 90% reduction in BOD, 30% reduction in phosphorous, and 50% reduction in total nitrogen, could be generally achieved.

(4) **Tertiary treatment:** This is the final treatment meant for "polishing' the effluents from the secondary treatment processes, to improve its quality further. The main objectives of tertiary treatment processes are:

(*a*) Removal of fine suspended solids

(*b*) Removal of dissolved inorganic solids

(*c*) Removal of final traces of organics, as desired

(*d*) Removal of bacteria

(*e*) Decrease the load of nitrogen and phosphorous in the effluents.

(*f*) Further purification of wastewater to enable its reuse.

The various processes employed in tertiary treatment include.

(1) **Precipitation:** Calcium compounds in the effluent from secondary treatment as calcium phosphate by adding lime.

(2) **Nitrogen Stipping:** Nitrogen is present in the effluent from secondary treatment in the form of ammonia, nitrites and nitrates. Ammonia is toxic to aquatic biota. Nitrogen compounds enhance *eutrophication.* Ammonia in the effluent is removed by trickling the effluent from the top of a baffle tower while it meets the air coming upwards.

(3) **Cholorination:** The residual microogranisms in the effluent are removed by chlorination before it is discharged.

(4) **Adsorption:** The undesirable tastes and odours are removed by adsorption on activated charcoal.

(5) **Coagulation and filtration:** The residual solids in the effluent are coagulated and removed by filtration.

(6) **Desalination:** The residual dissolved inorganic impurities may be removed by ion-exchange, reverse osmosis or electrodialysis.

(7) **Oxidation ponds:** Bacteria, particularly of faecal origin, can be removed by retaining the effluents from the secondary biological treatment plants in maturation ponds or lagoons for specific time periods. The final effluent which has very low BOD and very low suspended solids may be chlorinated before final disposal.

(8) **Anaerobic digestion :** Using digesters, septic tanks, Imhoff's tanks.

For large towns, a combination of aerobic and anaerobic treatment followed by irrigation may be ideal. The bio-gas produced in the anaerobic treatment can be used as domestic fuel.

Bhawalker Earthworm Research Institute (BERI), Pune developed a processe in which the waste water is passed through a vermifilter formed by enclosing earthworms and worm-castes which harbour cacoons and a variety of microflora in a specially developed medium. The impurities in the wastewater are converted into worm castes (i.e., earthworm excreta) which have strong absorption properties. After repeated filtration, clear water is obtained. The worm casts accumulated in the vermifilter may be harvested periodically for use as a fertilizer.

## EUTROPHICATION

Enrichment of a water body by nutrients is called, "eutrophication". The word eutrophication originated from two greek words- 'eu' = good or well, and "trophes" = food. Eutrophication thus means "well-fed" or "nutrient-rich". The enrichment of a water body with respect to nutrients may take place because of natural sources (eg., decomposition of plant and animal remains) or by anthropogenic sources (eg., man-made sources like domestic, industrial of modern agricultural practices.

A newly formed waterbody possesses a very low concentration of plant nutrients and hence little plant life grows in such water. Low primary prodction limits animal communities too. The nutrient content in it slowly increases due to surface run-offs, windborn dust and organic debris, excreta and exudates of animals which use the water. Bacteria and blue green algae fix atmospheric nitrogen. Phosphates present in the rocks and detritus at the bottom are solubilized by the microbial activity. Thus the nutrient status of the water body gradually increases. At this stage, a moderate population of plants, animals and microbes now develops in the system, which further increases with increasing nutrient enrichment with passage of time. Eventually, dense population of plants, phytoplanktons and animals appears. At this stage, the aquatic system becomes highly productive in terms of fish etc.

On the basis of nutrient status and productivity, anaquatic systems may be classified into the following three types:

(*i*) **Oligotrophic:** Water with poor nutrient status and productivity

(*ii*) **Masotrophic:** Water with moderate nutrient status and productivity

(*iii*) **Eutrophic:** Water with rich nutrient status and high productivity.

Oligotrophic waters gradually turns into mesotrophic and finally to eutrophic water. Further ageing causes over-abundance of nutrients wihc leads to profuse growth of rooted and floating greens plants and the water body loses its aesthetic and economic value. Organic debris and silt settles at the bottom. The water becomes useless. The boundaries of the waterbody turn into a marsh with only a small shallow pond in the middle. Organic debris and silt finally fill the depression and what was once a lake now converts into a dry land.

The *accelerated or cultural eutrophication* of several waterbodies is caused by human activity. Large quantities of mineral nutrients and organic matter are added to the waterbodies in the form of sewage effluents, organic wastes, agricultural run-offs, excreta and exudates of animals and humans, etc. These provide plenty of phosphates, nitrates (mostly from fertilizers applied to agricultural lands, domestic sewage, etc.) which lead to exuberant growth of algae and other water plants. A rich microbial and animal population also develps. If water from such a waterbody is to be used for domestic or industrial purposes, expensive cleaning operations will be required. The processes of natural eutrophication which is generally very slow, thus gets accelerated. Silt and organic debris accumulates at the bottom and the system turns into a shallow muddy pond, then to a marsh and finally into a dry land. Thus a waterbody which could have been useful as a reservoir of fresh water and could have helped the growth of fish, etc. for hundreds of year becomes totally useless within a span of a few years only.

Lake Washington and lake Mendota have undergone rapid eutrophication due to anthropological activities. Similarly, the recreational value of lakes in Kashmir is reduced. Nainital lake is undergoing accelerated eutrophication due to loading with sewage.

## UNDESIRABLE EFFECT OF EUTROPHICATION

1. **Algal blooms:** Dense population of Planktonic algae develops rapidly in eutrophic waters. The water turns green. Such waters are useless for human use because it is very difficult and expensive to remove the microscopic green plants. In due course of time, the entire mass of planktonic algae may die abrupty. The decaying organic matter cause bad tastes and odours. Further, the toxic chemicals released are fatal for fish, birds and other aquatic animals which causes stinking and repulsive smell.

   The decay and death of dense algae leads to biodegradation, cause sudden depletion of oxygen in the water, thereby destroying fish habitats and other desirable aquatic species.
2. The inability of the waterbody to replenish the oxygen results in suffocation and death of several aquatic organisms.
3. The layer of slime produce restrict the penetration of light and prevent atmospheric regeneration of water.
4. The decaying algae, fish Planktons and other organisms cause foul smell.
5. Anaerobic bacteria eg. Clostridium botulinum floroushing in such environment generate toxins which are fatal to livestock, birds etc.

6. Pathogenic microbes, bacteria, viruses, protozoa which flourish under the prevailing anaerobic conditions may result in causing water-borne diseases such as diarrhoea, dysentry, typhoid, viral hepatits, etc.
7. On depletion of oxygen level and on exhausting nitrate oxygen, sulphates are reuced as a last resort to yield hydrogen sulphide which results in bad smell and putrified taste of water.
8. Growth of very long filamentous weeds reduce the stream velocity and also trap solid particles along with them. In our Country, Dal lake Loktak lake, Hussain Sagar etc. are choked due to aquatic weeds thus affecting aesthetics, productivity of fish, and utlity of aquatic flora and recreational value.
9. Over fertilization results in over production of algae and diatoms which leads to clogging of filters in water treatment plants, retard water flow and affects water quality.
10. High population densities of hydrilla, potamogeton, myriophyllum, ceratophyllum and other macrophytes render the water body unsuitable for any useful purpose.
11. During eutrophication, growth of very large populations of tubicid worms and midge chironomous plumosus, etc occurs, thereby causing aesthetic and economic problems for maintaining the waterbodies.
12. Prolonged eutrophic conditions lead to "dystrophic" conditions when bog flora and large quantities of humic acid are produced while drastically reducing Plankton productivity.
13. The filamentous algae are washed into beaches during stroms and piled up. The rotting and stinking piles of organic matter render the beaches unsuitable for recreational uses such as swimming, boating and fishing.

## STEPS TO CONTROL EUTROPHICATION

(1) Effective wastewater treatment and removal or nutrients like nitrogen and phosphorous before discharging the sewerage into waterbodies.
(2) Controlling the recycling of nutrients through harvest.
(3) Effective disposal of organic matter as sludge.
(4) Removal of the algal blooms by dredging.
(5) Developing phosphate-free detergents for domestic use.
(6) Adopting effective physico-chemical methods for removal of disolved nutrients such as nitrogen and phosphorous compounds.
(7) Overcoming the temptation of over-fertilization.
(8) Controlling entrophication by applying algicides such as copper sulphate, chlorine, etc. on susceptible surface waterbodies.

## FLUORIDE PROBLEM IN DRINKING WATER

The word fluorine is derived from Latin word "Fluore" meaning "to flow". At room temperature it is pale yellow greenish gas. It is most reactive and electronegative elements. It is combined chemically to form fluorides. On our hand it is among the 14 physiologically essential elements for normal growth and development of human beings (as reported by WHO). On the other hand it is harmful too. The range of fluoride level in water varies in different parts of the world. In 1931 the fluoride was identified as "mysterious factor" responsible for motted enamel (Dr. Trendley H. Dean) He reported high concentration of fluoride in water is directly related to severity of enamel mottling. Enamel mottling was wide spread in areas with water having fluoride content of 3 ppm or more than that.

Though water is essential natural resource for sustaining life but the suitability of water depend upon chemical composition, geothermal or non thermal. Ground water forms a measure source of drinking water in urban as well as in rural areas. More than 90% of the rural population uses ground water for domestic purposes. Around 300 million people still live in absolute poverty. Mearly half of the population is illiterate not at all aware of the water borne diseases affecting their health. Measure problems and due to the presence of excess fluoride, arsenic and nitrate in ground water Fluoride problems are wide spread in almost the entire country.

## BASIC FACTS ABOUT FLUORIDE

Fluoride exists fairly abundantly in the earth's crust and can enter groundwater by natural processes; the soil at the foot of mountains is particularly likely to be high in fluoride from the weathering and leaching of bedrock with a high fluoride content.

According to 1984 guidelines published by the World Health Organization (WHO)[1], fluoride is an effective agent for preventing dental caries if taken in 'optimal' amounts. But a single 'optimal' level for daily intake cannot be agreed because the nutritional status of individuals, which varies greatly, influences the rate at which fluoride is absorbed by the body. A diet poor in calcium, for example, increases the body's retention of flouride.

Water is major source of fluoride intake. The 1984 WHO guidelines suggested that in areas with a warm climate, the optimal fluoride concentration in drinking water should remain below 1 mg/litre (1ppm or part per million), while in cooler climates it could go up to 1.2 mg/litre. The differentiation derives from the fact that we perspire more in hot weather and consequently drink more water. The guidelines value (permissible upper limit) for fluoride in drinking water was set at 1.5 mg/litre, considered a threshold where the benefit of resistance to tooth decay did not yet shade into a significant risk of dental fluorosis. (The WHO guideline value for fluoride in water is not universal: India, for example, lowered its permissible upper limit from 1.5 ppm to 1.0 ppm in 1998)

In many countries, fluoride is purposely added to the water supply, toothpaste and sometimes other products to promote dental health. It should be noted that fluoride is also found in some foodstuffs and in the air (mostly from production of phosphate fertilizers or burning of fluoride-containing fuels), so the amount of fluoride people actually ingest may be higher than assumed.

It has long been known the excessive fluoride intake carries serious toxic effects. But scientists are now debating whether fluoride confers may benefit at all.

## FLUORIDE: GOOD OR BAD FOR HEALTH?

Fluoride was first used to fight dental cavities in the 1940s, its effectiveness defended on two grounds:

- Fluoride inhibits enzymes that breed acid-producing and bacteria whose acid eats away tooth enamel. This observation is valid, but some scientists now believe that the harmful impact of fluoride on other useful enzymes far outweighs the beneficial effect on caries prevention.
- Fluoride ions bind with calcium ions, strengthening tooth enamel as it forms in children. Many researchers now consider this more of an assumption than fact, because of conflicting evidence from studies in India and several other countries over the past 10 to 15 years. Nevertheless, agreement is universal that excessive fluoride intake leads to loss of calcium from the tooth matrix, aggravating cavity formation throughout life rather than remedying it, and so causing dental fluorosis. Severe, chronic and cumulative overexposure can cause the incurable crippling of skeletal fluorosis.

## SYMPTOMS OF FLUOROSIS

Dental fluorosis, which is characterized by discoloured, blackened, mottled or chalky-white teeth, is a clear indication of overexposure to fluoride during childhood when the teeth were developing. These effects are not apparent if the teeth are already fully grown prior to the fluoride overexposure; therefore, the fact that an adult may show no signs of dental fluorosis does not necesarily mean that his or her fluoride intake is within the safety limit.

Chronic intake of excessive fluoride can lead to the severe and permanent bone and joint deformation of skeletal fluorosis. Early symptoms include sporadic pain and stiffness of joints: headache, stomach-ache and muscle weakness can also be warning signs. The next stage is osteosclerosis (hardening and calcifying of the bones), and finally the spine, major joints, muscles and nervous system are damaged.

Whether dental or skeletal, fluorosis is irreversible and no treatment exists. The only remedy is prevention, by keeping fluoride intake within safe limits.

Through there has been tremendous progress is rural water supply infrastructure after setting up of the Rajiv Gandhi National Drinking Water Mission in 1986, the goal to provide safe drinking water to all is still to be achieved. India's population has recently crossed the one billion. Ever-increasing population and the increased need for agriculture and industries has resulted in water scarcity. The country thus faces a series of threats to the management of water resources (2). This leads the rural population and even urban also to depend upon water from local tanks and tube wells and the consumption of untreated water for all purposes (1). In view to look into the aspects of water quality and related health problems, the water quality data from nine following states (a) Jammu and Kashmir (J & K), (b) Himachal Pradesh, (c) Rajasthan, (d) Haryana, (e) Bihar, (f) West Bengal, (g) Chattisgarh, (h) Orissa and, (i) Maharashtra, covering almost the entire nation has been collected and analyzed.

## RESULTS

The surface, subsurface and thermal water sample analysis indicate the fluoride concentration ranging from < 0.2 to 18 ppm in the States of Jammu & Kashmir, <0.2 to 6.5 ppm in Himachal Pradesh, > 1.5 ppm in Rajasthan, 0.2 to 0.6 in Haryana, 0.35 to 15 ppm in Bihar, on an average 12 ppm in West Bengal, 15 to 20 ppm in Chattisgarh, 8.2 to 13.2 ppm in Orissa and 0.7 to 6.0 in Maharashtra, indicating that except in Haryana, the concentration of fluoride is very high up to 20 ppm. The results of ch emical analysis of water are summarized in Table 2.11.

**Table 2.11. Average fluoride concentration detected in water from different parts of India**

| | 1 | 2 | 3 | 4 | 5 | 6 | 7 | 8 | 9 |
|---|---|---|---|---|---|---|---|---|---|
| Area State | Ladakh (J & K) | Manikaran (Him. Prad.) | Bhilwara (Raj.) | Sohna (Haryana) | Tantoli (Bihar) | Tattapani (Chattisgarh) | Bakreshwar (W. Bengal) | Khurda (Orissa) | Unkesher (Mahar) |
| Sample Nos. | 20 | 12 | 15 | 10 | 30 | 25 | 60 | 50 | 65 |
| Surface | | | > 1.50 | | 0.35 to 4.00 | | | 8.20 to 13.20 | 2.70 to 6.00 |
| Sub-surface | <0.2 to 18.00 | <0.2 to 6.50 | | 0.20 to 0.60 | | | 0.60 to 15.00 | | |
| Thermal | | | | | | 15.00 to 20.00 | | | |

**Source:** Diffuse Pollution Conference Dublin - 2003

## DISCUSSION

Probable source of high fluoride in Indian waters seems to be that during weathering and circulation of water in rocks and soils, fluorine is leached out and dissolved in ground water. Tile fluoride content of ground water varies greatly depending on the type of rocks from which they originate. Among the various minerals responsible for high concentration of fluoride, the Flour - apatite 3Ca-3 (PO-4)-2, CaF-2 and Fluorite, CaF-2 are important. However, the most important being the Fluorite, CaF-2 and the leaching of fluoride from the metamorphic rocks hornblende gneiss of Proterozoic age (3).

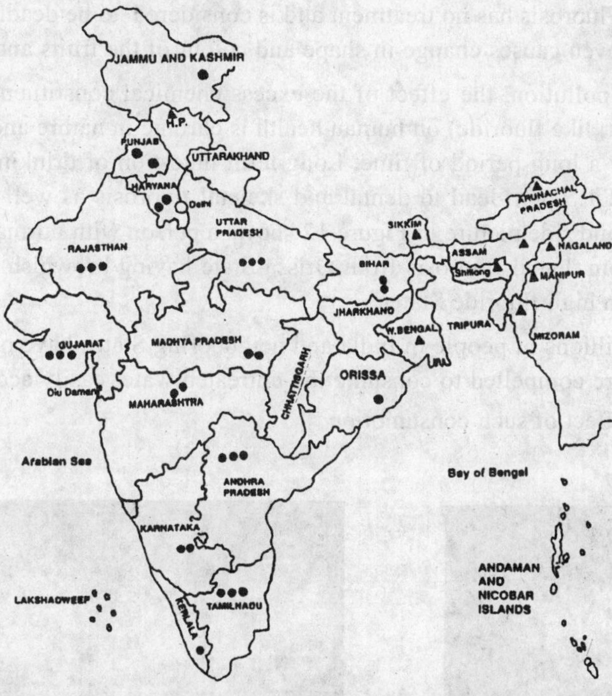


(●) Less than 30% of districts affected: 4 states
(●●) 30–50% districts affected: 5 states
(●●●) 50 – 100% districts affected: 5 states
(▲) disease not detected

**Fig.11.** Location map showing areas of high fluoride content in water

## ILL AFFECTS OF HIGH FLUORIDE CONTENT IN WATER

Endemic fluorosis is a public health problem in India. Around 25 million people of 150 Districts are affected by this disease (Survey report – Rajiv Gandhi National Drinking Water Mission, 1993). Medical advice recommends the drinking water should not contain more than 1.5 ppm of fluoride (WHO, 1994). Concentration of fluoride below 1.5 ppm are helpful in prevention of tooth decay, and such level of fluoride also assists in the development of perfect bone structure in human and animals. However, doses of fluoride above 1.5 ppm increases the severity of tooth mottling and induces the prevalance of osteoporosis and collapsed vertebrae. The disease resulting from excessive consumption of fluoride. Fluorosis has to treatment and is considered to be deadly disease.

High fluoride content in water even causes change in shape and colour of the fruits and vegetaion.

Ill affects of high fluoride content in water are manifested in the form of 'Endemic fluorosis' which is an acute public health problem in India. Around 25 million people of 150 Districts are affected by this disease (Survey report – Rajiv Gandhi National Drinking Water Mission, 1993). Medical advice recommends the drinking water should not contain more than 1.5 ppm of fluoride (5). Concentration of fluoride below 1.5 ppm are helpful in prevention of tooth decay, and such level of fluoride also assists in the development of perfect bone structure in human and animals. However, doses of fluoride above 1.5 ppm increases the severity of toothy mottling and induces the prevalence of osteoporosis and collapsed vertebrae (4). The disease resulting from excessive consumption of fluoride. Fluorosis has no treatment and is considered to be deadly disease. High fluoride content in water even causes change in shape and colour of the fruits and vegetation.

Unlike bacteriological pollution, the effect of the excess chemical constituents (that may be present in the groundwater, like fluoride) on human health is chronic in nature and manifest after consuming the water over a long period of time. Long term ingestion of drinking water having fluoride beyond a limit of 1.5 ppm lead to dental and skeletal fluorosis as well as non skeletal manifestations. The left hand side picture in Figure 12 shows a person with normal teeth whereas the right one suffering from dental fluorosis from Orissa State having brownish Yellow mottled teeth, a common feature in high fluoride States.

It is unfortunate that millions of people in India and neighboring States have no access to safe drinking water and they are compelled to consume the untreated water easily accessible to them without knowing the ill affect of such consumption

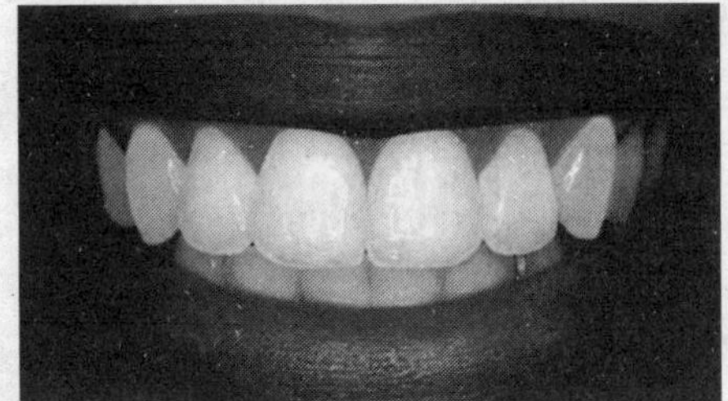

Normal Teeth

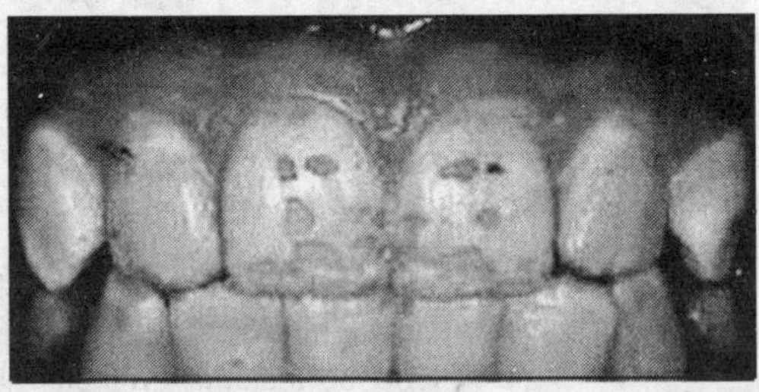

Dental Fluorosis mottled teeth

**Fig. 12.** Normal and Mottled teeth

A much elevated concentration of fluoride, randing from more than 1.5 ppm to 20 ppm in surface, subsurface and thermal waters in nin States in India, is beyond the permissible limit fixed by the WHO for human beings, the consumption which is bound to yield the deadly Fluorosis disease and if used for irrigation, may cause harm to the ecosystem and vegetation.

UNICEF (1996) has published a simple guide to identify a fluorosis diseased person for they can not perform some basic exercise related to body posture. The following Figure 13 clearly explains that if a person is not able to do the three body exercises illustrated in the figure, then the non-skeletal manifestation of the disease confirm that person to be chronic patient of the Fluorosis.

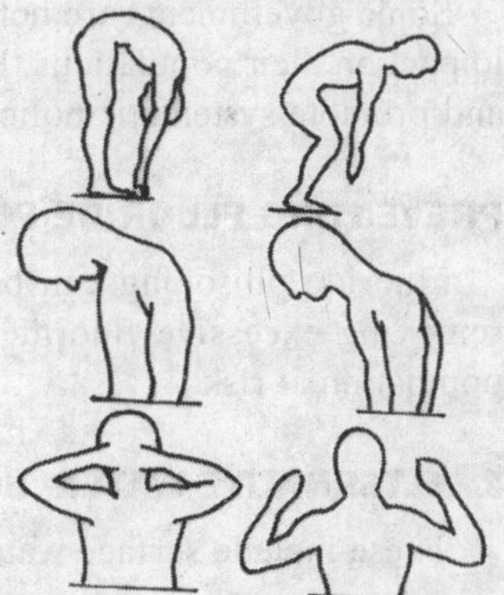
Fig. 13. A simple guide to identify a fluorosis diseased person from body posture.

Left Column: Normal capacity person

Right Column: Fluorosis diseased person

Top- The person is unable to bend completely from the waist.

Middle- Unable to bend the neck to touch the chin on the chest and

Bottom- Unable to bend arms to touch the back of the head.

High fluoride consumption leads to the fluorosis of the bones and affects the vegetation too. Hence, the possibility of reducing the fluoride content in water or effluent by defluorination process or by dilution with the surface water appears to be the feasible solution to mitigate the human sufferings from these affected areas. Solubility experiments carried out on fluorite in distilled water by Rao, *et al* (1993), revealed that addition of Ca ions to solution in contact with fluorite caused decrease in fluorine concentration. The fluoride reduction may be achieved by calcium ion treatment or by dilution.

## FLUOROSIS WORLDWIDE

The latest information shows that fluorosis is endemic in at least 25 countries across the globe (see map). The total number of people affected is not known, but a conservative estimate would number in the tens of millions. In 1993, 15 of India's 32 states were identified as endemic for fluorosis in Mexico, 5 million people (about 6% of the population) are affected by fluoride in groundwater[3]. Fluorosis is prevalent in some parts of central and western China, and caused not only by drinking fluoride in groundwater but also by breathing airborne fluoride released from the burning of fluoride-laden coal[4]. Worldwide, such instances of industrial fluorosis are on the rise.

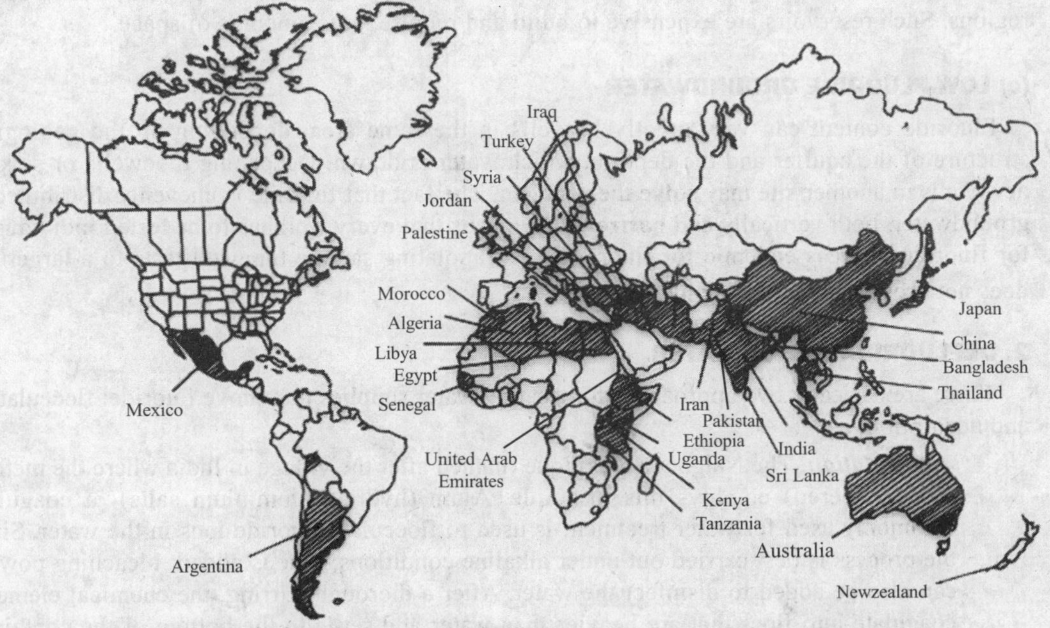

Fig. 14. Countries with endemic fluorosis due to excess fluoride in drinking water

Some governments are not yet fully aware of the fluoride problem or convinced of its adverse impact on their populations. Efforts are therefore needed to support more research on the subject and promote systematic policy responses by governments.

## PREVENTING FLUORIDE POISONING

Fluoride poisoning can be prevented or minimized by using alternative water sources, by removing excessive fluoride from drinking water, and by improving the nutritional status of population at risk.

### 1. ALTERNATIVE WATER SOURCES

These include surface water, rainwater, and low-fluoride groundwater.

#### (a) SURFACE WATER

Particular precaution is required when opting for surface water, since it is often heavily contaminated with biological and chemical pollutants. Surface water should not be used for drinking without treatment and disinfection. Many water treatment technologies are available, but the most effective are usually too expensive and complex for application in poor communities. Simple and low-cost technologies, such as sand filtration, ultraviolet water disinfection or chlorine water disinfection, are adequate in some but not all cases. Community capacity is an essential factor in ensuring successful utilization of these technologies. Water chlorination at household level is widely used only in emergencies.

#### (b) RAINWATER

Rainwater is usually a much cleaner water source and may provide a low-cost simple solution. The problem, however, is limited storage capacity in communities or households. Large storage reservoirs are needed because annual rainfall is extremely uneven in tropical and subtropical regions. Such reservoirs are expensive to build and require large amounts of space.

#### (c) LOW-FLUORIDE GROUNDWATER

Fluoride content can vary greatly in wells in the same area, depending on the geological structure of the aquifer and the depth at which water is drawn. Deepening tubewells or sinking new wells in another site may solve the problem. The fact that fluoride is unevenly distributed in groundwater, both vertically and horizontally, means that every well has to be tested individually for fluoride in areas endemic for fluorosis ; extrapolating sample tubewell tests to a larger are does not provide an accurate picture

### 2. DEFLUORIDATION OF WATER

There are basically two approaches for treating water supplies to remove fluoride: flocculation and adsorption.

- ***Flocculation***. The Nalgonda technique (named after the village in India where the method was pioneered) employs this principle. Alum (hydrate aluminium salts)- a coagulant commonly used for water treatment-is used to flocculate fluoride ions in the water. Since the process is best carried out under alkaline conditions, lime is added; bleaching powder can also be added to disinfect the water. After a thorough stirring, the chemical elements coagulate into flocs that are heavier than water and settle to the bottom of the container. The operation can be carried out on a large or small scale, and the technique is suitable for both community or household use. One household version uses a pair of 20-litre buckets, with a settling time of one hour and not more than two hours: after coagulation and settling are complete, the treated water is withdrawn through a tap 5 cm above the bottom of the first bucket, safely above the sludge level, and stored for the day's drinking in the second bucket.

*Adsorption*. The other approach is to filter water down through a column packed with a strong adsorbent, such as activated alumina ($Al_2O_3$), activated charcoal, or ion exchange resins. This method, too, is suitable for both community and household use. When the adsorbent becomes saturated with fluoride ions, the filter material has to be backwashed with a milk acid or alkali solution to clean and regenerate it. The effluent from backwashing is rich in accumulated fluoride and must therefore be disposed of carefully to avoid recontaminating nearby groundwater.

Both the community and household defluoridation systems have pros and cons. Defluoridation equipment connected to a community handpump is theoretically cheaper per capita than a household unit because of economies of scale ; but ensuring proper maintenance of a commonly owned facility is often problematic, so good community organization is necessary. The household units are more convenient for filtering the small amounts of water intended for drinking only, and people usually take better care of them ; but an extensive and efficient service system is required to ensure that the filters are replaced or regenerated at the capabilities, can be a far more critical and difficult task.

## 3. BETTER NUTRITION

Clinical data indicate that adequate calcium intake is clearly associated with a reduced risk of dental fluorosis. Vitamin C may also safeguard against the risk. In consequence, measures to improve the nutritional status of an affected population-particularly children-appear to be an effective supplement to the technical solutions discussed above.

## DEFLUORIDATION AND UNICEF

UNICEF has worked closely with the Government and other partners is defluoridation programmes in India, where excessive fluoride has been known for many years to exist in much of the nation's groundwater. In the 1980s, UNICEF supported the Government's Technology Mission in the effort to identify and address the fluoride problem; the Government subsequently launched a massive programme, still under way, to provide fluoride-safe water in all the areás affected.

Over the past five years, UNICEF's focus in the India programme has been on strengthening systems for monitoring water quality, facilitating water treatment by households, and advocating alternative water supplies when necessary. Education-both of households and communities- is key to strategy. A number of demonstration projects have been initiated in fluorosis-affected areas, with the emphasis currently on introducing household defluoridation. UNICEF has also sponsored research and development on the use of activated alumina for removal of fluoride from water.

Since fluoride must now be considered an issue of worldwide importance, the years of experience in India should help UNICEF and its partners provide four types of assistance towards an eventual solution:

- Promoting a better understanding of the problem and its impact on children;
- Raising the awareness of relevant governments and the public on the fluoride issue is particular and the importance in general of monitoring water quality;

- Demonstrating, through pilot projects, the efficacy of low-cost fluoride removal technologies;
- Strengthening community and government capacity for fluorosis prevention, including a credible system for risk assessment the comprises both water quality monitoring and health monitoriong.

# NOISE POLLUTION

## INTRODUCTION

The term "noise" may be defined as an unwanted sound at a wrong time and a wrong place. Although noise is undesirable, it could be meaningful or meaningless. A meaningful noise is generally meant for inviting attention or expecting a consequent response such as the cry of a baby or a screaming of a person for help. On the contrary, an irresponsible or meaningless noise is disturbing and annoying. Whether a given sound is wanted or unwanted may depend upon the person involved, the loudness, the rhythm, and the length of time for which one is exposed to it. A sound may be music to one person but noise to another; acceptable when soft, rhythmic or for short time, but unacceptable when loud, random or prolonged; reasonable when made by himself but unreasonable when it is made by others. However, prolonged and lound sound is generally considered as noise which is mostly caused because of industries, vehicles, aeroplanes etc. Since such a noise is found to cause some kind of physical, physiological or psychological harm or stress to human beings, it is also considered as a pollutant. Noise is a by-product of human activity and the background noise levels have been found to be increasing year by year for the last few decades. Excessive noise can result in loss of hearing, increase in accidents and decrease of productivity. Hence, it is essential to take suitable precautions to reduce noise pollution particulary in urban and industrial areas.

Sound has several physical properties among which "Frequency" and "Intensity" are the most relevant for the present discussion. Sound frequency is the rate at which compression waves arrive at or pass a fixed point. "Pitch" is the human perception of sound frequency (and also intensity to some extent). Sound intensity is the acoustical power (*i.e.*, the energy delivered by sound) per unit area. "Loudness" is the human perception of the sound intensity (and also frequency to some extent). "Hertz" (Hz) or cycles per second is a measure of sound frequency. Human beings can hear only sounds ranging from 20 Hz to 20,000 Hz. The range of frequencies of human speech is 200 to 3000 Hz, which is best heard by humans. Thus, we can hear only sounds of certain frequencies and even among them, we are more sensitive to some than to others. Sounds too high in frequency (above 20,000 Hz) is called "ultrasound" and that which is too low in frequency (below 20 Hz) is called " infrasound".

The response of ear to sound is proportional to the logarithm of its intensity or pressure. The loudness of two sounds is judged subjectively by the ear by the ratio of their intensities or pressues. The loudness is expressed in terms of a unit called "decibel" ("deci" comes from the Latin word for ten, and a "bel" is "the logarithm of a ratio" of any two acoustical or electrical intensities. In terms of sound, a "decibel" (dB) is ten times the logarithm of the ratio of two sound intensities, one being the intensity of any sound of interest and the other being a reference sound ($I_o$).

$$\text{Decibel (dB)} = 10 \log \frac{\text{Sound Intensity measured}}{\text{Reference sound intensity}}$$

in the United States, the Reference sound intensity is $10^{-12}$ watts per square meter, which is the sound intensity barely audible to human beings. This zero dB is the threshold of hearing. Thus, a sound which is 100,000 times louder (more intense) than the reference level would be called a 50-dB sound (because the logarithm of 100,000 is 5 and 10 times 5 is 50). Similarly, sound with 10 times the intensity of the reference level would be a 10 db sound. A normal conversion is done at 60 dB sound levels. A jet plane during its takeoff produces a 150 dB sound. The decibel scale for some familiar sound sources is given in Table 2.12. The psychological and physical harm at different decibel levels is shown in Table 2.13.

**Table 2.12. The Decibel Scale**

| Sound Source | Decibel, dB |
|---|---|
| Launching of space rocket | 170 |
| Jet plane at take off | 150 |
| Threshold of pain | 140 |
| Pneumatic riveter | 130 |
| Running motor cycle | 118 |
| Jet fly over at 150 m | 115 |
| Rock band | 111 |
| Jet fly over at about 300 | 103 |
| Farm tractor | 98 |
| Motor cycle at 25 ft | 90 |
| Heavy city traffic | 85 |
| Alarm clock | 80 |
| Average city traffic | 70 |
| Normal conversation | 60 |
| Business office or light city traffic | 50 |
| Living room | 45 |
| Library | 35 |
| Broadcasting studio or A quiet room at night | 20 |
| Rusting of leaves | 10 |
| Threshold of hearing | 0 |

**Table 2.13. Psychological and Physical Effects at Different Decibel Levels**

| | | |
|---|---|---|
| 135 dB | .. | Painful |
| 110 dB | .. | Discomfort |
| 88 dB | .. | Hearing impairment on prolonged exposure |
| 80 dB | .. | Annoying |
| 65 dB | .. | Intrusive |

## EFFECTS OF NOISE

**(a) Physiological Effects :**

The acute effects caused by noise depend upon the pressure and frequency. At high levels of about 150 dB, immediate permanent hearing impairment may be caused. At sound levels in the range of 120-150 dB, effects on respiratory system, dizziness, disorientation, loss of physical control, other physiological changes resulting from stress, nausea and vomiting may be caused. Even sounds of the order of 70 dB can have measurable physiological effects, although they may not result in any immediate impairment.

Nerve fibers leaving the inner ear carry impulses elicited by sirens, trumpets, etc. to the medulla of the brain stem, where they meet other fibers going to other parts of the brain. Nerve pathways permit both ears to communicate with numerous parts of both sides of the brain, including the centres of consciousness and the control centres that regulate breathing, blood pressure and other bodily functions below the level of consiousness.

Loud sounds can cause an increased secretion of many hormones of the pituitary gland *e.g.*, adrenocorticotropic hormone (ACTH). ACTH in turn stimulates the adrenal gland, which secretes several other hormones. Through a variety of influences, these hormones in turn trigger various effects such as (a) enhancement of the sensitivity of the body to adrenalin, (b) increase of blood-sugar levels (c) suppression of immune system and (d) decreasing the efficiency of liver to detoxify blood.

The major *chronic effect* of noise is the so-called noise-induced hearing loss. A person exposed to high noise level goes deaf more quickly than the one who is exposed to a relatively noise-free environment during his day to day activities. This effect depends upon and increases with the pressure, frequency and the time of exposure. Studies on a group of youngsters listening to rock music generated at over 92 dB throughout the 500- to 8000-Hz range and sustained over about an hour showed a "threshold shift" of 40-dB in about 10% of the listeners, and 20 to 30% "threshold elevation" in the others. Most of this "threshold shift" is usually temporary but repeated exposure can make it permanent (A "threshold shift" is an elevation in the threshold of hearing *i.e.*, the quietist sound that can be heard, becomes a louder sound).

**(b) Psychological and other effects**

Although there is little specific evidence regarding the onset of mental or nervous illness caused by noise, some reports are available to indicate temporary effects such as deterioration in concentration and even mental disorientation at high noise levels.

Loud continuous noise reduces the working efficiency, interferes with communication, increases the frequency of errors which may, at times, cause accidents. Noise reduces the mental capability. Noise has psychological effects on humans ranging from mild distress to complete unhinging.

Noise interferes with deep sleep and interrupts sleep. Because sleep is important to emotional stability, noise may contribute to distress and emotional disturbance. Noise also aggravates any existing psychological conditions and mental illness.

Noise has some behavioural effects on birds and animals also.

## NOISE POLLUTION CONTROL PROGRAMME IN INDUSTRIES

The modern technology for achieving greater productivity necessarily deals with greater speeds, higher fluid pressures and velocities and more powerful machinery. This creates the undesirable noise pollution. This results in speech communication interference, loss of productivity, hearing impairment and other physiological and psychological effects for the personnel working in the plant. Further, it also gives rise to community noise problems such as annoyance to the residential population in the vicinity of the plant.

Although the technology for noise pollution control is available, the necessary social motivation for it seems to be lacking.

The first step in implementation of any noise control programme in any industry is to determine the existing noise levels in the plant and to decide upon the noise control criteria to be achieved.

**Measurement of noise levels :** Measurement of noise levels in decibel is carried out with an instrument called sound level meter. It consists of three internationally accepted weighting networks, namely, A, B and C. Industrial noise levels are measured with weighting network A and accordingly the readings are designated as dBA (decibels measured on network A). In this scale, the frequencies to which human beings are most sensitive are given more weightage in the assessment of effects on human hearing. Since engineering noise control procedure depend on frequency* the noise levels are usually examined with the help of an octave meter attached to the sound level meter.

Noise surveys have to be made in the entire plant area to assess the overall noise level and noise contours (in the vicinity of individual noise sources) under different meteorological conditions. Noise emitted by individual sources must be carried out at specified distances as per the prescribed standards (*e.g.*, IS: 4758 - 1968). When the noise sources are grouped, the composite noise level may mask the individual sources. Then, it is difficult to identify a significant noise source. In such cases, special measuring devices such as directional microphones are to be used.

**Noise Control Criteria :** With increasing noise levels, annoyance, interference with communication and hearing damage can occur progressively. Noise control criteria are prescribed to provide standards for judging the acceptability of noise under various conditions.

**Annoyance :** The degree of annoyance depends upon the intensity and frequency of the noise and its variation with time. High frequency noise and impulsive or intermittent noise are much more annoying and irritating as compared to low-frequency and continuous noise. Table 2.14 shows the acceptable noise levels for different outdoor and indoor situations.

**Table 2.14. Acceptable noise levels for different outdoor and indoor situations**

| Location | | Acceptable noise level range, dBA |
|---|---|---|
| **Outdoor :** | | |
| Rural | .. | 25-35 |
| Residential (urban) | .. | 35-45 |
| Business (urban) | .. | 45-50 |
| City | .. | 45-55 |
| Industrial area | .. | 50-60 |
| **Indoor :** | | |
| Radio & TV studios Hospitals, classrooms, | .. | 25-30 |
| Apartments, hotels and conference rooms | .. | 35-40 |
| Private offices and court rooms | .. | 40-45 |
| Public offices, stores and banks | .. | 45-50 |

**Interference with Communication :** In plants where audible warning signals are used, the background noise level must be maintained low enough to enable the warning signals audible. Since the speech frequencies lie between 200 and 7000 Hz, the background noise must be restricted to the level that it does not mask the frequencies in this range. For judging the interference effects by the background noise, the arithmetical average of the noise levels in the

*Frequency is the number of times that a sound source vibrates in a unit interval of time and it is expressed as hertz (Hz) or cycles per second (CPS).

three octave bands *viz*., 600-1200 Hz, 1200-2400 Hz and 2400-4800 Hz are usually calculated. This average level is called "*Speech Interference Level, (SIL)*".

The SIL values in decibels for rendering different types of communication possible are given in Table 2.15.

**Table 2.15. Speech Interference Level and Voice Levels Required for Communication**

| Speech Interference Level (SIL), dB | Voice level and distance within which communication is possible |
|---|---|
| 45 | Normal Voice at 3 m (Relaxed Conversation) |
| 55 | Normal Voice at 1 m (Continuous conversation in work areas) |
| 65 | Raised Voice at 0.6 m (Intermittent communication) |
| 75 | Very loud voice at 0.3 m (Minimum communication) |
| 85 | Shouting at 0.3 m (Unvoidable communication) |

**Hearing Loss Criteria :** The extent of hearing impairment depends upon the noise intensity, frequency and time of exposure.

Table 2.17 given below indicates the permissible duration of exposure to various noise levels.

**Table 2.16. Permissible Exposure Limits to Various Noise Levels**

| Noise levels, dBA | Permissible duration of exposure (hours per day) |
|---|---|
| 90 | 8 |
| 95 | 4 |
| 100 | 2 |
| 105 | 1 |
| 110 | 1/2 |
| 115 | 1/4 |

Table 2.17 shows the American National Standard Specification for sound level meters.

**Table 2.17. Noise Threshold Limit Values as per American National Standard Specifications for sound level meters**

| Sound levels, dBA | Perimissible duration of exposure (hours per day) |
|---|---|
| 80 | 16 |
| 85 | 8 |
| 90 | 4 |
| 95 | 2 |
| 100 | 1 |
| 105 | 1/2 |
| 110 | 1/4 |
| 115 | 1/8 |

Further limits for exposure to impulsive or impact sound have also been prescribed as shown in Table 2.18. An impulsive or impact sound includes the variation in noise levels that involve maxima at intervals greater than one per second.

**Table 2.18. Threshold Limit Values of Impulses or Impacts per day**

| Sound level, dB* (Decibels peak sound pressure level) | Permissible number of impacts per day |
|---|---|
| 140 | $10^2$ |
| 130 | $10^3$ |
| 120 | $10^4$ |

Noise Exposure Index (NEI) is helpful to indicate whether a given noise exposure is excessive or not. It can be calculated as follows :

$$NEI = \frac{t_1}{T_1} + \frac{t_2}{T_2} + ... + \frac{t_n}{T_n}$$

where $t_1$ to $t_n$ are the actual limits of exposure at the corresponding noise levels, and $T_1$ to $T_n$ are the permissible times of exposure at the corresponding noise levels. If the value of NEI thus calculated is greater than 1, the noise exposure level is deemed to be excessive.

## EQUIPMENT USED FOR NOISE MEASUREMENT

The various types of noise measuring equipment commonly used are given below.

**(1) Sound Level Meter :** A simple sound level meter basically comprises of a non-directional microphone coupled to an amplifier to transform sound pressure level impinging on the atmosphere to a proportional electric signal indicated by a meter. The two main types of microphone used in this instrument are the piezoelectric and condenser types. The sound level meter is used for evaluating sound pressures on linear or weighted scale. It indicated the RMS value of the noise, but can be calibrated to read the peak level sounds as well. The sound level meters are available with various dynamic characteristics such as slow, fast, impulse, impulse hold, peak and peak hold. Sound level meters which are cheap and robust with limited accuracies (*e.g.*, IEC 123) as well as those which are highly accurate but are delicate and expensive (*e.g.*, IEC 179) are available. Additional facilities may be provided on industrial and precision grade instruments. These may include a wider measuring range, A-, B- and C- weighting networks, a linear frequency response, provision for the use of filters, etc.

**(2) Octave Band Analyser :** This can be built in or used as an accessory with sound level meter to determine the average sound pressure of the standard octave bands. This gives information regarding the frequency content of a noise which helps in identifying the noise source as well as in selecting the ear protectors.

**(3) Cassette Recorder :** This is used for storage of noise and for rough evaluation. It can be used for digital recording of sound pressure levels for computer evaluation.

**(4) Magnetic Tape Recorder :** This is used for recording the noise more accurately for storage or for subsequent evaluation. This instrument is more expensive than cassette recorder and is fed through a sound level meter or a microphone.

**(5) Pen Recorder :** This instrument can either be fed from a sound level meter or directly or from the magnetic tape recordings on any of the following two systems :

(*i*) The one which gives a continuous record of sound pressure level against time. This data can be used either for storage or for obtaining approximate statistical levels by inspection. Instruments with different dynamic levels and different degree of accuracies are available.

(*ii*) The one which gives a discontinuous record in the form of instantaneous sound pressure levels sampled from the noise received. The statistical levels can be found by inspection.

The dynamic range of the device is limited and only a single sound level can be determined accurately.

**(6) Statistical Analysers**

(*a*) Mechanical units which are used in conjunction with pen recorders to give any statistical level or an $L_{eq}$.

(*b*) Electronic units which can be fed either directly or through a sound level meter or from a magnetic tape recording to give the desired statistical analysis commensurate with the limitations of the equipment used. It can also be coupled to a print-out system to give a permanent record of the data collected. The dynamic range can be 60 dB or more and the accuracy complies with IEC 179.

(*c*) Digital tape processor, which is fed from digital tape recordings to provide any statistical analysis or an $L_{eq}$ as desired. This system is expensive. It can give a good dynamic range and accuracy.

**(7) Noise Average Meters ($L_{eq}$ meters) :** Equivalent continuous noise level ($L_{eq}$) is at present a widely accepted parameter for "noise dose" measurement. Although the $L_{eq}$ can be obtained from formula calculations applied to measurements with a conventional sound level meter taken over the specified period, the accuracy of the result depends upon the amount of variation of the signal and the number of results summed. Hence, Noise Average Meters have been developed to measure $L_{eq}$ levels directly and accurately. These $L_{eq}$ meters are more complicated and more expensive than the normal sound level meters. Since $L_{eq}$ measurements are taken over long periods of time (over 24 hrs) and perhaps in highly impulsive environments, a large dynamic range for the meter is absolutely essential to accommodate full range of expected levels and to ensure that these are accurately averaged. The noise average meter can be fed directly or from magnetic tape recordings to give $L_{eq}$ as a continuous digital display or as a print-out.

Noise Exposure Meters are similar to Noise Average Meters in that they register elapsed time of exposure to sound pressure levels above a predetermined level. Thus, basically, these may be considered as noise exposure monitoring devices but do not actually measure noise dosage specifically. These were originally developed to conform to earlier industrial noise dosage recommendations but are equally applicable to quasi equivalent continuous noise level monitoring.

The Noise Dosemeters or Dosimeters also function on the same principle excepting that the information is recorded as a percentage of a present dose instead of $L_{eq}$. These are small and compact and are intended to be worn by operators exposed to hazardous noise environments.

**Noise Analyser :** This instrument is fed either directly or from magnetic tape recordings. It basically consists of a sound level meter (IEC 179) which feeds a Noise Average Meter and a Statistical Analyser. The instrument may be coupled to provide an analogue or digital recording of the processed data, with both periodic and cumulative evaluations for print-out. It can also provide a digital display.

**(8) Noise Limit Indicators :** These are basically Sound Level Meters with or without filters and/or analyzers coupled to instrumentation or display units to indicate whenever a received noise level exceeds a predetermined value. These instruments are of particular use for quality control of a production line output with respect to noise and vibration of the product.

**(9) Impulse Noise Meters :** These are modified sound level meters with a response time reduced to the order of 1/10 ms to be capable of indicating or recording impulsive sounds. The circuit may be designed to measure peak instantaneous value, continuous peak level, RMS average level or all the three.

**(10) Vibration Meters :** Vibration measurements can often complement sound level measurement. A vibration meter for simple measurements comprises of a Sound Meter in conjunction with a vibration adapter. Such adapters generally utilize an accelerator type of vibration pick-up which provides an output signal proportional to acceleration. These pickups are small, light and robust. The signal from the accelerometer is fed to the Sound Level Meter via a network of passive circuity comprising of attenuators and integrating networks which provide acceleration, velocity or displacement conscious signals according to the parameter that is desired to be measured. Measuring range selection is achieved by the range attenuator on the Sound Level Meter and information is provided on how to translate the indication on the dial of the Sound Level Meter in to a vibration reading. Some modern instruments have been designed to provide a direct read-out of the virbation measurement.

Various codes, standards and guidelines have been laid down for noise measurements and monitoring in indoor and outdoor situations.

## APPROACHES FOR NOISE CONTROL

The following four approaches are available for noise control :

(1) Modifying some of the present practices and procedures in order to minimise the noise.

Ex: Reducing automobile traffic, outlaying sirens, discouraging stereos without headsets, using glue instead of rivets, etc.

(2) Shielding the sources of noise generation.

Ex: Use of sound-absorbing motor mountings, better installation, better design, use of motor enclosures, use of vibration damping or absorbing materials in automobiles and dishwashers, etc.

(3) Shielding the noise receiver.

Ex: Using earplug, control booths, etc.

(4) Shifting noisy sources and things away from people.

Ex: Isolating airports, industrial complexes, etc.

Obviously, some of the above measures can be implemented successfully only if they are mandatory.

## NOISE CONTROL IN INDUSTRIAL ESTABLISHMENTS

The problem of noise in industries is tackled by effective implementation of administrative and engineering controls.

**(A) Administrative Controls :** These are essentially short-term controls which are aimed at protecting the workers overexposed to noise levels of about 90 dBA before the Management can plan and implement long-term solutions. "Worker rotation" is an effective administrative control which aims at management of work schedule in such a way so as to ensure that no worker is exposed to high noise levels beyond permissible limits.

**(B) Engineering Controls :** In order to control high noise levels in industrial plants, one or more of the following three methods may be necessary :

(*i*) suppression of the noise at the source,

(*ii*) path control, and

(*iii*) protection of the personnel.

***(i)* Suppression of the noise at the source :** This is an effective and economical approach particularly in situations where the noise levels are very high. Such situations arise with devices such as compression, cooling fans, pneumatic equipment, valves, flares and pumps which are commonly used in various industries. If the noise is "inherent" in the operation itself, nothing much can be done about it. However, if the noise is "incidental" to the operation, then it can be minimised by proper design, proper location, proper installation, proper operation and regular maintenance.

**(*ii*) Path Control :** After taking all possible steps to suppress the noise at the source, further engineering controls are directed at the treatment of the major noise transmission paths from the source. The important paths of noise transmission from a source include (*a*) the structure which supports the entire machinery under consideration, (*b*) surface and panels attached to the source (*c*) cables, conduits, belts, wires, ducts, pipes, etc., and (*d*) fluids flowing through or surrounding the source (as in case of chemical plants).

Sound absorption, sound insulation and vibration control are the three important parameters in noise control.

**Sound absorption :** Sound absorption is mainly intended to reduce the loudness of reflected sounds in a room or enclosure, and to decrease reverberation. A sound absorbent surface is one which absorbs a proportion of the sound energy incident on it so that the level of sound reflected from the surface is substantially reduced. The absorption of sound energy by a sound absorbing material or composite structure is achieved by one or more of the following mechanisms :

(*a*) friction between the fibres of a porous fibrous material

(*b*) absorption in the voids of a porous, non-fibrous material, and

(*c*) absorption within narrow entries to an air space.

The performance or effectiveness of a sound absorbent is usually expressed by its "absorption coefficient" which is an arbitrary scale between 0 and 1.0. An absorption coefficient of 0 represents total reflection and no absorption, whereas the absorption coefficient of 1.0 represents total absorption and no reflection. The coefficient multiplied by 100 represents the percentage of sound energy absorbed. However, it should be emphasized that the absorption coefficient is not an absolute constant quantity for any material or composite structure. It varies with frequency and is also affected by size and position, method of mounting, etc.

Sound absorbent materials are generally light-weight, porous or fibrous which reduce reflection of sound. Examples of typical sound absorbing materials are mineral wool, glasswool, felts, porous panels, porous blocks, perforated panels and tiles with or without absorbent backing, carpets, drapes, etc. It is interesting to note that sound absorbents are, in general, poor sound insulators and hence, a high proportion of sound energy may be transmitted through the absorbent surface.

Sound absorbents play an important role in reducing the general noise level within a room and minimizing excessive background noise. They can be particularly beneficial in large offices, workshops, etc. Sound absorption treatment can prevent noise being reflected back into other quieter areas. In addition to the sound absorbers, use of sound insulating barriers may be necessary to reduce the noise radiated in a direct path from noise sources.

Although *sound absorbent materials* can contribute to *sound insulation* by reducing the sound level generated within a room or an enclosure, that does not justify them being called *sound insulating materials.*

**Sound insulation :** Sound insulation involves reduction of sound transmission with the help of barriers. Sound insulating materials are dense, heavy weight materials, their effectiveness as a sound barrier being directly related to their mass per unit area. The performance of a sound insulating material is expressed in terms of sound reduction index (R), which is given by the following equation :

$$R = 10 \log_{10} P_1/P_2 \text{ dB}$$

where

$P_1$ = sound pressure level on sound source side, and

$P_2$ = sound pressure level on the opposite side (*i.e.*, the receiving side).

The ratio $P_1/P_2$ is thus the ratio of transmitted sound energy to incident level and is called the transmission coefficient. The above formula can be written as follows :

$(P_2)^2 = (P_1)^2 \times TS/\alpha$, where

T = the transmission coefficient of the panel or structure

S = surface area of the panel, and

$\alpha$ = total sound absorption of the receiving side volume.

The transmission coefficient, and thus the sound reduction index, is frequency dependent, low frequencies are much more readily transmitted than higher frequencies. Concrete, glass, lead, steel, etc. are used as sound insulating materials.

**Vibration control :** Vibration isolation comprises of protecting a supporting structure from force-excited vibrations generated in an associated mass or protecting a mass from motion-excited vibration generated in the supporting structure. If the two are rigidly connected, then the vibration excited in one will be transmitted directly to the other and with the result, the entire system vibrates.

Free space is the ideal insulator, where no transmission of vibration is possible. However, this is a purely theoretical concept which defines a perfect vibration insulator with a "transmissibility value" of zero. In practice, a vibration isolation treatment involves the use of resilient mountings and mounting system designed to limit the transmitted force or motion.

The performance of an isolater is usually expressed as its transmissibility, which is a function of the ratio of the disturbing frequency to the natural frequency of the system :

$$\text{Transmissibility (T)} = \frac{1}{1-(f/f_n)^2}$$

where

f = disturbing frequency, and

$f_n$ = natural frequency of the mounted assembly.

Isolation efficiency is then expressed as (1 – T)%.

A transmissibility of less than 1 implies some degree of isolation.

Isolators may be used to mount a machine directly on to a base-plate, floor or foundation, or to a substantial inertia block (such as a concrete block) which is itself isolated.

"Damping" is a treatment aimed at "absorbing" vibrational energy present and is thus quite distinct from vibration isolation. Discrete dampers may, however, be used in conjunction with vibration isolation systems to compensate for low damping provided by the spring material, or to provide stability of controlling or reducing vibration at or near resonance.

The various techniques available for path control are as follows:

**(a) Acoustic Enclosures :** In the context of noise control, the term "enclosure" broadly refers to any method used for artificially impeding the free flow of acoustic energy away from the source. As a rough guide, it may be taken that whenever an overall sound reduction of about 15 dBA is desired, a complete enclosure is needed.

In industrial plants, the following two types of noise control problems are encountered which warrant acoustic enclosures :

The first one is where the source and the receiver are in the same room *e.g.*, operators working in machine shops. In such cases, a localised and close-fitting enclosure around the machine is recommended. Access to the machine may be provided via hinged or quick release panels and doors.

The second one is where the source is in one room and the receiving personnel are in adjacent rooms, *e.g.*, offices next to a mechanical services plant room. In such a case, the source room itself becomes the enclosure.

In cases where the size of the noisy equipment is too big or when the working area and the operation do not permit close-fitting enclosures, the equipment has to be housed in a separate room of its own. In this case, the operator inside the room may not get any relief, but other workers in the plant will be benefited from the reduced sound level.

Containment of a noise source in an enclosure such as close-fitting acoustic box is an effective method used for the path control of noise. The wall thickness of the enclosure depends on the depends on the material of construction, as for instance, 20 mm in case of plywood or 3 mm in case of sheet metal. The walls of the enclosure are acoustically lined inside with materials like glass fibre or mineral wool of thickness 20 to 120 mm depending on the frequency of the noise. Thicker linings are necessary for lower frequencies. The linings are fixed to the container with the help of a retainer such as chicken wire, expanded metal hardware cloth or fibre cloth which also protects the lining from weather or attrition. A good leak-proof enclosure can achieve a noise attenuation of the order of 50 dBA and thus can offer protection to the operator and other workers in the plant.

**(b) Noise Barriers :** Construction of a thick and high (> 2 m) barrier, covered with an acoustic material on the source side, blocks the path of the directly transmitted sound from the source and provides appreciable protection (of the order of 15 dBA) particularly from high frequency noises. A barrier may not give relief to the operator from the noise from his machine but prevents noise from other sources reaching him.

Any solid medium substantially normal to the ground acts as a barrier to sound and affords an appreciable sound reduction. Depending on the nature of the medium, it may "reflect" sound (as in the case of hard and dense surfaces), or "absorb" sound (as in the case of porous media). The degree of sound reduction achieved depends also on the physical form of the barrier, specifically on its effective height and length.

On an open site, buildings, earth embankments, stored materials and other solid obstructions can also act as sound barriers. However, unless suitably sited, such barriers may only transfer a noise problem from one area to another.

A noise barrier will be most effective only when it is placed as close as possible to the noise source or to the receiving point. A barrier should not have any gaps or joints through which sound can "leak". The length of the barrier should be at least 10 times its height. Wherever this is not possible, the barrier should curve around the noise source.

In order to act as an effective sound "insulator", the barrier material should have a superficial mass of at least 7 kg/m$^2$. A number of common building materials *e.g.*, brickwork, asbestos cement boards, fibre board, woodwool slabs, clinker blocks etc., satisfy this requirement.

Sound "absorbent" materials are generally poor sound insulators in that they readily transmit sound through them. They are useful as linings to reduce the sound reflections incident on particular surfaces. They are particularly useful for the lining of screens or semi-enclosures where the objective is both to reduce the sound generated by a source (*e.g.*, noisy machine or tool) and at the same time reduce the overall noise level in the vicinity of the source which would otherwise be reflected by a sound barrier and cause discomfort to the machine or tool operator. Typical sound absorbing materials include glass fibre, woodwool slabs, straw slabs, mineral wool etc.

By and large, a simple sound barrier can give a noise reduction of the order of 10 to 15 dBA.

**(c) Mufflers or Silencers :** A muffler or a silencer is a pipe or duct treated or shaped for reducing sound transmission while at the same time allowing the free flow of a gas. Mufflers or silencers are broadly classified into two categories: the *dissipative* and the *reactive*. The *dissipative silencers,* which are also known as absorptive silencers, work on the principle of absorbing noise. They make use of acoustical absorbent linings. The *reactive silencers* work on the principle of reflecting, and thus containing, sound within them. They can be of various geometrical shapes to provide impedence mismatch where the acoustic energy is reflected back towards the noise source. Some mufflers combine both these effects. The performance of a muffler depends on the frequency of the sound. In terms of general performance, *dissipative silencers* normally provide sound attenuation over a broad frequency range, and hence are more popular than the reactive variety, in industrial process plant applications. On the other hand, *reactive silencers* are tuned to provide maximum attenuation at specific frequencies and thus work as narrow band silencers. Some special designs of silencers incorporate both the above principles of working. For instance, a simple expansion chamber is essentially a tuned or reactive silencer, attenuating noise in a narrow frequency region. However, if it is lined with an acoustically absorbent material (*e.g.*, fibre glass, mineral wool, etc.), it would also exhibit dissipative characteristics and provide attenuation over broad frequency range.

Pure dissipative silencers are based on the use of flow-resistive materials, normally in the form of porous acoustic linings. Such materials may be used to cover the interior surface (as in lined ducts) or be in the form of parallel baffles. The latter arrangement may be considered as splitting the volume involved into a series of lined ducts. This effectively increases the absorptive length of the silencer without increasing its actual length, although its diameter may have to be increased. Mufflers of parallel or undulated baffles may be required where large quantities of air pass through openings between quiet and noisy spaces.

The *reactive silencer*, in its simplest form is a plenum chamber or expansion chamber, normally cylindrical, with or without baffles. Industrial silencers of this type commonly have inlet and discharge ducts offset to minimize the direct transmission of sound across the chamber.

The principal advantages offered by reactive silencers are relatively small size and high mechanical stability. Plenum chambers can provide high degree of sound attenuation at low frequencies, but when used for industrial silencers, large volume is required.

Expansion chamber lined with acoustically absorbent material is a basic type of reactive absorption silencer. Such a system, with suitable design, can achieve an overall attenuation much greater than the simple sum of the attenuation of the lining and that of the "tuned volume".

An *acoustic plenum* is essentially a combination of a muffler and enclosure used to admit low velocity cooling and combustion air without leaking excessive noise. The interior surface of the plenum chamber is lined with sound absorbing material, such as fibre glass. An expended metal screen is used as retainer. The inlet and outlet should not be located directly opposite to each other. In fact, it is essential to provide at least two 90 degree changes in the direction of the acoustically lined air flow path with no direct line – of – sight to the noise source from the outside. Fig. 14 shows a simplified diagrammatic representation of a single chamber plenum and multiple chamber plenum.

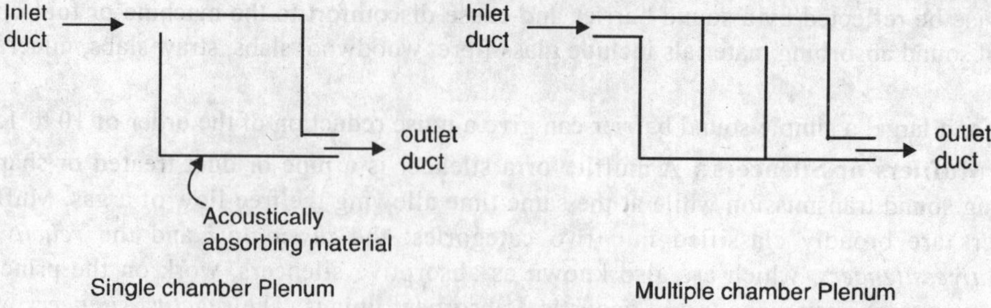


**Fig. 15.** Acoustic plenums.

**Automotive type silencers :** Until very recently, the design of silencers used for automobiles and transport vehicle engines has been based on reactive and/or dissipative principles described above. Reactive silencers work on the principle of reflecting sound energy back towards the source, but may also provide some energy dissipation by cross flow, depending on the design. On the other hand, dissipative silencers incorporate sections or chambers lined with sound absorbent materials in which sound energy is absorbed and converted into heat. Fig. 16 shows four basic designs of automobile engine silencers in simplified form:

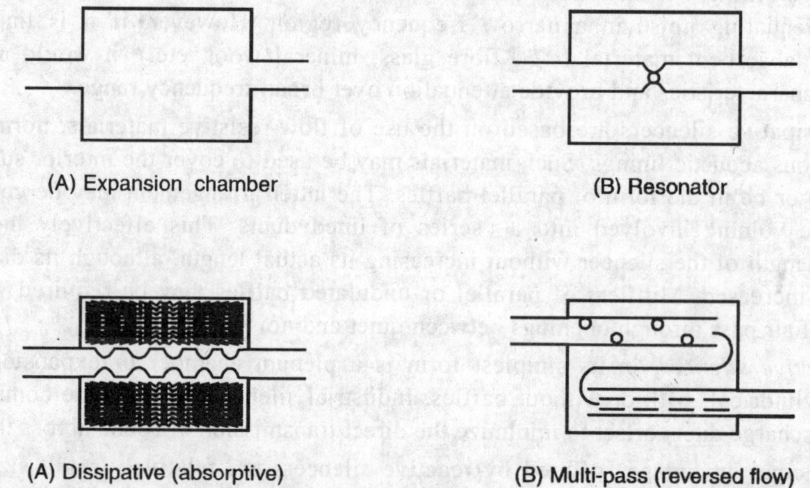


**Fig. 16.** Simplified diagrams of the four basic types of automobile engine silencers.

The expansion chamber and resonator are basically tuned filters with narrow band attenuation. The dissipative or absorbent type is capable of providing much broader band attenuation while still retaining the "straight through" geometry and thus minimum back pressure loss. The reversed flow or multipass type of silencer is capable of even wider band attenuation, particularly at low frequencies, but usually at the expense of higher back pressure (because of the tortuous gas flow path) on which the performance of the silencer largely depends. The multi-pass silencer is produced in a variety of forms, some of which are quite complex, which may include cross-flow chambers (Quincke tubes) and a Helmholtz resonator section. These are also called baffle silencers.

The present trend in the design of automobile silencers is to employ a combination of multi-pass or (reversed flow) chambers with acoustic absorbent chambers.

There are a wide variety of ingenious designs of reversed-flow silencers with minimum back pressure loss. In some designs, an exhaust pipe and tailpipe are employed in parallel configuration inside a silencer chamber, the ends of these pipes being blanked off. The slub pipes are then drilled and/or slotted, to provide a continuous gas path to atmosphere, at the same time expanding within the chamber. The main object is to reduce both the pressure and speed of gas flow. Expansion will also result in some cooling of the gases. Modern tailpipes also include resonator devices, pulsed-tuned baffling, or even sound-absorbing elements to provide additional effect.

Various materials used for the construction of silencer and exhaust systems include mild steel, zinc-proofed steel, tinplate, aluminium-coated steel, galvanised steel and stainless iron and stainless steel.

**Vibration isolation :** This is intended to reduce the magnitude of vibration between two sources attached to each other. When a sound source is attached rigidly to a large thin plate, the noise radiated from the plate will be very high. The larger the area of the radiating surface (*i.e.*, the plate), the greater would be the intensity of the noise from the source. When a machine is installed in a building, the vibration produced by the machine is transmitted to the floor and then to the entire building structure and thus propagating the noise to large distances. In such cases, the radiation of noise can be controlled by "impedance mismatch" between the source and the structure on which it is mounted. The "acoustic impedance" is a measure of the rate at which unit mass of a material receives the acoustic energy. The acoustic impedance of a material is given by the product of the density of the material and the velocity of sound in the material. Thus, a material having a low impedance is an inefficient sound transmitter or a good sound isolator. Rubber has a low impedance as compared to steel. Hence, a block of rubber between two blocks of steel can provide a very good sound isolation. For the same reason, a noisy machine should be floated on a soft and low-impedance mounting rather than bloting it to the floor mounting. Properly designed, soft and resilient spring or elastomer mounts act as very efficient sound isolators and also reduce wear and maintenance problems. Compressors, pumps, fans, etc should be installed on antivibration mounts to reduce noise from a few dB to 30 dB. Other examples of vibration isolators are flexible couplings to conduits, ducts and pipes, gaskets on ducts, etc.

**Damping :** Damping is a method used to control resonant response to vibration, the process involved being that of dissipating energy by converting it into heat and thus controlling the amplitude of the response to excitation at resonance. The greater the damping, the less the vibration motion at resonance. Thus, it is a basic method used for controlling vibration, shock or noise.

Damping can be applied to mechanical systems by the following three ways :

(*a*) viscous damping

(*b*) dry friction damping

(*c*) hysterisis damping.

The damping properties of materials themselves result from inner molecular action in transferring mechanical energy into heat. Visco-elastic materials have *inherent damping* properties whereas metals and rigid structural materials have little internal damping.

In *layer damping*, a visco-elastic material is coated or inserted in structural members, usually a beam or plate. As the base structure vibrates, the layer (or layers) of visco-elastic material undergoes stretching and compression, dissipating a large proportion of mechanical energy. For effective damping, the viscoelastic layer must be capable of storing a large amount of energy before dissipating it, which in practice means that an optimum layer of stiffness exists for minimum energy dissipation. This in turn is a function of type of structure, the stiffness of the elastic layer (or layers) and the boundary conditions. *Layer damping* may be of two types, *viz., constrained* and *unconstrained.* A constrained layer of *damping material* consists of a composite laminate made up of a damping core material sandwiched between the skins of beams or plates of structural material in a three or multi-ply layer. The damping action results directly from the shearing of the viscoelastic core material as the laminate is flexed. The amount of damping achieved depends upon the material properties of both the skin and the damping media, the geometric properties such as shape and size, and other factors such as excitation frequency and temperature. *An unconstrained or free-layer damping treatment* consists of a simple two-ply construction, where a highly damped material is applied directly to the structure to be controlled. Composite damping is obtained by alternate extension and compression of the damping layer as the structure is flexed. As in the case of constrained-layer treatment, the effectiveness of the damping treatment depends upon forcing the damping material to move and dissipate energy.

**Lagging :** Lagging is a process in which noise radiating surfaces are wrapped with acoustic materials. This process is generally applied to piping, valves, vessels and equipment which radiate excessive noise. Wrappings are generally made of one or more layers or porous material covered and separated by impermeable layers. The porous material attenuates the sound waves partly by reflecting them back into the structure to which it is wrapped and partly by converting the acoustic energy into heat. Blankets of fibre glass or mineral wool of thickness of about 300 mm can achieve attenuation of the order of 60 dB in case of sound wave frequency of the order of 1000 Hz. Lagging of ducts with 50 mm glass fibre quilt or polyurethane foam, and wrapping it with a mass skin of lead foil, Keene cement, or PVC sheets can achieve a sound attenuation of about 10 dB. Acoustic lagging treatment is also eminently suitable for noise control from axial flow fans. Fig. 17. shows the diagrammatic representation of lagging of a noisy pipe.

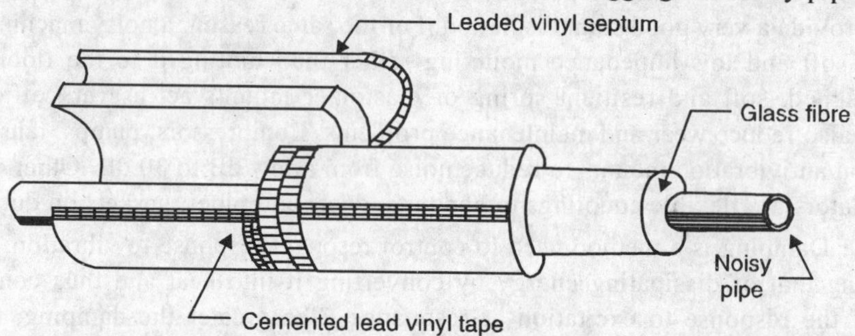


**Fig. 17.** Lagging of a noisy pipe.

***(iii)* Protection of the Personnel :** Important personnel protection methods are (a) use of ear protectors and (b) isolating the personnel by total or partial enclosures. Wherever the noise levels cannot be reduced to safe levels in industrial area by the engineering controls described above, the exposed workers should be protected by the employers by providing suitable hearing protectors. The choice of a suitable protector depends on the specific situation and the sound pressure level to which the wearer is subjected. An ear protector can provide an attenuation of 25 to 35 dB.

Ear protectors used for general use can be divided broadly into the following two types : (a) ear-plugs and (b) ear-muffs. However, there are other specialized types too. Any ear protector that works on the basis of an insertion loss between the sound source and the ear drum cannot be entirely effective in excluding the sound, since sound vibrations are transmitted through the skull by bone conduction as well as through auditory canal.

**Ear-plugs :** These are simple to use, inexpensive and can be relatively comfortable. The degree of sound attenuation depends upon the type of ear-plug used. For instance, a simple cotton wool plug can produce only very low attenuation, although this can be improved substantially by soaking the cotton wool in oil or wax.

Ear-plugs can be broadly cassified into the following five groups :

**(a) Disposable, mouldable or malleable ear-plugs** made from low-cost mateials such as cotton, wax, glass-wool, plastic foam or mixtures of these or other materials.

**(b) Reusable, pre-fabricated or moulded ear-plugs** which are permanent and washable. These can be designed in different forms ranging from simple barrier (plain plugs) to hollow plugs, which are filled with air or wax.

**(c) Individually moulded ear-plugs** which are usually made from silicone rubber and are actually moulded in a permanent form within the ear canal. The ear-plug material is usually supplied with a curing agent, the two are mixed to a puttylike consistency and then this is inserted into the ear canal of the person to be protected. The material gets cured within about 10 minutes. The ear-plug is then in a permanent form which may be removed or reinserted any number of times. This type of ear-plugs are comfortable and provide a high degree of protection.

**(d) Semi-inserts** which are essentially short ear-plugs mounted on a head band. High attenuation is possible with such designs, which can offer some of the advantages of an ear-plug as well as an ear-muff. Semi-inserts are less comfortable to wear than ear-plugs but are lighter and less obtrusive than ear-muffs.

**(e) Selective ear-plugs** which are permanent form of ear-plugs to provide selective attenuation. These designs generally incorporate a limited-movement diaphragm in the plug. Ordinary sounds at low to moderate pressure levels are transmitted via the diaphragm thus facilitating speech communication during quiet periods. At higher sound pressure levels, the diaphragm vibration reaches the limit of its free movement and thus can only transmit sound at reduced volume.

**Ear-muffs :** The ear-muff provides a complete "cushion" entirely surrounding and isolating each ear. The muffs are held in position with the help of a head-band which can readily be adjusted to fit any person. It is heavy and expensive. Ear-muffs give good protection for persons who move in and out of high-level noise areas. In such cases, ear-muffs may also be used in conjunction with ear-plugs to give better protection. For maximum protection, the protector cups of the ear-muffs should be made from a rigid, dense non-porous material. The volume enclosed within the muff is proportional to the attenuation provided at low-frequencies. Ear cup should be partially filled with an absorbent material to reduce the high-frequency resonances that may otherwise occur within the shell. Ear-muff seals may be liquid filled or plastic foam filled. In general, attenuation obtainable with ear-muff at frequencies lower than about 1000 Hz is usually lower than that of ear-plugs, but its performance at higher frequencies is generally superior to ear-plugs. The main disadvantage with ear-muffs is the over-heating of the skin because of the large surface area of contact between the compliant sealing surface of the muff and the skin.

**Helmets :** A protective helmet normally consists of a shell of reinforced glass fibre lined with sound-proofing material formed into a close fitting seal around the face and the neck, together with internally fitted ear-pad or ear-muffs. The total sound-proofing is only very little better than which would be possible with ear-muffs alone. The particular value of a helmet as a protector is due to its ability to protect the skull and other more vulnerable soft areas of the head against transmission of damaging pressure waves acting through bone conduction, particularly where ultra-high noise levels are encountered.

**Personnel Isolation :** In automated plants where only a few workers attend the processing operations, it is better to isolate the personnel in complete air-conditioned enclosures or in three-sided partial booths with a top and a viewing window. The inner surfaces of the booth may be lined with acoustic absorbents. In addition, provision of quiet rest areas where the workers can withdraw periodically can help in minimising the hearing damage and reversing the threshold shift of the exposed workers. Further, educating the workers regarding the adverse effects of undue exposure to excessive noise and the use of proper protecting techniques will also go a long way in protecting the personnel.

## ACOUSTICAL ABSORPTIVE MATERIALS

An ideal acoustical absorbent should have the following properties: (1) High noise reduction coefficient (2) adequate strength (3) smooth surface for low fluid friction (4) fireproof, rot- and vermin-proof (5) free from odours (6) good light reflectivity (7) good and decorative effect.

Acoustic absorbents used in the path control of noise are light weight, fibrous or of open cellular construction with highly porous surfaces to permit sound waves. Acoustic materials may be broadly classified as follows :

**(1) Acoustical plasters and spay-on-materials :** These are applied to the wall and ceiling surfaces in a wet state with a towel or by means of a sprayer. Plasters are mostly composed of an aggregate of perlite or vermiculite and a binder such as gypsum. Sprayed-on materials are composed of a mineral fibre or a mixture of mineral and asbestos fibre as the dry ingredient, which is mixed with quick-setting binder and adhesive during spraying with the help of a special spray gun. These materials provide high sound absorption and thermal insulation.

**(2) Tiles and Boards :** These are generally made up of materials such as compressed cane, wood fibre and mineral wool and are available in prefabricated units of definite size and thickness. Because of their high absorptivity, they are used for surfaces where relatively small areas are available for acoustic treatment.

**(3) Composite or assembled units :** These are comprised of different combinations of acoustic absorbing materials such as rock wool, glass fibre, slag wool, etc. and are faced with acoustically transparent materials like perforated hard-board or metal (steel or aluminium) plates. These are particularly useful in places of high humidity as they do not retain moisture.

**(4) Functional absorbers :** These are used in situations where separate enclosures for noise sources cannot be constructed or where acoustical treatment of the ceiling is not feasible. These may be in the form of panels (made of various acoustic materials) or in various shapes of hollow cubes, cylinders, spheres, pyramids etc. fabricated of perforated aluminium sheets lined inside with the mineral wool fibre. Both the panels and hollow type absorbers are suspended from the ceiling at a convenient height about the head and as near to the noise source as possible.

## NOISE CONTROL METHODS IN INDUSTRIAL PLANTS

Excessive noise is produced from various types of machines, petrol and diesel engines, electric motors, construction site equipment, pumps and pumping systems, compressed air systems, hydraulic systems, air distribution system, industrial fans, etc. It is always advantageous, economical and effective to identify the noise sources and noise problems right in the design and erection stages and incorporate the necessary noise control measures rather than attending to the problems at a later stage.

The various noise sources in industrial plants and the methods available for noise control are summarized in the Table 2.19 given below :

**Table 2.19. Typical Noise Sources in Industrial Plants and Methods of Noise Control**

| Equipment | Noise Source | Methods for noise control |
|---|---|---|
| Bulldozer, crane, compactor, excavator, dumper, shovel, scraper, etc. | Engine. | Fitting of more efficient silencer or exhauster; closing the enclosure panels, if fitted. |
| Rotary drills, diamond drilling drilling and boring. | Drive motor and bit. | Use of machines inside acoustic covers. |
| Riveters. | Impact on rivet. | Enclose the working area in acoustic screen. |
| Pumps. | Engine pulsing. | Enclosure in acoustic screen, allowing for engine cooling and exhaust, use of antivibration mounting, flexible coupling and hoses, maintaining adequate inlet pressure. |
| Motors. | Cooling fans. | Intake muffler unidirectional fan. |
| | Cooling systems. | Absorbent duct liners. |
| | Electrical motors. | Enclosure. |
| Engines. | Air intake and exhaust. | Use of mufflers |
| | Cooling fan. | Enclosing intake and discharge lines, use of quieter fan. |
| Vibrating screens | — | Stiffening and damping. |
| Pneumatic equipment. | Air and steam vents. | Use of mufflers; quieter valves. |
| | Air jets. | Use of mufflers, limiting of air velocities, improving orifice design. |
| | Ducts. | Lagging, adsorbent lining, limiting of air velocity. |
| Furnaces. | Combustion. | Use of acoustical plenum; seals around control rods. |
| | Ducts. | Lagging, use of mufflers. |
| Turbines. | — | Use of enclosure, use of intake and outlet mufflers. |
| Compressors. | Discharge piping and expansion joint. | Lagging and use of in-line muffler. |
| | Air intake and exhaust. | Use of mufflers |
| | Intake piping and suction drum. | Lagging. |
| | Speed changers. | Enclosure or constrained damping on case. |

## SOIL POLLUTION

Soil is a very important constituent of the lithosphere. The word "Soil" is derived from a Latin world "Solum" which means earthy material in which growth of plants takes place. "Soil" may be broadly defined as the weathered layer of the earth's crust with living organisms and their products of decay. It is a complex physio-biological system containing water, mineral salts, nutrients and dissolved oxygen. It provide anchorage to plants. The study of Soil science is called pedology or edaphology.

Earth's crust comprises of rocks and the loose material derived from these rocks, the disintegration of which results into the formation of grave sand, silt and clay. About 99% by weight of earth's crust is constituted by about 12 elements only, whereas the other elements exist in traces. The basic composition of the earth's crust is shown in table 2.20.

### Table 2.20: Elementary composition of the earth's crust

| Elements present | Percentage by weight |
|---|---|
| Oxygen | 49.85 |
| Silicon | 26.03 |
| Aluminium | 7.28 |
| Iron | 4.12 |
| Calcium | 3.18 |
| Sodium | 2.33 |
| Potassium | 2.33 |
| Magnesium | 2.11 |
| Hydrogen | 0.97 |
| Titanium | 0.41 |
| Chlorine | 0.20 |
| Carbon | 0.19 |
| Others | 1.00 |

The earth's crust basically consists of the following three rock types.

**1. Igneous rocks:** These are formed by cooling and solidification of molten rock material called 'Magma'. Ex:- Basalt and Diorite.

**2. Sedimentary rocks:** These are developed as a result of gradual accumulation, consolidation and hardening of products of weathering of mineral materials brought about by wind or waters. These rocks are characterised by the presence of distinct sedimentary layers. Ex:- Lime stone, sand stone and shale.

**3. Metamorphic rocks:** These are formed as result of metamorphosis of igneous and sedimentary rocks under the influence of high pressure and intense heat. Ex.:- Quartzite, Slate, Marble and Schist.

Out of about 2000 minerals known, nearly 99% of the earth surface is made of only about 10 to 12 minerals. The following our minerals groups account for over 90% mass of terrestrial rocks.

| | |
|---|---|
| Feldspars | 58% |
| Pyroxenes | 16% |
| Quartz | 13% |
| Mica | 4% |

The common minerals found in the earth crust are shown in Table 2.20.

**Table 2.21. Common minerals found in earth crust and their composition**

| Name of the mineral | Chemical composition |
|---|---|
| Feldspars | $K_2 Al_2 Si_6 O_{18}$, $Na Al Si_3O_6$, $Ca Al_2 Si_3 O_8$ |
| Pyroxenes | (Mg, Fe) $Si O_3$ |
| Quartz | $Si O_2$ |
| Micas | $K Al_2 (Al Si_3O)_{10} (OH)_2$, |
| K Mg Fe-Al Silicates | |
| Olivine and serpentine | (Mg, Fe) $Si O_4$ |
| Amphiboles | (Mg, Fe) $(Si_4 O_{11})_2 (OH)_2$ |
| Calcite | $CaCO_3$ |
| Magnetite | $Mg CO_3$ |
| Dolomite | ($Ca CO_3$, $Mg CO_3$) |
| Oxides of Iron | |
| (Halmetite, Magnetite, Limnite) | $Fe_2 O_3$, $Fe_3 O_4$, FeO (OH), $XH_2O$ |
| Montmorillonite | (Ca, MgO) $Al_2O_3$. $5SiO_2$. $5H_2O$ |
| Kaoline | $Al_2O_3$, $2SiO_2$, $2H_2O$ |

## FORMATION OF SOIL

Owing to constant battering of the earth's crust by wind, rains and changes of temperature, the constituent rocks are deposited ionto small fragments. This modification of earth's crust as a result of its interaction with atmosphere and hydrosphere is called weathering. The final products products of weathering are sand, silt, gravel and clays.

Depending on the nature of agencies which causes it, three mains types of weathering can be distinguished:

***(a)* Physical weathering :** Important processes involved in physical weathering include:

(1) Wetting - Drying (2) Heating - Cooling
(3) Freezing (4) Glacination
(5) Solution (6) Sand - blast

Physical weathering is mainly brought about by changes in temperature, water and wind. Differential thermal expansion and contraction due to change in temperature produces internal tensions in the rock body which leads to cracks and splits. Freezing of water in cracks and fissures expand and cause intense pressure within the rock body. This pressure, being far more than the tensile strength of the rocks, tear the rock fragments apart. Slow disintegration of the rocks also occurs due to abrasion produced by wind, rapidly flowing waters and glaciers.

***(b)* Chemical Weathering:** Chemical weathering involves processes such as (*i*) Hydration (*ii*) Hydrolysis (*iii*) Carbonation (*iv*) Chelation and (*v*) Oxidation – reduction. These changes are brought about by air, water, oxygen, $CO_2$, and other materials carried by these agencies which lead to gradual transformation of the rock fragments into gravel, sand, silt and clay. The mineral constituents of the rock also may be altered in this process.

***(c)* Biological Weathering:** Organisms such as bacteria, lichens, algae and some fungi in presence of moisture produce metabolites which aid in the process of disintegration of rocks. Roots of plants growing in cracks and fissures creates considerable pressure and tears rock fragments apart.

**Pedogenesis :** The modification of mineral matter through interaction between climatic, biological and topographic effects that leads to the formation of a "*Soil Profile*" is called Pedogenesis.

**Sequence of Soil formation due to varius factors**

**(*i*) Active factors:** The active factors involved in soil formation include temperature, wind, humidity, rainfall and evaporation.

**(*ii*) Passive factors:** The texture, aeration and chemical characteristics of the soil depend upon the parent materials and topography.

**(*iii*) Biospheric factors:** The living organisms acceleate and modify the Physico-Chemical and thus are important in the soil development.

## COMPOSITION OF THE SOIL

Soil is a complex biological system comprising of organic and inorganic material including water and gases in variable proportions. It is an important part of biosphere and is a dynamic entity on which the entire mankind depends for its food, shelter, etc. The major constituents of soil are as follows:-

**(*i*) Mineral matter:** The principal constituents of soil are sand, silt and clay. The distinction between these components is rather arbitrary and is based on the particle size:

| Type of the mineral particle | Particle diameter, mm |
|---|---|
| Stone and gravel | 2 and above |
| Coarse sand | 0.2 –2.0 |
| Fine sand | 0.02 – 0.2 |
| Silt | 0.002 – 0.02 |
| Clay and colloids | less than 0.002 |

Particle larger than 2mm diameter are called gravel sand particles in the range 0.02 – 2.0 mm in diameter are mainly composed of quartz and silica. Particles of slit in the diameter range 0.002 – 0.02 mm are mainly made up of quartz and other silicate minerals. Particles of diameter less than 0.002 mm consists of a variety of silicates, aluminosilicates and colloidal substances. Depending on the proportions of sand, silt and clay present, a soil is described as sandy, sandy loam, clayey loam, clayey etc.

**Organic matter:** The organic matter content of a soil is generally 2-5%. Partially decomposed organic matter present in a soil is referred to as "*humus*" and the process of formation of humus from plant and animal remains after decomposition by micro-organisms is called "*humification*". During this process, cellulosic materials disappear, lignins are modified while proteins are retained. In spite of the small quantity present in the soil, humus play a very vital role in maintaining the porosity of the soil, its nutrient and water retaining capacity and fertility of a soil. Humus also helps to bind the soil particles in aggregates of smaller or larger size which helps in restricting soil erosion. The top layer of soil consists of organic matter such as plant litter, decaying exreta and remains of a variety of organisms such as bacteria, algae, fungi, protoza, nematodes, worms, moluscs, arthropods etc. The organisms present in the soil are called "*Edaphons*". The amount of organic material in the soil depends on climatic conditions, the type of inorganic constituents present and the topography. In temperature regions of the world where decomposition is slow, plenty of organic debris collects in the soil. However, in tropical or subtropical regions, warm and humid conditions cause quick decomposition and mineralization of organic matter takes place.

**(*iii*) Soil water and Soil atmosphere:** Soil water and soil air fill up the space in between the soil particles. The quantity and the movement of water and gases in the soil is determined by the size of the soil particles. Soils with a higher percentage of smaller particles (eg. Clayey soils) have more interparticle space, in spite of the smaller pore size. Soils with a higher proportion of larger particles (eg. Sandy soils) have little interstitial space, in spite of the larger pore size. The water-holding capacity of sandy soils is low although most of the water present is available to the plants. On the other hand. the water-holding capacity of clay soils is higher but most it is held tightly by the minerals present in the microscopic pores and is unavailable to the plants. The water present in the soil acts as a solvent for dissolving nutrients and transporting the same to plants. The amount of dissolved nutrients in water depends upon the pH of the soil solution. It is due to this reason that pollutants that effect pH of the soil water indirectly affect the availability of nutrients to the plants. Soil water also helps in maintaining soil texture, arrangement and compactness of soil particles and makes the soil as a perfect habitat for plants and animals. Soil solution is dilute solution of various dissolved solids, liquids and gases. It is held in the soil by capillary or absorptive forces between soils/layers and soil particles. The soil water may be gravitational water, capillary water, hygroscopic water or combined water.

Some space between soil particles is also occupied by the "*Soil air*" which is very vital for the oxidation of organic material in the soil. Roots of higher plants and some soil microbes also required oxygen for their growth and activity. In case of water-logged soils, absence of oxygen promote anaerobic activity. Soil atmosphere consists of more cabon dioxide and moisture and less oxygen content, which is due to decay of organic materials. A loam soil containing 66% of water and 34% of air is preferred for several crops.

## IMPORTANCE OF SOIL TO THE BIOSPHERE

Soil plays a vital role in determining the quality and composition of the biosphere which develops over it. The multifarious functions of soil are as follows:

(*i*) Soil provides mechanical support to the plants.

(*ii*) Owing to the porosity and water-holding capacity of soil, it serves as a reservior of water and supplies water to the plants through their roots even when the land surface is dry.

(*iii*) The ion-exchange capacity of the soil ensures availability and supply of macro- and macro-nutrients for the growth of plants, animals and microbes. It also helps in preventing excessive leaching of nutrients ions, while maintaining proper pH.

(*iv*) The colloidal component of soils which comprises of clay micelles and humus particles (less than 0.002 mm) tightly adsorb a number of nutrient ions and supply them evently to the plants.

(*v*) Soil contains organotrophic bacteria, nitrifying bacteria, nitrogen-fixing bacteria, fungi, protozoans and other microbes which help in decomposition and mineralisation of organic matter and regeneration of nutrients.

## SOIL PROFILE

A display of vertical section from the surface of the ground down to the parent rocks, is called *Soil Profile*. It comprises of distinct layers of soil horizons as follows:

(*i*) A – horizon (Top soil)

(*ii*) B – horizon (Sub soil)

(*iii*) C – horizon (weathered parent rocks)

(*iv*) Bed rocks

## TYPES OF SOIL-TEXTURES

The following types of soil texture are generally distinguished.

(*a*) **Clayey Soils :** The particle size of clays is less than 0.002 mm. They have very small interspaces and hence inhibit free circulation of water and air. They tend to become water-logged and are unsuitable for growth for plants.

(*b*) **Sandy Soils:** They consist of loose and dry and particles. They have poor water-holding capacity and low nurient content.

(*c*) **Loam Soils:** They contain sand, clay and silt in nearly equal proportion. These soils are the best for copious plant growth. Depending upon the proportions of sand, clay and silt, they are further categorized as sandy-loam soils, clay-loam soils and silt-loam soils. Unlike sandy-loam soils, clay-loam soils and silt-loam soils. Unlike sandy-loam and clay-loam soils, the silt-loam soils are not suitable for plant growth.

## MAJOR TYPES OF SOIL IN INDIA

The major types of soil occurring in different parts of our country are given in Table 2.22. below:

**Table 2.22. Major types of Soil in our country**

| Type of Soil | | Rermarks |
|---|---|---|
| 1. Alluvial Soil | - | Occur in the great-northern plains of India and deltas of rivers in peninsular India. |
| (*a*) Khadar | - | Very fine and new alluviam. Very fertile. |
| (*b*) Bangar | - | Relatively coarse and old alluviam. Relatively less fertile |
| 2. Mountain or hill soils and forests soils. | - | Rich in organic matter (humes), occur in estern and western ghats, hills in Central India and northern hilly regions of Himalayas. |
| 3. Black Soils | - | Originates from volcanic rocks. Highly fertile clayey soils. Occur in parts of M.P., Tamilnadu, Gujarat and Deccan trap regions of Maharashta. |
| 4. Red soils | - | Relatively less fertile. Deficient in organic and nitrogenous matter. Occur in plateau of Kerala, Karnataka, A.P., Tamilnadu, Orissa and Southern parts of Bihar. |
| 5. Lateritic soils | - | These are poorly fertile but can support pastures and scrub forests. These soils occur in regions of tropical rainy climate such as western ghats, chota Nagpur plateau, Assam, Orissa, A.P., Tamilnadu and Kerala. |
| 6. Desert Soils | - | These are arid sandy soils with low moisture and low humus content. They occur mostly in Rann of Cach, Southern Punjab, Western Rajasthan and Haryana. |

## ACCUMULATION OF SALTS IN SOILS

Soils may consist of soluble salts such as chlorides, nitrates, sulfates, carbonates and bicarbonates of sodium, calcium and magnesium.

Accumulation of exchangeable sodium in soils is called "*Alkalization of Soils*", Build-up of large quantities of sodium makes the soil alkaline, which is unsuited for plant growth.

Concentration of soluble salt in the soil is called "*Salinization of soil*". In arid regions of Rajasthan and Gujarat, leaching out of soluble salts is not so complete as compared to that in humid regions. Higher rate of evaporation and restricted drainage further adds to the salt concentration in the soils.

Soils affected by salts may be of the following three types:

(*i*) **Saline Soils:** These contain a high concentration of soluble neutral salts and small quantities of sodium. The excess soluble salts comprising of sulphates of chlorides of sodium, calcium and magnesium can be readily leached out of these soils.

(*ii*) **Saline alkaline or sodic soils:** These soils are formed as a result of both alkalization and salinization of soils. They contain high concentration of neutral soluble salts and moderate quantity of absorbed sodium ions. Their pH may be about 8.5.

(*iii*) **Non-saline alkaline (Sodic) soils:** These contain high concentration of sodium ions. Their pH may be 8.5 to 10 due to hydrolysis of $Na_2CO_3$ present.

Reclamation of Sodic Soils is achieved by replacing sodium ions by calcium ions with the help of gypsum followed by leaching to remove the sodium sulphate formed. Other soil amendments include application of calcium chloride, phosphogypsum, basic slag, sulphur, sulphuric acid, iron pyrites, aluminium sulphate etc. Salt resistant crops such as cotton, barley, sugar beat, paddy and alfa-alfa are recommended for those soil.

## SOIL INDICATOR PLANTS

Some plants species can indicate or reveal the nature of the soil on which they grow, such are called "*Soil indicator plants*".

Plant species exhibit characteristics tolerance limits to soil pH. Many vascular plants show amphitolerance and can grow over a broad range of pH (viz., 3.5 to 8.5). Certain species of sphagun are strongly acidophilic and prefer strongly acidic soils such as bog soils for their growth. Grasses such as *Deschampsia flexuosa*, calluna vulgaris and sarothamus scoparis are indicators of acidic soil, although they can grow in neutral range and also can tolerate weakly alkaline soils. Hence they are called *acidophilic-basitolerant*. On the other hand, plant species such as Rhodomitrium, Pinacea, Polytrichum and Ericaceae decrease the pH of the soil on which they grow.

Glycophytes cannot grow well on strongly alkaline soils. Basiphylic acido-tolerant species like Tussilage ferfare can grow in neutral soil and also exhibit good tolerance down to pH 4. Plant species such a Bryum celluloze and barbura gracilena that grow on basic or neutral soil containing some calcium ions are called Calciphytes.

Saline soils having pH below 8.5 usually do not support the growth of non-halophytes and some species e.g. Kochia vestita, Capparis asphylla and Salicornia rubra. The presence of Saline soils is indicated by the growth of chenopodiacea (in weakly saline soils) to highly saline mangrove vegetation.

Alkaline soils are indicated by the presence of plant species containing more sodium ions and potassium ions than chloride, sulphate and other anions. Ex- Anabasis saisa suaeda microphylla, and other alkali halophytes.

On the other hand, acid halophytes contain more of chloride, sulphate and other inorganic anions than sodium and potassium ions. Growth of plant species such as Kalidium capicum and Salicornea herbaceae indicates chloride salinity whereas plant species such as Karelinia caspia and anabasis aphylla are sulphate indicators.

Some species of mosses associated with copper-bearing substrates indicates enrichment of the soil with copper. These species are called "*Copper Mosses*" which proved to be helpful in locating copper deposites in Europe.

## PLANTS AS POLLUTION INDICATORS

Some plant species have been used as indicators of pollution of soil, air and water. Certain plant species such as chara utricularia and Wolffa were found to grow well in polluted water. Species like agrotis, festuca, anthoxanthum and impatiens are used as metallic tolerance plant indicators for Cu, Pb, Zn and Cd respectively. Presence of diatoms indicate sewage pollution. Escherichia coli bacteria indicate water pollution. Leaf cabbage indicate accumulation of polycyclic hydrocarbons in the soil. The growth of lichens was found to decrease when the soil is polluted. Soil pollution inhibits plant growth and reduce productivity. Plant growth is inhibited or destroyed in areas near smelters.

## SOURCES OF SOIL POLLUTION

Soil pollution differs from water pollution on air pollution, because the pollutants remain in direct contact with the soil for relatively longer periods and hence alter the chemical and biological properties of the soil. The hazardous chemical can also enter the human food chain from soil or water plants.

The major sources of metallic contamination of soils include mining, smelting, sludge, fertilizers, pesticides, composted town refuse etc. Metals such as Cd, Pb, Hg, Ni, Mo, Ni, Cr etc. are toxic to plant and animal life.

Indiscriminate dumping of industrial wastes and municipal wastes leads to the leaching and/or seepage of toxic substances into the soil and pollution of ground water. Further due to some modern agricultural practices, obnoxious pesticides, fungicides, insecticides, biocides, bacteriocides, etc. contaminate land. Direct pollution of soil by dangerous pathogenic organisms is also important.

Fly ash generated from thermal power plants, industrial waste discharged into streams of dumped into the surrounding land, mining wastes, non-biodegradable organic pollutants industrial sludges such as flue gas desulphurization sludge, heavy metal sludges etc. cause serious water and soil pollution problems.

Commercial and domestic urban wastes consisting of dried sewage sludge as well as garbage and rubbish materials such as plastics metal cans, glasses, street sweepings, waste paper, fibres, rubber etc contribute to soil pollution.

Human and animal excreta, farm wastes, soil conditioners, soil fumigants, radioactive wastes etc. also cause soil pollution.

## EFFECTS OF SOIL POLLUTANTS

"*Soil Pollution*" was originally defined as the contamination of the soil system by considerable quantities of chemical or other substances, resulting in the reduction of its fertility or productivity with respect to the qualitative and quantitative yield of the crops. However, if some of the contaminants are such that if they are taken up by the plants (with or without any detrimental effects on them), and enter into the food chain and impart detriminatal or toxic effect on the consumers, then that also should be treated as soil pollution.

Soil pollution is receiving greater and greater attention due to its direct impact on public health. The major effect of various types of pollutants are given below:

**(a) Effects of modern agricultural practices:**

***Synthetic Fertilizers***: Synthetic fertilizers are employed to increase the soil fertility and crop productivity. These fertilizers concentrate the essential nutrients in layer of top soil. However, the soil enriched by chemical fertilizers cannot support the microbial flora which are so essential to enrich the humus that helps in plant growth. Excessive and indiscriminate use of chemical fertilizers may result in the following undesirable effects:

(*i*) Wheat, maize, corn, etc grown on soils fertilized with NPK fertilizers may result in considerable reduction in protein content of the crops.

(*ii*) Excessive use of nitrogenous fertilizers leads to the accumulation of nitrates in the soil which may contaminate the ground water. Nitrate concentrations exceeding 90 ppm in drinking water may lead to diarrhoea, blue Jaundice (Cyanosis) in children, "*methemoglobinemia*" (or blue baby syndrome) in infants. Further, the nitrates and nitrites enterning the human body may be eventually converted nitroso amines and other nitroso-compounds which are suspected nitroso amines and other nitroso-compounds which are suspected to cause stomach cancer. Surveys in Rajasthan and other parts of the country indicated much higher nitrates levels than the permissible 45 ppm levels.

(*iii*) Vegetation growth in nitrate-rich soils may exert toxic effects in cattle.

(*iv*) Excessive use of chemical fertilizers may enter the water bodies and contribute to "*eutrophication*". (Eutrophication is the excessive growth of algae and aquatic plants to undesirable levels).

(*v*) Excessive use of chemical fertilizers may reduce the ability of plants to fix nitrogen.

(*vi*) Excessive quantities of potassium fertilizers in soils may reduce the quantities of valuable ascorbic acid (vitamin C) and carotene in fruits and vegetables grown in such soils.

(*vii*) The large-sized fruits and vegetables grown in highly fertilized soils may be more vulnerable to attacks by pests and insects.

## PESTICIDES

As per the reports of the World Health Organization (WHO), about 50,000 people in developing countries are poisoned and about 5,000 people die because of improper use of pesticides and other chemicals in modern agricultural practices.

Pesticides pose potential hazard to animals, humans and aquatic life. They also cause deleterious effect on soil fertility and crop productivity. Pesticides applied to crops are retained in the soil in considerable quantities. They enter into cyclic environmental processes such as absorption by soil, leaching by water, etc. and contaminate both lithosphere and biosphere. Pesticides including herbicides, fungicides and rodenticides, are persistent pollutants. Owing to interactions between lithosphere and biosphere, pesticides may enter the food chain and pose serious health hazards. Some of them undergo metabolic transformation and bio-degradation. The degradation products of some of the pesticides are more dangerous than their respective parent compounds. Some of the pesticides residues are carcinogenic while their metabolic products too are toxic. The rate of degradation of pesticides depend upon their properties and structural characteristics.

The following types of pesticides are commonly used:

(*a*) Chlorinated hydrocarbons (eg. DDT, Aldrin, Dieldrin, Lindane, BHC etc.)

(*b*) Carbamate compounds (eg. Carboryl or Sevin, Zectrion etc.)

(*c*) Organo-Phosphorous compounds (eg. Methyl or ethyl parathion, melathion, guthion etc.)

(*d*) Inorganic compounds (eg. $As_2O_3$, $PbO_2$, $NiCl_2$, $CuSO_4$ etc.)

(*e*) Miscellaneous compounds (eg, Organic mercurials, 2, 4D; 2, 4, 5T etc.)

Some of the adverse effects of pesticides are given below:

(*i*) Some arsenic pesticides may render the soil permanently infertilile.

(*ii*) Pesticide residues in soil may be taken up by plants and cause phyto-toxicity. They may enter the aquatic environment and enter the food chain.

(*iii*) Pesticides such as, endrin, dieldrin, DDT, hepachlor etc. may seep through the soil and contaminate groundwater and surface waters. They may eventually contaminate drinking water supplies.

(*iv*) Fruit, vegetables, rice, wheat, barley, maize etc. are known to contain considerable quantities of toxic pesticide residue such as of DDT, BHC and other organochloro pesticides.

(*v*) Polychlorinated biphenyls (PCB) having half-life periods of about 25 years in soil are among the most hazardous soil pollutants. They may accumulate in soil and plants and when they eventually enter the animal or human body, they may severe health disorders, including eye damage, skin problems, nervous disorders, foetus deformities and liver or stomach cancer.

(*vi*) Irrigated water from pesticide contaminated soils may evaporate and spread the toxic pesticide vapours in the atmosphere.

(*vii*) DDT can enter the food chain and accumulate in human fats and may lead to disorders such as impotency.

(*viii*) Persistant pesticides can damage human tissues and interfere with the normal metabolic activities by disturbing enzymatic functioning.

(*ix*) Chlorinated pesticides and herbicides are hazardous soil pollutants which can affect the soil texture and damage the ecosystem.

(*x*) Herbicides such as dioxan may cause congential birth defects in offsprings.

(*xi*) Hunting birds feeding on grains contaminated with DDT are threatened of extinction.

(*xii*) Organophosphate pesticides may cause muscular disabilities, tremors and dizziness.

(*xiii*) Excessive use of synthetic pesticides may lead to defoliation of forests and adverse effect on fauna and flora.

(*xiv*) Farm animals drinking stagnant water in fields sprayed by pesticide developed toxic symptoms and some mortalities were reported.

(*xv*) Farmers and farm workers are particularly prone to pesticide poisoning because of greater exposure while handling and spraying.

(*xvi*) Accidental spillages and leakages in pesticide manufacturing industries cause disastrous effects on the people residing in nearby areas due to pollution of air, water and soil. The Bhopal tragedy on 3$^{rd}$ December, 1994 is a lingering example.

(*xvii*) Contaminated soils may act as potential carriers of pathogenic bacteria and other dangerous organisms which may endanger human health.

(*xviii*) Volatile pesticides may cause pollution of air in the surrounding area.

**(*b*) Effects of Industrial Effluents :**

Solid, liquid and gaseous chemicals from various industries such paper and pulp, iron and steel, fertilizers, dyes, automobiles, pesticides, tanneries, coal-based thermal power plants etc. contain a variety of pollutants such as toxic heavy metals, solvents, detergents, plastics, suspended particulates, and refractory/non-bio-degradable/recalcitrant chemicals. If they are not properly treated at source, they give rise to water-, air- and soil pollution. Fly ash resulting from coal-based thermal power plants is one of the alarming and continuously increasing source of soil-pollution leading to degradation of soil, apart from water- and air-pollution in the nearby areas. Some trade wastes such as tannery wastes may contain pathogenic bacteria.

Indiscriminate dumping of untreated or inadequately treated domestic, mining and industrial wastes on land is an important source of soil pollution.

Fall-out of gaseous and particulate air-pollutants from mining and smelting operations, smoke-stacks, etc. is a major source of soil pollution in nearby areas.

**(*c*) Effects of urban wastes:**

Millions of tonnes of urban waste are produced every year from critically polluted cities. The inadequately treated or untreated sewage sludge not only pose serious health hazards but also pollute soil and decrease its fertility and productivity. Other waste materials such as rubbish, used plastic bags, garbage, sludge, dead animals, waste medicines, hospital wastes, skins, tyres, shoes, cans, etc. also cause land and soil pollution. Some solid wastes may cause clogging of ground water filters. Suspended matter present in sewage can act as a blanket on the soil and interfere with its productivity.

Apart from the above major sources, radioactive waste dumped in the soil from natural and man-made sources, soil erosion due to deforestation, unplanned irrigation and unscientific agricultural practices also result in land and soil pollution.

## CONTROL OF SOIL POLLUTION

As discussed in earlier sections, the major sources of soil pollution are the domestic wastes, industrial wastes, and agricultural wastes including those toxic chemicals (eg. Pesticides) arising from modern agricultural practices. The various approaches to control soil pollution are as follows:

(1) Implementing stringent and pro-active population control programmes.

(2) Launching extensive afforestation and community forestry programmes.

(3) Implementing deterrent measures against deforestation.

(4) Formulation of stringent pollution control legislation and effective implementation with powerful administrative machinery.

(5) Imparting informal and formal public awareness programmes to educate people at large regarding the health hazards and undesirable effects due to environmental pollution. Mass media, educational institutions and voluntary agencies should be involved to achieve these objectives.

(6) Banning the use of highly toxic and resistant synthetic chemical pesticides or atleast regulating/restricting their use only for special purposes under thorough monitoring.

(7) Encouraging the use of bio-pesticides in place of toxic chemical pesticides.

(8) Conservation of soil to prevent the loss of precious top soil from erosion and to maintain it in a fertile state for agricultural purposes.

(9) Transforming intensive agriculture into a sustainable system by measures such as

   (*a*) maintaining a healthy soil community in order to regenerate soil fertility by providing organic manures, increasing fellow periods, avoiding excessive use of chemical fertilizers and pesticides.

   (*b*) Infusing bio-diversity in agriculture by sowing mixed crops, crops rotation etc.

(10) Effective treatment of domestic sewage by suitable biological and chemical methods and adopting modern methods of sludge disposal.

(11) Municipal wastes have to be properly collected by segregation, treated and disposed scientifically in land fills. Recycling and reuse of materials should be done wherever possible, such as recycling of glass, plastics, paper and production of bio-gas.

(12) Industrial wastes have to be properly treated at source, by segregation of wastes and/or adopting integrated waste treatment methods. Proper care should be taken in treating heavy metal wastes and other obnoxious waste materials.

(13) Security land-fills have to be constructed for permanent disposal of hazardous and recalcitrant industrial wastes.

(14) Sponsoring more intensive R & D efforts on bio-fertilizers, bio-pesticides, utilization of wastes by recovery, reuse and recycling processes, and safer treatment and disposal of hazardous waste.

(15) Enforcing environmental audit for industries and promoting ecolabelled products.

(16) Avoiding excessive use of chemical fertilizers and insecticides and providing more organic manures to the fields and thereby maintaining healthy biota. This in turn regenerates soil fertility. A soil rich in organic matter also helps in controlling soil erosion.

# MARINE POLLUTION

## INTRODUCTION

Seas are the unlimited source of water for man. Secondly they are the main source of food and earnings for persons living in coastal areas. When the marine water is polluted it effects the animals and other food chain components. Researches shows that many marine animals secrets the medicinal chemicals which are useful to mankind and other living organisms. When water will be polluted it will effect the animals present in seas. Generally drainage from rivers, industries, human activities from coastline area, disposal of radioactive wastes and toxic materials, leakages from ships are main source of marine pollution. When ships and other animals take such polluted water get affected. When man take them dies. Leaked oil other pollutants from ship spread over in sea, get absorbed on sediments. It effects the marine life. This is due to dumping of wastes material from outside, which is harmful too, in ocean affect their ecosystem. This is known as marine pollution.

## SOURCES OF MARINE POLLUTION

The main sources of marine pollution are:-

1. Rivers are the main source of marine pollution. They carry wastes in their drainage and joints the sea/ocean. The drainage include sewage sludge, industrial effluents, detergents, agrochemicals, plastics, metal scraps etc.
2. Catchment area-like India and other countries too, many big cities and industries are situated along the coast line. Every large amount of wastes from hotels, wastes effluents mixed with detergents, sewage from corporations and industries, other wastes from human activities are mixed in sea water.
3. Ships which carry toxic substances, lubricating oil, paints heavy oils, fuels, automotive materials and other chemicals from one place to another, some times by accident or by leakages pollute the marine water.
4. Testing of atomic weapons, space aircrafts, missiles (generally developed country do this) and other radioactive wastes when dumped in seas, causes heavy loss to aquatic biota.

5. Harmful effluents from nuclear power stations or from other scientific organizations like BARC in India, chemical industries, fertilizers, Pesticide and insecticide industries when mixed in marine water causes harmful effects to marine life.
6. Marine pollution also caused by oil drilling in seas, tourism activities and heat released from industries. etc.

## EFFECTS OF MARINE POLLUTION

The major effects of marine pollution are as follows:-

1. Oil is most dangerous pollutant when afloat on sea or mixed with water a great threat to marine life specially fish, birds, invertebrates and algae. Thousands of birds killed every year because once they oiled, seldom survived despite efforts to clean themselves.
2. Oil of sea also effects sensitive flora and fauna, phytoplankton, zooplankton and other animals. In Alaska, Brittany (France), Elbe (Germany) thousands of birds died by oil spillage.
3. Plastic or plastic materials when dumped into sea by commercial ships or from drainage, animal take it through their food in stomach. It causes ulcer and reduces hunger.
4. Marine pollution effects the food chain in seas. Serious diseases like cancer are the caused when affected animals are taken by man from ocean.
5. Detergents, either from cleaning up the spills or from drainage, also responsible for high mortality of marine life.
6. Heavy metals (like lead and mercury), fectory materials, mineral oils, acids and other biocides are also measure threat to marine life when mixed with sea water.

Apart from these major effects, there is a heavy loss of economy after getting polluting animals and chemicals from marines.

## CONTROL OF MARINE POLLUTION

The control of marine pollution can be studied in following two steps:-

**1. Steps already in operation**

(*i*) Port authorities are alert and introduced antipollutant measures by creating pollution cell. But deeper in sea coastal guards are doing this job.

(*ii*) Various research organizations, institutions are working in this field to check the marine pollution.

(*iii*) In most of the countries (India too), the monitoring and survey in operation to control the marine pollution.

(*iv*) Authorities are taken care of effective measures to check the oil leakage from ships and tankers.

(*v*) Urban and coastline corporations are trying to check the dumping of wastes from human activities & Municipal etc. solid waste management is helping to recycle or reuse.

## SUGGESTING STEPS TO CONTROL MARINE POLLUTION

(1) Dumping of oil ballest, hazardous and toxic substances, gases from radioactive labs into sea, should be banned or should be properly treated before dumping.

(2) Drainage, sewage sludge and effluents from industries should not be discharged in to rivers which joint sea.

(3) Developmental activities on coastal areas should be minimised.

(4) Toxic pollutants from industries and treatment plants should not be discharged into sea.

(5) Ships and ports should have certain facilities for reducing pollution.

(6) Certain biological and other methods should be followed to restore species diversification and ecobalance in the water body to prevent pollution.

(7) Effective measures should be developed to check the leakage in ships and oil tankers.

(8) Nuclear explosions and other nuclear activities in sea should be minimized.

(9) Wastes from municipal, industries, sewage and thermal power stations should be recycled for reutilization. Such plants should be developed. Some are in operation.

(10) We should develop awareness in people to reduce the amount of waste in their daily life.

(11) Drilling should not be allowed in coastal areas.

# THERMAL POLLUTION

## INTRODUCTION

The term Thermal Pollution has been used to indicate the detrimental effects of heated effluents discharged by various power plants. It denotes the impairment of quality and deterioration of aquatic and terrestrial environment. Various Industrial plants like thermal, atomic, nuclear, coal fired plants oil field generators and mills utilize water for cooling purposes.

The heated effluents are discharged at a temperature 8 to 10°C higher than the temperature of intake waters, which reduces the concentration of D.O. (Dissolved Oxygen).

## THERMAL POLLUTION: IT CAN BE DEFINED AS:

1. The warming up of an aquatic system to the point where desirable organisms are adversely affected.

2. Addition of excess of undesirable heat to water that makes it harmful to man, animal, plant or aquatic life or other wise causes significant dangers to the normal activities of aquatic communities in water.

3. Heated effluents either from natural or man made sources, contaminated with water supplies, may be harmful to life because of their toxicity, reduction in Dissolved Oxygen (D.O.), aesthetically unsuitable and spread diseases.

4. It reduces the number of aquatic species and destroys the balance of life in streams as is evidenced by the biological indices of community and diversity.

5. It is a by-product of rapid and unplanned industrial progress and over population.

## SOURCES OF THERMAL POLLUTION

The accelerated pace of development, rapid industrialization and extensive population density have increased demand of thermal power plants. Human activities, today, are constantly adding pollutants to air and water at an alarming rate. The following sources contribute to thermal pollution:

**1. Nuclear Power Plants**: Nuclear power plants, including drainage from hospitals, institutes, nuclear experiments and explosions, discharge a lot have unutilized heat and trapped radionuclides into nearby water streams. Emissions from nuclear reactors and processing instruments are also responsible for increasing the temperature of water bodies. Heated effluents from power plants are discharged at 10°C higher than the coolant receptor and severely affect the aquatic flora and fauna.

**2. Coal-fired Power Plants:** Some thermal power plants ultimately discharges effluent having temperature difference of 15°C between effluent and water body. The Thermal power plants utilize coal as fuel and they constitute the major source of thermal pollutants. The heated coils are cooled with water from nearby like or river and discharge the hot water back to the receptor water body and thereby increasing the temperature of the nearby water. The heated effluent decreases the dissolved Oxygen content of water. It results into killing of fish and other marine organisms.

**3. Industrial Effluents:** Industries generating electricity, like coal as fuel and Nuclear powered thermal plants, require huge amounts of cooling water for heat removal. Other industries like textiles, paper and pulp as well as sugar also release heat in water but to a much lesser extent. The heat from the turbo-generators installed in industries have temperature of effluent as 5°C to 9°C more than the normal temperature of stream. To cope with the increased demand of electricity and rapid industrialization the number of installations are raised which results in discharge of more volume of water/heated effluent and above the receptor water body temperature.

**4. Hydro-electric Power:** The generation of hydroelectric power, sometimes, results in negative loading in water systems. Apart from electric power industries, various factories with cooling contribute to thermal loading. It has been reported that about 18% more heat is given to cooling ponds in nuclear power plants than any other plant of equivalent size.

**5. Domestic Sewage**: Domestic sewage is commonly discharged into rivers, lakes and canals with or without waste treatment. The municipal sewage normally has a higher temperature than receiving water. The discharged water not only raises the stream temperature to a measurable extent but also creates numerous deleterious effects on aquatic biota. The organic matter present in the sewage utilizes the dissolved oxygen present in the surface water for oxidation. With the increase in the temperature of the water, the D.O. content decreases and the demand of oxygen increases. Hence, the anaerobic conditions will result in the release of foul and offensive gases. The marine life dependent upon the D.O. will die out and the quality of water is also adversely affected.

## EFFECTS OF THERMAL POLLUTION:

The various effects of the thermal pollution are:-

**1. Reduction in Dissolved Oxygen**: Concentration of dissolved oxygen decreases with increase in temperature of water. For example, the D.O. content is 14.6 ppm in water at a temperature of 32°F and 6.6 ppm at 64°F. Thus cold-water fish, which requires about 6 ppm to survive, would not tolerate the high water temperatures. If they remained in the area they would die of oxygen starvation. Since the aquatic biota live acrobically, so a healthy stream should have an adequate supply of dissolved oxygen.

**2. Change in Water Properties:** A rise in temperature changes the physical and chemical properties of water. The vapour pressure increases sharply, while the viscosity of water decreases. The decrease in density, viscosity and solubility of gases increases the settling speed of suspended particles, which seriously affect the food supply of aquatic organisms.

**3. Increase-in Toxicity:** The rising temperature increases the toxicity of the poison present in water. A 10°C rise in temperature doubles the toxic effect of Potassium cyanide, while an 80°C rise in temperature triples the toxic effect of O-Xylene causing massive mortality of fish.

**4. Interference with biological Activities:** Temperature is considered to be of vital importance to physiology, metabolism and biochemical process in controlling respiratory rates, digestion, excretion and overall development of aquatic organisms. The temperature changes totally disrupt the entire ecosystem. Sharp changes in temperature are often destructive. Because, the life of aquatic animals involves several chemical reactions and the rate of these reactions vary according to changes in temperature.

**5. Interference with Reproduction:** In fishes, several activities like nest building, spawning, hatching, and migration and reproduction etc. depend on some optimum temperature. For instance, the maximum temperature at which lake trout will spawn successfully is 8.9°C. The warm water not only disturbs spawning but also destroys the laid eggs.

**6. Variations in Reproductive Rate:** The increase in temperature triggers deposition of eggs by female. The triggering is particularly dramatic in estuarine fish, which spawn in four hours after the water temperature reaches critical level.

**7. Changes in Metabolic Rate:** Fishes show a marked rise in basal rate of metabolism with temperature to the lethal point. The respiratory rate, oxygen demand, food uptake and swimming speed in fishes increase.

**8. Increased Vulnerability to disease:** Activities of several pathogenic microorganisms are accelerated by higher temperature. Hot water causes bacterial disease in certain fishes such that they fail to develop eggs above critical temperature.

**9. Invasion to destructive organisms:** Thermal pollutants may permit the invasion of organisms that are tolerant to warm waters and highly destructive e.g. invasion of ship worms into New Jersey's Oyster Creek.

**10. Undesirable Changes in Algae Population:** The life in an ecosystem is greatly influenced by the algal growth. Excess nutrients from the washout waters from farmlands, thermal plants cause an excessive algal growth with consequent acceleration of eutrophic and other undesirable changes.

**11. Destruction of Organisms in Cold Water:** The volume of water required for cooling purposes from a stream is enormous. Unfortunately many of plankton, small fish, insect larvae that are sucked into the condenser along with cooling water are killed by the thermal shock, increased pressure and water viscosity.

**12. Biochemical Oxygen Demand**: When the temperature of stream carrying biodegradable organic matter rises, the intensified action of aquatic organisms causes B.O.D. to be accomplished at a lower temperature. When the temperature of stream carrying biodegradable organic matter rises fish death may occur due to synergistic action, which is caused due to accelerated chemical or biochemical action.

**13. Effect on Marine Life:** Temperature plays an important role in affecting the physiology, metabolism, growth and development of marine animals. Sea organisms are poikilothermic i.e. their body temperature varies with the surrounding water. Some marine creatures cannot tolerate wide changes of temperature, so they die at higher temperature.

**14. Effect on Bacteria:** Due to the heated discharges from the industries and plants (industrial), the bacteria are severely damaged. The effect includes coagulation of body Protein, melting of cell fats, toxic action of metabolic products etc.

## CONTROL OF THERMAL POLLUTION

Heat must be removed from the condenser cooling waters prior to their disposal into water bodies. The major principles involved in the process of heat loss are:-

1. Conduction 2. Convection 3. Radiation 4. Evaporation

The following methods can be adopted to control high temperature caused by thermal discharges:

**(1) COOLING PONDS:** The cooling towers are beneficially used in dissipation of heat as shown in fig. 18.

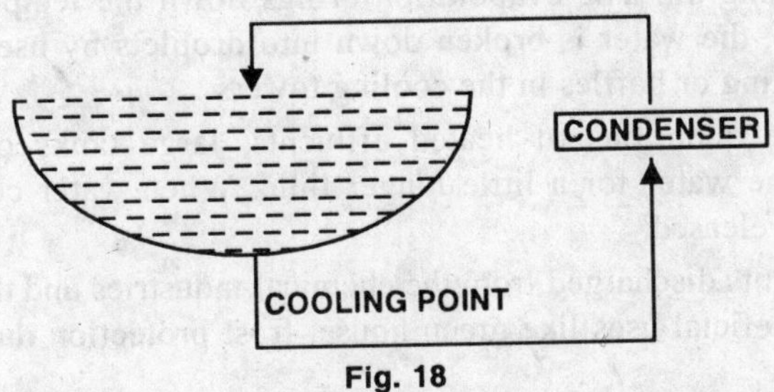


Fig. 18

The water from the condensers is stored in the earth like ponds where natural evaporation brings down the temperature. The water is re-circulated again. Another method for installation of cooling ponds is shown in fig.19.

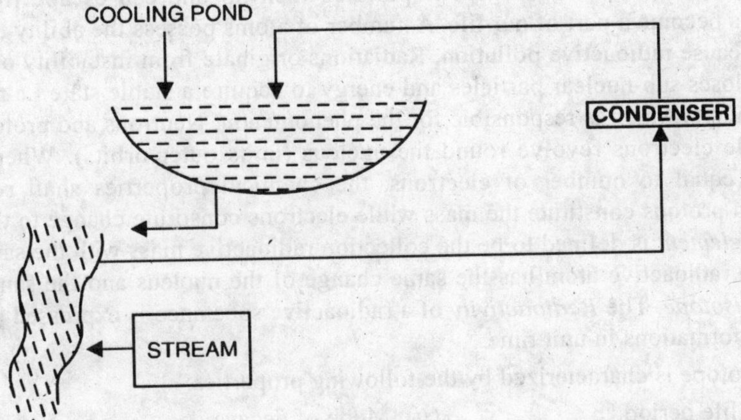


Fig. 19

**(2) SPRAY PONDS**: In spray ponds, the water is sprayed in the cooling ponds with the help of spray nozzles to convert it into fine droplets which provide more surface area to facilitate efficient heat transfer to atmosphere.

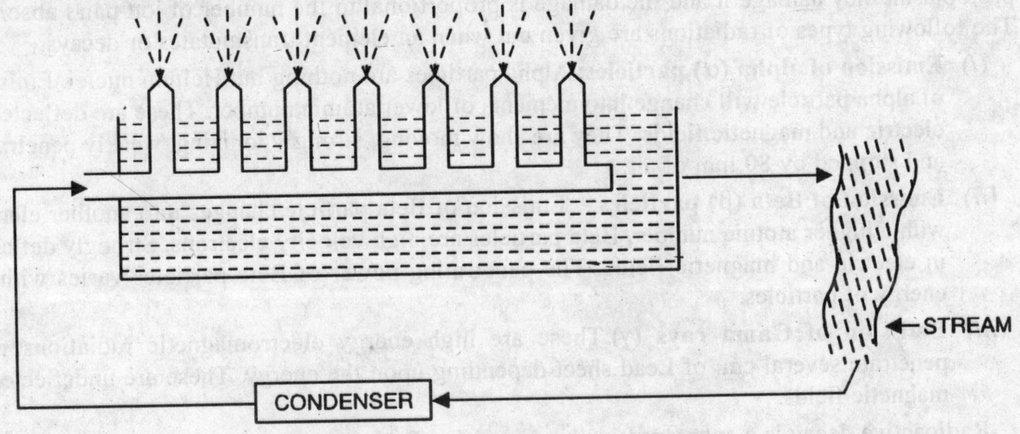


Fig. 20

(3) **COOLING TOWERS:**

**Wet Cooling Towers:** In wet cooling towers, the heated water is brought in direct contact with continuously flowing air. The evaporation brings down the temperature. To increase the surface area of contact, the water is broken down into droplets by use of spray nozzles or by splashing it on the packing or baffles in the cooling towers.

(4) To handle large quantities of heated effluents, large tanks or reservoirs should be constructed to retain the water for a little longer time. When water cool down to a tolerable temperature, it may be released.

(5) The heated effluents discharged from the chemical industries and thermal power plants can be put in to certain beneficial uses like green house, frost protection during colds, aquaculture, heating the buildings etc.

## NUCLEAR HAZARDS

Hazard mean dangerous to human being or so by external source. This external source is from environment. When our environment is polluted then no one can escape from the pollution hazard. It has become a part of our life. A number of atoms possess the ability to limit radiations and thereby cause radioactive pollution. Radiations originate from instability of the nuclei of an atom which loses sub-nuclear particles and energy to acquire a stable state i.e radioactivity. It is the state of nuclei which is responsible for the phenomenon. Neutrons and protons constitute the nucleus while electrons revolve round the nucleus (units outer orbits). When the member of protons are equal to number of electrons, the chemical properties shall remain the same. Neutrons and protons constitute the mass while electrons constitute change to the element. Thus *Radioactive element* is defined to be the collection radioactive mass with the same change of the nucleus. The radioactive atom has the same change of the nuçleus and the same mass is called *Radioactive isotope*. The *Radioactivity* of a radioactive substance is expressed by the number of nuclear transformations in unit time.

A radio isotope is characterized by the following properties:-

(*i*) Half life period. (*ii*) Mode of decays.

(*iii*) Energy of radiations (*iv*) Definite energy state

Radiation is the emission of rays and particles or release of energy from the source (atom). There are two types of radiations ionizing and non-ionizing radiations. These radiations destroy the organic molecules of which the body cells are composed. If ion pairs enter into a living protoplasm, they damage it and the damage is proportional to the number of ion-pairs absorbed. The following types of radiations are given out when an element transmutates or decays.

(*i*) **Emission of alpha ($\alpha$) particles.** Alpha particles are nothing but Helium nuclei. Emission of alpha particle will change into elements of lower atomic number. These are deflected by electric and magnetic fields. They are show moving, strongly ionizing, weakly penetrating and stopped by 80 mm of air.

(*ii*) **Emission of Beta ($\beta$) particles**:- Emission of Beta particle changes into another element with a higher atomic number. Beta particles are high velocity electrons. Strongly deflected in electric and magnetic fields. The penetrating power of Beta particles varies with the energy of particles.

(*iii*) **Emission of Gama rays ($\gamma$).** These are high energy electromagnetic radiations. Can penetrate several cm. of Lead sheet depending upon the energy. These are undeflected in magnetic fields.

Radioactive decay is a spontaneous process arising from nuclear instability.

## SOURCES OF RADIO ACTIVE POLLUTION

The two main sources of radioactive pollution are, natural and manmade.

## NATURAL SOURCES

The natural sources of radioactivity are considered mainly of the cosmic radiation received from the space, and the naturally occurring radioisotopes present in the environment and those contained within the body of the organisms. The cosmic radiations are of extraterrestrial origin, which probably arise from the sun or even beyond it. They are consisted of particles of very high energy, primarily of protons and some heavy nuclei. These cosmic particles collide with the gas molecules of the upper atmosphere bringing about intense ionization in gases accompanied with the formation of secondary cosmic rays composed mainly of neutrons, mesons, and gamma rays. Eventually a complex mixture of particles reaches the earth as cosmic rays. These particles also form substantial quantities of $^{3}H$ and $^{14}C$ in the atmosphere.

Another source of natural radiation is the presence of radionuclides in the lithosphere, hydrosphere and atmosphere. All the elements above atomic number of 82 (Lead) are radioactive in nature and emit a variable quantity of radiations. The most abundant naturally occurring radionuclides on the earth are Uranium, Thorium and Potassium-40. Soils, rocks and even building material contain small quantities of $^{40}K$; and Uranium and its daughters.

### Man-Made Source

Man causes radioactive pollution by testing of nuclear weapons, establishment of nuclear power plants, mining and refining of plutonium, and thorium, and preparation of radioactive isotope.

### 1. Nuclear weapons

Testing of nuclear arms comprises:

(*a*) The use of Uranium 235 and Plutonium 239 for fission.

(*b*) Hydrogen or lithium as fusion material.

Atomic explosions are uncontrolled chain reactions. They give rise to very large neutron flux conditions that cause other materials in the surrounding environment to become radioactive.

Huge clouds of fine radioactive particles and gases are thrown up in the environment and are carried away to distant areas by the agency of wind. Gradually they settle down on earth as fall out or are brought down by rain.

### 2. Atomic Reactors and Nuclear Fuel

The most common fuel used for fission in the nuclear power plants are uranium, thorium and plutonium. Uranium undergoes several processes, right from its mining to its inception into the reactors. The spent materials obtained from the reactors, after the energy has been utilized, are reprocessed to recover unburnt uranium, plutonium and some other important isotopes, which can be used in medicine or for some other useful purposes. The whole operation from the mining of the fuel to its final disposal is called "nuclear fuel cycle".

At almost all stages of the nuclear fuel cycle, liquid, gaseous and solid radioactive wastes are released having a tremendous potential to contaminate the environment and hence, a great care is to be taken for the environmental safety during the nuclear operations.

### 3. Radioactive Isotopes

Radioactive isotopes such as $^{125}I$, $^{14}C$ and $^{32}P$ and their compounds find wide usage in scientific research institutions contain varying amounts of radioactive materials. When this waste water reaches the different water sources such as rivers, streams, lakes etc. through the sewers they cause water pollution. Radioactive iodine and phosphorus also enter the food chain through water and may finally reach man through fish etc.

**4. Other Sources**

During different medical treatments, varying concentrations of radiations enter the human body for instance, X-rays are common for detecting skeletal disorders, and therapy for cancer patients often includes radium and other isotope radiations.

It has been reported that about 240 million dental and medical X-rays are taken annually and that 15 million tests using radioactive materials as tracers in the human body are also made.

A common type of ionizing radiation is X-ray which is produced by radiographic equipment. X-ray therapy equipment, dental X-rays that can operate at a voltage about 10 KV produce more penetrating radiation and may be more hazardous, if not properly shielded.

## DAMAGES TO A BIOLOGICAL SYSTEM

Most of the damages caused by radioactive pollutants stem from their capacity to produce high energy radiations, which are very harmful to a living system. There are two main modes in which radioactive pollution can be dangerous to a biological system.

(*i*) Damages caused by radiations from outside source.

(*ii*) Damages caused by radiations from sources inside the body.

**Damages caused by Radiations at different levels**

**(*i*) Damages at Molecular level**

Damages to macromolecules such as enzymes, DNA, RNA etc. through ionization cross-linkages within and between two affected molecules.

**(*ii*) Damages at sub-cellular level**

Damages to cell-membranes nuclei, chromosomes such as fragmentation, mitochondria etc.

**(*iii*) Damages at cellular level**

Inhibition of cell division death, decay and transformation to malignant state.

**(*iv*) Damages to Tissues and Organs**

Disruption of such systems as central nervous system, loss of sight, inactivation of bone marrow activity resulting in blood cancer malignancy and ulceration of intestinal tract.

**(v) Damages to an Individual and whole population**

Death or shortening of life due to radiations changes in characteristics due to mutations. In human beings exposure of radiations results in little visible effects in early stages. But after 12-24 hours injury symptoms manifest themselves. This includes reddening of skin, anemia, anorexia. Vomiting, and diarrhoea and with heavy doses, blister formation, pigmentation of skin, burning sensation all over the body, loss of sight etc. It must be noted that for all this there is no cure available. Once a person is exposed to radiation he has to bear its consequences. Medical aid can do little.

## HAZARDS ASSOCIATED WITH RADIO ACTIVE POLLUTION

Radio-active pollutants are not like other pollutants which are sooner or later converted into harmless material and degraded into simpler constituents to be recycled into the ecosystem and used again. It is not the element itself, but the instability of its nuclei which is responsible for damages caused by these pollutants. As long as radiations continue, these wastes are dangerous for the living beings. After the emission of radiations nuclei attain stable state and behave like any other element in the environment or the biosphere. Life on earth's crust could evolve only when the nuclear activity ceased, atoms acquired stable state and radiations were reduced to the level which could be tolerated by living beings. Major hazards associated with radio-active pollution can be summed up as follows:

(1) No physical, chemical or biological process can influence the process of radio active emissions. The unstable nuclei have to decay and acquire a stable state.

(2) A number of radio-active isotopes have a very long half-life. Thorium-232 ($_{90}Th^{232}$) takes 14,0000,000,00 years to lose half of its radio-activity. Half of Uranium-235 ($_{92}U^{235}$) takes 710,000 years to disintegrate. Half of Neptunium-237 ($_{93}Np^{237}$) decays in 21,00,000 years. This makes these radio-active wastes almost a permanent hazard for the biosphere.

(3) Most of the radiations have a high penetrating power. Thick sheets of steel, cement concrete walls etc, can not contain them. They can easily penetrate to deep seated organs and cause injury.

(4) Nucleic acids (DNA and RNA) effectively absorb these radiations. Even low level radiations which do not cause any visible damage are completely absorbed by nuclear material which causes carcinogenic, mutagenic and teratogenic effects.

(5) A biological system is unable to distinguish between a radio-active and a normal isotope of an element as their physical and chemical properties are similar. Radio-active isotopes are therefore, absorbed and incorporated within the bodies of living organisms as normal isotopes are. This lodges a radio-active source within the body of the organism itself.

(6) Like any other element radio-active isotopes are also absorbed, accumulated and bio-magnified thousands of times. Thus the entire food chain becomes contaminated. Organisms at higher trophic levels may, therefore, receive a highly concentrated source of radio-active material through their food supply.

(7) There is no other way to dispose of these hazardous wastes except to store them for thousands or millions of years away from living beings. This is too long a period on human scale of time. Even the safest burial places for radio-active wastes, which represent the best of human efforts, have shown signs of leakage. At present it appears very difficult, though not impossible to store radio-active wastes away from the biosphere for such long periods.

(8) In spite of all these hazards, nuclear reactors and tests are still continuing and an increasingly large amount of radio-active wastes is accumulating every day while no solution to the problem of their safe disposal is in sight till date.

The *harmful effect of radiation* upon human beings is due to its ability to wise and ultimately destroy the organic molecules of which body cells are composed of. The damage depend upon, the energy and the type of radiation. The energy is expressed in **Rads** (i.e. absorption of 100 ergs or $10^8$ joules of energy per gram of tissue). The total biological effect of radiation expressed in **Rems.**

Number of rems = n × Number of rads

Where n = 1 for $\beta$, $\gamma$ and x-rays.

= 10 for $\alpha$ radiations or high energy neutron.

## CONTROL OF RADIOACTIVE POLLUTION

Control of natural radioactive pollution may not be possible. Out of all the sources, only artificial radioactivity is the scope of intervention, wherever controls can be thought of. Radioactive pollution can be controlled by strict enforcement of the following safety measures.

All low or high level wastes have tremendous capacity to pollute the environment. As low level wastes are often produced in large quantities, their containment is not possible. They are visually subjected to a treatment for removal of radioactivity and then discharged in the water bodies or on land in usual way. High level wastes, on the other hand, cannot be disposed of freely in the environment, but have to be concentrated, contained and stored out of the reach of human's environment.

The radioactive wastes concerned with water pollution are usually in liquid or solid state. These different kinds of wastes pose various problems, as disposal techniques suitable for one kind may be risky for other. All techniques however have a single goal that radioactive constituents of wastes are not allowed to cause harm to organisms and in particular humans.

## QUESTIONS

1. How are air-pollutants classified? State five common air-pollutants, their source and effects on the man and his environment.
2. What are the sources of $NO_x$ and $SO_2$ in the atmosphere? What are the undesirable effects manifested by them? How can they be controlled?
3. What is the acid rain? How is it caused? What are its bad effects on the man and his enviroment? What steps are needed to control acid rain?
4. What is photochemical smog? How is it formed in the atmosphere? What are its consequences?
5. Discuss the sources and effects of particulates in the atmosphere.
6. Describe the sources of the following in the atmosphere and their effects of human beings.
   (*a*) As (*b*) Be (*c*) B (*d*) Cr (*e*) Hg
7. Describe the sources and effects of the following in the atmosphere
   (*a*) Zn (*b*) Ni (*c*) Pb (*d*) Se (*e*) Cd
8. Describe the manifestation of air-pollution on man and his environment giving examples.
9. Explain the terms "Lapse rate" and "temperature inversion". What are their consequences in the atmosphere?
10. Discuss with the help of diagram the plume characteristics under different lapse conditions.
11. Write informative notes on any two of the following:
    (*a*) Atmosphere sampling
    (*b*) Significance of air quality standards
    (*c*) Instrumental techniques used for estimation of atmospheric pollutants.
12. Discuss the various techniques available for atmospheric sampling.
13. Describe briefly how the following pollutants in the atmosphere are determined?
    (*a*) Suspended particulates (*b*) $H_2S$ (*c*) $SO_2$ (*d*) Hydrocarbons
14. How are the following pollutants estimated?
    (*a*) Dust (*b*) CO (*c*) NO*x* (*d*) $SO_2$
15. Describe the major pollutant emissions from thermal power plants, cement and leather industries. How can they be controlled?
16. Discuss the various techniques available for air-pollution control.
17. What are the various types of pollutants from automobiles? How can they be controlled?
18. How can you control the emission of the following pollutants into atmospheric air?
    (*a*) NOx (*b*) SOx (*c*) Hydrocarbons (*d*) Particulates (*e*) CO
19. Discuss the sources of Pb in the particulate matter in the atmosphere. How can it be controlled?
20. How can the following pollutants in air be monitored?
    (*a*) CO (*b*) $NO_x$ (*c*) $SO_x$ (*d*) heavy metals.
21. State the characteristic and biochemical effects of the following air pollutants:
    (*a*) $SO_x$ (*b*) $NO_x$ (*c*) CO (*d*) $O_3$ (*e*) hydrocarbons
22. What are the physiological and psychological effects of noise ?
23. Discuss the different types of equipment used for the measurement of noise levels ?
24. Write informative notes on any two of the following :
    (a) The Decibel Scale
    (b) Approaches for noise control
    (c) Noise Exposure Index
25. Write short notes on any three of the following :
    (a) Threshold shift
    (b) Speech interference and voice levels
    (c) Vibration meters
    (d) Weighted decibel scale
26. Discuss the various measures needed for effective noise control in industrial establishment.
27. Write short notes on any three of the following :
    (a) Acoustic enclosures and noise barriers
    (b) Vibration control devices
    (c) Sound absorbents and sound insulators

(d) Acoustic plenum
(e) Mufflers

28. Discuss the principle involved in the following :
    (a) Automative silencers
    (b) Damping
    (c) Lagging
29. What are the various methods available for the protection of the personnel exposed to high noise levels ?
30. Describe the various acoustical absorptive materials used for noise control and state their uses and limitations.
31. Indicate briefly the sources of noise and the noise control methods used in the following equipment :
    (a) Bulldozer
    (b) Electrical
    (c) Industrial cooling fans
    (d) Engines
    (e) Compressors
32. What are the sources of soil pollution? What are the detrimental effects of soil pollutants?
33. Discuss the various sources of soil pollutants. How can the soil pollution be controlled?
34. Discuss the role of urban waste and modern agro-technology in soil pollution.
35. How do industrial wastes contribute to soil pollution? What measures are to be taken to minimise soil pollution from this source?
36. Write about the role of an individual in prevention of pollution.
37. Write an essay on Radioactive Pollution.
38. What are the main sources of marine pollution? Give effects and methods to control marine pollution.
39. What do you know about thermal pollution? Give their effects and methods to control.
40. Write short notes on–
    (i) Thermal pollution (ii) Marine pollution (iii) Bhopal Gas Tradegy
    (iv) Decible (v) 3R
    (vi) Fluorosis (vii) Nuclear hazards
41. Write about fluoride problem in drinking water.
42. How are water pollutants classified? Give suitable examples.
43. Discuss the important characteristics of waste waters.
44. How are the following determined in wastewater? (*a*) COD (*b*) BOD.
45. Discuss the various methods available for primary waste water treatment.
46. What are the major objectives of the secondary wastewater treatment? Describe any one of the biological wastewater treatment process in detail.
47. Write informative notes on the following:
    (*a*) Water treatment for municipal purposes
    (*b*) Sewage treatment
    (*c*) Eutrophication
48. Discuss with the help of a diagram trickling filtration process for waste-water treatment. What are its merits and limitations?
49. Describe the activated sludge process. State the advantages and disadvantages of this process.
50. Write short notes on the following:
    (*a*) Oxication ditch process (*b*) Coagulation (*c*) Equalisation.
51. Describe the anerobic digestion process of wastewater treatment. What are its advantages and limitation?
52. Write informative notes on any two of the following:
    (*a*) Oxidation pond (*b*) Teritary treatment methods (*c*) Aerated lagoons

# Solid Wastes-Pollution, Treatment and Disposal

## INTRODUCTION

Solid wastes can be broadly defined as those wastes which have been rejected for further use and which can neither be transported by water into streams nor can readily escape into the atmosphere. They include all the discarded solid materials from municipal, agricultural and industrial activities.

## CLASSIFICATION AND ORIGIN

Solid wastes can be broadly classified into the following three categories:

**(1) Domestic Wastes:** These include wastes from household preparation, cooking and serving of food; waste paper and plastics; cloth; rags; etc.

**(2) Municipal Wastes:** These include garbage and rubbish from households, offices, hotels, markets, etc. and also the street refuse such as street sweepings, dirt, leaves, contents of litter receptacles, etc. The term "garbage" is generally used to putrescible or biodegradable food wastes. The term "rubbish" is used to denote nonputrescible or non-biodegradable solid wastes which include combustible materials (*e.g.*, paper, cloth, plastic, tyres, etc.) as well as non-combustible materials (*e.g.*, broken crockery, metals, glass, masonry wastes, used metal cans and containers, etc.) Apart from these, the following types of special wastes are also encountered: ashes, cinders, dead animals, abandoned vehicles, construction and demolition wastes, septic tank sludges, pipes, wires, conduits, insulations, etc.

**(3) Industrial wastes :** These include

(*a*) Non-process wastes such as office and cafeteria wastes, packing wastes, etc. which are common to all industries and

(*b*) Process wastes which depend upon the type of the products being manufactured, such as tannery wastes, weaving and dying wastes, food-processing wastes, plastic wastes, rubber wastes, metal scraps, etc., from the respective industrial establishments.

Mineral wastes from mining and mineral processing units also fall under this category.

**(4) Agricultural wastes:** These wastes result from farms, feed lots and livestock yard. The agricultural wastes include paddy husk, baggasse from sugarcane, tobacco and corn residues, slaughter house wastes, manures, etc.

**(5) Special wastes:** These include hazardous wastes from different sources e.g.

(*a*) Radioactive wastes from nuclear power plants, laboratories, hospitals, etc.

(*b*) Toxic substances such as heavy metal sludges, pesticides, pharmaceuticals, etc.

(*c*) Biological product such as enzymes, antibiotics, pathogenic and pathological wastes, etc.

(*d*) Miscellaneous products such as inflammable substances, corrosive materiais, explosives, security wastes, etc.

## MAGNITUDE OF THE PROBLEM

Owing to increasing industrialization and ever increasing population, the production of paper, leather, rubber, metals, plastics and ceramics has been steeply increasing over the last few decades. This in turn resulted in steep increase in the generation of solid wastes both in the developed and developing countries. For example, the annual urban refuse of USA is estimated to be over 400 mega tons (*i.e.*, 400 × $10^9$ kg). The national average of solid waste generated in North America is 1 tonne per capita per year (*i.e.*, about 2.7 kg per person per day).

The annual cost of collection and disposal of urban waste in USA is estimated to be 6 billion dollars.

The quantity of domestic and municipal solid wastes generated in some major cities of India is 300 to 600 g per person per day. Since the quantity of waste generated depends on the affluence of the society, the per capita waste produced in India is naturally lesser as compared to that produced in affluent western countries. The amount being spent in municipalities in India for solid waste management is estimated to be 850 to 1200 million rupees annually which accounts for about 10% of their total budgetary provisions. During the seventies, the quantity of solid waste generated was increasing at a rate of 50% per year and in eighties, the annual rate of increase was 2 to 3%. Further, the new products such as cans, bottles, plastic containers, household appliances, tyres, food packing materials and several other items are considered to be cheaper to throw away than to reclaim. Thus, with the advent of "throw-away" concept in the present civilized society, the generation of solid waste has been considerably increasing both in quantity as well as complexity.

## CHARACTERISTICS OF SOLID WASTES

**(1) Composition :** Apart from variations in quantity, wide variations in the composition of domestic and municipal solid wastes may also occur depending on the following factors;-

(*a*) Degree of urbanization and industrialization of the area

(*b*) Per capita income

(*c*) Social customs

(*d*) Climatic conditions of the area

(*e*) Acceptability of packaged foods

(*f*) Frequency of collection by the municipality, etc.

**(2) Density:** The density of municipal solid wastes varies from 150 kg/$m^3$ to 8000 kg/$m^3$ depending upon the waste composition and degree of compaction for land-filled, pulverized refuse.

**(3) Energy content:** Municipal solid wastes contain about 50% of combustible matter. Hence the waste is often burnt as a means of disposal and occasionally used as a source of energy. The average calorific value of the urban solid wastes produced in India is about 1500 Kcals/kg. Since they contain a large quantity of compactable organic matter, these wastes are mostly disposed of by composting (aerobically) or for generation of bio-gas by anaerobic treatment.

## OBJECTIVES AND CONSIDERATION IN SOLID WASTE MANAGEMENT

**Objectives:** The principal objectives of solids waste management are to control, collect, treat, utilize and dispose of the solid wastes in an economical manner consistent with the protection of public health.

**Considerations:** The major considerations of concern with solid waste management apart from the economics are (*a*) public health, (*b*) waste separation for recycling, and (*c*) energy recover.

**(*a*) Public Health:** Under warm and moist conditions, and particularly with the help of vectors (carriers) like water, air, food, rodents, mosquitoes and flies, organic solid wastes are ideal breeding places for pathogenic organisms. Potentially hazardous substances like solvent and pesticide cans, medical wastes and asbestos debris present in solid wastes, air pollution from gaseous and particulate emissions from land-fill sites and municipal incinerators, etc., also present some additional environmental concerns related to solid waste disposal. Further, the solid waste management strategies should also take into account the possible deterioration of ground water quality because of land-filling of solid wastes, residues from incinerators and leachates from the decomposing refuse.

**(*b*) Waste separation for recovery and recycling:** Recovery and recycling of some of the resources in solid wastes, although a very appealing idea, is rather difficult in practice. Expensive materials such as some metals are found to be economical to recycle by industries. Returnable bottles and refundable cans may be recycled from municipal wastes.

Separation of wastes at sources is warranted and feasible only when the reclaimed materials find reasonable market. However, this concept of waste separation at source is receiving increasing attention in some developed countries due to the dwindling landfill capacity, economic incentives, improving markets for the reclaimed materials, environmental concerns and political will.

Separation of mixed municipal refuse for reclamation at a centralised facility is another approach in waste management which is particularly suitable for dense metropolitan areas. Hand sorting and machine sorting processes are used depending upon the labour costs and other sanitary considerations. Methods for the separation of dry, pulverized mixed refuse are based on characteristics such as size, diversity, inertia, conductivity, etc. Air or water classifiers for light materials like paper, and plastic, magnetic separation for separation of ferrous metals, screening for separation of non-ferrous materials, inertial classifiers for separating inorganic contaminants from compost or for separating heavy, resilient particles from light, inelastic ones, are some of the unit processes employed.

**(*c*) Energy recovery:** Recovery of energy from municipal solid wastes can be achieved by the following two ways:

(*i*) Solid wastes can be burnt directly in incinerators or converted to more efficient "refuse-derived fuel" (RDF). Pyrolysis and anaerobic decomposition of organic matter in solid wastes are the other methods available for recovering the fuel value of solid wastes.

(*ii*) Reuse of the recovered materials from solid wastes is the other principal mode of energy conservation. Obviously, mining and manufacture of ferrous and nonferrous metals starting from mining of the ores is so energy-intensive that reuse of these metals is certainly justified from the standpoint of energy conservation.

## METHODS OF SOLID WASTE TREATMENT AND DISPOSAL

The most commonly used solid waste treatment and disposal methods are : (1) Composting (2) Sanitary landfills or controlled tipping (3) Thermal processes (*e.g.*, incineration and pyrolysis), and (4) Recycling and Reuse.

### A. Auxiliary operation involved in solid waste treatment and disposal

Before actually taking up the discussion on the disposal methods, it will be useful to consider some of the essential auxiliary operations involved in these processes. These processes include.

(*a*) **Collection, transport and handling:** Solid wastes should be collected at least once a week. However, in high population density areas and for wastes with high putrescible content, particularly during warm weather, more frequent collection of refuse is desirable. The modern method of transporting and handling the refuse is by packer trucks or container trucks (with carrying capacity of 4 to 5 tonnes each) provided with compaction facility, which are loaded manually or mechanically. The wastes are then collected at transfer stations equipped with trailer units with higher carrying capacity (about 20 tonnes each). The wastes are then compacted to high density and then transported to the disposal sites. Hydraulic and pneumatic waste-transport systems also are being developed.

(*b*) **Pulverisation:** Solid wastes are pulverised with the help of gyratory crushers or jaw crushers or other similar equipment prior to transfer loading, compacting, landfills or incineration in order to facilitate these processes. Pulverisation helps in getting homogeneous material which helps both in the greater initial settlement of the solids as well as in future reclamation of the land.

(*c*) **Compaction and Baling:** Compaction and baling of solid wastes using hydraulic or pneumatic presses is carried out at the site or at a central facility. The advantages achieved by this process are (*i*) Reduction in refuse value (*ii*) Reduction in collection time (*iii*) Reduction in transport time and cost (*iv*) Lesser storage area (*v*) Lesser safety hazards (*vi*) Cleaner storage area.

## B. MICROBIOLOGY INVOLVED IN SOLID WASTE TREATMENT AND DISPOSAL

Micro-organism play a vital role in water and wastewater treatment, composting and sanitary land-filling. Some types of these macro-organisms of interest from the point of view of environmental science and engineering are shown in Fig 3.1-A.

Most living things were originally classified as belonging to either the plant kingdom or the animal kingdom. In 1866, a third kingdom called the "*protista*" was proposed to fit in such micro-organisms which did not unequivocally fit into the plant or animal kingdom, *e.g*, protozoa, bacteria, fungi and algae viruses were discovered later.

**Bacteria** are the most important group of micro-organisms which are essential to the nutrient cycle of the ecosystem. Pathogenic (disease causing) bacteria received greatest attention from the point of view of epidemiology. Many other types of bacteria are important in natural self purification of streams and lakes, in the decomposition of solid wastes in landfills, solids and in composting. Bacteria also play an important role in water an wastewater treatment processes (vide Chapter 2).

**Viruses** are smaller than bacteria and may also cause diseases in plants, animals and human beings.

**Algae** belong to a group of photosynthetic plant-like micro-organisms. The impart odours and tastes in water. They may also interfere with water supply by clogging water filters. Algae provide oxygen for low-cost wastewater treatment in oxidation ponds. (vide Chapter 2) However, excessive amounts of nutrients in water can lead to algae blooms which leads to "*eutrophication*", removing dissolved oxygen from lakes when they decompose.

**Fungi** are unicellular or multicellular micro-organisms which can survive under low pH conditions. They are non-photosynthetic "*protista*". Fungi are useful in the biological treatment of some industrial wastes and also in composting of solid organic wastes.

**Protozoa** are, by and large, an order of magnitude larger than bacteria. They also play a role in biological wastewater treatment processes described in Chapter 2.

**Rotifers** are multicellular micro-organisms which are sometimes present in the effluents of some biological waste treatment processes. They consume organic colloids, algae and bacteria and thus perform a "Polishing" function.

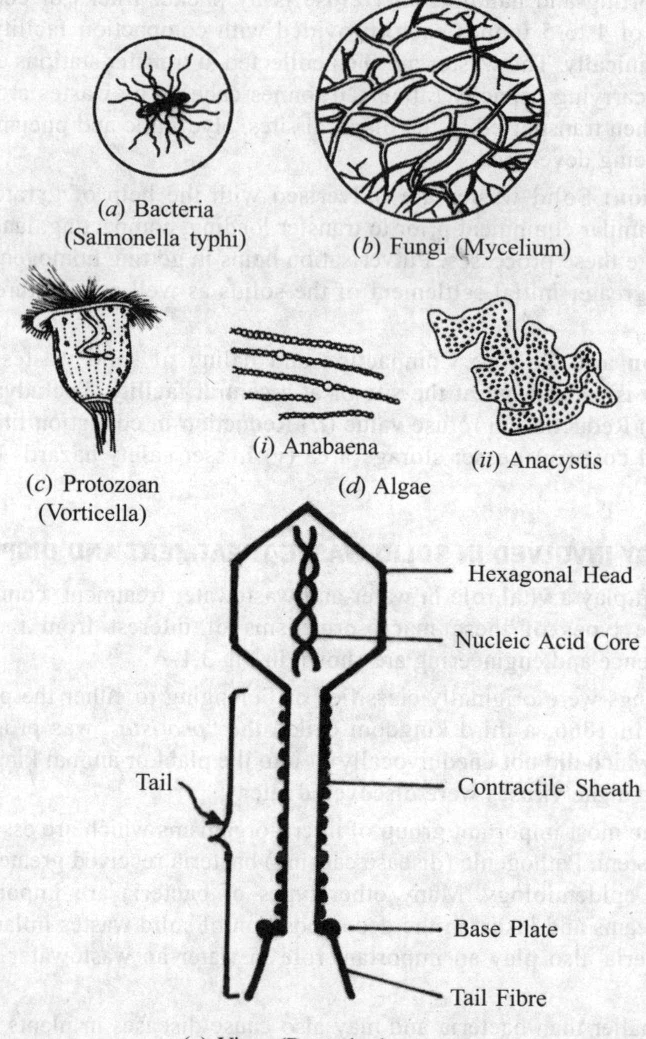


**Fig. 3.1-A.** Soil micro-organisms of interest from the point of view of Environmental Science and Engineering.

**Crustaceans** are multicellular organisms having a hard shell. Some of them are microscopic in size and serve as food for fish. They are also considered as indicators for unpolluted condition in receiving waters.

**(a) Bacteria:** These are unicellular microscopic organisms usually found in water, wastewater, soil, plants, animals, milk and intestinal tract of humans. Bacteria reproduce by binary fission. The are characterized by their size, shape, structure and cellular arrangement. Bacteria with spherical shape are called "cocci", those with spiral shape are called "spirilla", and those having cylindrical or rod-like shape are called "bacilli". Bacterial cells may be arranged in pairs, clusters, or chains.

The cells of all types of bacteria have a rigid cell wall of about 0.25 μm that protects their contents from osmotic pressure and also maintains their shape (Fig. 3.1-B). Just beneath the cell

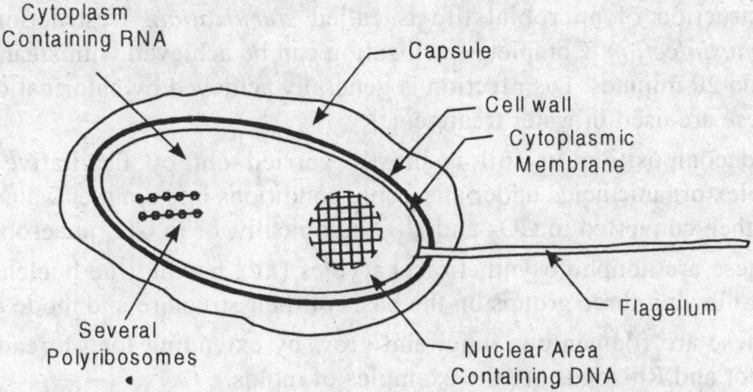


**Fig. 3.1-B.** Schematic representation of a typical bacterial cell.

wall is the semipermeable "cytoplasmic membrane" having a thickness of about 0.0075 μm. This semipermeable membrane separates and protects the protoplasm (*i.e.*, the internal content of the cell) from the external environment by allowing the nutrients to pass into the cell and the waste products out of the cell. Any damage to this cytoplasmic membrane by physical or chemical agents cause the death of the cell. The protoplasm contains three different areas: (*i*) the cytoplasm, containing dissolved nutrients and RNA (ribonucleic acid) which helps in the biosynthesis of proteins, (*ii*) the nuclear area containing DNA (deoxyribonucleic acid) which is responsible for the genetic stability of the species, and (*iii*) the polyribosomes, which produce enzymes specific to certain biochemical reactions.

Some types of bacteria, such as bacillus and clostridium, form "spores" which represent a dormant of resting phase of the cell and which are very resistant to adverse physical or chemical environments. When the active cell (called the vegetative cell) is exposed to adverse growth conditions, "spores" may form within the cytoplasm. When conditions conducive for growth develop, the "spores" germinate and new vegetative cells emerge. Spore-forming bacteria commonly occur in air, water and soil and are difficult to destroy.

Many bacteria are capable of rapid movement in liquids with the help of their whip-like "flagella" and hence are said to be "motile". The existence and the type of "flagella" help to differentiate between various bacterial groups.

Bacteria are divided into two broad groups on the basis of their source of energy and carbon source. Bacteria which obtain both their energy and carbon from organic matter are called *Heterotrophic Bacteria.* Bacteria which require carbon dioxide as their energy source and obtain their energy from sunlight or by the oxidation of inorganic substances are called *Autotrophic Bacteria.*

On the basis of their need for oxygen, bacteria are also classified as follows;

(*a*) **"Aerobic bacteria"** which require free oxygen for their growth.

(*b*) **"Anaerobic bacteria"** which can grow without free oxygen.

(*c*) **"Facultative bacteria"** which can grow with or without oxygen.

(*d*) **"Microaerophilic bacteria"** which grow in the presence of minute quantities of free oxygen.

Most bacteria grow between pH 4 to 10 but exhibit optimum growth in the pH range 6.5 to 7.5. All bacteria require moisture for their growth for facilitating the dissolution of nutrients and to penetrate the cell membrane. Growth and reproduction of the cell occur as the nutrients are taken up into the cell to process new cell material.

Complete destruction of microbial life is called *sterilization.* Destruction of pathogenic bacteria is called *disinfection.* Complete sterilization can be achieved with steam at about 210 K pa (15 psig) within 20 minutes. Disinfection is generally achieved by chlorination or with iodine or ozone; and these are used in water treatment.

Solid waste decomposition in soil is mostly carried out by facultative bacteria which decompose complex organic acids under anaerobic conditions into simpler water-soluble organic acids. These are then converted to $CO_2$ and $H_2O$ aerobically, or to $CH_4$ anaerobically.

***(b)* Fungi:** These are nonphotosynthetic eukaryotes (*i.e.*, having true nucleus). They may be divided into the following three groups on the basis of their structure and mode of reproduction:

***(i)* Molds:** These are filamentous fungi and grow by extending long thread-like structures: Aspergillus, Mucor and Rhizopus are the examples of molds.

***(ii)* Yeasts:** These are non-filamentous fungi and are unicellular. Raspore and Candida are the examples. They are unicellular and are considerably larger than bacteria. They are widely distributed in nature. They are generally spherical, ellipsoidal and egg-shaped cells. Unlike molds, yeasts are facultative, i.e., they can grow under both aerobic and anaerobic conditions. The reproduce asexually by binary fission or by budding. Yeasts are useful in many fermentative processes (e.g., preparation of bread, beer and wine) and also for synthesis of certain vitamins, fats and proteins from simple sugars and ammonia nitrogen. However, some types of yeasts (*e.g.* Candida) can cause hazardous infections in humans.

***(iii)* Mushrooms:** These are macroscopic fungi belonging to the Basidiomycetus division. Several types of edible and non-edible mushrooms are known.

Fungi are mostly aerobic and feed on decaying organic matter. They are more resistant to acidic condition. They reproduce by either sexual or asexual spores.

Soil fungi play a vital role in stabilizing solid wastes in composting and landfilling processes by decomposing the plant tissues like cellulose and lignin.

***(c)* Protozoa:** These are highly specialized unicellular organisms. They reproduce asexually by binary fission. Most of them are non-photosynthetic and are motile. They are of varying size from a few to several hundred microns. They survive adverse conditions by forming cysts with thick walls. Protozoa may be saprophytic (e.g., sarcodina). As most of the protozoa are predators on bacteria, they are found wherever bacteria are prevalent. Thus, protozoa help in maintaining the equilibria of microbial flora in solid waste disposal systems.

**Entamoeba histolytica** is common pathogenic protozoa which causes amoebic dysentery in humans. Some protozoa belonging to mastigophora group are blood parasites which cause sleeping sickness in humans. Some species of parasitic protozoa belonging to sporozoa group cause malaria in humans. The female *Anopheles* mosquitos are the carriers of these parasites to human hosts.

***(d)* Algae:** These are microscopic plants which are capable of photosynthesis and hence contain chlorophyll. Unicellular algae are mostly spherical, spiral or cylindrical while multicellular ones can grow in filaments, long tubes or simple masses of single cells that cling together. Algae are the important primary producers in the aquatic food chain. However, algae impart tastes and odours in water and cause high chlorine demand for water treatment. They clog water intakes and shorten filter runs. Excessive growth of algae, known as "algal blooms" interferes with recreational uses of water. The algae are present in fresh water, marine water as well as polluted water. They are generally classified on the basis of appearance in water, *e.g,* green yellow, red brown and blue-green.

***(e)* Viruses:** These are the smallest micro-organisms in size range $1 \times 10^{-2}$ μm to $25 \times 10^{-2}$ μm. They do not contain any internal enzymes and hence cannot grow or metabolize on their own. They are obligate parasites infecting the tissues of bacteria, plants, animals and humans. Examples of humans pathogenic viruses include those which cause mumps, influenza, smallpox, infectious hepatitis and poliomyelitis.

Although viruses are not living organisms, they can reproduce or replicate themselves within their specific host cells.

### C. Waste treatment and disposal methods

#### (1) Composting

Composting is the aerobic and thermophilic decomposition of organic matter present in the refuse by microorganisms, primarily bacteria and fungi. The organic matter is transformed into a stable humus like substance during this process. The reactions taking place during composting generate heat and hence the compost temperature raises during the process. Depending upon the composition and nature of the waste, the waste volume is reduced by about 30 to 60%.

For an optimum composting operation, the following control parameters are usually adhered to:

| | | | |
|---|---|---|---|
| 1. | Temperature | : | 40 to 50°C (If the temperature goes beyond 66°C, biological activity will be reduced) |
| 2. | pH | : | 4.5 to 9.5 (It is better to maintain pH below 8.5 to minimise the loss of nitrogen in the form of ammonia as gas). |
| 2. | Moisture | : | 40 to 70% (The optimum value is about 55%). |
| 4. | Particle size | : | 0.63 to 2.54 cm |
| 5. | Air | : | 0.5 to 0.8 $m^3$/day/kg of volatile compost solids |
| 6. | Carbon to Nitrogen Ratio | : | (35 to 50) : 1 |
| 7. | Carbon to Phosphorus ratio | : | 100 : 1 |

Composting may be carried out naturally under controlled condition or in mechanized composting plants. In natural systems, the garbage (which is ground after removing glass and metallic materials) is mixed with a nutrient source (*e.g.*, sewage sludge, animal manure or night soil) and a filler (*e.g.*, wood chips or ground corn cobs) which permits the air to enter into the pile. The mixture, which is maintained at about 50% moisture content, is kept in windrows having a width of about 2.5 m. The mixture is turned over twice a week. Within about 4 to 6 weeks, the temperature falls, the colour darkens and a musty odour develops. This indicates completion of the process. The filler may then be removed and the remaining humus-like material is used as soil conditioner. With mechanical systems, the composting time is reduce to half of that required in natural systems, because of continuous aeration and mixing.

The composting process usually consists of the following three steps:

**(*i*) Waste preparation:** The solid waste is placed on slow moving conveyor belt. Materials like corrugated paper are hand-picked and then the ferrous materials are removed by magnetic separation. The waste is then ground in hammer mills or wet pulpers to the desired size range (0.6 to 2.5 cm). Then, it is mixed with nutrient source, filler and water (to provide 50% moisture).

**(*ii*) Digestion :** The mixture is placed in the windrows for 4/6 weeks, while turning it once or twice a week. The waste is decomposed by thermophilic micro-organisms during this period. The material is then allowed to stabilize for another 2 to 5 weeks.

(*iii*) **Product upgradation:** In order to ensure quick and better marketing prospects, the product is sometimes upgraded by operations such as curing, grinding, screening, pelletising and bagging.

In Western Europe, Japan, Israel and some Third-World countries which are committed to land reclamation, many successful composting plants have been operating for several years. India has made considerable progress in organising compost plants in cities like Bombay, Baroda, Calcutta, Nagpur, Delhi, Bangalore and Ahmedabad, Mechanical compost plants are encouraged by the Government in our country as a national programme. The National Research Development Corporation is now in a position to offer the India know-how to other countries. The agricultural research institutes of our country are developing the technology to produce blue green algae coated granulated compost.

**(2) Sanitary Landfilling**

Landfilling is the most common and economic method of solid waste disposal in many countries.

Disposing solid wastes into "open dumps" was the normal practice by municipalities from good olden days. The "town dump" generally used to be in a low-lying area near a watercourse. This unsanitary dumping of the waste used to result in water pollution, bad odours, blowing papers, fires, flies, rats, etc. These problems are mostly reduced by burial of the waste. Further improvement was achieved by compacting the waste in layers and covering it with earth at the end of the disposal of the waste every day. In order to distinguish from the earlier practice of unsanitary "open dump", this method is called "sanitary landfill". Although compacting and covering still continue to be the basic operations today, several further improvements in the process have taken place during recent years. These include careful and scientific site selection, controlled deposition, better method of compaction, reduced cover, leachate collection to avoid water pollution and site monitoring to ensure environmental protection.

In a sanitary landfill, complex organic wastes are slowly degraded or decomposed by the soil micro-organisms, primarily by aerobic or facultative bacteria and fungi. Since air cannot enter through a compacted land-fill, the aerobic bacteria decompose the organic solid waste by utilizing whatever oxygen is present inside the landfill. Then, decomposition by the anaerobic micro-organisms begins, and this accounts for the degradation of most of the solid waste present in the landfill. The water-soluble organic acids generated in this process enter the water media and diffuse through the landfill soils. The bacteria and fungi present in the soils aerobically metabolise these organic acids into $CO_2$ and water.

Occasionally, anaerobic methane bacteria accumulate in landfill systems and generate appreciable quantities of methane gas. A portion of this gas may be utilised by aerobic bacteria as it diffuses through the landfill. But, however, most of the methane escapes into the atmosphere and may pose a potential fire hazard.

An ideal sanitary landfill site should satisfy the following criteria:

(*i*) It should be cheap, accessible and at a reasonable distance.

(*ii*) It should be atleast 1½ km downwind from the commercial and residential neighbouring area.

(*iii*) It should be reasonably levelled, clear and well drained, with capacity of use for at least 3 years.

(*iv*) Its soil should be a low permeability so that it can be used as satisfactory cover material.

(*v*) It should be well above the ground water table so that the underground water supplies are not polluted.

(*vi*) The site selected for landfilling should not be deleterious or offensive to the surrounding environment. It should be consistent with the topography, climatic conditions, hydrogeological requirements and economical considerations.

The planning of a sanitary landfill operation should be based on the following considerations:

(*i*) The quantity and nature of the solid waste to be treated.

(*ii*) Overall suitability of the land for such operation.

(*iii*) Economic considerations of the landfilling process as compared to other available solid waste treatment processes.

(*iv*) Public health criteria.

(*v*) Proper design of the sanitary landfill operation with respect to the appropriate method of the landfill suitable for the chosen site and efficient organisation of the possible pollution problems such as leachate generation, escape of gases, etc.

Preparation of the landfill site involves fencing, grading, stockpiling of the cover material, construction of berms, landscaping and the installation of leachate collection system, gas collection system and monitoring system.

A cross-sectional view of an ideal sanitary landfill is shown in Fig. 3.2.

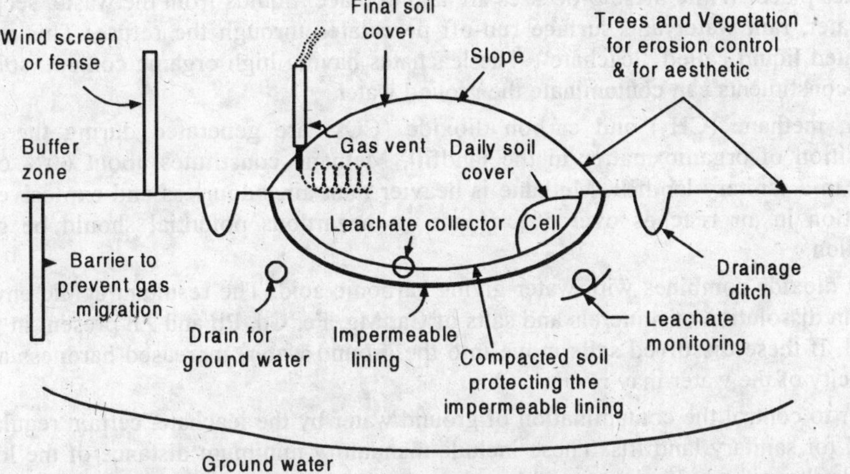


**Fig. 3.2.** Cross-sectional view of a typical sanitary land-fill.

Mixed solid wastes with varying degrees of compaction are delivered to the landfill site by packed trucks or trailer units. After hand sorting the wastes, pulverising or high pressure compaction and baling for volume reduction may be done before placement. Loose material is placed in the lower part of the pit or trench. It is then spread and compacted by machines in layers of about 0.5 m thickness. After the end of each day's operation and when the depth is about 3 m, the refuse is covered with 15 to 30 cm of earth. This consolidated solid waste enclosed by earth at the end of a day's operation is called a "cell".

Landfilling by low density refuse may not be economical particularly at a site with limited capacity. Reduction of waste volume extends the life of the landfill and will also prove to be economical. Volume reduction by incineration is expensive and hence physical methods such as pulverising (by a hummer mill) or high-pressure compaction into bales or blocks (of about 1 $m^3$ weighing about 900 kg) are usually preferred. Pulverisation brings about a volume reduction by about 50% and also helps in reducing the cover material required, problems of odours, rodents, insects, fires, blowing papers, etc. High-pressure compaction into solid bales also provides all the above advantages in addition to provide resistance to infiltration by rain water.

## PROBLEMS WITH LANDFILLING

**(1) Economic problems:** The land use as landfill cannot be used in future as a productive farm land. Even after closure of the landfill site, its further use should be restricted only to some type of open development such as a park or recreational area. Building construction in the site must be controlled.

**(2) Aesthetic problems:** Poorly operated landfill operations may cause problems due to bad odours, insect, vermin, blowing papers, rats, and scavenger birds, apart from the dust and noise from waste transporting vehicles and compacting operations. Continuous field compaction of the loose refuse and proper covering with earth at the end of each day's operation reduce these problems. Volume reduction by pulverisation or high-pressure compaction may ensure an aesthetically acceptance operation.

**(3) Environmental problems:** During landfilling operations of municipal solid wastes, aerobic degradation occurs initially for periods varying from a few weeks (in wet areas) to a few months (in dry areas). After all the oxygen is exhausted, anaerobic degradation of the organic wastes takes place. While these processes are taking place, liquids from the waste, seepage from ground water, rain water and surface run-off percolates through the refuse. This produces a contaminated liquid called "leachate". The leachates having high organic content, soluble salts and other constituents can contaminate the ground water.

Further, methane ($CH_4$) and carbon dioxide ($CO_2$) are generated during the anaerobic decomposition of organic matter in the landfill. Methane constitutes about 60% of the gas generated in a sanitary landfill. Methane is heavier than air, odourless and explosive when its concentration in air reaches over 5%. Hence its hazardous potential should be given due consideration.

Carbon dioxide combines with water giving carbonic acid. The resultant acidic environment may help in dissolution of minerals and salts of Ca, Mg, Fe, Cd, Pb and Zn present in the refuse or the soil. If these dissolved salts move into the ground water, increased hardness and heavy metal toxicity of the water may result.

In order to control the contamination of ground water by the leachate, certain regulations are prescribed for sanitary landfills. These include mandatory minimum distance of the location of the landfills above the ground-water table, minimum distance from the nearest point of water use, use of soil of low permeability, preventing the build-up of the leachate within the landfill, extensive hydrogeological investigation of the actual landfill site including topography, soil stratigraphy, drainage and groundwater characteristics (*e.g.*, depth and movement), etc. Supplementary measures such as inclusion of clay and/or membrane covers or liners for the landfill, leachate collection, removal and treatment facility, installation of monitoring system are also suggested to ensure that the buried refuse remains as dry as possible and the leachate does not reach the ground water. In spite of these precautions, if some leakage of the leachate still occurs, the soil beneath serves as an additional barrier to attenuate the contaminants before reaching the groundwater by mechanisms such as filtration, adsorption, precipitation and biological activity.

The gases produced in sanitary landfills by the anaerobic digestion or organic waste are generally vented to the atmosphere through gravel-packed seams or wells. In some landfills, the escaping gases are burnt off with the help of burners installed at the top of the vents. These measures are meant for preventing undue accumulation and lateral movement and migration of the gases under the landfill cover to nearby residential areas which may cause fires and explosions.

**(3) Thermal Processes**

The important thermal processes used in solid waste treatment are incineration and pyrolysis.

*Incineration* : In this process, the solid organic wastes are subjected to controlled combustion so as to convert them into incombustible residue and gaseous products. Incineration process is considered when suitable site for landfilling is not available within economic haul distance from the sources of solid wastes.

The advantages of this process are:

(*i*) The volume of the waste is reduced to more manageable levels thereby reducing the transportation costs to the ultimate disposal site.

(*ii*) Waste from a bigger community of people can be accommodated for a given acreage available for a landfill. Incineration reduces the land requirement to one-third of that required if the refuse is to be landfilled.

(*iii*) The residue after incineration, if properly carried out, is free from any degradable materials and hence is no longer a source of pollution. Moreover, the stabilised residue thus produced, minimises or even eliminates the need to transport the cover material to the landfill site.

The disadvantages of this process are.

(*i*) high capital and operation costs.

(*ii*) the possibility of air pollution if not carried out properly.

(*iii*) ordinary incinerators cannot be used for radioactive wastes.

Successful incineration process involves the following steps:

(*i*) handling and feeding the wastes to the incinerator.

(*ii*) combustion of the waste within the incinerator.

(*iii*) removal of the incombustible residues to the disposal site.

(*iv*) removal of gaseous and gas-borne residues.

The design of an incinerator should be based on the following considerations:

(*i*) the quantity of the waste to be handled, its composition and characteristics.

(*ii*) measures to ensure that there will not be any pollution of air, water or land resulting from the operation of the incinerator.

On economic considerations, the size of the incinerator is determined of the basis of the weekly quantity of the waste to be incinerated. For unsorted wastes, two types of incinerators are used:

**(*i*) The batch-type incinerator** which is manually stoked and has a relatively small rated capacity. These plants have several disadvantages:

(*a*) owing to the intermittent operation, the burning temperature cannot be maintained in a uniform manner. This many result in an inadequate and irregular combustion of the waste.

(*b*) the output of particulate matter is more.

(*c*) the volume reduction of the waste is lesser than the optimal value expected.

(*d*) this may end up with an unstable residue still containing some putrescible matter. Thus, it may still possess some pollution potential.

(*e*) the intermittent incinerator plants are unsuitable for large urban centres.

**(*ii*) The continuous type of incinerators** are equipped with large storage bins, automatic feed hoppers, varied types of moving grates and ash discharging systems. These units are capable of maintaining uniform temperatures for combustion and can be equipped with pollution control devices such as gas scrubbers and electrostatic precipitators. These units are capable of yielding a stable residue which is non-polluting. Although the capital and operating costs are very high, these units, which provide controlled furnace temperatures of 760° to 980°C, can remove odours and also bring about a substantial reduction in waste volume, in an environmentally acceptable form. Since the final residue is stable, the cost of cover material required to ultimately dispose it in the landfill will be substantially minimised or even eliminated in some cases.

The modern municipal incinerators are of continuosly burning type. Many of them have "waterwall" construction in the combustion chamber instead the older and more common refractory lining. The "waterwall" consists of joined vertical boiler tubes (made of steel) through which water is circulated. These water tubes adsorb heat to provide not water for steam generation and at the same time control the furnace temperature. Thus, the heat recovery is simpler and the waste heat of combustion can be utilised for space heating, power generation, desalination, etc., as has been done in some recent installations.

Further, the "waterwall" units help in eliminating the costly maintenance of refractory-lining and also reduce the pollution control requirement (by reducing the volumes of quench water and gas, that require treatment). However, corrosion of the "waterwall" unit may be a serious problem, as experienced by some installations in European countries.

Putrefiable organic solid wastes from slaughter house have to be dried in special driers before being incinerated. Hospital wastes are generally disposed of by pathological incineration.

If the wastes are burnt only for volume reduction, there is no need for an auxiliary fuel, except at the start-up. When steam production is the objective, supplementary fuel must be used with the pulverised refuse. Ferrous metals are usually recovered from the ash.

High temperature incineration is a recent innovation where temperatures of the order of 1,650°C are attained using supplementary fuels. In this process, noncombustible fractions of the refuse (*e.g.*, metal and glass) are melted in a bed of high-temperature coke in the refractory lined incinerator and are drained of as molten slag. This technique can achieve volume reduction of the refuse by 97%. In spite of the fact that the first high temperature incinerator was installed on a pilot plant scale as early as in 1966, application to full-scale commercial units was not followed up enthusiastically for a long time due to the high costs involved.

The Volkswagen Works in Germany reportedly operated the so called melt zit slag-tap process. In this process, the glass in the soil waste is liquefied at about 980°C and the ash is melted at about 1300°C. The resultant magma is either molded into large blocks or run into water to produce a coarse type of sand useful for preparing road or concrete aggregate.

Another technological innovation in the field of incineration is the so called CPU-400 which is designed as a turboelectric generator plant utilising the solid waste itself as fuel. The refuse is first passed through a shredder and a drier (where the hot gases from the combustion unit themselves are used for drying) and then fluidised in the combustion chamber. The hot gases produced are made to turn a turbine which will drive an electrostatic precipitator.

The major problems associated with incineration process are:

(*i*) Air pollution control, particularly the removal of the fine particulates and toxic gases *e.g.*, dioxin, are the most difficult.

(*ii*) disposal of liquid wastes from the floor drainage, quench water and scrubber effluent.

(*iii*) disposal of ash containing heavy metal residue in landfills.

## PYROLYSIS

The chemical constituents and chemical energy of some organic wastes can be recovered by destructive distillation (or pyrolysis) of the solid waste. In this process, the combustible constituents of the solid waste are heated in a special retort like chamber known as a pyrolysis reactor at 600 to 1000°C in a low-oxygen or an oxygen-free environment. This is an endothermic process and thus differs from the conventional incineration.

Pyrolysis of the solid wastes yields the following components:

(*i*) Tar or oil phase containing methanol, acetone, acetic acid, etc.

(*ii*) Gaseous phase containing $H_2$, $CH_4$, $CO$, $CO_2$, etc.

(*iii*) Solid phase containing pure carbon char and inert materials like glass, rock metal, etc.

The advantages of the pyrolysis process include:

(*i*) Volume reduction by about 90%

(*ii*) Possibility of handling potentially hazardous plastic *e.g.*, PVC in a safe way

(*iii*) Absence of pollution problems

**(4) Recycling and Reuse**

Recycling and reuse of the waste helps to reduce the problem of waste disposal. Resource recovery is a method to turn wastes into resources by recovering usable products-both materials and energy. As the disposal costs are expected to rise continuously due to increasing land prices and pollution controls, resources recovery is becoming more common and finding more favour.

About 70% by weight of municipal solid waste from domestic and commercial areas is combustible. One ton of such a waste is approximately equal to 9 million British Thermal Units (BTU) of heat or 65 gallons of fuel oil or 9000 cubic feet of natural gas. But still only a small percentage of the resources is being recovered.

In some municipal solid waste-processing facilities, the combustibles are separated from the non-combustibles. The combustibles are then shredded and burnt in utility boilers or industrial boilers as a primary fuel or as a supplement to fossil fuels. This type of solid waste processing operation is known as a refuse-derived fuel (RDF) system. RDF is used to supplement other fuel sources in a ratio of 20% RDF to 80% fossil fuels. It was reported that 29 RDF systems were operated in USA for power generation as early as in 1983, processing 1250 to 18,000 tons of waste per week. Incineration and RDF technologies seems to be competitive in cost.

A ton of solid waste processed by pyrolysis is believed to yield an energy equivalent of one barrel of oil. The city of Baltimore reportedly operated a commercial scale facility in 1975 to produce 4.8 million pounds of steam daily from the low-BTU gas generated by pyrolysis of municipal solid waste. Thus, saving of 357,000 barrels of oil annually was accomplished, in addition to the revenue earned from ferrous metals sorted out and the sale of glassy aggregate for use in concrete manufacture and street paving. Economic viability of full-scale commercial pyrolysis facility has still not been proved beyond doubt. The advantage of pyrolysis is that it produces a more generally useful and transportable form of energy.

Rubber tyres continue to pose disposal problems. They do not decompose well in landfills. Incineration of rubber must be done is specially designed facilities to check air pollution and to accommodate the intense heat produced by burning rubber. The BTU content of burning rubber is nearly equal to that of coal. Some systems burning rubber as fuel were successful. Production of fuels from rubber by pyrolysis was also successful. However, these were necessarily small-scale operations due to limited supply of tyres.

Reclamation of reusable materials such as paper, glass, metal, cardboard and plastics from municipal wastes, though received considerable attention, has not been very successful due to economic considerations.

Paper constitutes nearly 50% by weight of the solid wastes. Total recycling of paper has not been possible so far. Unrecoverable Paper is generally disposed by controlled incineration and the gaseous products obtained are $CO_2$ and water vapour which are non-polluting. Paper has a high heat of combustion and hence is disposed by converting it to energy. Refuse derived fuel (RDF) can be used as supplementary fuel for use in boilers or in other combustion processes.

From the standpoint of energy conservation there is every justification of the recycling and reuse of ferrous metals, aluminium and some other non-ferrous metal because the mining of virgin material is energy intensive. As far as glass is concerned, it was found that the energy spent on reuse of old glass is more than that required to process new material. Hence, energy savings can be realized only by reuse of the glass containers than to reprocess them. Reynolds Metals Company (USA) was reportedly recycling used aluminum cans from packaging industry.

Recycling of plastics received considerable attention primarily because of ever increasing use of plastics and also because of their non-biodegradable nature. For a satisfactory recycling of plastics the following two requirements are to be satisfied:

(*i*) The plastic material should be made up on only one type, *i.e.*, it should be homogeneous.

(*ii*) The plastic scrap or waste should be collected from the consumers or intercepted on its way from consumers to the municipal refuse site.

Recycling of plastics may be carried out in any of the following ways:

(*i*) Primary recycling where the same plastic product is manufactured again.

(*ii*) Secondary recycling where the material is reprocessed to a new product with different composition and in some cases may be inferior in properties.

(*iii*) Tertiary recycling where the plastic material is completely processed to a new form as in pyrolysis (where some chemicals are recovered). In USA, high density polyethylene bottles used for supplying milk, are collected from consumers and are converted to flake powder by grinding. This can be used for manufacturing plastic drainage pipes or as inert fill material or an aggregate for low weight concrete.

## CO-DISPOSAL

Co-disposal involves simultaneous disposal of solid waste and sewage sludge. However, the use of sewage sludge as a soil fertilizer is questionable due to its volume and the presence of persistent chemicals such as heavy metals. Moreover, due to the high water content, the sewage sludge does not burn readily.

**Thermal co-disposal** involves the use of energy from solid waste to dry the sludge to a point at which it will bu without supplemental fuel or in combination with more refuse. The technique is receiving greater attention because of the following factors: (*i*) increasing cost of land which makes land disposal of sludge more and more expensive, and (*ii*) increasing cost of fossil fuels, which make incineration of sludge more expensive. The process is used in European countries.

## BIOCONVERSION

As has been stated earlier, organic solid wastes decompose under landfill conditions to $CH_4$ and $CO_2$. This process can be accelerated artificially by means of an anaerobic digester. Sewage sludge containing decomposers is mixed under anaerobic condition with shredded combustible components of municipal solid waste. The mixture is heated in a digester for about a week. A demonstration bio-conversion plant has been in operation in Pompano Beach, Florida since 1978, in which $CH_4$ is produced from a mixture of RDF and sewage sludge.

## QUESTIONS

1. How are solid wastes classified? What are the characteristics of solid wastes? What are the important objectives of solid waste management?
2. Describe briefly the various methods of solid waste treatment and disposal.
3. What are the important control parameters for an optimum composting operation? Discuss the different steps in composting operation. What are the merits of composting as a solid waste treatment process?
4. Discuss the advantages and problems associated with sanitary landfilling. Give a diagrammatic representation of the cross-sectional view of an ideal sanitary landfill.
5. Describe the thermal processes used in solid waste treatment. Discuss the merits and demerits of these processes.
6. Give an informative appraisal on recycling and reuse of solid wastes.
7. Write informative notes on any two of the following:
   (*a*) Continuous type of incinerators and their advantages
   (*b*) Microbiology of solid wastes treatment
   (*c*) Pyrolysis
8. Write short notes on any three of the following:
   (*a*) Bio-conversion
   (*b*) Co-disposal
   (*c*) Environmental problems involved in sanitary landfilling
   (*d*) Pyrolysis
   (*e*) Recycling of plastic wastes.

# Hazardous Wastes

## INTRODUCTION

As per the abbreviated definition by the United States Environmental Protection Agency (US EPA), hazardous waste means "a waste or combination of wastes which because of the quantity, concentration or physical, chemical or infectious characteristics may cause, or significantly contribute to increase in mortality or an increase in serious irreversible, or incapacitating reversible illness, or pose a substantial present or potential hazard to human health or the environment when improperly treated, stored, transported or disposed off, or otherwise managed. The hazardous wastes include wastewater treatment sludges from electroplating, spent halogenated solvents used for degassing (*e.g.*, trichloroethylene, methylene chloride, etc.), spent unhalogenated solvents such as xylene, acetone, ethylbenzene, ethyl ether, etc.), sludges from petroleum refineries, dewatered air pollution control scrubber sludges from coke ovens and blast furnace sludges, radioactive materials and medical wastes.

In this chapter, we mainly concentrate on the management of the following three types of wastes; (*i*) Radioactive wastes (*ii*) Bio-medical wastes and (*iii*) Non-radioactive industrial liquid wastes.

**(*i*) Radioactive Wastes**

Nuclear power production is the result of radioactive waste problem. The environmental effects of nuclear power generation mainly arise from

(*a*) the nuclear fuel cycle

(*b*) low-level dose radiations from nuclear-power-plant effluents, and

(*c*) low- and high-level dose radiations from radioactive wastes.

All nuclear reactors rely on uranium as fuel. The natural uranium contains 99.2% of U-238 and 0.72% of U-232. It can be used as such in heavy water reactors. For use in light water reactors, it should be enriched to about 3 to 3.6% with respect to U-235 content in an isotope enrichment plant. The enriched uranium is fabricated into fuel rods and are inserted in nuclear reactors. During operation of the reactor (for 1 or 2 years), the fuel rods become intensely radioactive due to the generation of fission products and trans-uranium elements such as plutonium americium and curium. After the fuel rods are removed from the reactor, they should be cooled in water-filled pools (known as spent-fuel storage bay) to remove the heat liberated due to radioactive decay. The fuel cycle is then completed by disposing of the fuel in an underground repository or separating the residual plutonium from uranium and fission products in a chemical reprocessing plant. The fission products are then immobilized by incorporating into ceramic or glass matrix and disposed off into and underground repository. The enriched uranium fuel cycle involving light-water reactor is shown in Fig. 4.1 given below.

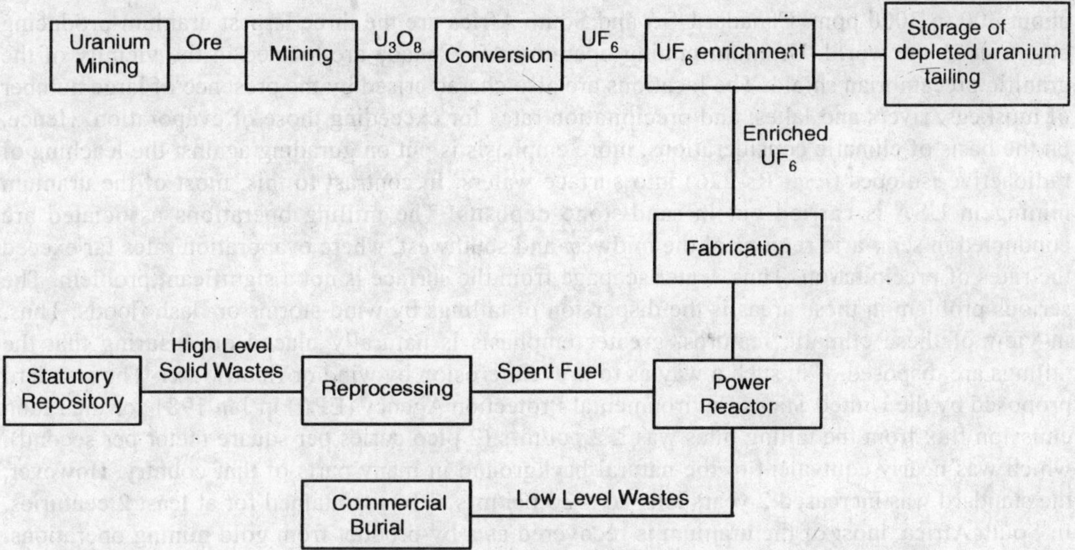


**Fig 4.1.** Fuel cycle for light-water reactor which require enriched fuel.

Low-level radioactive wastes arise from a broad spectrum of radioisotopes produced by mining facilities, and by medical and research activities *e.g.*, radio-isotopes used for cancer treatment, contaminated glass, clothes, filter paper and other laboratory appliances.

High-level radioactive wastes are generated in nuclear reactor fuel cores as spent fuel and also from nuclear fuel reprocessing.

## EFFECTS OF RADIATION ON LIVING CELLS

The particulate and electromagnetic radiations emanating from radioactive materials inflict deleterious effects on the living cells. These effects are classified as "somatic" and "genetic". "Somatic effects" are caused on the exposed individuals and the cell damage caused may manifest in malignancies such as leukemia or cancer. "Genetic effects" are transmitted to the descendants of exposed individuals and thus can affect unexposed generation too. The radiation-induced changes in the genes may manifest themselves in (*a*) gene mutations, (*b*) chromosome aberrations and (*c*) changes in the number of chromosomes. Such changes can result in abnormalities in the offsprings which may be mild or lethal.

The extent of damage from radioactive materials depends on

(*a*) physical properties such as half-life of the radioisotope involved and the type and energy of the radiations emanating.

(*b*) their ability to enter the food-chain, and

(*c*) their tendency to become concentrated in the living tissues.

The extent of damage is maximum in the reproductive organs, block forming tissues, the digestive tract and developing embryos which are called radiosensitive organs.

## ENVIRONMENTAL PROBLEMS AND MANAGEMENT OF NUCLEAR WASTES

### Wastes from uranium mining and milling

Uranium is present in the earth's crust in a concentration of about 3 parts per million (ppm) by weight. However, economic recovery is possible only if the uranium content of host rock is about 300 to 3000 ppm. Canada, USA and South Africa are the three largest uranium-producing countries in the world. Uranium mining operations in Canada are located in the vicinity of the granitic precambrian shield. The locations are also characterised by the presence of large number of muskegs, rivers and lakes, and precipitation rates for exceeding those of evaporation. Hence, on the basis of climatic considerations, more emphasis is put on gurading against the leaching of radioactive isotopes (*e.g.*, Rs-226) into surface waters. In contrast to this, most of the uranium mining in USA is carried out in sand-stone deposits. The milling operations associated are conducted in semi-arid regions of the midwest and southwest, where evaporation rates far exceed the rates of precipitation. Thus, water seepage from the surface is not a significant problem. The serious problem in these areas is the dispersion of tailings by wind-storms or flash floods. Thus, in view of these climatic features, greater emphasis is naturally placed on ensuring that the tailings are disposed of in such a way as to prevent erosion by wind or flood water. The standard proposed by the United States Environmental Protection Agency (EPA) in Jan 1981 for the radon emission flux from the tailing piles was $\leq 2$ pci/m$^2$s (2 pico curies per square meter per second), which was nearly equivalent to the natural background in many parts of that country. However, the standard was increased 2 years later to $\leq 20$ pci/m$^2$s to be maintained for at least 2 centuries. In South Africa, most of the uranium is recovered as a by-product from gold mining operations. The uranium concentrations are generally below 300 ppm and hence the radioactive hazard is far less than that from the richer North American ores.

The uranium recovery from the ore involves crushing and grinding the ore to a finely divided state, followed by dissolution of the uranium containing minerals in an acid or alkali. The solution is filtered to separate it from the bulk of the inert ore and is then treated further to recover the uranium. The large mass of the residual solids in then piped as a slurry to the tailing pond for disposal.

The tailing pose problems due to the presence of chemical, mineral and radioactive contaminants. For instance, if the ore contains pyrites ($FeS_2$), the fine particles of pyrites formed during crushing and grinding of the ore go with the tailings. These particles of pyrite are most susceptible for the attack by oxygen, water and bacteria than the original massive form of the ore underground. Thus, sulphuric acid is formed which can percolate slowly into the rivers and lakes nearby which is undesirable. Further, the tailings also contain toxic metals such as As, Cd, Hg, Fe, V, Zn and molybdenum are large quantities of ammonium, chloride and sulfate ions which may pose a threat to the nearby environment. However, the radioactive isotopes present in the tailings include Ra-226, Th-230 and Rn-222 (and its radioactive daughters) which are of greater environmental concern. The Ra-226 found in freshly discharged tailings can undergo slow dissolution by the water percolating through the deposit and contaminate the surface waters. When ingested by humans, the Ra-226, which is similar to Ca, gets deposited into the bone structure. Even when 1 microgram of Ra-226 is retained in the body, it can cause serious bone damage. The radio concentration above the tailing piles can be several hundred times higher than the normal background level but it quickly gets diluted by the surrounding air mass and does not pose any treat beyond 1 km distance from the tailings piles. Th-230 has a long half-life of $7.7 \times 10^4$ years and hence has a relatively low specific activity. But, it is the continuous generation of its daughter element *viz*. Ra-226 (with a half-life of 1600 years and having higher chemical mobility) which poses a low-level radioactive hazard over thousands of years after disposal.

The method for storage and disposal of the tailings are site-specific and depend upon parameters such as the local climatic features, topography, geology, the nature of mining operation (i.e., open pit or underground) and the proximity of populated areas. In the arid regions of the south western regions of U.S., the tailings are disposed in a depression below the ground level and covered with 1 to 3 meters thick layer of soil, sand gravel or crushed rock. In areas where precipitation (*i.e.*, rainfall) is higher such as at Alliot Lake in Canada, the design of the tailings disposal facility is based on ensuring minimum dissolution of Ra-226, toxic metal ions and other contaminants such as chlorides and sulphates.

## WASTES FROM REFINERY AND FUEL FABRICATION

In the refineries where uranium concentrates from mines are purified, small quantities of uranium bearing wastes are generated. These wastes are usually accepted back into the milling circuit at mine-mill sites at no charge because those wastes are equivalent to high-grade ore from many mines.

In fuel fabrication such as pressing, sintering and grinding, small amount of uranium oxide waste is produced. Since uranium is so valuable, this waste is recycled through the fabrication process by dissolving it in nitric acid, precipitating as ammonium diuranate and converting to uranium dioxide powder.

Thus uranium refining and fuel fabrication process are closed circuit operations and no significant uranium wastes escape into the environment.

## HIGH-LEVEL WASTES FROM SPENT NUCLEAR FUEL

When spent fuel is discharged from light-water reactors, it consists a mixture of uranium, plutonium and fission product. Since the heat flux from the decaying fission products is too high during the first one or two years, underwater storage is the most desirable option. However, as the time passes, interim dry storage in concrete canisters seems to be a more economical option, even for longer storage times.

The ultimate waste disposal process envisages the use of multiple barrier system. The spent fuel bundles are loaded into containers made up of corrosion-resistant material such as titanium-nickel- or copper-based alloys. The interstices between the fuel bundles and the container are filled with a particulate support (*e.g.*, glass beads) or by a lead-antimony alloy. The sealed containers would then be placed in caverns drilled in a geologically stable granite rock repository, around the container to separate it from the rock. After filling of the repository, all rooms, boreholes and shafts would be backfilled and plugged so that no further supervision is required.

The possibility of disposal into underground salt domes has also been studied in USA. Another possibility that is being explored is the disposal of radioactive waste into the deep sea bed, where tectonic movement would slowly propel the wastes downwards into the earth's crust.

Technology is also available for incorporating high level radioactive wastes into borosilicate glass or ceramic matrix or polycrystalline "synrock". Thus, the radioactive waste is fixed or immobilized into the solid matrices of proven stability so that it can then be safely disposed in suitably designed and constructed underground repositories that remain free from any disruptive seismic or tectonic activity for millions of years.

The general public still seems to believe that there are serious unresolved environmental problems associated with the nuclear fuel cycle. This view is fostered because of incomplete, incorrect and exaggerated information. But, despite the perceived problems, the fact remains that technical solutions are now available for the safe, long-term management of uranium mine and mill tailings and of high-level wastes from nuclear power reactors.

**(*ii*) Biomedical Wastes**

Biomedical wastes originate from hospitals, clinics, research and testing laboratories and drug companies. These include pathological and surgical waste: experimental animals and cadavers, drug and chemical residues and their containers; discarded bandages, linens and other infectious wastes; disposable syringes, needles and surgical instruments; contaminated equipment, food and other waste materials. Chemical and chemotherapy wastes, organic solvent wastes and radioactive wastes are not included in the biomedical wastes because they are governed by different regulations.

In countries like USA and Canada, biomedical wastes are controlled as per USEPA's 1985 guidelines for Infectious Waste Management and the Ontario Ministry of the Environment's Guidelines for the Handling and Disposal of Biomedical Wastes respectively. On-site handling of wastes and their off-site disposal are considered as the responsibility of the waste generators themselves. The normal practice for handling hospital wastes is the on-site incineration of the combustible solids in a specially designed high temperature incinerator provided with after-burners to heat the gases leaving the chamber to 700°C for controlling odours. The ash thus produced is disposed in a sanitary landfill. Wastes from such hospitals which do not have facilities for incineration or sterilization are segregated, packed in specially coded and labelled containers, and transported to a place where such facilities exist.

**(*iii*) Chemical Wastes**

Modern civilization has resulted in large scale-manufacture of products such as TV sets, aerosol cans, pesticides, plastics, preservatives, automobiles, etc. These industries generate a multitude of hazardous industrial wastes, toxic substances such as heavy metals. Proper management of these hazardous industrial wastes is mandatory from the point of view of public health and environmental protection. Incidents like **"Love Canal Episode"***, which festered for 30 years before its dangers to humans were exposed, emphatically demonstrated the consequences of improper disposal of hazardous wastes.

***Love Canal Episode:**

Hocker Chemical company (Niagara, New York) disposed its hazardous wastes for several years until 1952 by burying them in 55-gallon drums in a ditch known as the Love Canal. After the ditch was filled, it was capped with clay. Later the land was sold for a throw away price to the Board of Education of Niagara Falls. A school and some residences were built on the site. The clay cap was cracked due to construction, water got infiltrated into the trenches containing the wastes. After 25 years, the waste began to leach out. The leachate penetrated the basement walls in the area, surfaced in backyards and volatalized in the air. Several people fell ill. Over 300 different chemicals, including some carcinogens, were identified in the ambient air. The President of USA had to declare in August 1978 that Love Canal was a Federal Emergency Disaster area.

Several families were evacuated from the area. The clean-up of the canal and stabilization of the area (by claycapping and drainage ditches) costed over 40 million dollars. Hooker Chemical Company was sued for claims amounting to about 2 billion dollars.

The disaster obviously was the result of improper disposal of hazardous waste.

In 1985, it was found that Love Canal Experience is only the tip of an ice-berg and several other waste disposal sites in USA need federal cleanup assistance over the next 50 years at a cost of about 100 billion dollars.

## ENVIRONMENTAL EFFECTS

All chemicals are toxic to humans beyond some concentration. From the environmental protection point of view, the important point to be considered are

(*a*) at what concentration a particular substance becomes toxic, and in what form?

(*b*) What are its environmental pathways? That is, what are the entry routes for these hazardous chemicals and wastes into the environment?

(*c*) With what persistence and at what biological concentrations that toxicity is manifested in the environment?

The various entry routes for hazardous chemicals and wastes (for dumps and industries) into the environment include

(*a*) leaching into adjacent land,

(*b*) leaching into streams,

(*c*) leaching down into the water table,

(*d*) evaporation,

(*e*) incineration, and

(*f*) spillage during transit.

## TOXIC CHEMICAL

The inorganic chemicals of concern are heavy metals such as Hg, Cd, Pb and As which act as biological poisons even at parts per billion (ppb) levels. These toxic elements accumulate in organic matter in soil and sediments and are taken up by growing plants, thus entering the food chain. Since these chemicals are poorly excreted by humans, they accumulate in organs and tissues to toxic levels in the body. Toxic metals enter the atmosphere by burning of fossil fuels (*e.g.*, coal and leaded gasoline), reach receiving water through atmospheric fall-out and leaching from mines and landfills, and contaminate land as a result of sewage sludge application. Low pH caused by acid rain or the generation of carbon dioxide increases the transportability and hence availability of these metals by rendering them more soluble.

The organic chemicals of concern are those that persist (*i.e.*, those that degrade slowly) in the environment and are fat soluble, since they can accumulate in the food chain. Polychlorinated biphenyls (PCB's) and pesticides fall under this category which cause immediate toxicity as well as long term effects such as carcinogenicity and mutagenicity. Many of these persistent pollutants are generated from the degradation of primary substances of from burning of chlorine compounds. Also, incineration of municipal solid wastes, and burning of organic materials and fossil fuels also release toxic chlorinated organics into the atmosphere. Some of these polluting organics are also absorbed on dust and fly ash, thus enabling wide spread distribution in the atmosphere. Other important organic pollutants include pesticides (*e.g.*, DDT), biocides (*e.g.*, Dioxin), insecticides (*e.g.*, Fenitrothion and Mirex), $CHCI_3$ and trihalomethanes.

## IDENTIFICATION OF HAZARDOUS WASTES

A simple approach to identify hazardous wastes is to consider these under general categories, such as toxic, inflammable, radioactive etc. Such a classification system helps the Fire Service to properly deal with the hazardous wastes under emergencies.

Many countries (*e.g.*, UK, Germany, France and Netherlands) supply a supplementary classification system for hazardous substances along with the general classification system and some countries also set maximum concentration limits for different contaminants. However, in Japan, the following four types of wastes, *viz.*, sludges, slags, acidic wastes and alkaline wastes, which contain any of the 9 toxic materials (*viz.*, As, Cd, Cr (VI), Pb, Hg, CN, PCB's and organic phosphates) beyond permissible limits, are considered to be hazardous.

It is best to identify the hazardous wastes on the basis of algorithms that consider toxicity, reactivity, persistence, corrosivity, inflammability, quantity of waste involved, the extent of hazard to the environment and the ultimate effects on living organisms.

The hazardous wastes are identified in the U.S. as per the Federal Register of May 19, 1980. The US EPA (United States Environmental Protection Agency) considers the following five basic criteria for identifying a hazardous waste:

(*i*) Whether the material falls under the category of solid wastes as defined by the Resource Conservation and Recovery Act (RCRA) of 1976? As per this definition, the term "solid waste" included not only just solids, but liquids, semi-liquids and contained gaseous materials as well.

(*ii*) Whetner the waste has been legally discarded? This stipulation is applicable to the wastes that are stored or treated prior to disposal but not to those to be recycled.

(*iii*) Whether the waste is specifically excluded by the regulations? For instance, agricultural wastes, municipal solid wastes, animal manures, etc., are excluded.

(*iv*) Whether the waste has toxic or hazardous characteristics? That is, whether it has the potential to increase the mortality or illness or whether it poses a substantial threat to human health or environment on the basis of hazardous characteristic such as flammability, corrosivity, toxicity or explosiveness?

As regards the human toxicity, in the absence of adequate toxicity data, a waste is considered to be hazardous if

(*a*) it has an oral LD50 toxicity equal to or greater than 50 mg/kg (*i.e.*, LD50 is the lethal dose of the toxic waste at which 50% of the experimental animals die as a result of oral ingestion or dermal penetration), or

(*b*) it has an inhalation LC50 toxicity of 2 mg/kg (LC50 is the lethal ambient concentration of the toxic material in mg/L of air causing 50% mortality to test rats during 4 hours-inhalation).

As regards the hazard to health or the environment, the following standard tests which have been quantified by EPA, are considered:

(*i*) Ignitability (*i.e.*, whether the waste causes or enhances fires),

(*ii*) Corrosivity (*i.e.*, whether the waste destroys the tissues or metal),

(*iii*) Reactivity (*i.e.*, whether the substance reacts violently or causes explosion, and

(*iv*) Toxicity (*i.e.*, whether the substance polluted water supplies threatens the health).

As far as the toxicity to the groundwater is concerned, limits for 14 common contaminants have been considered as the basis for determining human toxicity. These 14 toxic substances include 8 metals (*viz.*, Hg, Cd, Cr, As, Pb, Ba, Se and Ag), 4 Pesticides (*viz.*, Endrin, Lindane, Methoxychlor and Toxaphene and 2 herbicides ( viz., 2, 4–D and 2,4,5–TP). When the concentration of these contaminants exceed 100 times their respective permissible concentration in drinking water, the waste is designated as hazardous. This toxicity of the waste is determined by the "Extraction Procedure (EP) Toxicity Test" which consists of agitating not less than 100 grams of the "solid waste" in 16 times its weight of de-ionised water at a controlled pH of 5 for 24 hours. The leachate is then analyzed for the 14 toxic substance listed above. If their concentrations exceed 100 times those allowed in drinking water, the waste is designated as "hazardous". Tests for radioactivity, phytotoxicity, mutagenicity, infectiousness, etc. are also being developed.

(*v*) Whether the waste is listed as hazardous? Wastes which are known to be hazardous (carcinogenic, mutagenic, etc.) but not amenable to the EP Toxicity Test described above are listed in the following three categories: (*a*) spent solvents used for degreasing etc., (*b*) specific sources such as process waste from wood preservation or manufacture of materials such as halogenated hydrocarbons, and (*c*) discarded product such as mercury compounds, toluene, xylene, etc. and their containers.

## MANAGEMENT OF HAZARDOUS WASTES

The primary objective of a hazardous waste management plan, is to eliminate or reduce the hazardous waste through processing changes or resource recovery. Having accomplished this objective, the remaining hazardous wastes must be accounted for from its origin to ultimate disposal. This "cradle-to-grave" concept is followed by countries like USA and Canada.

The important components of a hazardous waste management plant are as follows:

**(1) Inventory:** All the industries producing more than 100 kg of hazardous wastes per month should be registered. An inventory of such waste generating agencies and the quantity of the waste produced per month should be prepared.

**(2) Storage:** The hazardous waste generating industries should equip themselves with special on-site tanks or basins for storage of large quantities of hazardous wastes. Chemically resistant drums should be used for temporarily storing small quantities of corrosive materials.

**(3) Transport :** The wastes stored as above must be collected at regular intervals by licensed haulers and transported by suitable tanker trucks or flat-bed trucks (for the drums) or rail cars (for large volume of waste) to the approved disposal sites.

**(4) Spillage:** A well-publicized emergency plan should be prepared to deal with unexpected spillage or accidental release of contaminants during transport to prevent environmental damage or public health hazard.

**(5) Disposal:** The wastes collected as above should be transported to a physical and chemical treatment plant (PCT) for processing or concentration; or directly hauled to an approved hazardous waste treatment facility for final disposal.

**(6) Waste-exchange:** Where on-site recycling or recovery of waste is not feasible, the possibility of waste exchange should be explored. Transferring wastes from big industries to smaller ones which are able to reuse low-purity oils, acids, alkalis, solvents, catalysts, etc. or recover valuable metals and other materials from concentrated wastes, can also be considered.

## TREATMENT AND DISPOSAL OF HAZARDOUS CHEMICAL WASTES

### (A) Treatment and Disposal by Industry

Where recovery or reuse of the waste is not economically feasible or waste exchange is not possible of off-site waste disposable facility is not available, the industry concerned should determine a suitable wastes pre-treatment procedure before safe disposal to an on-site facility or municipal sewers. Some types of wastes may be disposed in deep wets provided ideal underground conditions occur. In North America and Canada, 80 per cent of the least hazardous wastes are disposed on-site, in lagoons, evaporation ponds and landfills. However, in France, West Germany, Netherlands and Denmark, 73 to 80 per cent of the hazardous wastes is transported to centralized hazardous waste treatment and disposal facilities.

When wastes are unsuitable for disposal, a suitable method or combination of methods described below, for pretreatment of the waste before safe disposal must be selected.

**(A) Physical method:** A number of physical processes are available for pretreatment of hazardous wastes before safe disposal. Solid-liquid separation can be achieved by centrifugation, sedimentation, filtration or flotation. Adsorption of toxic organic components can be accomplished with activated carbon. Removal of specific components can be accomplished by stripping, distillation or membrane process such as dialysis, electrodialysis and reverse osmosis.

**(B) Chemical method:** Chemical process play an important part in hazardous waste-treatment operations. Some of the important examples are given below.

***(i)* Oxidation** of cyanides to cyanates or to $CO_2$ and $N_2$ by alkaline chlorination.

***(ii)* Reduction** of hexavalent chromium to less toxic trivalent chromium compounds using suitable reducing agents such as $SO_2$.

***(iii)* Precipitation** of toxic heavy metal salts as sulfides.

(*iv*) **Ion-exchange** for removing dissolved metallic or nonmetallic inorganic compounds.

(*v*) **pH-adjustment** of lime slurries with acidic pickle liquor.

(*vi*) **Immobilization of Stabilization or chemical fixation** of toxic inorganic sludge using ceramic, glass, cement or lime based inert matrices.

**(C) Biological Process** The various biological processes used in wastewater treatment have been discussed in the chapter on Water Pollution. Some toxic organic materials *e.g.*, phenols, oils, refinery wastes, etc., can be successfully treated and over 60% removal of many metals (*e.g*, Cd, Pb, etc.) can be achieved if the concentration of these toxic materials is not high and the biological process can be acclimatized to these materials. These results are usually accomplished in municipal wastewater treatment plants in combination with sanitary wastes which provide dilution and buffering capacity along with organics and nutrients required for the growth of micro-organisms. Aerobic biological processes where rapid microbial growth occurs are less sensitive to toxic materials as compared to the anaerobic processes which have slower microbial growth. Further, other factors which improve the ability of conventional biological treatment plants to accept toxic wastes are high active solids concentration, long sludge retention time and a high organic concentration (*i.e.*, a strong sewage). However if sufficient sewage is not available for combining with the industrial wastes, degradation of toxic organics or removal or metals by the conventional biological processes may not be feasible. In such cases, use of other biological methods may be explored. For instance, land farming of oily wastes offers and efficient and economic treatment method if thin layers of waste are spread over longer intervals of time. Another interesting but distinct possibility seems to be the use of bacteria for removing more concentrated metals from wastes. It has been reported that 15 to 20% of the copper production in USA is by bacterial leaching. A uranium mine in Canada is reported to rely totally on bacterial leaching for its uranium production.

**(B) Off-site hazardous waste disposal**

When an industrial establishment is not able to dispose of its hazardous waste in an on-site facility or into municipal sewers, the following three alternatives for offsite disposal are considered.

(*i*) **Hazardous waste treatment facility:** In a hazardous waste treatment facility, organic wastes are either incinerated or treated (physically and/or biologically) to produce an acceptable liquid effluent and a concentrated sludge that can be landfilled. Inorganic wastes are detoxified, neutralized and concentrated to produce an acceptable liquid effluent and a sludge which can be further concentrated and solidified by chemical fixation for landfill disposal. Generally, a combination of these or other process may be required for treatment and disposal of hazardous waste product and by-product.

(*ii*) **Co-disposal:** The hazardous waste may sometimes be co-disposed with municipal refuse. In such a case, the relatively small quantities of hazardous inorganic liquid wastes and some organics will be absorbed by large quantities of the refuse so that the contaminants are attenuated by the waste and the surrounding soil.

(*iii*) **Secure landfill:** The practice of disposing some detoxified hazardous wastes under carefully controlled conditions was followed in Japan. But, however, most of the hazardous waste systems rely on the land disposal at some stage or the other. In places like California, New York and Illinois landfills have been the usual repository for untreated or inadequately treated industrial wastes. In contrast, European countries used security landfills only for the residues from physical and chemical treatment (PCT) facilities, incineration and solidification, thereby reducing the waste and leachate volumes to be stored, treated and monitored indefinitely. In Denmark, 20% of the hazardous waste are landfilled with an option for heavy metal recovery from them in future.

Secure landfills are meant to accept and retain concentrated inorganic and organic hazardous wastes for indefinite period, perhaps, in perpetuity. If necessary, leachate can be periodically removed for treatment and disposal.

The secure landfill is the ultimate repository for all hazardous waste residues and it should be considered as the last resort when all other efforts to eliminate the waste problem fail. In designing a security, landfill, protection of groundwater from pollution is the main concern, apart from subsidence, fires, explosion, erosion and migration of organics. In the absence of adequate data regarding the migration, degradation and synergistic effects of hazardous pollutants in the soil, the North American practice is to design secure landfills for total containment. In contrast to this, the approach in the UK considers in its design the attenuation of the contaminants in the soil.

A secure landfill must be located in a land of marginal agricultural potential in a rural area, at least 8 km from populated areas and 750 m from the nearest neighbour. It should have 1 to 5% surface slopes overlying deep impermeable clay and the maximum ground water table should be at least 1.5 m below the bottom of the landfill. Other hydrogeological parameters, surface drainage, wild life, flooding potential, access to vehicles to the area and connection to major water resources and community water supplies are the other important considerations in selecting the land for security landfills.

Although many designs are possible for a secure landfill, a typical one used in Niagara Falls, New York is described here to illustrate an example. An area of about 10 hectares and located in heavy clay, partly above the partly below the original ground, to provide an overall depth of over 10 meters. Three lifts were allowed for several cells, with each lift and cell being separated by a clay or absorbent barrier. The clay bottom is excavated and sloped to instal an underdrain monitoring system below the landfill for future surveillance. A synthetic membrane liner (made up PVC, Hyapalon or High Density Polyethylene) sandwiched between protecting clay layers, was then placed over the sloped bottom. Sump pipes of about 900 mm dia were installed vertically as the filling progressed for leachate monitoring. If there is a probability of burial of organic wastes, provision for gas venting is also made. Finally, when the filling was completed, sloping clay cap with a synthetic membrane was placed and finished with top soil and grass cover. Some of the features of a Secure landfill are shown Fig. 4.2.

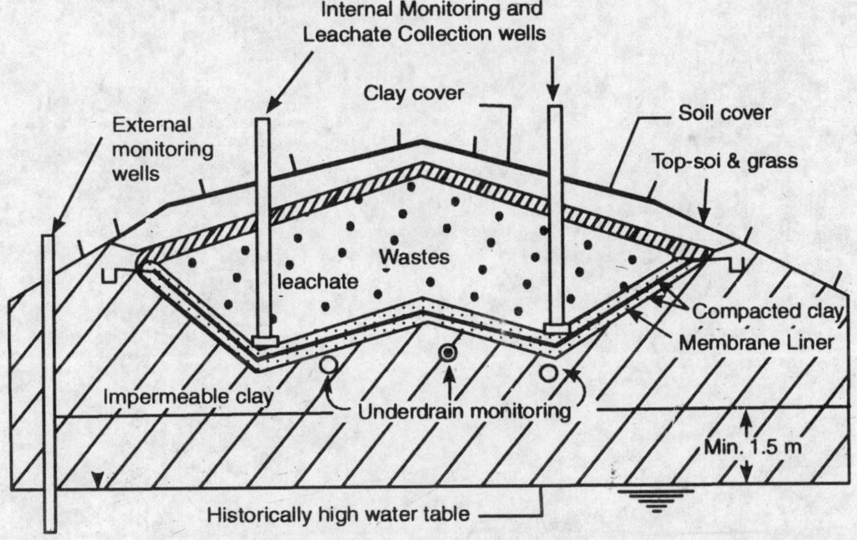


**Fig. 4.2.** Some special features of a secure landfill.

A secure landfill should not be considered as the sole means to hazardous waste disposal. It should be a part of an overall system which minimizes the volume and hazard of the waste by physical and chemical treatment (PCT), incineration and solidification.

Since the leachate characteristics are mostly site-specific and waste-specific and vary considerably, the type of the leachate treatment, disposal, recycling or interim storage have to be decided accordingly. So far, no standardized approach to leachate treatment could be made. However, recent researches on the use of upflow anaerobic filter and anaerobic fluidised bed systems for treating organics present in leachate seem to be very promising.

To sum up, a secure landfill is an essential component of a hazardous waste management plan. It should not be used as a receptacle for untreated hazardous wastes. It should serve only as the ultimate resting place for the unrecyclable, detoxified and solidified hazardous waste residues. Even then, the leachate formed should be collected and treated to specified requirements before disposal.

## QUESTIONS

1. What are hazardous wastes and how are they classified?
2. Discuss the environmental problems associated with nuclear wastes. How can they be managed?
3. Write informative notes on biomedical waste management.
4. What are the major types of chemical wastes that we are generally confronted with? What are the various entry routes for them to escape into the environment?
5. Discuss the important factors that should be considered for a hazardous waste management plant.
6. Discuss the various methods used for the treatment and disposal of hazardous chemical wastes from industrial processes.
7. Describe with the help of a diagram the essential features of a secure landfill.
8. Write informative notes on any two of the following:
   (*a*) Considerations in the design of a security landfill
   (*b*) Co-disposal
   (*c*) Biological processes used in treatment of wastewaters containing hazardous chemicals
   (*d*) Basis for the identification of hazardous waste.

# Trace Elements-Pollution & Control

## INTRODUCTION

About more than 90 elements are found in the earth's crust out of which a mere 9 elements (*viz.*, Al, Fe, Ca, Mg, O, Si, Na, K and Ti) only account for over 99% by weight. The remaining 81 elements together, which account for hardly 0.14% by weight, constitute the so called "trace elements". These elements, notwithstanding their low abundance in nature, play a vital role in plant and animal nutrition. These elements occur in living tissues in such minute amounts (or traces) that their precise concentration could not be measured by the early workers with the analytical techniques then available. This is another reason why the term "trace elements" was used to describe them. The term "trace elements" has remained in popular use still, despite the fact that today, virtually all these elements can be estimated with great accuracy and precision with the help of modern analytical gadgets, such as neutron activation analysis, chromatography, x-ray fluorescence spectrometry, atomic absorption spectrometry, flameless AAS, polarography, etc.

## MECHANISM OF DISTRIBUTION

In order to clearly understand and appreciate the role played by the trace elements in the environment, it is necessary to study the mechanism by which these elements attained the present level of distribution within the earth. The trace elements are mobilized by the following three types of mechanisms:

***(i)* Primary dispersion :** Trace elements are mobilized in the "primary environment" by a process called "hypogene mobility" (or igneous activity) whereby the liquid magma formed deep under the earth at high temperature and pressure, crystallizes and distributes the elements among the various rock types in a definite pattern. In the course of primary dispersion, the elements get concentrated in certain type of geological concentrations, known as "ores". These ores fall under the following three main geochemical categories of the elements :

***(a)* Siderophile elements**, which are concentrated in iron deposits and in the iron-nickel core of the earth. Ex :- Fe, Co, Ni, Cr, Pt, etc.

***(b)* Chalcopile elements**, which are concentrated in sulphide deposits.

Ex :- Sb, As, Cu, Cd, Pb, Hg, etc. These elements, though rare in overall abundance, are commercially important because they tend to accumulate in easily accessible sulphide deposits.

***(c)* Lithopile elements**, which are concentrated as silicate deposits,

Ex :- Ca, Mg, Cr, V, alkali metals, etc.

***(ii)* Secondary dispersion :** After the deposition of the rocks by the primary dispersion, the elements are then mobilized in the "secondary environment" by "supergene mobility" which occurs at relatively lower temperatures and pressures by the "weathering" process. This mobilization is strongly influenced by Eh, pH and the stability of the minerals. The factors which ultimately bring about the breakdown of the minerals during the weathering process are mechanical, chemical and biological in nature.

***(iii)* Tertiary dispersion :** This is caused by human activities in which the elements are re-distributed in such a manner that some of them pose pollution problems. In fact, what man is really doing is extracting metals that are present in high concentrations in the form of ores in some localised areas and redispersing them in the whole world in such a form and to such an extent that cause adverse effects on the environment. A classical example of this is the mining of the locally abundant but environmentally rare element, lead (Pb) and dispersing it throughout the world via heavy industry and as an anti-knocking additive in gasoline (as Tetraethyl lead) in internal combustion engines. This "tertiary dispersion" of trace elements by the anthropogenic activities is causing concern among scientists due to the potential environmental hazards.

## ESSENTIAL AND NON-ESSENTIAL TRACE ELEMENTS

A trace element is considered to be "*essential*" if it satisfies the following criteria :

(*i*) If it is present in all healthy tissues in all living things or within a botanical or a zoological family.

(*ii*) Its concentration is fairly constant among various species.

(*iii*) Its deficiency, depletion or withdrawal from the body induces reproducible structural and physiological abnormalities irrespective of the species studied.

(*iv*) The deficiency symptoms or abnormalities induced by the deficiency disappear or reverse on replenishment of the element under consideration.

(*v*) The abnormalities induced by the deficiency of the elements are always accompanied by pertinent, specific biochemical changes, and

(*vi*) These biochemical changes can be prevented or cured when the deficiency is prevented or cured.

Out of about 40 naturally occurring elements detected in living organisms, about 25 of them seem to be highly essential for the higher animals and humans. These include the following metals and non-metals:

Metals:- Na, K, Ca, Mg, V, Cr, Mn, Fe, Co, Cu, Mo, Zn and Sn.

Non-metals:- H, C, N, O, Si, P, S, Se, F, Cl, Br and I.

Some trace metals such as Pb, Cd and Hg and metalloids like As, Sb and Se are considered to be toxic. However, some of the metals listed as environmental hazards also seem to play some useful role in the development and growth of animals and humans. Elements like nickel, fluorine, bromine, arsenic, vanadium, cadmium, barium and strontium are considered to be possibly essential on the basis of suggestive but not completely convincing evidence.

About 25 trace elements which do not meet the above six exacting criteria described above occur more or less constantly in highly variable concentrations in living tissues. These include Al, Sb, Hg, Cd, Ge, V, Si, Rb, Ag, Au, Pb, Bi, Ti etc. These elements are believed to be acquired by the animal body as environmental contaminants due to contact of the organism with its environment. "Skewed" (*i.e.* log-normal) distribution patterns have been reported for the concentrations of these elements in human organs, whereas the essential elements have a "normal distribution" pattern. In fact, the shape of the distribution curve for a trace element in tissue has been proposed as a method to determine whether an element is essential or not. Elements with "normal distribution" in plants and animals are considered to be essential, whereas those elements which tend to be distributed "abnormally" are considered as non-essential elements.

The term "normal distribution" implies that if a histogram is plotted of the frequency of occurrence of each elemental abundance value, the plot will be a perfectly symmetrical Gaussian curve. In the case of "log-normal distribution" the data plot will be asymmetric with a positive or negative "skew". If the logarithm of the concentration data is taken, the resultant histogram will again follow a symmetrical Gaussian curve. This has been observed both for plants as well as humans.

The reason for the "normal distribution" of essential elements is that the concentrations of these elements in organisms are "internally" controlled and tend to be constant irrespective of the amount present in the environment. On the contrary, non-essential elements tend to be "externally" controlled, and hence the organisms accumulate the element to an extent proportional to its concentration in the environment.

The tendency of organisms to accumulate non-essential elements to an unlimited extent is particularly serious where pollution is involved, because toxic levels can soon be attained. For this reason, excesses of the non-essential elements tend to be more serious than excesses of essential elements. However, even if the essential elements are supplied to the organism in too great an excess, the regulatory mechanism can break down and toxicity symptoms or even death can result.

The trace elements are further classified into a sub-group known as "toxic" elements. This term may be justified for a small number of elements such as As, Pb, Cd and Hg, the biological significance of which is at present largely confined to their toxic properties at relatively low concentrations. However, this classification has little general application because all the trace elements are toxic if ingested at sufficiently high levels and for a long enough period. In fact, every element has a whole spectrum of actions, depending upon the dose and the nutritional state of the animal or system in respect to the element in question. The dose response curve for each of these elements shows two maxima. The first part of the curve, which shows an increasing effect with increasing concentration until a plateau is reached, expresses the biological action of the element. The plateau expresses optimal supplementation and normal function. The width of the plateau is determined by the homeostatic capacity of the animal or system. With further increase in the dose, the element enters a phase of irritation and stimulation of some function, expressing what may be called its *pharmacological* action. In this phase, the element acts as a drug independent of a deficiency state. If the dose is still further increased, the signs of toxicity appear and even death may result, thus expressing the toxic action of the element. The margin between the phases of biological and toxic actions differs widely among different trace elements.

## PHYSIOLOGICAL ROLE OF TRACE METALS

The physiological role of some essential trace metals is shown in Table 5.1, from which it is clear that these have a specific function in acting as enzyme activators. In fact, it is this function that carries with it the greatest danger to the well-being of living organisms, because other non-essential elements such as Cd, Ni or Ag can replace an essential element and cause toxicity symptoms or even death in the organism. Further, it is also possible for some essential elements present in excess to displace other essential elements from their rightful function and thus producing toxic symptoms.

**Table 5.1. Physiological Role of some Essential Trace Metals in Plants and Animals**

| *Element* | *Physiological function* |
|---|---|
| Zinc | Activator for several mammalian enzymes |
| Copper | Activator for several enzymes |
| Chromium | Phosphoglytamase |
| Cobalt | Methymalonyl GA mutase in livers, constituent of Vitamin $B_{12}$ |
| Manganese | Activator for pyruvate carboxylase and arginase in liver |
| Molybdenum | Found in metalloflavin enzymes and in xanthine oxidase in liver |

Interactions between essential elements which induce nutritional imbalances can take place at any of the following sites :

(*a*) In the soil, thus preventing the uptake by the plants.

(*b*) At the site of the metabolic action

(*c*) At some other site within the body by promoting release from storage, combining together to form a non-metabolizable structure, or by altering the binding sites carrying proteins.

(*d*) At the gastro-intestinal tracts of animals.

(*e*) At the excretory pathway by promoting abnormal (*i.e.*, excessive or low) excretion.

Trace elements find their way into humans either by direct absorption via the air or drinking water, or via the food chain. An indispensable link in the food chain is the terrestrial or marine plant life from which humans receive their quota of trace elements directly, or indirectly by feeding upon herbivorous animals which depend on plants for their nutrition. Plants absorb trace elements either via the root system or by foliar absorption.

Wherever trace elements have abnormally low- or high-level concentrations, adverse effects on animals and plants in that area can be identified, and the region is said to be a "*biogeochemical province*". Seleneferous soils of the western United States and iodine-deficiency-areas of Switzerland etc., provide convincing examples. Sometimes, if a particular element is present in excess, this may result in reduced uptake of other essential elements. A striking example of this is the prevalence of deficiency diseases such as hypocuprosis in cattle living in areas with a high molybdenum content in the soil.

Uptake of trace elements via the roots of plants is generally the more important pathway of absorption and must be appreciated in order to understand the effects of pollutants upon the environment. The uptake of trace elements by plants is illustrated by the so called "*biogeochemical cycle*". The extent of the absorption of trace elements by plants mostly depends upon the "*ionic potential*" of the element concerned. The ionic potential is given by the ratio of the charge and the ionic radius (that is, Z/r). Elements of low ionic potential form easily soluble cations such as $Na^+$ and $Ca^{2+}$. On the other hand, elements of high ionic potentials, such as $P^{5-}$ and $N^{5-}$, form soluble anions. Elements with intermediate ionic potentials such as $Al^{3+}$ and $Fe^{3+}$ form insoluble hydrolyzates.

The ionic potential values also help in understanding the probable effects of these pollutants upon the environment. For instance, Be with an intermediate ionic potential of 5.9 has a tendency to get precipitated whereas Cd having a lower ionic potential of 2.0 has a greater tendency to migrate. Cd is also accumulated by vegetation to a greater extent and is more likely to be leached from industrial wastes. Thus, the ionic potential values help in predicting the migration of a pollutant in the environment and assessing its probable danger to humans.

## TRACE ELEMENTS IN THE MARINE ENVIRONMENT

Unlike many terrestrial organisms, many marine organisms can concentrate pollutants to an alarming degree. The process is called bio-accumulation. For example, the levels of Cu and Cd in shell-fish in New Zealand were respectively 3000 and 2 million times higher in scallops (pectin novae-Zelandiae), and 14,000 and 30,000 times higher in oysters (ostrea sinuata), than in environmental sea-water. Enrichment coefficients of the order of $> 10^5$ have been documented for trace elements in the marine biosphere. Further, the stabilities of the trace metals with organic ligands and also their enrichment coefficients increase with increasing basicity of divalent metal ions in the following order.

$$Cu > Ni > Pb > Co > Zn > Fe > Cd > Mn > Mg > Ca > Sr > Ba > Ra$$

This order is particularly disturbing from the environmental pollution point of view because the most strongly absorbed ions such as Ni, Pb and Cd are also among the most toxic to living organisms. The situation is all the more difficult because the organisms where bio-accumulation was detected were also the lower members of the food chain. Progress through food chain successively increases the concentrations of some elements such as Hg to such an extent that the "natural" levels in some commercial fish are close to, or even exceed sometimes, the lowest levels now set by the Public Health Authorities in many countries. It is thus obvious that even a small additional "pollution" component can pose a public health hazard under some circumstances. Such situation has already reached for Hg and may soon reach for Cd too.

## HEAVY METALS

Out of the 105 elements discovered and confirmed so far, over 65 are metals. Owing to their high thermal and electrical conductivity, high density, high melting and boiling points, malleability, ductility and other distinctly useful properties, metals find extensive use in human civilization. According to a rough estimate, 0.5 million tonnes of Zn and 310 million tonnes of Cu have been mined so far, used for domestic and industrial purposes, and thus dispersed into the biosphere.

From the point of view of environmental pollution, metals may be broadly classified into the following three categories :

(1) Non-toxic but accessible

(2) Toxic but non-accessible

(3) Toxic and accessible.

It is the third category of potentially toxic and relatively accessible metals that have attracted more attention from the points of view of environmental pollution and public health. These toxic metals occur in very small quantities in the earth's crust (less than 1000 ppm) and hence are called "*trace metals*". These are further arbitrary sub-divided on the basis of their densities : those having densities below 5 g/cm$^3$ are called "light metals" and those with densities above 5 g/cm$^3$ are designated as "heavy metals". Thus, metals like Hg, Pb, Cu, Cd, Zn, Ni, etc. are generally known as "*toxic heavy metals*". Surprisingly, even metalloids like As and Sb are also generally considered under this category, thus making it appear as if the term "heavy metals" is a misnomer. Although the toxic heavy metals are present in the earth's crust in trace levels, anthropogenic activities such as industrial processing and use of metals, alloys and metallic compounds disperse them into the environment, thus adding up to their natural background levels. Table 5.2 shows the various heavy metals found in the effluents from some major industries.

**Table 5.2. Heavy metals found in the effluents from some major industries**

| *Industry* | *Heavy metals present in the effluents* |
|---|---|
| Chlor-alkali | Cd, Cr, Cu, Pb, Zn, Hg, Se |
| Electroplating | Cu, Cr, Zn, Ni |
| Leather | Cr |
| Batteries | Pb, Cd, Hg |
| Paints & Dyes | Cd, Cr, Cu, Pb, Hg, Se |
| Textiles | Cr |
| Paper and pulp | Hg |
| Petroleum refining | Zn, Cd, Cu, Cr, Pb |
| Fertilizers | As, Cd, Cr, Cu, Pb, Mn, Hg, Zn |
| Motor Vehicles | Cd, Cr, Pb, Hg, Se, Zn |
| Pharmaceuticals | As, Hg |
| Mining and Metallurgy | As, Cd, Cr, Cu, Hg, Se, Zn |
| Explosives | As, Hg, Pb, Cu |
| Pesticides | As, Hg, Se, Pb. |

As stated earlier, some of the heavy metals in traces are considered to be essential for plant and animal nutrition, and they serve some useful biological function. Thus, Cu, Co, Mn, Mo, Se and Zn are essential to both plants and animals. Minute traces of As, Cr, Ni and Sn have been found to be essential for animals but not for plants. Although some essential physiological roles have been inferred for As, Cd and Pb recently, Hg and Ag have not yet been confirmed to be essential both for plants and animals since they do not seem to serve any useful biological function. However, all the heavy metals, when present in excessive quantities have been found to be toxic for animal and vegetable life. In fact, the toxicity of all the essential trace elements follow the following general trend: under-supply leads to deficiency, optimum supply helps in healthy growth, and over-supply leads to toxicity and even death to the organisms.

The adverse effects of heavy metal toxicity in biological systems may result from the following reasons :

(*a*) interaction of the metal with protein leading to denaturation
(*b*) interaction with DNA leading to mutation
(*c*) effect on cell membranes
(*d*) effect on regulatory enzymes.

These adverse effects in mammals may manifest in the following disorders :

(*a*) Retardation of growth
(*b*) Decrease in longevity
(*c*) Detrimental changes in reproductive cycle leading to mortality of the offspring
(*d*) Morbidity
(*e*) Pathological changes
(*f*) Symptoms of chronic disease
(*g*) Formation of tumors.

The extent of heavy metal toxicity in mammalian systems is influenced by the following factors:

(*a*) Electrochemical character of the metal

(*b*) State of oxidation of the metal

(*c*) The solubility, stability and reactivity of the heavy metal or its compound or its complex under the existing conditions.

(*d*) The rate of transport of the heavy metal compound in blood and its distribution and retention in the body tissues

(*e*) The ability of the heavy metal to chelate with the various ligands in the body tissues and the stability of the metal chelates thus formed

(*f*) The rate of absorption of the metal in the body as against the rate of clearance

(*g*) The efficiency of the enzymatic and homeostatic mechanism which controls the absorption, distribution, retention and excretion of the heavy metal ions or compounds or complexes.

Traces of Cu, Zn, Mn, Cr, Mo and Co have been found to be highly essential for human for human beings. These metals have been detected in different body organs but relatively higher levels of these metals were found in the liver. As stated earlier, higher levels of even the essential metals may lead to toxicity. The toxicity due to the metals and their compounds is mainly determined by (*i*) the delivery of the metal to the cell to attain a critical concentration level at the site of action, and (*ii*) the cellular biochemical defence mechanisms. Although the toxic action of different metals may be different, most of them involve binding to the metabolically-active groups such as amino–, imino–, sulphydryl—, carboxyl–, phenolic– or phosphoryl groups. The toxicity of a metal is mainly determined by its solubility, stability, physical form at the site of its action, biological reactivity and the presence or otherwise of a homeostatic mechanism. Among the metals studied, Cr, Ni, Cd and Be have been found to be potentially *carcinogenic* (*i.e.*, causing cancer). Be, Cr, radionuclides such as $^{90}Sr$ and $^{239}Pu$ have been found to be potentially *mutagenic* (*i.e.*, causing chromosomal aberrations and thus altering generic characteristics of suceptible cells). Further, Cd, Cu, Pb, Hg, Mo, Ni and Se in excessive amounts have been found to be potentially *teratogenic* (*i.e.*, embryotoxic, inducing anatomical birth defects) on the basis of studies on test animals. Some arsenic compounds have been found to be potentially carcinogenic, mutagenic as well as teratogenic.

The heavy metal pollutants, like some other pollutants, on "*acute*" exposure (*i.e.*, a single but relatively high dose of the toxicant) or "*chronic*" exposure (*i.e.*, repeated long term doses of the toxicant) severely affect different body organs. Some of the body organs are affected more severely in a poisoning situation by different chemical pollutants. The "target organs" affected by different heavy metal pollutants are given in Table 5.3 below:

**Table 5.3. The target organs affected by heavy metal pollutants**

| *Heavy Metal Pollutant* | *Target Organ* |
|---|---|
| (As), Hg, Mo, Se | Liver |
| (As), Cd, Hg, Pb | Blood |
| (As), Pb, Hg | Brain |
| (As), Cd, Hg | Lungs |
| (As), Pb, Hg, Cd | Kidney |
| Cd, Se | Bones and teeth |

Excepting in case of occupational and accidental exposures, we are constantly exposed in nature to low levels of different trace metal and other pollutants.

For evaluating the toxic effects of any pollutant and or comparing the information regarding the toxicity of different chemicals, a measure called $LD_{50}$ is generally used. $LD_{50}$ represents the median lethal dose of a chemical that kills 50% of the population of experimental animals. These values are specific to the species studied, duration of exposure, mode of exposure and generally under certain dietary and environmental conditions. Obviously, the smaller the $LD_{50}$ values, the more toxic the chemical is. The $LD_{50}$ values are not extrapolated from the experimental animals to calculate the $LD_{50}$ values for humans. These tests are mainly meant to identify the relative toxicity of a chemical and to predict the mode of attack and the specific toxic effects produced by a chemical.

**Physiological role:** Heavy metals form stable coordination complexes with a variety of ligands. They tend to form more stable complexes with sulphur and nitrogen ligands than with those of oxygen. Although most of the heavy metals are known to exert biological effects through combination with sulfhydryl groups, they can also combine with other groups, especially in chelate configurations. For example, Hg and Ag can combine with amino-, imidazole, phosphate and carboxylate and sulfhydryl groups. Pb can combine with carboxylate and phosphate groups. The biological properties must be sought in the stability, steriochemistry and magnetic susceptibility.

Several heavy metals exert toxic effects on rapidly proliferating tissues such as the gastrointestinal mucosa and bone marrow and on highly specialized cells such as neurons and renal tubular cells.

Heavy metal ions such as $Pb^{2}$, $Hg^{2+}$, and $Cd^{2+}$ act as enzyme inhibitors. They attack ligands like -SH and $SCH_3$ in cysteine and methionine amino acids in the enzyme structure as follows :

$$\text{Enzyme}\left\langle\begin{matrix}SH\\ SH\end{matrix}\right. + Hg^{2+} \longrightarrow \text{Enzyme}\left\langle\begin{matrix}S\\ S\end{matrix}\right\rangle Hg + 2H^{+}$$

Heavy metal ions can also replace the right metal ion of similar charge and size from metalloenzymes and inhibit its normal activity or induce toxicity. For example $Cd^{2+}$ can displace $Zn^{2+}$ in some metalloenzymes and lead to cadmium toxicity. $Cd^{2+}$ can inhibit activity of enzymes such as amylase, carbonic anhydrase, adenosine triphosphatase, alcohol dehydrogenase and carboxy peptidase. $Pb^{2+}$ can inhibit activity of enzymes like carbonic anhydrase, cytochrome oxidase, alkaline phosphatase, adenosine triphosphatase and some important enzymes in the synthesis of "heme".

A lot of more is yet to be learned about the physiological role of heavy metals. The future progress obviously depends upon further advances in analytical instrumentation and their judicious application with due regard to the characteristics of the analyte and the chemistry of the various biological matrices.

The sources, uses, environmental levels, ecological, biochemical and toxicological effects of some typical trace elements are reviewed in the following sections.

## MERCURY (Hg)

**Introduction :** Mercury is a non-essential trace metal constituting 0.00003% by weight of the earth's crust. It is extracted mainly from its sulfide ore, cinnabar (Hgs). Mercury, both in the metallic form (quicksilver) and as the sulfide (Hgs) played prominent role in therapeutics, alchemy and folklore. Mercury chloride or calomel enjoyed popularity as a laxative. Mercuric oxide, salicylate and chloride salts were used as antiseptics. Red mercuric sulfide, mercuric benzoate, mercurous acetate and mercurous iodide were used for treating syphillis. Organic mercurials such as metaphen and mercurochrome exert a weak bacteriostatic action and are used as local antiseptics. Mercurin is still used as diuretic to a limited extent. Mercury is now mostly used as filling materials for dental cavities as silver amalgam. Mercury was used in the past in felt-hat manufacture and fur processing but its use for this purpose is minimal.

**Industrial uses and pollution sources :** Mercury finds extensive use in chlor-alkali industry (for the manufacture of $Cl_2$ and NaOH), electrical and electronic industries (in manufacture of mercury vapour lamps and fluorescent tubes, batteries, electric switch gears, etc.) plastics industry (in the manufacture of vinyl chloride), paper and pulp industry and pharmaceutical industries. Hence the effluence from these industries pose environmental hazard. Organomercurials used as fungicides for seed dressing in agriculture are also a widespread source of pollution. Combustion of fossil fuels is the main source of air pollution by mercury and about 3000 tonnes of mercury is annually released into the atmosphere by this source. The mercury contents of bituminous coals, anthracite coals, crude oil and residual tars are in the range, 1–25 ppb; 1100–2700 ppb, 1900–21,000 ppb and 520,000 ppb respectively. Natural weathering processes and submarine vulcanism release about 5,000 tonnes of mercury annually into the oceans. Although natural agencies are predominant suppliers of mercury in the environment, the proportion contributed by man-made sources is also increasing rapidly. It is estimated that the mercury electrodes used in chlor alkali industry all over the world themselves disperse 1 million pounds of mercury into the environment.

Mercury poisoning has been prominent at times among goldsmiths and mirror makers and the expression "mad as a hatter" is derived from the symptoms shown by workers engaged in the treatment of furs with mercuric nitrate. Industrial exposure to mercury is now relatively well controlled. The present concern mostly centres upon the biological consequences of the dispersion of mercury into the environment by the anthropogenic activities such as burning of coal and oil, use of mercury containing ointments, dental amalgams, other medical formulations and fingerprint powders. Organic mercury fumigants and fungicides in industry and agriculture and the consequent increase in mercury concentrations in the marine life of lakes, rivers and estuaries as well as in the plant and animal products in the local environment have also become a matter of considerable concern in recent times.

**Environmental levels :** Information regarding environmental levels of mercury came only after the discovery of flameless atomic absorption analysis technique. On the basis of mean concentrations of mercury in cereals, vegetables, pulses, milk and fish, the average intake of mercury through the normal diet of an adult is estimated to be 7.3 μg/day. The mercury concentration in ambient air in some industrial areas in our country is found to be 0.7 μg/$m^3$. The mean mercury concentration in water varies from 2 μg to 33 μg/L. The US EPA limits of mercury in drinking water is 2 μg/L.

**Biochemical effects, Toxicity and Toxicology :** The extreme toxicity of mercury, though suspected some centuries back, was fully appreciated in the year 1953, when 52 persons living in fishing villages along Minamata Bay, Japan died of a mysterious disease. Investigations revealed that the victims had eaten shellfish contaminated with mercury containing effluent from a nearby plastic factory, Shin-Nihon Chisso Hiryo Co., operating since 1949. The unit was producing vinyl chloride and acetaldehyde by the catalytic conversion of acetylene using $HgCl_2$ as catalyst. The methyl mercury compounds present in the effluent wastes discharged into the Minemata Bay were gradually bioconcentrated by fish and shell-fish in the bay, which provided the sea-food for many families along its shores. The ailment became known as "Minamata disease", characterised by peculiar neurological disorders and was ultimately attributed to the organomercurials, in the year 1958. Another major environmental crisis relating to mercury poisoning came to be known in the year 1972 when 450 Iraqi villagers died after consuming grain which was treated with a mercury containing pesticide. Emergence of such episodes and the discovery of sensitive analytical techniques helped in further research on mercury poisoning.

Mercuric ions act as potent enzyme inhibitors, protein precipitants and corrosive agents. Mercury has great affinity for sulfhydryl groups, although it combines with phosphoryl, carboxyl, amide, and amino-groups also. The biochemical toxicology of mercury depends on the chemical form and entrance route into the body. Elemental mercury, being lipid-soluble, can be absorbed through the intact skin. However, negligible amounts of elemental mercury are absorbed through the gastric mucosa. Hence, elemental mercury swallowed during accidental breakage of thermometer in one's mouth is of lesser concern than swallowing the pieces of glass concurrently, because elemental mercury is fairly inert and is excreted without any serious damage.

Mercury has a fairly high vapour pressure (0.001 mm at 18°C and rises markedly to 0.27 mm at 100°C) and the vapour is quite toxic if inhaled. Vapour can diffuse through the lungs into the blood and then to brain resulting in severe damage. This is the reason why mercury should be handled only in well-ventilated areas and any spilled Hg should be cleaned promptly. Mercury vapour, on inhalation, enters the brain through the blood stream and causes severe damage to the central nervous system.

Mercurous ions react with chloride ions forming insoluble $Hg_2Cl_2$. Since our stomach contains fairly high concentration of chloride, mercurous compounds do not pose any hazard. Inorganic mercuric compounds mainly attack liver and kidney. Mercuric chloride is corrosive and, when ingested, precipitates proteins of the mucous membrane causing ashen appearance of the mouth, pharynx and gastric mucosa. The affected areas cause intense pain. Profuse vomiting, diarrhoea and shock may also be experienced. $Hg^{2+}$ ions, on account of their high affinity for sulfur atoms, easily attach themselves to the sulfur-containing amino acids of proteins. $Hg^{2+}$ ions can also attach to sulphydryl groups present in haemoglobin and serum albumin. $Hg^{2+}$ ions replace the hydrogen atom from the sulfhydryl group to form mercaptides of the type *x*-Hg-SR and Hg $(SR)_2$ where *x* is an electronegative radical and R is a protein. Organic mercurials form mercaptides of the type R Hg-SR'. Mercurials even in low concentrations are capable of inactivating sulfhydryl enzymes and thus interfering with cellular metabolism and function. The affinity of mercury for thiols provides the basis for treatment of mercury poisoning by reagents such as dimercaprol and penicillamine.

Organic mercurials such as methyl mercury ($CH_3$ $Hg^+$) are the most toxic substances because of their solubility in fat, the lipid fraction of the membrane and brain tissue. The covalent Hg-C bond is quite stable and the alkyl mercury is retained in the cells for prolonged periods of time. Further, the $CH_3$ $Hg^+$ can pass through the placental barrier and enter the foetal tissues. This can cause irreversible damage to the central nervous system in the babies born to mothers exposed to poisoning by $CH_3$ $Hg^+$. The symptoms of mercury poisoning may set in at blood levels of as low as 0.5 ppm of $CH_3$ $Hg^+$ and may even lead to chromosomal segregation, chromosome disruption and inhibition of cell division. A concentration of 6 ppm in brain cells can cause irreversible brain damage.

**Environmental fate :** Although mercury may enter the hydrosphere initially in the form of $Hg^{2+}$, it is readily scavenged by organic matter and converted by anaerobic bacteria such as "clostridium cochlearium" (which is facilitated by Co (III)-containing vitamin $B_{12}$ coenzyme) to produce more toxic organic mercurials such as methyl mercury ($CH_3$ $Hg^+$) and dimethylmercury ($CH_3$–Hg–$CH_3$). Dimethyl mercury is stable at high pH values, but dissociates to $CH_3Hg^+$ in the low PH range which normally exists in the bottom muds of streams and lakes. Since methyl mercury is soluble, it is readily incorporated into organisms in the aquatic environment and ultimately finds its way into the higher members of the food chain such as man. Dimethyl mercury is formed from mono-methyl mercury under alkaline conditions. Dimethyl mercury being relatively volatile can contaminate the air. The biological half-life of methyl mercury is about 74 days as compared to 5 days for that of inorganic forms. A daily exposure to 1 mg of mercury vapour per $m^3$ of air can be fatal for humans. Mercury is teratogenic and can induce abortions and embryotoxicity.

**Control of mercury pollution :** Most of the environmental mercury is derived from natural sources such as weathering processes and terrestrial and submarine volcanic activity. Hence mercury pollution is a fact of life and we will have to live with it. However, what we can do is to restrict the mercury pollution from man-made sources. Although it may be possible to substitute mercury with other materials in some industries, it will continue to be used in the fluorescent lamp industry, in chloralkali plants, in dentistry and in electronic industry. Stringent legislations are being enforced in many countries to control mercury pollution from man-made sources and the results are encouraging. Other measures such as banning of alkyl mercury pesticides, restricted use of other mercurial pesticides in only selected areas and encouraging new technology to replace the use of mercury in chlor-alkali plants go a long way in controlling mercury pollution. Decontamination of mercury rich bottom sediments of rivers and lakes by covering with inert or adsorbent materials is also being experimented. However, mankind will never be able to control mercury pollution from natural sources with its serious problems in the fishing industry. In the world suffering from hunger, it is impossible to neglect or prohibit tuna and shark as food sources merely because their present mercury levels are currently higher than the current safety limits.

**Treatment of Mercury Poisoning :** Acute mercury poisoning usually results from the oral ingestion of highly dissociated inorganic mercury compounds or inhalation of vapours of elemental mercury, or from organic mercurials, or even from mercurial ointments applied topically. Treatment consists of introducing in the stomach, a source of sulfhydryl-rich protein such as milk or raw eggs and then performing copious lavage. Sodium formaldehyde sulfoxide, if available, provides excellent local antidote. As soon as mercury intoxication is suspected, intra-muscular injection of dimercaprol or penicillamine is given to chelate the mercury and accelerate its excretion. In the absence of the metal-chelating agents, isotonic NaCl solution is infused in amounts as much as 10 liters per day to produce a copious diuresis so as to protect the kidney from high concentrations of mercury.

Chronic mercury poisoning can be treated by removing the patient from the exposure area, supportive thereby, and the use of appropriate metal chelating agents such as dimercaprol, penicillamine or N-Acetyl-D, L-pencillamine.

$$\begin{array}{ccccc} & H & H & H & \\ & | & | & | & \\ H- & C- & C- & C- & H \\ & | & | & | & \\ & SH & SH & OH & \end{array}$$

Dimercaprol

$$\begin{array}{cccc} & CH_3 & & \\ & | & & \\ H_3C- & C- & CH- & COOH \\ & | & | & \\ & SH & NH_2 & \end{array}$$

Penicillamine

## CADMIUM (Cd)

Cadmium as a distinct element was discovered by Stromeyer of Germany in 1817. Owing to the similarity in atomic structure and chemical behaviour, Cd and Zn often occur together in nature. However, while Zn is an essential element, Cd is considered as a toxic element. They are found together in sulfide deposits. Igneous rocks contain about 0.03 ppm Cd and 80 ppm Zn on an average. These elements are mobilized in the weathering process and find their way into oceans, where their respective concentrations reach to about 0.1 ppb of Cd and 20 ppb of Ni. Soils contain about 4.5 ppm of Cd because of its biogenic enrichment in humid materials. Natural Cd levels in air are about 0.002 $\mu g/m^3$ although values as high as 0.3 $\mu g/m^3$ may be found near zinc smelters.

**Industrial uses and pollution sources:** Cadmium is widely used in various alloy formulations, electroplating, paint-pigments, plastics and silver-cadmium batteries. Cadmium stearates are used as stabilizers in the manufacture of PVC (to stabilize the double bonds in the polymer by displacing the labile allyl chlorine atoms). Cadmium is a constituent of fusible alloys and aluminium solder. It is used in varied applications such as process engraving, amalgam in dentistry and nickel-cadmium storage batteries. Cadmium acetate is employed for producing iridescent effects on porcelain and potteries. Cadmium bromide and iodide are used in photography, process engraving and lithography. Cadmium oxide in used in glass manufacture, in ceramic glazes, in electroplating and in the manufacture of ceramic alloys. Cadmium selenide is used in photoelectric cells, photoconductors, rectifiers and in phosphors. Cadmium sulphide, which is known as "cadmium yellow" or "juane brilliant", being a pigment fast to light, is used in colouring glass, soaps, textiles, paper, rubber, printing inks, ceramic glazes and fireworks. Cadmium oxide and iodide were used to a limited extent as anthelminthics in swine and poultry. Cadmium salts were used in the early period of 1900 sporadically in treating human syphilis and tuberculosis.

Over 1 million kg of Cd are released into the atmosphere every year from smelters and factories processing cadmium which represents about 45% of the total cadmium pollution. About 52% of total cadmium pollution comes from the incineration or disposal of cadmium-bearing products such as automobile tyres, motor oils, fungicides, plastics and coal. Atmospheric pollution by heating or burning cadmium containing materials is due to cadmium oxide. The tendency of Cd to be distributed into the atmosphere is partly due to its relatively high volatility. The melting point of pure Cd is only 765°C as compared to that of 907°C for Zn and 1744°C for Pb. Discharge of Cd into natural waters is mainly from electroplating industry and nickel-cadmium battery industry. Fertilizers containing Cd (about 8 ppm), and fungicides containing Cd are other sources of cadmium pollution.

## BIOCHEMICAL EFFECTS, TOXICOLOGY AND TOXICITY

Cadmium acts as inhibitor of sulphydryl enzymes. It has also got affinity for other ligands in cells, such as hydroxyl, corboxyl, phosphatyl, cysteinyl and histidyl sidechains of proteins, purines and porphyrin. It can disrupt pathways of oxidative phosphorylation.

Cadmium interacts or competes with other metals such as Cu, Fe and Zn and induces the deficiency symptoms of these essential metals.

Cadmium is highly toxic because of the absence of homeostatic control for this metal in the human body. The general population is exposed to cadmium via ingestion or inhalation. The workers in mining and smelting industries may be occupationally exposed to higher levels of Cd in the body by steady accumulation over a long period. About 1 to 2% of ingested and about 11% of inhaled cadmium is retained in the body and in mammals, the absorption increases which there is deficiency of Ca in the diet. About one-third of the absorbed cadmium is stored in the kidney, which is the target organ. When excessive amounts of $Cd^{2+}$ are ingested, it replaces $Zn^{2+}$ at key enzyme sites and induce metabolic disorders.

The symptoms of cadmium toxicity produced by enzymatic inhibition include hypertension, respiratory disorders, damage to kidney and liver, aminoaciduria (urinary excretion of aminoacids), hypecalciurea (urinary excretion of excessive Ca) glucosuria (excretion of blood sugar in the urine) proteinuria (urinary excretion of proteins), osteoporosis (decalcification of the skeleton), formation of kidney stones, etc. Carcinogenic and teratogenic effects have also been observed in epidemiological studies on animals.

A disease specifically associated with cadmium poisoning was recognised in Japan. People residing along the banks of Janstsu river in Northern Japan were consuming the rice which was contaminated with cadmium and they were ingesting 100 to 1000 μg of Cd every day. The pollution was due to a nearby mining complex and by the end of 1965, about 100 deaths were reported. The victims accumulated about 500 to 600 mg of Cd in their body over several years of consuming the contaminated rice as their staple food and finally fell fatal due to the so called itai itai (or ouch-ouch) disease. The disease was characterized by kidney malfunction, drop in the phosphate level of the blood serum, loss of minerals from the bones and osteomalacea (which is characterized by bone fracture and intense pain) or severe osteoporosis. Concentrations of cadmium in the rice grown in polluted areas were found to range from 0.6 to 1 μg per gram.

## ENVIRONMENTAL LEVELS AND ECOLOGICAL EFFECTS

Natural cadmium levels in air are in the range of 0.002 to 0.048 $\mu g/m^3$ but near smelters, the value may be as high as 0.3 $\mu g/m^3$. The safety level of Cd in drinking water in many countries is 0.01 ppm but many surface waters show higher Cd levels. The maximum permissible concentration of Cd in water is 5 ng/g as per the W.H.O. guidelines.

Cadmium in blood reflects current exposure level but cadmium in urine may reflect body burden when the exposure is low. Normal Cd levels are less than 1 μg % in blood.

The main problem with Cd in human nutrition is that the body does not completely excrete whatever Cd is absorbed. The estimated daily intake of Cd in different countries ranges from 25-60 μg/day while the tolerable daily intake (TDI) of cadmium is about 57 to 72 μg/day. Normal cadmium content of rice is about 29 ppb. Cadmium content of sea water varies from 0.05 to 0.12 μg/g.

Cadmium in water at 10 ppm level can kill fishes in one day while at 2 ppm level they will be killed in 10 days. Hardness and salinity of water provides some degree of protection. Amongst the aquatic invertebrates, crustaceans are more sensitive to Cd toxicity as compared to molluscs and polychaetes.

Plants exposed to Cd at toxic levels exhibit "chlorosis" and reduced growth. Cd in agricultural soils is mainly derived from the Cd present in fertilizers and sewage sludge applied in the crop fields. When grown in cadmium contaminated soils, the roots of corn, oats, soybeans, tomatoes and alphalpha accumulate highest levels of Cd, whereas the aerial parts of carrot, potato, lettuce and tobacco accumulate highest levels of Cd. Thus, water, food and smoking are the major sources of Cd for general population. Tobacco contains 1 ppm of Cd. Hence the tissues of smokers contain more Cd concentration than non-smokers. On an average, a smoker inhales 0.1 to 0.2 μg of Cd from each cigarette he smokes.

Cadmium may be methylated in the environment. Although the transient formation of methyl cadmium has been inferred under laboratory conditions, it has not been detected in sediments or biota, nor has it been isolated in the natural environment. Cd along with Pb and Hg are considered to be the three major trace elements having the greatest potential environmental hazard. Much greater efforts are needed to control them.

## LEAD (Pb)

**Introduction :** Lead is one of the seven metals of antiquity and was perhaps discovered by accidental dropping of galena (pbS) into a campfire. The average lead content of igneous rocks is about 15 ppm and hence it is categorised as a rare element. Fortunately it is concentrated as sulfide deposits in many parts of the world which is easily mined and smelted. In nature, lead occurs mainly as galena (Pbs), although carbonate and chloride ores also exist. Ancient people used lead for ornaments, dishes, trays, as a core for bronze statues, as sinkers for fishing nets, etc. Greeks and Romans used lead extensively for sweetening the bitter taste of the food cooked in bronze pots, storing olive oil, preparing wine, etc. Incidentally, it is believed now that one of the main reasons for the fall of the Roman empire was endemic lead poisoning due to over exposure to Pb. Sweetening of wine with lead salts caused an endemic disease called "lead colic".

**Industrial Uses and Pollution Sources :** Lead is used in building construction, preparation of various alloys with Cu, Sn and Sb for use as accumulator plates in storage batteries, paints, pigments and varnishes, cable coatings, ammunition, etc. Lead arsenate is used as pesticide. Lead borate is used in plastic industries. Lead is used for making lead pipes, containers for corrosive liquids, and lead chambers. White lead, red lead and litharge are used in pigments and paints. Lead is used for preparing alloys like solder, pewter (for making utensils), woods metal, Rose metal, Lipowitz alloys, Newton metal, etc. These fusible alloys are used for making soft solders, electric fuses, safety plugs for automatic water sprinklers, etc. Organic forms of lead *e.g.*, tetraalkyl leads are used as antiknocking additives to gasoline in internal combustion engines along with dichloroethane or dibromoethane as scavenging agents. In fact, the tetra alkyl lead accounts for 20% of lead used by mankind which accounts for the main environmental lead pollution problem. Street dusts and roadside soils on busy streets may contain Pb in the range of 1000 to 4000 mg $kg^{-1}$.

Apart from natural weathering processes, the main sources of lead pollution are mining and smelting of lead ores, emissions from automobile exhausts, effluents from storage battery industries, use of glazed earthenware-containers, lead pipes and lead pigments. Hair dyes, eye-paints, canned foods, sea foods, painted toys, wicks in decorative candles, coloured pages in magazines, putty on the window frames, pewter ware etc., are the important sources of lead exposure.

## BIOCHEMICAL EFFECTS, TOXICOLOGY AND TOXICITY

Lead is considered as a general protoplasmic poison which is cumulative, slow-acting and subtle. It produces a variety of symptoms. Similar to other heavy metals, it has an affinity for sulphur. Although Pb exerts much of its activity through sulphydryl inhibition, it also interacts with carboxyl and phosphoryl groups. The major biochemical effect of lead (Pb) is its interference with heme synthesis and leading to hematological damage. Pb inhibits several important enzymes involved in the overall process of heme synthesis.

Lead inactivates delta-aminolevulinic acid (ALA dehydrase) and thus obstructs its conversion to porphobilinogen (PBG), which is an important step in heme synthesis. It also impairs the activity of porphobilinogen decarboxylase.

$$HOOC-CH_2-CH_2-\overset{\overset{\displaystyle O}{\|}}{C}-\underset{\underset{\displaystyle NH_2}{|}}{C}-COOH$$

**(delta-aminolevulinic acid)**

$$HOOC-CH_2-C-C-CH_2-CH_2COOH$$

$$\begin{array}{c} \| \quad \backslash\backslash \\ C \quad C \\ H_2N-H_2C \quad N \quad H \end{array}$$

(Porphobilinogen)

These effects lead to the disruption of the synthesis of haemoglobin and other respiratory pigments such as cytochromes which require heme. Further Pb obstructs the utilization of oxygen and glucose for the life-sustaining energy production. The interference with normal metabolic functions starts when the blood-lead level reaches about 0.3 ppm. When the blood-lead level reaches about 0.8 ppm, symptoms of anaemia will be observed due to the deficiency of haemoglobin. Higher levels of Pb in the blood can cause kidney dysfunction (leading to Fanconi syndrome) and brain damage because it is toxic to the central and peripheral nervous system.

One of the most insidious effects of inorganic lead is its ability to replace Ca in bones and accumulate there as a reservoir for long-term release. This lead is subsequently remobilized along with phosphates from the bones and exert toxic effects when transported to soft tissues.

Air-borne lead of small particle size is readily absorbed through the lungs. On an average only about 10% of inorganic lead that is ingested is absorbed, and is distributed in the various tissues - liver, bones, kidneys and brain. About 90% of this lead is immobilised in bones. The bone-lead concentration goes on increasing and accumulating with age until it reaches toxic levels. Chronic exposure to lead causes weight loss, constipation and loss of teeth.

Organic lead, as tetra ethyl lead and tetramethyl lead is even more acutely poisonous (10 to 100 times) than inorganic lead. Tetraethyl lead (TEL) is non-polar and lipophilic and hence is less soluble in water than inorganic lead. Inorganic lead is mainly absorbed by food or inhalation of particular lead whereas TEL is absorbed by inhalation of the volatile compound or by its dermal entry into the body. Inorganic lead mostly causes gastrointestinal and hematological disturbances whereas tetraethyl lead selectively attacks the central nervous system. Inorganic lead can bind to complexing agents such as calcium-EDTA or to thiol groups but organic lead does not.

Organic lead compounds such as tetraethyl lead can penetrate the skin and absorb into the body tissues more rapidly as compared to inorganic lead compound. It is believed that the metabolism of tetraethyl lead involves sequential dealkylation process that forms triethyl lead, diethyl lead and then inorganic lead. It is the formation of triethyl lead from tetraethyl lead which accounts for the severe toxic effects of tetraethyl lead; further dealkylation renders this compound progressively more polar and less toxic. Brain is the target organ for tetraethyl lead. The organo-lead compounds, being lipophilic, and selectively localized in the central nervous systems and result in toxicity. The biochemical effects of tetraethyl lead include uncoupling of oxidative phosphorylation and inactivating and obstructing its conversion to porphobilinogen as explained earlier. Organo lead compounds are suspected to cause genetic modifications.

## ENVIRONMENTAL LEVELS AND ECOLOGICAL EFFECTS

The lead-levels of urban-dwellers may be as high as 200 mg. Owing to the continued use of leaded-gasoline, the blood-lead levels of population in major cities may vary between 2 to 35 µg/dl. The World Health Organisation suggests a provisional tolerance of 7 µg/kg body weight per day for adults. The degree of inorganic lead poisoning can be indicated by determining lead content in blood, hair and urine.

Leafy vegetables, potatoes and beans are likely to absorb more lead than fruiting crops like tomatoes, beats, etc. It is better not to grow carrots, turnips, beetroot and sprouts near highways where contamination by vehicular exhausts is more. The movements of lead into the vegetables grown in soil contaminated with lead can be controlled by adding large amounts of compost and maintaining the soil pH near 6 to 7.

Human activities have seriously upset the natural cycle of lead and urgent action is needed to prevent continuing and cumulative poisoning of the environment with lead. Improved plant control in smelting operations and improving the automobile engine design are suggested so that lead-free gasoline can be used and at the same time emission of other pollutants like CO in the exhaust should be controlled by using suitable catalytic control devices. Immediate attention is imperative for controlling Pb pollution because in no other case the "normal levels" in humans are so close to the lowest prescribed safety levels.

**Treatment of lead poisoning :** Treatment of lead poisoning consists in supportive measures and chelation therapy. A combination of intramuscular dimercaprol and calcium disodium edetate (Ca $Na_2$ EDTA) is used, sometimes followed by a course of oral penicillamine for long-term prophylactic chelation. In treating acute poisoning by alkyl lead compounds, only chelating agents are of no avail. Sedative, anti-convulsant and fluid and electrolyte therapy may be required.

## CHROMIUM (Cr)

**Introduction :** Until about forty years ago, interest in chromium was mainly confined to its toxic effects. However, today the element is considered to be essential for maintaining the normal glucose metabolism and glucose tolerance in human beings. Perhaps, Cr (II) acts as a co-factor with insulin at the cellular level through the formation of ternary complex between membrane sites, insulin and chromium. The chromium concentration in normal blood is found to be 0.5 to 5 $\mu g\ l^{-1}$, in urine, 5-10 $\mu g\ l^{-1}$ and in hair, over 500 ng $g^{-1}$. The chromium content in all tissues decreases with age.

Chromium occurs in a variety of foods in the form of a complex with nicotinic acid and possibly glycine, glutamic acid and cysteine, chromium in this complex form, termed "glucose tolerance factor" is absorbed-better than in inorganic form. Animal protein is the best and most reliable source of chromium. Both the trivalent and hexavalent forms are biologically important. Cr (VI) is far more toxic than the Cr (III).

Chromium is distributed in the earth's crust at about 100 ppm levels. Chromite is the important ore of chromium which contains about 68% of chromium oxide and 32% iron oxide.

**Industrial uses and pollution sources :** Chromium is used in the manufacture of stainless steel and other alloys. The trioxide is used in chrome-plating, copper stripping, photography and as a corrosion inhibitor. Chromium sulphate is used as a mordant in textile manufacture, in leather tanning, in manufacturing green varnishes, paints, inks, and glazes for porcelain. Chromic potassium sulphate and oxalate are used in colouring glass. Chromic acetate is used in dyeing and tanning. Chromium is also used in explosives and for preservation of wood. Wastes from all the above industrial units can be a source of water pollution from soluble chromate salts. Spent chrome liquors and wastes from tanneries contain chromium to the tune of 2900 to 4500 mg/L and 10-50 mg/L respectively. Textile mill wastes may contain 32 mg/L of chromium. Fertilizer materials are also a source of chromium pollution in water.

The presence of chromium in air is mostly due to burning of fossil fuels, which contributes about 1450 metric tonnes of Cr every year. Soil contamination by Cr results from the land disposal of sewage sludge, land disposal of by-products from ferrochrome and chrome-steel industries and from the use of phosphatic fertilizers.

## BIOCHEMICAL EFFECTS, TOXICOLOGY AND TOXICITY

Trivalent chromium is found to be essential to humans and animals. It is found in small quantities in ribonucleic acid (RNA) of some organisms. It plays a vital role in insulin metabolism as the Glucose Tolerance Factor (GTF). The supplementation of Cr (III) was found to help normalizing or improving glucose tolerance in diabetics, malnourished children and older people. May mammalian species tolerate about 100 times their normal body burden of Cr (III) without any apparent adverse effects. Chromium deficiency is characterised by impaired growth and longevity and by disturbances in glucose, lipid- and protein metabolism. Rats maintained on diets low in protein and Cr also develop a corneal lesion manifested as a pronounced opacity of the cornea and congestion of the iridal vessels. Chromium supplementation prevents the appearance of the lesion but does not cure the full developed defect.

Orally administered trivalent chromium is poorly absorbed to the extent of about 1% only, regardless of dose and of dietary chromium status. Cr (VI) is better absorbed than Cr (III). Absorbed anionic hexavalent chromium readily passes through the membrane of the red blood cells and gets bound to the globin fraction of the hemoglobin. Cationic trivalent chromium is unable to pass through this membrane and combines with the β-globulin fraction of the plasma proteins and, in physiological quantities, is transported to the tissues bound to siderophilin.

Absorbed or injected Cr is excreted mainly in the urine, with small amounts being lost in the bile and small intestine and possibly through the skin. Tissue uptake of Cr is quite rapid and the plasma is cleared of it within a few days. The chromium entering the tissues is distributed among the subcellular fractions in rather unusual proportions.

Chromium is mobilized from the body stores in response to glucose administration.

Trivalent chromium has a low order of toxicity. A wide margin of safety exists between the normal quantities of Cr (III) ingested and those likely to induce deleterious effects. Cr (III) showed that $LD_{50}$ value is about 1 mg per 100 g body weight. This is many times greater than the 0.05 to 0.10 μg Cr/100 g required to correct the impairment of glucose metabolism "in vivo".

Hexavalent chromium is found to be much more toxic than the trivalent chromium. Chronic exposure to chromate dust has been correlated with increased incidence of lung cancer. Oral administration of excessive levels (50 ppm) has been associated with liver and kidney damage and depression in experimental animals.

Exposure to hexavalent chromium causes allergic skin irritations, dermatitis, irritation to mucous membranes and conjunctiva and gastro-intestinal ulcers. Chrome holes *i.e.*, penetrating ulcers which occur around the fingernails, the surface exposed finger joints, eyelids and occasionally on forearms. Lesions on the nasal mucosa may lead to perforation of the septum.

A higher rate of bronchitis among chromate workers has been reported and a condition called "chronic chromate lung" has also been described. Alkali bichromates can be absorbed through skin lesions in sufficient quantities to cause renal damage. Hexavalent chromium may be carcinogenic, teratogenic and mutagenic.

Fortunately, the more toxic hexavalent chromium is reduced to less harmful Cr (III) due to the stomach acidity.

## ENVIRONMENTAL LEVELS AND ECOLOGICAL EFFECTS

Chromium concentrations in river and sea waters are about 0.04 ppm and 0.00005 ppm respectively. Chromium concentration in potable water supplies varies from 0.002 to 0.1 ppm and in rain water it is about 0.003 ppm. The U.S.E.P.A. suggests 0.05 ppm as the chromium limit for domestic water supply. The environmental standards are usually set on the basis of hexavalent chromium which is more toxic. The daily intake of Cr per person averages in the range 150 μg to 280 μg in different countries.

The toxicity of Cr to aquatic life is less as compared to Cd, Hg and Pb. Cr at 10 ppm levels in water is considered to be toxic to several species of algae. Fishes are less susceptible to the toxic effects of Cr than other aquatic species such as invertebrates. The rate of uptake is higher in young fish but the Cr body burdens decline with age due to rapid elimination.

## ZINC (Zn)

**Introduction :** Zinc and cadmium, being chalcophile elements tend to be found in sulfide deposits. Igneous rocks contain 80 ppm of Zn and 0.13 ppm Cd. During weathering process, these elements are mobilized easily and find their way into oceans, where their concentrations average about 20 ppb for Zn and 0.1 ppb for Cd. Soils contain about 50 ppm Zn. Since Zn is essential to plant nutrition, they contain relatively high concentration of Zn. For instance in plant ash its concentration can be as high as 1400 ppm. Zn forms 0.004% of the earth's crust. It occurs as sulphide, carbonate, oxide and silicate minerals in nature.

Although zinc was found to be essential in the nutrition of "Aspergillus niger" over a century ago, the essentiality of zinc for higher forms of plant life was established only in 1926. Since then, lot of original work on the function of Zn in the nutrition and growth of higher animals and Zn metabolism followed.

## INDUSTRIAL USES AND POLLUTION SOURCES

Zinc has many industrial applications, the most ancient being the manufacture of bronze and brass. Zn sheets are used for building purposes. Zn is used for galvanising iron and steel. It is used in construction materials, dry batteries, pigments and printing processes. Zinc salts, being astringent, mildly antiseptic and corrosive, have limited use in medicine. The world production of zinc was about 7 million tonnes as early as 1976. Smelting of zinc ores is the main source of pollution from zinc. Municipal refuse, automobiles (tyre wear, brake linings, fuel additives, etc.) and agricultural use of pesticides and fungicides containing $ZnSO_4$ are the additional sources of environmental pollution due to zinc.

## BIOCHEMICAL EFFECTS, TOXICOLOGY AND TOXICITY

The role of Zinc as an essential nutrient is well established now. A number of enzymes including aldolase, alkaline phosphatase, alcohol dehydrogenase, carboxy peptidase, carbonic anhydrase, lactic acid dehydrogenase are dependent on zinc. Zinc is also present in co-factors of other enzymes such as arginase and diaminase. Thus, it takes part in the synthesis of DNA, proteins and insulin. Hence, zinc is essential for the normal functioning of the cells including protein synthesis, carbohydrate metabolism, cell growth and cell division.

A normal human body contains 1.4 to 2.3 g of Zn and it is present in all body cells. Highest Zn concentrations are found in eye, prostate, liver, kidney, muscle, heart, skin and pancreas. It is present in plasma and in all the cellular elements of blood. It is also present in the hair, nails and bones.

Zinc influences growth rate and bone development, the integrity of the skin, the development and function of the reproductory organs. It also helps in wound healing capacity.

Recommended daily dietary intake of Zn is about 15 mg. Zinc absorption occurs mainly in the duodenum. Zinc excretion occurs through the intestines, urine and skin.

The zinc deficiency syndrome manifests itself by retardation of growth, anorexia, lesions of the skin and appendages, impaired development and function of reproductive organs.

Many of the toxic effects attributed to zinc may have been due to other associating metals such as Cd, Pb, Sb and As. However, if the zinc concentrations in air are over 15 mg/$m^3$ in the work environment, "metal fume fever" may be caused in the workers. This is mainly attributed to ZnO fumes and dusts. Inhalation of Zn fumes causes fever, depression, malaise, cough, vomiting, salivation and headache. Toxic quantities of Zn might go into solution where acidic foods such as fruit juices, meats, salads, etc. are prepared, or stored in galvanized utensils. Gastro-intestinal symptom and fever have been reported within 200 minutes to 12 hours following such ingestion. Some of the symptoms might be due to the presence of other metals like Cd, As and Fe. Some cases of occurrence of acute renal failure following ingestion of $ZnCl_2$ have been reported and in such cases the possible co-existence of other metals was not documented. The emetic action of Zn may be a protective mechanism. However, dehydration, electrolyte imbalance, dizziness, lethargy and muscular uncoordination may follow.

Zinc, unlike Cd, As and Sb, is not cumulative. Some workmen exposed for 2 to 35 years to zinc oxide dust poor in Pb and Cd, did not show any acute or chronic illness attributable to zinc.

## ENVIRONMENTAL LEVELS AND ECOLOGICAL EFFECTS

The concentration of zinc in cow's milk lies between 3 to 5 μg/ml. Normal hen's egg contains 0.7 to 1.0 mg Zn, almost all of which is present in the yolk. Wheat germ and bran contain 40–120 ppm and oysters contain over 1000 ppm of Zn. These represent the richest sources of Zn. White sugar, pome and citrus fruits are among the lowest in zinc, which hardly contain 1 ppm of fresh edible portion. Between these extremes, in ascending order, are white flour and bread, milk, leafy vegetables, meat, fish, eggs, whole cereals, nuts and leguminous seeds. Adult human diets supply 5 to 22 mg Zn per day.

Zinc plays a vital role in plant nutrition, it being a component of a number of metalloenzymes. Paddy grown in soils containing less than 20 ppm suffers from zinc deficiency, whereas if the Zn content is about 460 ppm in soil, 10% reduction in the paddy yield was observed. Soil Zn-content of about 1000 ppm may be harmful to the rice paddy plants. Zn is less toxic to fishes as compared to Hg, Cd, Cr and As. They accumulate Zn in gulls, gut and lever. Zinc in benthic invertebrates in unpolluted waters may lie in the range 20 to 200 mg/kg. Zn at about 40 ppm levels in water may impart a metallic taste.

## COPPER (Cu)

**Introduction :** Copper is one of the earliest known metals. Bronze, an alloy of Cu and Sn was invented as early as 500 B.C. The presence of copper in animal and plant tissues was recognized more than one and half centuries ago. It is present in all land and main organisms. In 1847, copper was shown to be present in combined form with the blood proteins of snails. Later, it was demonstrated that Cu, in addition to iron, is necessary for blood formation. Soon after this discovery, several reports appeared regarding disorders of grazing sheep and cattle due to deficiencies of copper and their recovery after providing copper supplements.

Several enzymes such as tyrosinase, cytochrome oxides, uricase, ascorbic acid oxidase, monamine oxidase and d-amino-levulinic acid dehydratase contain copper. Many of the manifestations of copper deficiency in animals seem to be related to the decreased tissue concentrations of some of these enzymes.

## INDUSTRIAL USES AND POLLUTION SOURCES

Copper is a chalcophile element which is mostly found in sulfide deposits along with Pb, Cd and Zn. It is lustrous, malleable, ductile, soft, tough, corrosion resistant and a good conductor of heat and electricity. Hence it is widely used in industry. It is used in the manufacture of alloys, paints, ceramics and pesticides.

Contamination of air with copper usually occurs near industrial smelters used for producing Zn and other non-ferrous metals. Cu is carried away to long distance (about 8 km) from these industrial areas by air and water. Iron and steel industry, fertilizer industry and burning of wood are other sources of copper in the atmosphere. Water pollution due to copper also results from the discharge of mine tailings, disposal of fly ash and disposal of municipal and industrial wastes.

## BIOCHEMICAL EFFECTS, TOXICOLOGY AND TOXICITY

Laboratory studies have shown several physiological functions involving copper. Defects in pigmentation, bone formation, reproduction, myelination of the spinal cord, cardiac function and connective tissue formation, in addition to defects in growth and hematopoiesis were found to be the manifestations of copper deficiency.

Many copper protein compounds were isolated from living tissues, most of which were found to be enzymes with oxidative function. These copper enzymes help in catalyzing the reduction of molecular oxygen to water. Tyrosinase, laccase, ascorbic acid oxidase, cytochrome oxidase, uricase, monoamine oxidised, -aminolevulinic acid dehydrase, and dopamine-**b**-hydroxylase were all identified as copper enzymes. Whenever the tissue concentrations of these enzymes are low, manifestations of copper deficiency are observed, thus revealing the basic biochemical defects.

As stated earlier, copper is essential for the normal biological activities of amino-oxidase and tyrosinase enzymes. The former is involved in the formation of the elastin and collagin both of which are important proteins. Elastin is a constituent of large blood vessels and collagen is the constituent of tendons and bones. Tyrosinase is required for the catalytic conversion of tyrasine to melanin, the vital pigment located beneath the skin which protects the skin from dangerous radiations.

A daily dietary intake of 2 to 3 mg of Cu is recommended for human adults. Copper is mainly absorbed from the duodenum in man and chicks. Copper absorption and retention depends on factors such as the chemical form in which the metal is ingested, by the dietary levels of other minerals and organic substances, and the acidity of the intestinal contents in the absorptive area.

Ingestion of 15-75 mg of Cu causes gastro-intestinal disorders. Excessive intake of Cu may cause hemolysis, hepatotoxic and nephrotoxic effects. Inhalation of air-borne copper causes irritation of the respiratory tract and metal fume fever. Workers involved in use of fungicides containing copper sulphate develop a respiratory disorder called "vineyard sprayer's lung" which is characterised by the development of interstitial pulmonary lesions and nodular fibro-hyaline scars. This may lead to even lung cancer.

Copper poisoning can occur as an industrial hazard in workers engaged in copper mining or processing. Continuous ingestion of Cu iron from food or water at intakes sufficient to induce chronic copper poisoning in man is unlikely. "Wilson's disease" (or hepatolenticular degeneration), which is characterised by excessive concentrations of Cu in the tissues, arises from metabolic defects involving absorbed copper and not from the ingestion of excessive amounts of Cu. Other conditions which manifest increased Cu content in liver include thalassemia (Mediterranian anemia), hemachromatosis, cirrhosis, atrophy of liver, tuberculosis and carcinoma.

## ENVIRONMENTAL LEVELS AND ECOLOGICAL EFFECTS

Crustaceans and shellfish, particularly oysters and the organ meats (liver, kidney and brain) are the richest sources of Cu. These are followed by nuts, dried legumes, dried vine and cocoa. These contain Cu in the range 20 to 400 ppm. The poorest sources of Cu are milk, butter, cheese, white sugar and honey which consist of about 0.5 ppm Cu. The non-leafy vegetables, fresh fruits, refined cereals, including white flour and bread contain about 2 ppm Cu.

The concentration of Cu in air is in the range 0.2 to 0.83 $\mu g/m^3$ in urban areas like Bombay. Soluble copper levels in water lie in the range 0.5 to 2 μg/L. Water containing more than 1 mg/L gives undesirable taste and odour. The level of Cu in sea water lies in the range 1 to 3 μg/L.

Copper is essential to plants for synthesizing chlorophyll and for the activity of some enzymes. However, at higher levels (about 0.1 mg/L) Cu can inhibit growth. Blue green algae are particularly susceptible to Cu because it inhibits the nitrogen-fixing properties of these algae. In animals, the toxic effects of Cu result from the complexation of essential carboxylic acids.

## ARSENIC (As)

Arsenic is a metalloid which is widely distributed in the biosphere. It was used in medicine and as poison in ancient times. It is another chalcopile element and it accumulates in sulphide deposits. It is distributed in earth's crust at an abundance level of about 2 ppm. It exists in nature mostly as sulfide ore. Copper acetoarsenite is another feed additive. Arsenic is found in concentration of about 1% in Pb-Zn concentrates used is smelters and because of its volatility (sublimation point of As in 650°C), it can spread into the atmosphere as a pollutant. However, the greatest source of pollution from arsenic is its use in agricultural chemicals such as weed killers, fungicides, insecticides, and rat poisons. Arsenious oxide, known as white arsenic, is produced as a byproduct in smelting of copper, lead and gold ores.

Aresenic occurs in the air wherever coal is burnt, particularly near smelters and refineries. Sea water contains 2 to 5 ppb of As and in public water supplies, its concentration may be higher. In normal soils, 1 to 40 ppm of As may be present but in areas where continuous arsenical sprays are used, high levels can be there.

## INDUSTRIAL USES AND POLLUTION SOURCES

Elemental arsenic is used in the manufacture of glass, particularly infra-red transmitting glass. It is also used in the manufacture of electrical semi-conductors and photoconductors. It is used in the manufacture of linoleum and oil cloth.

$As_2S_3$ is used as a depilatory for hides. $As_2S_5$ is a brownish yellow, glassy and highly refractive material and is used in pigments and in light filters as thin sheets. The hemiselenide of arsenic is used in glass manufacture. $As_2O_3$ is a byproduct in smelting operations. $As_2O_3$ is used in rodenticides, insecticides, herbicides, and as a mordant in textile industry. Lead arsenate, sodium arsenate and calcium arsenate were used as pesticides. Monosodium arsenite and dimethyl arsenic acid are specifically used as weed killers. Chromated copper arsenate (CCA) and fluorochrome arsenate phenyol (FCAP) are used as wood preservatives. Elemental arsenic is incorporated in some copper- and lead-based alloys to enhance their hardness and thermal resistance.

Arsenic released from natural agencies such as weathering processes on a global scale is estimated to be about $8 \times 10^4$ metric tonnes per year while man-made activities account for about $24 \times 10^4$ metric tonnes per year.

Environmental arsenic pollution occurs mainly from its release into air from smelting of As-containing ores, burning of coal and use of arsenic compounds in various applications such as fungicides, insecticides, herbicides, pesticides and preservatives.

**Arsine ($AsH_3$),** which is a hyproduct of many industrial processes involving arsenic, is the most poisonous of all arsenicals. It is toxic even at dilutions of 1: 20,000. Cases of chronic arsine poisoning were reported in workers involved in ore smelting and metal refining, etching, soldering, galvanising and lead-plating operations. Arsine may be liberated by the action of fungi in sewage plants. Accidental arsine poisoning was reported in petroleum industry.

## BIOCHEMICAL EFFECTS, TOXICOLOGY AND TOXICITY

Arsenic is a general protoplasmic poison and it affects all systems in the body. It is a cumulative poison. The major biochemical effects of arsenic are: (1) complexation with coenzymes (2) uncoupling of phosphorylation and (3) coagulation of proteins.

Trivalent arsenicals react with sulphydryl (–SH) groups in cells and thus inhibit the sulphdryl containing enzyme systems essential to cellular metablism.

$$\text{Enzyme}\begin{matrix}\diagup SH \\ \diagdown SH\end{matrix} + \begin{matrix}-O\diagdown \\ -O\diagup\end{matrix}As-O \longrightarrow \text{Enzyme}\begin{matrix}\diagup S\diagdown \\ \diagdown S\diagup\end{matrix}As-O+2OH^-$$

The pyruvate oxidase (or pyruvate dehydrogenase) system gets inactivated by complexation with As (III) and thus the generation of adenosine triphosphate (ATP), which is an important energy source, is inhibited. The enzymes generating cellular energy in the citric acid cycle are also adversely affected.

Arsenic uncouples phosphorylation. Arsenic also interferes with some biochemical processes involving phosphorous due to the similarity in their chemical properties. This is particularly noticed in the biochemical generation of ATP. A vital step in the generation of ATP is the enzymatic synthesis of 1, 3-diphosphoglycerate from glyceraldehyde 3-phosphate. As (III) interferes in this reaction by producing 1-arseno-3-phosphoglycerate instead of 1, 3-diphosphoglycerate. The former compound undergoes spontaneous hydrolysis forming 3-phosphoglycerate and arsenate. Thus instead of phosphorylation, arsenolysis takes place.

Arsine ($AsH_3$) combines with hemoglobin and is oxidised to a hemolytic compound that does not appear to act by sulphydryl inhibition.

Low chronic doses of arsenic ingested tend to accumulate in lipid rich tissues. High arsenic levels in man are usually found in hair, nails and skin. When arsenic is inhaled it is deposited in lungs and is retained in the lung tissues for a long time.

Soluble arsenicals are absorbed from all mucous membranes. Arsenic containing ointments or lipid soluble vesicants are absorbed through the skin. Non-allergic contact dermatitis and conjunctivitis are frequently suffered by workers exposed to arsenic containing dusts. Continued inhalation of arsenic dusts may cause perforation of the nasal septum, as is the case with chromium and other metallic dusts.

Chronic ingestion of inorganic arsenic causes peripheral arteriosclerosis, commonly known as 'black foot disease'. It may also cause peripheral neuritis resulting in motor and sensory paralysis of the nerve extremities. Arsenic is toxic to liver, and it produces fatty infiltration and causes central necrosis and cirrhosis. Arsenic poisoning also affects bone marrow and cellular elements of blood.

Arsenic can cross placental membrane and is known to be teratogenic to animals. Arsenicals are known to be carcinogenic to lungs in humans. They may also lead to skin cancer through the initial skin lesions.

## ENVIRONMENTAL LEVELS AND ECOLOGICAL EFFECTS

Most human foods like vegetables, fruits, dairy products and meats contain less than 0.5 ppm and rarely exceed 1 ppm of arsenic. On the other hand, foods of marine origin are much richer in As. For instance, mussels, oysters and bony fish contain As upto 120 ppm, 3 to 10 ppm and 2 to 8 ppm respectively.

The total amount of As in normal adult human body was estimated at 15 to 20 mg or 0.2 to 0.3 ppm. Analysis of As in human hair and blood gives an indication of As poisoning and the analysis is accomplished by neutron activation technique. Normal cow's milk contains 0.03 to 0.06 ppm of As.

Arsenites being more soluble in water are more phytotoxic than arsenates. The toxicity of arsenic to aquatic biota generally decreases with increasing pH of the medium due to the oxidation to higher valency state. Leguminous plants, rice, onion etc. are more susceptible to arsenic toxicity whereas grapes, carrots, apples, tomato, potato and tobacco are more resistant.

Sequential methylation of arsenic takes place by some soil microbes such as bacteria and fungi leading to the release of dimethyl and trimethyl arsine.

## TREATMENT FOR ARSENIC POISONING

Acute arsenic poisoning is usually diagnosed by gastrointestinal symptoms such as intense pain, projectile vomiting, diarrhea etc. followed by shock, hypoxic convulsious and coma. Sometimes, headache, vertigo, loss of memory followed by stupor and coma may occur without any gastrointestinal symptoms.

Acute arsenic poisoning is treated with the antidote, Dimercaprol (British anti-Lewsite, BAL). Restoration of fluid and electrolyte balance is mandatory.

## NICKEL (Ni)

Metallic nickel has known for over two centuries. Nickel constitutes about 0.008% of the earth's crust. Important nickel deposits occur as sulphides (chalcopyrite, pentlandite and pyrrhotite) or as oxides (Laterites).

Nickel exists in meteorites. Soil contains 40 ppm nickel on an average.

Nickel and its salts are used in several industrial applications such as in electroplating, storage batteries, automobile and aircraft parts, spark plugs, electrodes, coins, cooking utensils, pigments, paints, lacquers, cosmetics, water and printing fabrics. Raney nickel, made by fusing equal parts of Al and Ni, is employed as catalyst for hydrogenation of oils and of other organic compounds with gaseous hydrogen. Several nickel alloys are known. Constantan, German Silver. Nichrome. Raney Nickel are some examples.

Fairly recent studies indicated that nickel is essential in animal nutrition. Significant concentrations of Ni have been shown to be present in RNA and DNA. Nickel is shown to be essential for some micro-organisms and animals but not to plants. It is associated with the synthesis of Vitamin B-12. It is toxic at higher concentrations.

Nickel is found to be toxic for most plants and fungi. Growth of woody plants was hindered by high nickel concentrations in the soil. Land plant tissues contain about 4 times more nickel than in animal tissues.

Emission of nickel into the atmosphere by burning of fossil fuels is estimated to be 70,000 tonnes per year on global basis. Dietary intake of Ni is about 300 to 600 μg per day. Only 1 to 10% of the dietary nickel ingested is absorbed. Nickel is found in tea, cocoa, peanuts and several other food stuffs. Cow's milk contains about 0.03 ppm of Ni.

One of the most serious sources of nickel pollution is the air-borne effluent from industrial units using nickel carbonyl method for manufacturing nickel. Nickel carbonyl, $Ni(CO)_4$ is several times more toxic than CO. The complex breaks down and deposits finely divided Ni, in the lungs, which can cause cancer. Initial symptoms of toxicity due to nickel carbonyl are nausea, dizziness, headache, chest pain, etc. After 1 to 5 days, severe pulmonary symptoms, tachycardia and extreme weakness develop and even death may occur within 4 to 13 days. Dermatitis or nickel-itch is common among workers involved in making nickel-containing jewellery and those using nickel plated watches nickel-containing detergents. Nickel is believed to inhibit various enzymes such as cytochrome oxidase, isocitrate dehydrogenase and maleic dehydrogenase. Nickel dusts are reported to be carcinogenic. Allergic reactions to nickel-containing prostheses, surgical devices, kitchen ware, etc. have been reported and some of them could be serious at times.

## SELENIUM (Se)

**Introduction :** Selenium is a chalcopile element. It is a non-metal similar in properties to those of sulphur. It has been studied extensively in recent years because of its association with animal health. Selenium is an essential trace element. It is a natural component of the enzyme glutathione peroxidase. Selenium deficiency in cattle has been recognized in many areas, such as New Zealand. Some soils contain abnormal levels of Se and as such they are known as seleniferous soils, which occur mainly in the arid and semi-arid regions of Canada, western parts of USA, China, Columbia, Mexico, USSR, Israel, Hawaii, etc. Seleniferous soils do not occur in humid regions. Although Se is an essential element, excessive amounts of it are extremely toxic. One of the best known examples of Se toxicity of the poisoning of cattle in Colarado plateau because of their consumption is selenium-accumulating plants such as the "Astragalus" species, which may contain as high as 1% Se on dry weight basis.

Se is considered as more toxic than arsenic. Owing to its extreme volatility, it is easily lissipated into the atmosphere. It is emitted in ambient air by burning fossil fuels.

## NDUSTRIAL USES AND POLLUTION SOURCES

Selenium is used in electrical and electronic industries (*e.g.*, low-voltage rectifiers, photoelectric cells, etc.), ceramic industry, rubber production, inks, photocopies, in anti-dandruff shampoos, plastics, alloys, etc. Fossil fuels contain 1-10 μg of Se per gram. The atmospheric levels of Se in urban area may be 0.1 to 10 n/m$^3$. Element Se present in fossil fuels is transformed to $SeO_2$ during combustion and thus spreads in air as a pollutant. Water pollution by Se results from the Se present in rock phosphate and fly ash from coal-based thermal power plants. Most of the Se present in ambient air is due to the burning of fossil fuels. The use of Se in the above mentioned industries and refinery works processing ores of Cu, Ni, Ag and Au, may bring about exposure of the workers concerned to hazardous seleniferous dusts, fumes and vapours.

## CHEMICAL FORMS AND ENVIRONMENTAL AND METABOLIC FATE

Se exhibits variable valency as follows :

## VALENCY STATE

– 2 (as in hydrogen selenide, dimethyl selenide, etc.)
0 (as element selenium)
+ 2 (as in selenium dioxide)
+ 4 (as in selenites), and
+ 6 (as in selenates).

Hydrogen selenide is an unstable compound which is toxic and irritating at concentrations of the order of 1-4 μg/L in air, Selenium dioxide is formed by burning of fossil fuels but it may eventually get reduced to Se by the $SO_2$ which is concurrently present Elemental Se is mostly insoluble and hence is not available to the biosphere as such. Selenates mostly occur in alkaline soil and are water soluble. However, selenites are generally less soluble.

The metabolism of selenates and selenites involved reduction to hydrogen selenide, methylation to dimethyl selenide or trimethyl selenium ion which are exhaled or excreted in urine respectively.

## BIOCHEMICAL EFFECTS, TOXICOLOGY AND TOXICITY

Animals grazing in seleniferous areas suffer from chronic disease called "alkali disease" which is characterised by hair loss, soreness of joints, dullness, sloughing of hooves, etc. and sometimes cirrhosis and anemia. In an acute state, the selenosis may lead to a potentially fatal disease called "blind staggers" which is characterised by impairment of vision, paralysis, leading to death by respiratory failure, starvation and thirst. The mechanism whereby selenium induces toxicity is not fully established but inactivation of sulphydryl-containing enzymes is suspected.

Chronic Se poisoning was reported in the year 1925 in copper refinery workers and the symptoms observed were sore throat, coryza and gastro-intestinal irritation. Dermatitis and red staining of fingers, teeth and hair were also reported. Selenium oxyhalide compounds are strong irritants and vesicants and can cause pulmonary edema. Selenium is mutagenic and teratogenic. It has been implicated as a carcinogen in rats and as a possible protective against human cancer. However, in neither case is the evidence entirely convincing.

## ENVIRONMENTAL LEVELS AND ECOLOGICAL EFFECTS

As Se is an essential nutrient, an average daily intake of 50-200 μg of it is recommended. However, an intake of 1 mg of Se per kg, per day, is found to be toxic to humans.

Sea water contains 0.06 to 0.12 μg of Se per liter. The Se content of normal soils ranges between 0.1 to 2 ppm. According to the U.S. EPA, the maximum tolerable limit of Se in drinking water is 10 μg/g.

Selenium toxicity occurs mainly in neutral and alkaline soils, in which the soluble selenates are formed. On the other hand, in acid soils selenite ion ($SeO_3^{2-}$) formation is favoured which forms insoluble complexes with ferric ions, thus rendering the Se unavailable.

The Se concentration in human blood ranges from 0.10 to 0.34 μg/ml.

## TIN (Sn)

Tin occurs in the earth's crust at the levels of about 3 ppm. It is mostly found as oxide deposits such as cassiterite ($SnO_2$).

Tin is mainly used in tin plating, manufacture of alloys (*e.g.*, soft solder, bronze, bell metal, fuse metal, type metal, babbit metal, etc.), in medicine in anticaries preparations, as reducing agent in laboratories and in nuclear medicine for reducing $^{99}Tc$. Organo-tin compounds have several applications. For instance, TBTO (bis-tributyl tin oxide) and TPT (triphenyl tin) hydroxide are used as biocides. Tricyclohexyl tin is used as acaricide. TBTO is also used as antifouling agent in paint formulations. Dibutyl tin compounds are used as stabilizers in plastic industry and also as antihelminthics for poultry. In addition to the above applications, tin poisoning in humans can be from storage and cooking vessels made of tin and consumption of canned food products stored in tin-plated containers.

The toxic dose of tin for humans is 1.5 to 3 g. This causes symptoms like severe headache, nausea, cramps, vomiting, diarrhoea and fever. Tin inhibits the activities of enzymes like acid phosphatase and hepatic succinate dehydrogenase, diminishes the Ca of femur and interferes with heme metabolism. Organo-tin compounds are much more toxic because they tend to degenerate the bile duct, produce edeme in the central nervous system, inhibit the hydrolysis of ATP and uncouple the oxidative phosphorylation in mitochondria.

The tin content in air ranges from 0.0023 to 2.3 μg/g. Sea water contains about 3 μg/g of tin. The tin content of soils varies widely. Fruit juices stored in open cans may reach levels as high as 2000 ppm and hence should not be consumed.

## ANTIMONY (Sb)

The chemical and biological properties of antimony are similar to those of arsenic. It was used by ancient people as a medicine as well as a cosmetic but both the applications were abandoned after its toxicity and inefficacy were realized. Now it is mostly used as a parasiticide. The pentavalent antimonials are used in the treatment of the protozoal infection called leishmaniasis. The trivalent antimony compounds are primarily used for the management of helminthic infections. The emetic and expectorant uses of trivalent antimony compounds are now obsolete because of their toxicity.

On local application, antimony compounds are more caustic than those of arsenic compounds. Antimony salts act as emetics mostly due to their irritant action on the gastrointestinal mucosa. Trivalent antimonials act as parasiticidals by inhibiting essential sulphydryl enzymes such as pyruvic dehydrogenase.

The symptoms of actute and chronic antimony poisoning are similar to those observed in arsenic poisoning such as throat constriction, excruciating gastric pain, projectile vomiting, diarrhoea, shock, coma and death.

## FLUORINE (F)

**Fluorine occurs in combined form in Fluorspar ($CaF_2$),** Cryolite ($Na_3AlF_6$) and Fluorapatite [$Ca_{10}F_2(PO_4)_6$]. Fluorine is present in the earth's crust to the tune of 0.08%. Volcanic activity releases large quantities of fluorides into the environment.

Fluorides such as HF, $SiF_4$ and $H_2SiF_6$ are emitted in the industrial plants manufacturing phosphatic fertilizers (since phosphatic rock contains 3% by weight of fluoride) and in aluminium smelters where cryolite is used as a flux. Pollution due to fluoride emissions also occurs during burning of coal (coals contain 0.001 to 0.048% of fluoride) and from the manufacture of tiles, bricks, steel, pottery, glass and ferro-enamels. Water pollution due to fluorides is caused by contaminated domestic sewages and run-off from agricultural lands using phosphatic fertilizers.

HF is a gaseous pollutant which is highly toxic and irritating. Fluorides are mainly deposited on bones, teeth and further on skeleton and later in soft tissues. Overdose of fluorides causes mottling of enamels and bones. Crippling fluorosis was reported in some places of India and China where well waters contained as high as 10 ppm of fluoride. The disease is characterised by hyper-mineralization of skeleton, calcification of ligaments, painful joints leading to immobilization.

Fluorides in high doses can cause respiratory failure and paralysis.

Several cases of dental fluorosis were reported from Andhra Pradesh, Punjab, Tamilnadu, Karnataka, U.P. and Rajasthan. Poisoning due to toxic levels of fluorides was reported in 1937 in Andhra Pradesh and the people effected showed symptoms of pain, stiffness of spine and mottling of teeth.

Small doses of fluoride (about 1 ppm) in drinking water reduced the incidence of dental caries, but excessive doses can cause mottling of teeth in humans and domestic animals. As per the W.H.O. norms, the fluoride levels in water should not exceed 0.6 ppm. The fluoride levels in various fresh water streams lie at about 0.2 ppm, but in water sources contaminated by human activities, the fluoride levels of even 2-3 ppm are noticed.

Plants such as peaches and apricots are highly sensitive to HF even at 0.02 to 0.05 ppm levels. Orange and lemon trees are also sensitive to fluoride emissions and result in smaller yields. Plant species that accumulate fluorides suffer from a disease called "fluorosis". Livestock ingesting fluoride-contaminated vegetation also suffer from fluoride poisoning.

## BERYLLIUM (Be)

Be is present in the earth's crust to the extent of about 4 ppm. It resembles Al in its geochemistry. Small quantities of Be are released during combustion of coal.

Be is used in nuclear reactors, rockets, air crafts and in the manufacture of fluorescent lamps and durable alloys. In spite of its limited commercial applications, its inherent pollution potential should not be under estimated. Early complacency about the toxic effects of Be led to several industrial fatalities in the late 1960s. In 1949, the U.S. Atomic Energy Commission discovered that one of its Be manufacturing plants was contaminating the surrounding air to the extent of 0.1 $\mu g/m^3$ (which is the toxic limit) upto 1 km away from the plant. Ten cases of Be poisoning were reported out of which one person died. This experience showed that Be can pose a public health hazard in addition to being a mere occupational hazard.

Owing to the lack of homeostatic mechanism, Be tens to accumulate in mammalian systems. Symptoms of beryllium poisoning are acute pneumonitis and "berylliosis" which is a progressive disease affecting the heart. When inhaled as a dust, it can cause lung cancer. Be causes non-healing ulcers and heart diseases. It is carcinogenic and mutagenic. It increases the chances of occurrence of bone tumours.

In our country, there are only limited reserves of Be which are protected under the Indian Atomic Energy Act.

## COBALT (Co)

Its concentration in the earth's crust is about 23 ppm. It is an essential element for humans and animals because it is associated with the synthesis of vitamin $B_{12}$. Cobalt is used in the manufacture of alloys, permanent magnets, paint driers, and as industrial catalyst in processes such as manufacture of $NH_3$ and alcohol. Cobalt salts are used in the treatment of anemia and cyanide poisoning. Instant coffee and tea contain considerable amounts of cobalt. Bread contains about 1 ppm and beer about 0.93 of cobalt. Ingestion of excessive amounts of cobalt causes intercellular hypoxia and polycythemia. Chronic exposure of cobalt may lead to goitre.

## MANGANESE (Mn)

It is an essential trace element but is toxic at higher concentrations Pyrolusite ($MnO_2$) is the most common mineral of Mn. India is the second largest producer of Mn in the world.

Manganese is used in dry battery cells, ferro-manganese and other iron alloys, glass, ceramics, electric coils, etc. Naturally, these industrial processes are the major sources of Manganese pollution. Most of the manganese in the environment is due to burning of fossil fuels. Use of manganese bearing fertilizers also contributes to air and water pollution by Mn.

Manganese is an essential trace element at low concentrations. A daily intake of 2.5 to 5 mg of Mn by humans contributes to the well-being of the cells because it acts as a co-factor in some enzymatic reactions such as those involved in Phosphorylation, Synthesis of fatty acids and cholesterol. However, when exposed to higher levels of Mn, it gets accumulated in kidney, liver and bones and cause "Manganese psychosis", which is an irreversibly brain disease characterised by uncontrollable laughter, euphoria, impulsiveness, sexual excitement followed by impotency, etc. Mn is among the least toxic metals to plants. High concentrations of Mn may be toxic, particularly an acidic soils. It generally affects plant tops that roots. Accumulation of Mn in levels leads to Chlorosis and necrosis of leaves. Leuminous plants grown in acidic soils are sensitive to Mn. Calcium and Phosphorous of protection of plants from Mn-toxicity.

Human beings get their daily quota of manganese from vegetables, grains, fruits, nuts, etc.

## QUESTIONS

1. Discuss the various mechanisms by which rare elements have been dispersed or mobilized to the present level in our environment.
2. On what consideration a trace element in considered to be essential ? Give three examples of essential elements and their physiological roles in plants and animals.
3. Write short notes on the following
   (*a*) Biogeochemical cycle (*b*) Ionic potential and its significance
   (*c*) Trace elements in marine environment.
4. What are heavy metals ? Why have they attracted lot of attention in recent times ? Which of the heavy metals are considered to be essential for biological processes ?
5. Discuss the pollution sources, biochemical effects, toxicology and toxicity of any two to the following trace elements:
   (*a*) Arsenic (*b*) Cadmium (*c*) Lead.
6. Discuss the sources of pollution of mercury, its toxicology, toxicity and environmental fate.
7. Write short notes on
   (*a*) Minamata disease (*b*) Itai-Itai disease (*c*) Blind-staggers (*d*) Alkali disease.
8. Discuss the pollution sources, biochemical effects, toxicology and toxicity of the following:
   (*a*) Cr (*b*) Be (*c*) Ni

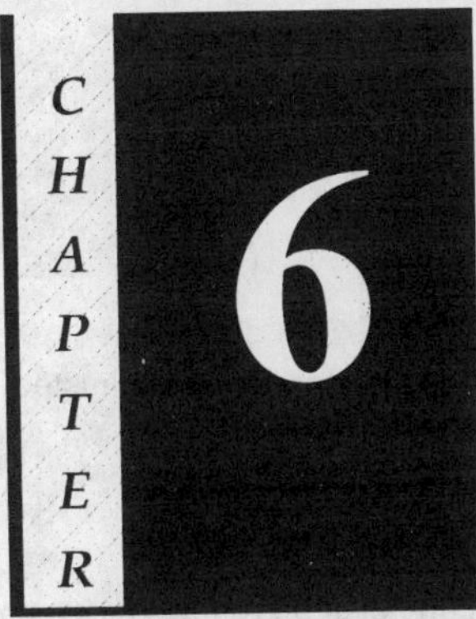

# Natural Resources

Man is the most highly evolved animal. Since its appearance on earth man and his associates has been dependent on nature. The development of new ideas enabled him to identify materials of nature and to put them in to different uses for his comfort and development. In this series water, air, sunlight, land, minerals, forest, coal, oil, metals etc. were available and were identified by man. As the development was in progress, he started to make use of these natural gifts. He converted most of the natural sources in to natural resources for his own interest. This was possible later, when development and research technology was developed to transform these natural sources or materials in to more valuable goods. Thus, natural resources can be defined as—**Things/materials of the nature, that can be put to some use by human beings for their growth, development, comfort and other necessities are called as "Natural Sources".**

For example, air, water, soil, forests, animals, minerals, metals, energy and other substances are some examples of natural resources that are utilised by human beings. At present, total global production is nearly enough to match the human demand for energy, materials etc., if we judiciously distribute the resources available to us. But for future, the situation appears pretty grim.

The resources are not equally distributed throughout the world. We can realise the value of resource only when it is rather scarce. Over exploitation threatens most of our natural resources now.

The primitive man was a part of environment. His basic requirement were limited i.e. food and shelter. Food was collected from the surroundings while caves, bushes, trees were enough to provide shelter. Those days natural materials, fields and living beings only were considered useful. But exploitation, destruction, development of natural resources began right from the time when man learned to use fire, metal copper for arms, tools, pans, domestic animals and grow plants for food. Now the boundaries of environment is from polar ice to equitorial mountain tops and from deep down ocean to ionosphere of sky. Everything present in this environment is regarded as natural resource.

## TYPES OF NATURAL RESOURCES

All the natural resources can be divided into two categories—

(*i*) Exhaustible natural resources

(*ii*) Inexhaustible natural resources

**Exhaustible natural resources** are soils, forests, water, coal, petroleum, natural gas, minerals etc. These are consumed or exhausted through continuous use or misuse. Exhaustible natural resources can be further divided into two—

(*a*) Renewable natural resources

(*b*) Non-renewable natural resources

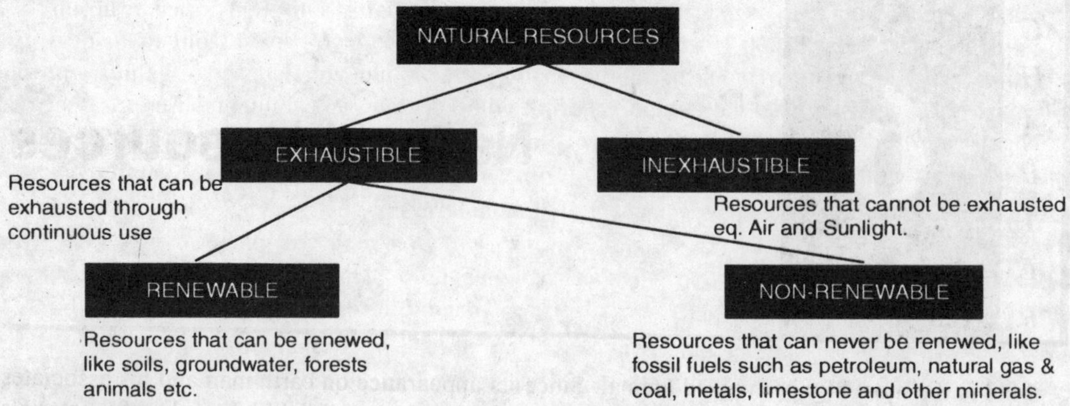


**Fig. 6.1.** Natural Resources

**Inexhaustible natural resources** are those which can not be exhausted through continuous use or misuse e.g., air and sunlight etc.

## RENEWABLE RESOURCES

The natural resources which are consumed/exhausted/depleted through continuous use and can be recovered by very hard efforts taken up for long periods are called **Renewable Resources**, for e.g., soils, forests, groundwater etc. In other words we can say that all renewable resources are replenished through natural cycles or manually. For example oxygen in air is replenished through photosynthesis. Forest is maintained themselves and manually. Similarly fresh water is available through cycles & manually too.

Most of the removable resources are interdependent to each other. Forests maintained the environment/climate, plants need to check soil erosion & soil is needed for plants. Air and insects are needed for pollination. Wood, fibre, fodder, fruits, vegetables, milk etc. are developed directly or indirectly by recent photosynthetic activity. Thus the renewal of these resources will continue at as long as photosynthesis continues on this planet. These resources are the life support system which can fulfill all human needs. But its productivity/renewability is limited or depend upon availability of water, nutrients and environmental conditions.

The natural resources are useful to human society in one way or other. Hence we should ensure a continuous yield of useful plants, animals and materials by establishing a balanced cycle of harvest and renewal (**Odum - 1971**).

## NON-RENEWABLE RESOURCES

Non-renewable resources are not replenishable or we cannot get back our coal and petroleum reserves in our life time, if ones they are consumed/exhausted completely. Non-renewable resources are metals (iron, copper, zinc etc.), coal, oil deposits, minerals, stone, mineral salts (Phosphate, nitrates, carbonates etc.) etc.

Minerals are often called the '*STOCK*' resources, because their new materials can only be extracted from the earth's crust once. But even in the transformed state in which they are used, they are not lost to the planet and so are ideally available for reuse. Metals/minerals come from a very slow process of geochemical concentration, which took millions of years to form. Therefore, these deposits which occur today can disappear at some point of time in future.

Coal, petroleum and natural gas are called as '*FOSSIL FUELS*' because they are formed from dead remains of plants and animals buried in the earth long long ago. They are called fuels because they are burnt to give off energy. Coal have a high heating value, hence it is a useful fuel. Since fossil fuels are non-renewable sources of energy, therefore it is essential to explore more and more alternatives. Today we are utilizing solar energy, wind energy, ocean, geothermal & atomic energy as an alternative sources of energy.

Minerals, rocks, salts and chemicals etc. are termed as '*abiotic resources*', as biological activity is not involved in their formation. Continuous over exploitation of these resources shall exhaust many of our valuable deposits, which took millions of years to form. They cannot be duplicated within the human scale of time.

## NATURAL RESOURCES AND ASSOCIATED PROBLEMS

Being most highly developed/evolved animal, man possesses certain special characteristics. Man apply all their power and intelligence for food and development. They adopt new ways to fulfil their needs and often make improvements in old ways to derive resources and to fulfil their desire more efficiently. This is how they develop new technologies for utilization of natural resources. As we have seen, natural resources are exhaustible and non-exhaustible the exhaustible resources are renewable and non renewable, therefore proper utilization of our natural resources is the need of today. Human beings utilize most of resources like air, water, land, minerals, flora, fauna, fuels, energy etc. for their growth and development.

Now the problem is, how and up to what extent human beings should utilize various resources. Resources are valuable gift of nature. Hence use of natural resources should depend on knowledge, availability, type, quantity, value, necessity etc. The use of resources should be in limit not to exhaust them so that ecological balance within the nature should also remain undisturbed.

In the past, the man was not so advanced and was satisfied with what he received from the nature due to his limited needs. Thus, there was a complete balance among all the components of the natural environment in the past. At present, the story has become very different due to human activities of consumption and misuse of natural resources and the explosion of population. Therefore, nature has lost its capacity of tolerance. Fast growth in human population has led to the fast rate of consumption of natural resources and creation of wastes. Human activities of cutting trees, killing animals, damaging and mixing poisonous wastes in air and water are creating severe crisis in the environment and natural resources. To some extent nature tries to recycle some of these resources. But, because of spurt in population these are used up faster than these are replenished, leading to their depletion.

Directly or indirectly there are some problems from the over exploitation of natural resources. They are, lowering of water level, extinction of wild animals, soil erosion, siltation of rivers, floods, climatic change, interruption of water cycle, loss of aquatic plants/animals, various diseases, ozone layer depletion, global warming, acid rains, green house effects, shortage of food, problem of housing, fossil fuels etc. Using wood as fuel not only cause severe damage to earth but also produce pollution.

Oceans are also provide different types of food materials and minerals. Large quantities of petroleum and natural gas are also obtained from oceans. Thus they are valuable contributors in the development and prosperity of human beings. But due to human activities, oceans have been put under heavy conditions of serious stress.

Now the conservation of natural resources should be the priority of everyone citizen. Methods of conservation management should be followed.

By making rules only, we cannot protect natural resources. Awareness programmes, social forestry, joint forest management, van panchayat, organisations of seminars, public programmes are some means by which we can protect and conserve natural resources. During the recent years many societies and NGO's have come forward in this direction.

Human activities should be refined using high technology which adversely affect the natural resources and environment under environmental resource management. Population should be controlled because it is the root cause of resource exploitation. We should plan for conservation of natural resources for their long term and maximum usage. Non-renewable resources should be used in as sustainable manner i.e. they should remain available for future generations.

## FOREST RESOURCES

India is one of the 12 mega diversity countries, commanding 7% of the world's biodiversity and supporting 16% of the major forest types, varying from alpine pastures in the Himalayas to temperate sub tropical, tropical forests and mangroves in the coastal areas. But nearly half of the country's area is degraded, affected by problems of soil degradation and erosion. Sixty percent forests are located in ecologically sensitive zones. These forests need to be managed in a way to ensure that they are ecologically protected and maintained, as well as sustained at the highest productivity level to meet the growing population's burgeoning demands for fuel, food, fodder, and timber.

**Country data**

| | |
|---|---|
| Total land area (thousand ha) | 297,319 |
| Total forest area 1995 (thousand ha)/% of total land area | 65,005/21.9 |
| Natural forest 1995 (thousand ha) | 50,385 |
| Total change in forest cover 1990-95 (thousand ha)/annual change (%) | 36/0 |
| Population total 1997 (million)/Annual rate of change 1995-2000 (%) | 960.2/1.6 |
| Rural population 1997 (%) | 72.6 |
| GNP per person 1995 in US$ | 340 |

*Source of data*: FAO - State of the World's Forest 1999

India has a large and diverse forest resource. Its forest types vary from tropical rainforest in the north-east, to desert and thorn forests in Gujarat and Rajasthan; mangrove forest in West Bengal, Orissa and other coastal areas; and dry alpine forests in the western Himalaya. The most common forest types are tropical moist deciduous forests, tropical dry deciduous forests, and wet tropical evergreen forests.

According to the Forest Survey of 1997, the country has 76.5 million ha of forest. The degraded area was 26.13 million ha and there was another 5.72 million ha of scrub; thus, in total 31.85 million ha of forests were degraded or open.

The land use outside for habitations (rural and urban), industries and infrastructure, such as roads, rivers, canals, railway lines, under permanent snow, rocks, desert, or not available for other reasons amounted to 264 million ha. It consists of cultivated land of 142 million ha, forestland of 67 million ha, fallows of 24 million ha, pastures of 12 million ha, tree groves of 3 million ha, and cultural waste of 16 million ha. Thus, in order to achieve the national goal of one third of the country under forest/tree cover, an area of 29.7 million ha has to be brought under plantations.

It was reported that the country's achievement in raising forest plantations, in terms of area, has been impressive. Up to 1998, the total area of tree plantations was 28.38 million ha, of which about 17 million ha were planted before 1990's. The current annual rate of plantation is 1.2 million. The quality of these plantations varies considerably. It should be noted that forest plantations are a means to meet the increasing demand for industrial raw material or for direct consumption, i.e. fuel wood, but not to justify deforestation or claim restoration of biodiversity and other environmental services.

There are other woodlands established in small blocks on non-forestry lands, which are not included in the forest survey because of limitations of interpretation of satellite data.

The performance of forest plantations, in terms of survival, growth and yield, has been poor caused by several factors, including inadequacies in site selection and site-species matching, poor planting stock, lack of proper maintenance and protection (from fire, grazing, pests and diseases), lack of timely tending/thinnings, delays in fund allocation, and inadequately trained staff. In this regard, some people are of the opinion that a master plan for tree plantations should be developed specifying categories, management regimes, utilisation and investment needs; and emphasis should be given to enhancing qualitative and quantitative productions.

Involvement of the private sector in plantation development has not been substantial and are not adequately supported by the government through relevant research, extension, technological packages, input delivery, market information or credit facilities. This sector is dominant in the area of harvesting and processing. It was noted that the needs and problems relating to this area are different from those producers of wood in rural areas.

According to the latest State of Forest Report, 1999, the total forest cover was 633.73 million ha or 19.39% of the geographical area, with dense forest accounting for 11%. The Report stated that the forest cover has increased by 4,000 km$^2$ since the last survey in 1997. Thus, the overall decline in the forest cover has been halted. In this regard, the Minister of Environment and Forests stated that a major constraint facing the afforestation programmes is funding, which requires Rs 66.95 billion per year (1US$ approximately equal to Rs 43.5 in September 2000) in order to achieve one-third forest cover within the next 20 years. Rs 16 billion per year is available from both the central and state budget together to be allocated for afforestation. Involvement and investment from various NGOs, corporate, public and private sectors to fund this sector is being approached. In this connection, consultation with several donor agencies, including international and bilateral banks, for possible support have also been carried out. In this regard, WB and EU have provided substantial support to several forestry programmes in some states.

In regard to national parks, sanctuaries and other reserves, the country's achievement in terms of area is substantial. The Protected Areas (PA) cover about 14.8 million ha, or about 14% of the forest area, consisting of 80 national parks, 441 wildlife sanctuaries and 8 biosphere reserves. However, the condition of several PAs is poor because of fire, grazing and inadequate management. The Management plans of some PAs are not comprehensive. Some are below the minimum size required to be effective.

Non-wood forest products (NWFPs) have a great potential to support the socio-economic development of the country and also the principles of sustainable forest management. These products are essential to local communities. Some products have great potential for export. Some products have also provided employment and income earning.

## DEFORESTATION

The conversion of forested areas to non-forest is deforestation. Historically, this meant conversion to grassland or to its artificial counterpart, grainfields; however, the Industrial Revolution added urbanization and technological uses. Generally this removal or destruction of significant areas of forest cover has resulted in a degraded environment with reduced biodiversity. In developing countries, massive deforestation has accompanied mankind's progress since the Neolithic, and has shaped climate and geography.

Deforestation (whether deliberate or unintended) is the result of the removal of trees without sufficient reforestation. There are many causes, ranging from extremely slow forest degradation to sudden and catastrophic wildfires. Deforestation can be the result of the deliberate removal of forest cover for agriculture or urban development, or it can be an unintentional consequence of uncontrolled grazing (which can prevent the natural regeneration of young trees). The combined effect of grazing and fires can be a major cause of deforestation in dry areas. In addition to the direct effects brought about by forest removal, indirect effects caused by edge effects and habitat fragmentation can greatly magnify the effects of deforestation.

While tropical rainforest deforestation has attracted most attention, tropical dry forests are being lost at a substantially higher rate, primarily as an outcome of slash-and-burn techniques used by shifting cultivators.

### Definition of Deforestation

Deforestation is the loss or continual degradation of forest habitat due to either natural or human related causes. Agriculture, urban sprawl, unsustainable forestry practices, mining, and petroleum exploration all contribute to human caused deforestation. Natural deforestation can be linked to tsunamis, forest fires, volcanic eruptions, glaciation and desertification. The effects of human related deforestation can be mitigated through environmentally sustainable practices that reduce permanent destruction of forests or even act to preserve and rehabilitate disrupted forestland. (See Reforestation and Tree planting)

The term deforestation is often the source of disagreement between various interest groups. Conservation groups often use broad definition while groups seeking to maintain the status quo often use a narrow definition.

Deforestation defined broadly can include not only conversion to non-forest, but also degradation that reduces forest quality - the density and structure of the trees, the ecological services supplied, the biomass of plants and animals, the species diversity and the genetic diversity. Narrow definition of deforestation is: the removal of forest cover to an extent that allows for alternative land use. The United Nations Research Institute for Social Development (UNRISD) uses a broad definition of deforestation, while the Food and Agriculture Organization of the UN (FAO) uses a narrow definition.

Definitions can also be grouped as those which refer to changes in land cover and those which refer to changes in land use.

*Land cover* measurements often use a percent of cover to determine deforestation. This type of definition has the advantage in that large areas can be easily measured, for example from satellite photos. A forest cover removal of 90% may still be considered forest in some cases. Under this definition areas that may have few values of a natural forest such as plantations and even urban or suburban areas may be considered forest.

*Land use* definitions measure deforestation by a change in land use. This definition may consider areas to be forest that are not commonly considered as such. An area can be lacking trees but still considered a forest. It may be a land designated for afforestation or an area designated administratively as forest.

### Use of the Term Deforestation

The term deforestation has been used to refer to fuel wood cutting, commercial logging and the slash and burn technique, a component of some shifting cultivation agricultural systems. It is also used to describe forest clearing for annual crops, for grazing, and establishment of industrial forest plantations. The meaning of the term is ambiguous enough and so charged with emotion that the use of a more precise term might be better suited in specific cases. Related terms are forest decline, forest fragmentation and forest degradation, loss of forest cover and land use conversion.

The term also has a traditional legal sense of the conversion of Royal forest land into purlieu or other non-forest land.

## Causes of deforestation

**Present causes:** While short-sighted, market-driven forestry practices are often one of the leading cause of forest degradation, the principal human-related causes of deforestation are agricultural and livestock grazing, urban sprawl mining and petroleum extraction. Causes include demand for farm land and fuel wood. Underlining causes include poverty, lack of reform. The causes of deforestation are complex and often differ in each forest and country. Government policies, such as ones in Brazil, make it a priority to resettle some of the country's numerous landless people. The largest cause as of 2006 is slash-and-burn activity in tropical forests. Slash-and-burn is a method sometimes used by shifting cultivators to create short term yields from marginal soils. When practiced repeatedly, or without intervening fallow periods, the nutrient poor soils may be exhausted or eroded to an unproductive state. Slash-and-burn techniques are used by native populations of over 200 million people worldwide.

**Prehistory:** Deforestation has been practiced by humans for thousands of years. Fire was first tool that allowed humans to modify the landscape. The first evidence of deforestation shows up in the Mesolithic. Fire was probably used to drive game into more accessible areas. With the advent of agriculture fire became the prime tool to clear land for crops. In Europe there is little solid evidence before 5000 BP. Mesolithic foragers used fire to create openings for red deer and wild boar. On Great Britain shade tolerant species like oak and ash are replaced in the pollen record by hazels, brambles, grasses and nettles. Removal of the forests led to decreased transpiration resulting in the formation of upland peat bogs. Widespread decreased in elm pollen across Europe between 6400–6300 BP and 5200–5000 BP, starting in southern Europe and gradually moving north to Great Britain, may represent and clearing by fire at the onset of Neolithic agriculture.

**Pre-industrial history:** The historic silting of ports along the southern coasts of Asia Minor (*e.g.* Clarus and the examples of Ephesus, Priene and Miletus, where harbors had to be abandoned because of the silt deposited by the Meander) and in coastal Syria during the last centuries BC, and the famous silting up of the harbor for Bruges, which moved port commerce to Antwerp, all follow periods of increased settlement growth (and apparently of deforestation) in the river basins of their hinterlands. In early medieval Riez in upper Province, alluvial silt from two small rivers raised the riverbeds and widened the floodplain, which slowly buried the Roman settlement in alluvium and gradually moved new construction to higher ground; concurrently the headwater valleys above Riez were being opened to pasturage.

A typical progress trap is that cities are built in a woody area providing wood for some industry (e.g. shipbuilding, pottery) which starts consuming it so fast – and without proper replanting – that it becomes impossible to obtain it close enough to remain competitive, leading to the city's abandonment, as happened repeatedly in Ancient Asia Minor. Especially the combination of mining and metallurgy went along this self-destructive path.

Meanwhile most of the population remaining active in (or indirectly dependent on) the agricultural sector, the main pressure in most areas remained land clearing for crop and cattle farming; fortunately enough wild green was usually left standing (and partially used, e.g. to collect firewood, timber and fruits, or to graze pigs) for wildlife to remain viable, and the hunting privileges of the elite (nobility and higher clergy) often protected significant woodlands.

**Industrial Pressure :** The massive use of Charcoal on an industrial scale was a new acceleration of the oslaught on western forests.

## Environmental effects

**Atmospheric pollution:** Deforestation is often cited as one of the major causes of the enhanced greenhouse effect. Trees and other plants remove carbon (in the form of carbon dioxide) from the atmosphere during the process of photosynthesis. Both the decay and burning of wood releases much of this stored carbon back to the atmosphere. A.J. Yeomans asserts in Priority One (http://www.yeomansplow.com.au/priority-one.htm) that overnight a stable forest releases exactly the same quantity of carbon dioxide back into the atmosphere. Others state that mature forests are net sinks of $CO_2$ (see Carbon dioxide sink and Carbon cycle).

**Wildlife:** Some forests are rich in biological diversity. Deforestation can cause the destruction of the habitats that support this biological diversity - thus causing population shifts and extinctions.

Hydrologic cycle and water resources

Trees, and plants in general, affect the hydrological cycle in a number of significant ways:

- their canopies intercept precipitation, some of which evaporates back to the atmosphere (canopy interception);
- their litter, stems and trunks slow down surface runoff;
- their roots create macropores - large conduits - in the soil that increase infiltration of water;
- they reduce soil moisture via transpiration;
- their litter and other organic residue change soil properties that affect the capacity of soil to store water.

As a result, the presence or absence of trees can change the quantity of water on the surface, in the soil or groundwater, or in the atmosphere. This in turn changes erosion rates and the availability of water for either ecosystem functions or human services.

The forest may have little impact on flooding in the case of large rainfall events, which overwhelm the storage capacity of forest soil if the soils are at or close to saturation.

## Soil erosion

Deforestation generally increases rates of soil erosion, by increasing the amount of runoff and reducing the protection of the soil from the tree litter. This can be an advantage in excessively leached tropical rain forest soils. Forestry operations themselves also increase erosion through the development of roads and the use of mechanized equipment.

China's Loess Plateau was cleared of forest millennia ago. Since then it has been eroding, creating dramatic incised valleys, and providing the sediment that gives the Yellow River its yellow color and that causes the flooding of the river in the lower reaches (hence the river's nickname 'China's sorrow').

Removal of trees does not always increase erosion rates. In certain regions of southwest US, shrubs and trees have been encroaching on grassland. The trees themselves enhance the loss of grass between tree canopies. The bare intercanopy areas become highly erodible. The US Forest Service, in Bandelier National Monument for example, is studying how to restore the former ecosystem, and reduce erosion, by removing the trees.

### Landslides

Tree roots bind soil together, and if the soil is sufficiently shallow they act to keep the soil in place by also binding with underlying bedrock. Tree removal on steep slopes with shallow soil thus increases the risk of landslides, which can threaten people living nearby.

### Controlling deforestation

**Farming:** New methods are being developed to farm more food crops on less farm land, such as high-yield hybrid crops, greenhouse, autonomous building gardens, and hydroponics. The reduced farm land is then dependent on massive chemical inputs to maintain necessary yields. In cyclic agriculture, cattle are grazed on farm land that is resting and rejuvenating. Cyclic agriculture actually increases the fertility of the soil. Selective over farming can also increase the nutrients by releasing such nutrients from the previously inert subsoil. The constant release of nutrients from the constant exposure of subsoil by slow and gentle erosion is a process that has been ongoing for billions of years.

**Forest management:** Efforts to stop or slow deforestation have been attempted for many centuries because it has long been known that deforestation can cause environmental damage sufficient in some cases to cause societies to collapse. In Tonga, paramount rulers developed policies designed to prevent conflicts between short-term gains from converting forest to farmland and long-term problems forest loss would cause, whilst during the seventeenth and eighteenth centuries in Tokugawa Japan the shoguns developed a highly sophisticated system of long-term planning to stop and even reverse deforestation of the preceding centuries through substituting timber by other products and more efficient use of land that had been farmed for many centuries. In sixteenth century Germany landowners also developed silviculture to deal with the problem of deforestation. However, these policies tend to be limited to environments with good rainfall, no dry season and very young soils (through volcanism or glaciation). This is because on older and less fertile soils trees grow too slowly for silviculture to be economic, whilst in areas with a strong dry season there is always a risk of forest fires destroying a tree crop before it matures.

**Afforestation:** Today, in China, where large scale destruction of forests has occurred, the government has required that every able-bodied citizen between the age of 11 and 60 plant three to five trees per year or do the equivalent amount of work in other forest services. The government claims that at least 1 billion trees have been planted in China every year since 1982. In western countries, increasing consumer demand for wood products that have been produced and harvested in a sustainable manner are causing forest landowners and forest industries to become increasingly accountable for their forest management and timber harvesting practices.

The Arbor Day Foundation's (http://www.arborday.org/) Rain Forest Rescue Program is a charity that helps to prevent deforestation. The charity uses donated money to buy up and preserve rainforest land before the lumber companies can buy it. The Arbor Day Foundation then protects the land from deforestation. This also locks in the way of life of the primitive tribes living on the forest land.

## CASE STUDIES

### Brazil

In Brazil the rate of deforestation is apparently driven by commodity prices. Recent development of a new variety of soyabean has lead to displacement of beef ranches and slash and burn farmers which in turn move further into the forest.

**Indonesia**

There are large areas of forest in Indonesia that are being lost as native forest is cleared by large multinational pulp companies and being replaced by plantations.

**United States**

Upon arrival European-Americans began clearing large areas of forest for wood and agriculture. Beginning in about 1850 farm land began to be abandoned because of soil exhaustion and competition from the mid-west. Also, mechanization allowed land formerly used as pastures for horses to revert to forest. From 1850 to about 1920 the amount of forest land in the United States actually increased. Today the trend in forest cover increase has reversed as urban sprawl causes conversion of forest as the forest is transformed to suburbs.

## TIMBER EXTRACTION

Once world bank study in 1989 has argued that tree crop estates are a better employment generating option than even forest plantations, notwithstanding the high density of usable timber in the plantations. In Malaysia, timber using industry i.e. timber extraction is on large scale. In Indonesia timber manufacturing employment is as high as 3.7 million. Newspaper Kompas (24 Oct., to 1989) reported that small regional sawmills alone were employ 200,000 people in Indonesia. Even in Sarawak, employment in the timber industry is about 55000 indirect timber extraction, plus a large number in associated tasks (ITTC 1990).

In 1980, there was a serious attempt to developed wood based industry in Sabah. Sawmills, veneer and plywood mills all increased sharply in number. Although sawn timber production increased by 40% during the same period, the total volume of exports fell by almost one third. The situation in Peninsula is less serious than that in Sabah only because of the much greater diversity of the economy. For the timber and timber using industries, it is worse.

With good prices, production expanded, and at the end of 1980s the Peninsula had 681 sawmills, 43 Veneer and plywood mills and more than 1200 small woodworking plants, furniture factories. There has been an increasingly severe shortage of timber for mills, for the building industry, for exporters of sawn timber and plywood and for the downstream factories. The future of the big plans for timber manufacture is not withstanding optimism, uncertain and is potentially in conflict with the conservationist goals now formally adopted by the Govt.

Unlike Sabah and Peninsula, Kalimantan and Sarawak have shortage of timber due to exploitation of forest. They have banned export for time being. If it is, then on very high rate. In Indonesia has been warned not to exceed an annual capacity of 10 million $m^3$, otherwise real shortage of raw material could arise.

In India also, millions of people are engaged in timber industries. Not only they are earning but solving the problem of employment to some an extent. Plywood industries are also increased. There is gap between demand and supply, this results import of loggs from other countries like Malaysia, Indonesia etc. Forest area is also receding. Day by day, the demand of loggs is increasing due to population growth. Timber shortage are clearly either real or in prospect, posing a severe threat to both entrepreneurship and employment. Given a low rate of success with enrichment planting, growing emphasis is now being placed on plantation forestry to ensure future supply. Indonesia has already stepped up its reforestation efforts using mainly Acacia mangium, with a view to the future establishment of pulp & paper mills.

A major conflict is emerging those whose interest lies from the production of roundwood, sawn timber, plywood and those whose depend upon the sustainable supply of timber to the sawmills, plywood mills and factories. Unauthorised invasion by cultivators too is now a days a problem. In 1991 the International Tropical Timber Organisation issued a document that how a sustainable tropical timber production could be achieved.

## Mining

Mining is the extraction of valuable minerals or other geological materials from the earth, usually (but not always) from an ore body, vein, or (coal) seam. Materials recovered by mining include bauxite, coal, diamonds, iron, precious metals, lead, limestone, nickel, phosphate, rock salt, tin, uranium, and molybdenum. Any material that cannot be grown from agricultural processes must be mined. Mining in a wider sense can also include extraction of petroleum, natural gas, and even water.

### History

The oldest known mine in the archaeological record is the "Lion Cave" in Swaziland. At this site, which by radiocarbon dating is 43,000 years old, paleolithic humans mined for the iron-containing mineral hematite, which they ground to produce the red pigment ochre. Sites of a similar age where Neanderthals may have mined flint for weapons and tools have been found in Hungary.

Another early mining operation was the turquoise mine operated by the ancient Egyptians at Wady Maghareh on the Sinai Peninsula. Turquoise was also mined in pre-Columbian America in the Cerillos Mining District in New Mexico, where a mass of rock 200 feet (60 m) in depth and 300 feet (90 m) in width was removed with stone tools; the mine dump covers 20 acres (81,000 $m^2$). Black gun powder in mining was first used in a mineshaft under Banska Stiavnica, Slovakia in 1627, in the same town in 1762 the first Mining Academy in the world was established.

Mining in the United States became prevalent in the 19th century. Mining for minerals and precious metals, such as in the California Gold Rush in the mid 1800s, was very important in westward expansion to the Pacific coast along with ranching and exploration of oil and gas fields. During this time period many white Americans and post-slavery African Americans, with the aid of railroads, travelled west for work opportunities in mining. Many western cities such as Denver and Sacramento originated as mining towns.

### Steps in the mining process

1. Prospecting to locate ore
2. Exploration to defining the extent and value of ore where it was located
3. Conduct resource estimate to mathematically estimate the extent and grade of the deposit
4. Conduct mine planning to evaluate the economically recoverable portion of the deposit
5. Conduct a feasibility study to evaluate the total project and make a decision as whether to develop or walk away from a proposed mine project. This includes a cradle to grave analysis of the possible mine, from the initial excavation all the way through to reclamation.
6. Development to create access to an ore body
7. Exploitation to extract ore on a large scale
8. Reclamation to make land where a mine had been suitable for future use

### Environmental effects and mitigation

Environmental issues can include erosion, formation of sinkholes, loss of biodiversity and contamination of groundwaters by chemicals from the mining process and products.

Modern mining companies in many countries are required to follow strict environmental and rehabilitation codes, ensuring the area mined is returned to close to its original state, or an even better environmental state than before mining took place. In some countries with pristine environments, such as large parts of Australia, this is impossible despite the best intentions. Past mining methods have had, and methods used in countries with lax environmental regulations can continue to have, devastating environmental and public health effects.

Mining can have adverse effect on surrounding surface and ground water if protection measures are not exercised. The result can be unnaturally high concentrations of some chemical elements over a significantly large area of surface or subsurface. Coal mining releases approximately twenty toxic release chemicals, of which 85% is said to be managed on site. Combined with the effects of water and the new 'channels' created for water to travel through, collect in, and contact with these chemicals, a situation is created where mass-scale contamination can occur. In well-regulated mines hydrologists and geologists to careful measures to mitigate any type of water contamination that could be caused by mines. In modern American mining, operations must, under federal and state law, meet standards for protecting surface and ground waters from contamination, including acid mine drainage (AMD). To mitigate these problems water is continuously monitored at coal mines. The five principal technologies used to control water flow at mine sites are : diversion systems, containment ponds, groundwater pumping systems, subsurface drainage systems, and subsurface barriers. In the case of AMD, contaminated water is generally pumped to a treatment facility that neutralizes the contaminants.

## CASE STUDIES

Some examples of environmental problems associated with mining operations are :

**Ashio Copper Mine,** Ashio, Japan was the site of substantial pollution at the end of nineteenth century

**Berkeley Lake,** an abandoned pit mine in Butte, Montana that has filled with water which is now acidic and poisonous. In 2003, a water treatment plant came on-line, initially treating "new" water entering the pit and thereby reducing the rate of rise of pit water. Treated water is currently used in the concentrator of the nearby Montana Resources Continental Pit, but it is clean enough to return to Silver Bow Creek. Eventually, water in the pit itself will be treated.

**Britannia Mines** a former copper mine near Vancouver, British Columbia. Copper from the abandoned mine washes into Howe Sound, polluting the water. No animal life remains there now - Latest reports are that after a water treatment plant was put in, fish are returning to Britannia Bay - may be for the first time ever. The name used by the First Nations tribes of Britannia Beach, even before mining started, means "The Place of No Fish".

**Scouriotissa,** a copper mine in Cyprus that has been abandoned. Contaminated dust blows off this site.

**Tar Creek,** an abandoned mining area in Picher, Oklahoma that is now an Environmental Protection Agency superfund site. Water in the mine has leaked through into local groundwater, contaminating it with metals such as lead and cadmium. [1]

(http://www.health.state.ok.us/PROGRAM/envhlth/sites/ottawa.html)

## BENEFITS OF LARGE DAMS

Water is essential for sustenance of all forms of life on earth. It is not evenly distributed all over the world and even its availability at the same locations is not uniform over the year. While the parts of the world, which are scarce in water, are prone to drought, other parts of the world, which are abundant in water, face a challenging job of optimally managing the available water resources. No doubt the rivers are a great gift of nature and have been playing a significant role in evolution of various civilizations, nevertheless on many occasions, rivers, at the time of floods, have been playing havoc with the life and property of the people. Management of river waters has been, therefore, one of the most prime issues under consideration. Optimal management of river water resources demands that specific plans should be evolved for various river basins which are found to be technically feasible and economically viable river basins which are found to be technically feasible and economically viable after carrying out extensive surveys. Since the advent of civilization, man has been constructing dams and reservoirs for storing surplus river waters available during wet periods and for utilization of the same during lean periods. The dams and reservoirs world over have been playing dual role of harnessing the river waters for accelerating socio-economic growth and mitigating the miseries of a large population of the world suffering from the vagaries of floods and droughts. Dams and reservoirs contribute significantly in fulfilling the following basic human needs:

- WATER FOR DRINKING AND INDUSTRIAL USE
- IRRIGATION
- FLOOD CONTROL
- HYDRO POWER GENERATION
- INLAND NAVIGATION
- RECREATION

**Water for drinking and industrial use**

- Due to large variations in hydrological cycle, dams and reservoirs are required to be constructed to store water during periods of surplus water availability and conserve the same for utilization during lean periods when the water availability is scarce.
- Properly designed and well-constructed dams play a great role in optimally meeting the drinking water requirements of the people.
- Water stored in reservoirs is also used vastly for meeting industrial needs.
- Regulated flow of water from reservoirs help in diluting harmful dissolved substances in river waters during lean periods by supplementing low inflows and thus in maintaining and preserving quality of water within safe limits.

**Irrigation**

- Dams and reservoirs are constructed to store surplus waters during wet periods, which can be used for irrigating arid lands. One of the major benefits of dams and reservoirs is that water flows can be regulated as per agricultural requirements of the various regions over the year.
- Dams and reservoirs render unforgettable services to the mankind for meeting irrigation requirements on a gigantic scale.
- It is estimated that 80% of additional food production by the year 2025 would be available from the irrigation made possible by dams and reservoirs.
- Dams and reservoirs are most needed for meeting irrigation requirements of developing countries, large parts of which are arid zones.
- There is a need for construction of more reservoir based projects despite widespread measures developed to conserve water through other improvements in irrigation technology.

**Flood Control**

- Floods in the rivers have been many a time playing havoc with the life and property of the people. Dams and reservoirs can be effectively used to control floods by regulating river water flows downstream the dam.

- The dams are designed, constructed and operated as per a specific plan for routing floods through the basin without any damage to life and property of the people.
- The water conserved by means of dams and reservoirs at the time of floods can be utilized for meeting irrigation and drinking water requirements and hydro power generation.

**Hydro power generation**

- Energy plays a key role for socio-economic development of a country. Hydro power provides a cheap, clean and renewable source of energy.
- Hydro power is the most advanced and economically viable resource of renewable energy.
- Reservoir based hydroelectric projects provide much needed peaking power to the grid.
- Unlike thermal power stations, hydro power stations have fewer technical constraints and the hydro machines are capable of quick start and taking instantaneous load variations.
- While large hydro potentials can be exploited through mega hydroelectric projects for meeting power needs on regional or national basis, small hydro potentials can be exploited through mini/micro hydel projects for meeting local power needs of small areas. Besides hydro power generation, multi purpose hydroelectric projects have the benefit of meeting irrigation and drinking water requirements and controlling floods etc.

**Inland navigation**

- Enhanced inland navigation is a result of comprehensive basic planning and development, utilizing dams, locks and reservoirs that are regulated to play a vital role in realizing large economic benefits of national importance.

**Recreation**

- The reservoir made possible by constructing a dam presents a beautiful view of a lake. In the areas where natural surface water is scarce or non-existent, the reservoirs are a great source of recreation.
- Alongwith other objectives, recreational benefits such as boating, swimming, fishing etc. linked with lakes are also given due consideration at the planning stage to achieve all the benefits of an ideal multipurpose project.

While dams provide a yeoman service to the mankind, the following impacts of the construction of dams are required to be handled carefully:

- Resettlement and Rehabilitation
- Environment and forests
- Sedimentary issues
- Socio economic issues
- Safety aspects

The above problems related to the construction of dams may be resolved successfully in case the approach of management is objective, dynamic, progressive and responsive to the needs of the hour.

**Problems from Dams**

We have seen the benefits of dams, but the otherside of coin is also not good when such projects are undertaken and hundreds crores of public money is spent, individual or organisations in the grab of PIL cannot be permitted to challenge the policy decision taken. For such developmental projects thousands of acre land is acquired, results the public become landless. As far as relief and rehabilitation are concerned, people are not given properly. Therefore, if the interests of any affected persons are not protected in terms of the approved mechanism, they certainly have the right to approach appropriate authorities or even the supreme court in case its directions are not complied with. Later on people go on strike, & other demonstrations. "Narmada Bachao Andolan", "Chipko movements" etc. are such movements, which work for the relief & rehabilitation of affected people.

Due to these dams, most affected people are tribals, poor, labour class people etc. They can not oppose of their own. As regards reliefs & rehabilitation detailed and exhaustive plans have been drawn by the states. The conditions of the rehabilitation are such that the affected families are to be offered choices that should, in fact, make the quality of their lives better than the conditions they encountered in their original inhabitation. Eleborate mechanisms have also been established to monitor and ensure compliance and even a grievances redressal mechanism has been put in place. Govt. also take more than adequate care to protect the interests of the oustees.

## CASE STUDY

### The Chipko Movement (1987) (*India*)

*"... for its dedication to the conservation, restoration and ecologically-sound use of India's natural resources."*

The forests of India are a critical resource for the subsistence of rural people throughout the country, especially in hill and mountain areas, both because of their direct provision of food, fuel, fodder and because of their role in stabilising soil and water resources. As these forests have been increasingly felled for commerce and industry, Indian villagers have sought to protect their livelihoods through the Gandhian method of *satyagraha* or non-violence resistance. In the 1970s and 1980s this resistance to the destruction of forests spread throughout India and became organised and known as the Chipko Movement.

The first Chipko action took place spontaneously in 1973 and over the next five years spread to many districts of the Himalaya in Uttar Pradesh. The name of the movement came from a word meaning 'embrace': the villagers hugged the trees and thus saved them by putting their bodies in the way of the contracts' axes. The Chipko protests in Uttar Pradesh achieved a major victory in 1980 with a 15-year ban on green felling in the Himalayan forests of that State by order of India's the then Prime Minister, Indira Gandhi. The movement later spread to Himachal Pradesh in the north, Karnataka in the south, Rajasthan in the west, Bihar in the east and to the Vindhyans in central India. In addition to the ban in Uttar Pradesh, the movement succeeded in halting clear felling in the Western Ghats and the Vindhyas, as well as generating pressure for a natural resources. Policy is more sensitive to people's needs and environmental factors.

The Chipko Movement was the result of hundreds of decentralised and locally autonomous initiatives. Its leaders and activists have primarily been village women, acting to save their means of subsistence and their communities. Men have been involved, too, however, and some of them have given wider leadership to the movement. One of the most prominent leaders have been Sunderlal Bahuguna, a Gandhian activist and philosopher, whose appeal to Mrs Gandhi resulted in the green-felling ban and whole 5,000-kilometre trans-Himalayan footmarch in 1981-83 was crucial in spreading the Chipko message.

In the late 1980s, Bahuguna joined the campaign which already for many years had been opposing construction of a proposed Himalayan dam on the river near his birthplace of Tehri. In 1989 he began the first of a series of hunger strikes to draw political attention to the dangers posed by the dam and in due course the Chipko Movement gave birth to the Save Himalaya Movement.

Bahuguna ended a 45-day fast in 1995 when the Indian government promised a review of the Tehri dam project. But the promise was not kept and the following year he committed himself to another fast, only broken after 74 days when the Prime Minister gave a personal undertaking to conduct a thorough review, largely on Bahuguna's terms. The veteran environmentalist, then in his 70th year, told the Prime Minister that the Himalayan glaciers were receding at an alarming rate. If this was not checked, the glacier feeding the Ganges would disappear within 100 years.

***"We in Himalaya are facing a crisis of survival due to the suicidal activities being carried out in the name of development... The monstrous Tehri dam is a symbol of this... There is need for a new and long-term policy to protect the dying Himalaya. I do not want to see the death of the most sacred river of the world - the Ganga - for short-term economic gains."***

*– Sunderlal Bahuguna*

## WATER-RESOURCES

Water is a crucial natural resource, its availability greatly influences the health of the people and development potential of the area. Water as a resource in relation to its needs is becoming increasingly scarce. Proper assessment of the availability of this resource from surface and sub surface sources is crucial for its proper planning, development and efficient management.

About 70% of the global surface is covered with water in the form of oceans, seas, river, lakes, ponds. Total quantity of water available on the earth is 1386 million cubic kilometers. 97.3% of the water available on earth is saline and only 2.7 is available as fresh water. Most of which lies frozen in polar regions or in deep aquifers, not available for use.

The mean annual rainfall, taking the country as a whole is 1170 mm. This gives an annual precipitation of about 4000 $km^3$. A significant part of this precipitation returns back to the atmosphere as evaporation. A large part of the remaining precipitation seeps into the ground and the balance flows through streams, rivers and collects in water bodies adding to the surface flow. A part of the water which seeps in to the ground remains as soil moisture in the upper layers and the rest adds to the ground water resource.

### INDIA'S WATER RESOURCES

India is a country of vast biological, geographic, and climatic diversity. It has total geographic area of 329 Mha; excluding bodies of water, India's total land area is estimated at 297 Mha.

India is bordered in the north by the 2,500-kilometer long Himalayan Mountains. Melting snow and glaciers provide a continuous flow for numerous rivers running south from the Himalayas into the vast Indo-Gangetic Plain, which is dominated by the Ganges River and its tributaries. Heavy rains are typical in the Himalayas during the monsoon months between June and October, causing frequent floods. Southern India consists largely of the Deccan Plateau, which is flanked by the Western Ghats running along the west coast and the smaller Eastern Ghats on the east coast. The Deccan rivers are rainfed and fluctuate in volume; many of these rivers are not perennial.

India receives average annual precipitation of 4000 $km^3$, out of which 700 $km^3$ is immediately lost to the atmosphere, 2150 $km^3$ soaks into the ground, and 1150 $km^3$ flows as surface runoff (CGWB 1996, NCA12). See Figure 6.2.

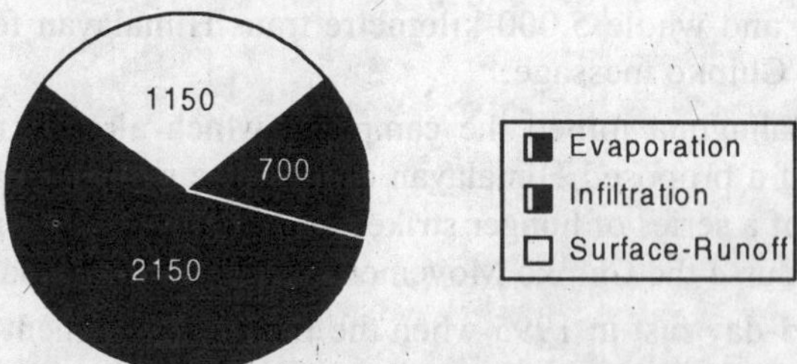


**Fig. 6.2.** Distribution of Precipitation (*Source:* Central Ground Water Board, 1965)

The basinwise details of water resources are given in Table 6.2. The total water resources in the country has been estimated by the National Commission as 1,953 $km^3$.

Nearly 62 per cent or 1,202 $km^3$ or total water resources lie in Ganga-Brahmaputra-Meghna basin. The remaining 23 basins have 751 $km^3$ of total water resources.

### River Basin

River basin is the basic hydrologic unit for planning and development of water resources. It is useful and convenient to make the assessment of water resources basin wise.

Earlier, the entire country was divided into twenty river basins by the Central Water Commission. These comprised of 12 major basins (having a drainage area exceeding 20,000 sq. km.) and 8 other river basins each combining a number of medium and minor rivers. However, it was seen that some areas such as the area of North Ladakh not draining into Indus and areas of Andaman, Nicobar and Lakshadweep islands are not covered in the 20 river basins. The Commission in consultation with CWC reviewed the grouping of the river basins. Based on this review, the drainage area of the country has now been divided into 24 basins, which includes 2 basins, namely Area of North Ladakh not draining into Indus and drainage area of Andaman, Nicobar & Lakshadweep Islands. The minor rivers of east coast from Mahanadi to Kanyakumari have been divided into five basins in place of the earlier two basins and the rivers draining into Myanmar and Bangladesh have been grouped under two separate basins, in place of the earlier one basin. The rivers of west coast south of Tapi are grouped into one basin in place of the earlier two basins. Area of inland drainage of Rajasthan which was one of the twenty basins, has been deleted. The list of basins alongwith their catchment areas and locations are given in Fig. 6.3.

During the review undertaken, it was noticed that there were some discrepancies in the reported drainage area of the Indus basin. Both the reports of the second Irrigation Commission (1972) and the Ravi & Beas Waters Tribunal (1987) had adopted the area of Indus Basin as 3,21,289 sq. km. on a watershed basis. However, as per the delineation of the basin boundary shown in the map of the basin in the CWC Publication (1989) "Major River Basins of India – An Overview" the area of the basin appears to be more than 3,21,289 sq. km. The Working Group has adopted the area of 3,21,289 sq. km. for the basin, since the same has been used by the earlier reports.

Based on the physiography, the river systems of India can be classified into four groups, viz. (i) Himalayan rivers, (ii) Deccan rivers, (iii) Coastal rivers, and (iv) Rivers of the inland drainage basin. The main Himalayan River Systems are those of Indus and Ganga-Brahmaputra-Meghna system. The Himalayan rivers receive very heavy rainfall in monsoon months and the rivers swell, causing frequent floods. The flows in the summer months are due to melting of snow and glaciers and therefore, have continuous flow throughout the year. The important river systems in the Deccan are West flowing rivers of Narmada and Tapi and the East flowing rivers of Brahmani-Baitarani, Mahanadi, Godavari, Krishna, Pennar and Cauvery. The Deccan rivers are rainfed and some of them are non-perennial. There are numerous coastal rivers which are comparatively small. Most of them are non-perennial. While only handful of such rivers drain into the sea near the deltas of east coast, there are as many as 600 such rivers on the west coast. The west coast rivers are short in length and have limited catchment areas. A few rivers in Rajasthan do not drain into the sea. They drain into salt lakes or get lost in sands with no outlet to sea.

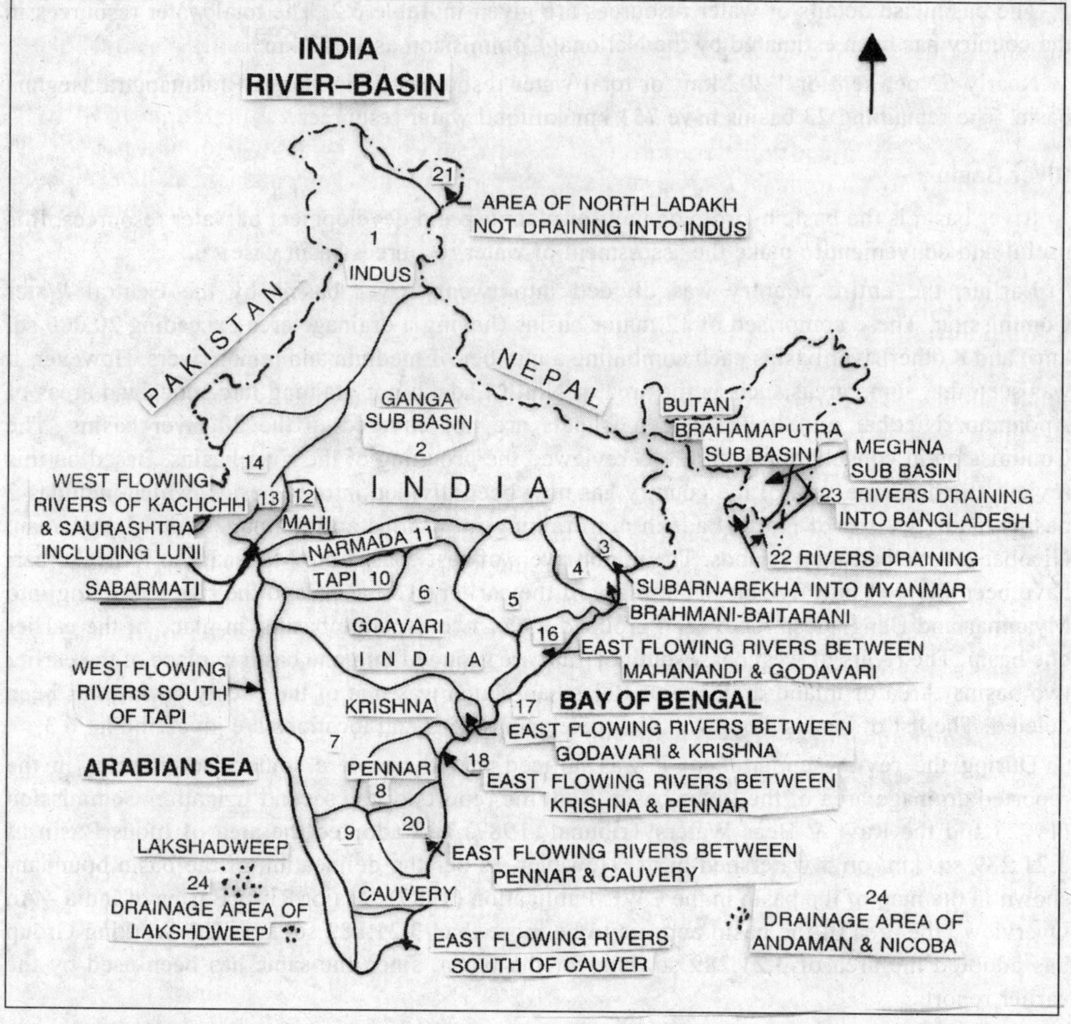


| *S.No.* | *Name of Basin* | *S.No.* | *Name of Basin* |
|---|---|---|---|
| 1. | Indus | 15. | West Flowing Rivers South of Tapi. |
| 2. | Ganga-Brahmaputra-Meghna Basin | 16. | East Flowing Rivers between Mahanadi and Godavari |
| 3. | Subarnarekha | 17. | East Flowing Rivers between Godavari and Krishna |
| 4. | Brahmani-Baitarani | 18. | East Flowing Rivers between Krishna and Pennar |
| 5. | Mahanadi | 19. | East Flowing Rivers between Pennar and Cauvery |
| 6. | Godavari | 20. | East Flowing Rivers South of Cauvery |
| 7. | Krishna | 21. | Area of North Ladakh not draining into Indus |
| 8. | Pennar | 22. | Rivers draining into Bangladesh |
| 9. | Cauvery | 23. | Rivers draining into Myanmar |
| 10. | Tapi | 24. | Drainage Area of Andaman and Nicobar and Lakshadweep |
| 11. | Narmada | | |
| 12. | Mahi | | |
| 13. | Sabarmati | | |
| 14. | West Flowing Rivers of Kachchh and Saurashtra including Luni. | | |

**Fig. 6.3.** India-River Basins

## AVAILABILITY OF WATER RESOURCES

India is one of the few countries in the world endowed with abundant land and water resources. Average annual precipitation including snowfall over the country is 4000 billion cubic meters (BCM). In addition, it receives another 200 BCM from rivers flowing in from other countries. Average annual water resources in various river basins are estimated to be 1869 BCM, of which the utilizable volume of water has been estimated to be 1086 BCM including 690 BCM of surface water and 396 BCM of ground water. The rest of the water is lost by evaporation or flows into the sea and goes waste. The utilization of water is expected to be 784 to 843 BCM by the year 2025. Though the present utilization level is only about 50%, the availability of water is highly irregular. It is not available in places of need, at times of need and in required quantities.

In the major part of the country, rainfall is the only sources for water which is available mainly during the monsoon season lasting for less than 3 months. Due to tropical climate and its geographical, location, the country experiences vast spatial and temporal variation in precipitation. About one-third of the country's area is drought prone. The south and western parts comprising the states of Rajasthan, Gujarat, Andhra Pradesh, Madhya Pradesh, Maharashtra, Tamil Nadu and Karnataka are the drought prone states. On the other hand, north and north eastern regions including states of Uttar Pradesh, Bihar, West Bengal and Assam are subjected to periodic floodings. The recent drought that the country is facing has not only created water scarcity but also led to migration of people and cattle from water scarcity areas to adjoining areas causing hardships to them but also social and economic problems in the areas where they migrate. In such areas, women have to fetch water from wells, ponds, lakes etc. from long distances which consume lot of time, energy and affect the health of the women folk. Therefore, concentrated and dedicated efforts are needed not only to harness the rain water but also conserve it so that it can be utilised in the time of need. The demand is maximum in summer when the available water resources dwindle. In the time of crisis, ground water is the most dependable source. The problem is more serious in urban areas, where natural recharge has been reduced due to reduction in permeable areas as a result of increased urbanisation.

### Surface Water

There are twenty river basins of which 12 are major basins having drainage areas of 20,000 km$^2$ (Fig. 6.3). The annual average runoff estimated to be 1952.87 BCM. The storage built up in various river basins through major and medium irrigation projects is about 173.73 BCM. The major and medium irrigation projects under construction and identified would account for 75.42 BCM and 132.3 BCM respectively taking the total to Rs. 381.45 BCM. If minor irrigation structures are included the total storage goes upto 420 BCM. However it is estimated that about 49.5 BCM and 75.2 BCM capacities of the existing reservoirs will be lost through siltation by the year 2020 and 2025 AD respectively. Total live storage capacity of all reservoirs would be about 462 km$^3$.

The average annual utilisable water resources through conventional schemes of all the 20 river basins taking into account the uneven nature of distribution of water resources and the topographic constraints is estimated to be 690.3 BCM which is about 35% of the total surface water resources. This indicates that vast amount of runoff is going as waste to sea.

The interbasin transfer proposals investigated by the National Water Development Agency (NWDA) envisage utilization of additional 200-250 km$^3$ of surface water. According to another study, carried out by the Central Ground Water Board (CGWB), an additional quantity of 160.5 km$^3$ water can be artificially recharged by utilizing surplus monsoon runoff. Many of these proposals are at conceptual stage only and would require further investigations and studies – and broad agreements between concerned State Governments – before a reasonable estimate of additional utilisable surface water can be made. In view of this any substantial use of surface

**Table 6.1. Rainwater availability in India**

| *Monsoon Activity* | *Mham* | *Percentage Share* |
|---|---|---|
| South West | 296 | 74 |
| North East | 12 | 3 |
| Pre monsoon | 52 | 13 |
| Post monsoon | 40 | 10 |
| Total | 400 | 100% |

*Source:* Fertiliser Statistics, FAI, New Delhi (1994)

water through interbasin transfer and/or artificial recharge by utilising surplus monsoon runoff is not for seen till the middle of the 21st century.

### Ground Water Resources

Replenishable ground water resources assessed is of the order of 431.32 BCM and utilisable ground water resources as 395.6 BCM of which 325.6 BCM is available for irrigation and the rest for domestic and industrial uses (Central ground water Board 1995). Thus the total utilisable ground water resource amounts to about 92% of potential replenishment.

The gross available and utilisable water resources of the country, are, therefore 2384.5 BCM and 1086.0 BCM respectively. The available and utilisable water resources per capita based on 1991 population are 2830 $m^3$ and 1288 $m^3$/capita/year respectively. However, there is a wider variation in its availability in basins. Out of the 20 basins, 4 basins had more than 1700 $m^3$/capita/year utilisable water resources while 9 basins had between 1000 and 1700 $m^3$/capita/year, in 5 basins it is between 500 $m^3$ to 1000 $m^3$/capital/year and 2 basins it is less than 500 $m^3$/capita/year. With increase in population which has crossed one billion the availability of land and water resources gets further reduced.

According to a standard definition for water availability if it falls below 1000 $m^3$/capita/year water supply begins to hamper health, economic development and human beings resulting in scarcity conditions requiring to take up immediate remedial measures. At less than 500 $m^3$/capita/year, water supply is primary constraint to life. So many river basins are already facing water stress requiring remedial measures.

According to the initial estimates of the Central Ground Water Board an additional quantity of about 10,081 $km^3$ of static ground water can be exploited. Generally static water is not regular replenished on annual basis, its one time use is only possible as a short time strategy. Further experience indicates that large scale use of static water or its mining is usually associated with surface settlement which may cause heavy damages to properties of land and soils and may, possibly, trigger seismic activities. Therefore, any appreciable use of static ground water on regular basis, is not foreseen till the middle of the 21st century.

## CONTRIBUTION OF RAINFALL TO WORLD'S WATER SUPPLY

Rain is the main source of fresh water. Its contribution is as

| | |
|---|---|
| 1. Water in sea & oceans | 97.2% |
| 2. Water in glaciers and ice caps | 2.14% |
| 3. Underground water | 0.16% |
| 4. Surface water | 0.009% |
| 5. Soil moisture | 0.005% |

Rainfall is seasonal and is not evenly distributed. On an average, annually about 112 cm rain is received over the plains of India. Highest rainfall in India is about 250 cm while lowest is 15-20%. In Cherrapunji (Meghalaya) rainfall is about 1087 cm annually.

## WATER DEMAND

During 1990 the total utilisation of water for all uses, was about 518 BCM or 609 $m^3$/capita/year.

The water demand to meet the requirement of domestic, industrial and irrigation is given in Table 6.2.

**Table 6.2. Requirement of water for various uses in BCM**

| *Sl. No.* | *Category* | *Year 2010* | | | *Year 2025* | | | *Year 2050* | | |
|---|---|---|---|---|---|---|---|---|---|---|
| | | *Low* | *Medium* | *High* | *Low* | *Medium* | *High* | *Low* | *Medium* | *High* |
| 1 | Irrigation | 489 | 536 | 556 | 619 | 688 | 734 | 830 | 1008 | 1191 |
| 2 | Domestic | 39.4 | 41.6 | 61 | 47 | 52 | 78 | 59 | 67 | 104 |
| 3 | Industries | 37 | 37 | 37 | 61 | 67 | 79 | 69 | 81 | 116 |
| | | 555.4 | 614.6 | 654 | 727 | 807 | 881 | 958 | 1156 | 1411 |

The total medium water requirement of the country would be for the above purposes is about 614.6, 807 and 1156 BCM by the years 2010, 2025 and 2050 respectively.

Water demand for irrigation to maximise agricultural production with the maximum possible level of irrigation to achieve self sufficiency in food grains is given in Table 6.3.

**Table 6.3. Water demand for irrigation**

| *Year* | *Low demand scenario million tons* | *Water requirement BCM* | *Middle demand scenario million ton* | *Water requirement BCM* | *High demand scenario million ton* | *Water demand BCM* |
|---|---|---|---|---|---|---|
| 2010 | 249 | 489 | 265 | 536 | 271 | 576 |
| 2025 | 322 | 619 | 349 | 688 | 365 | 734 |
| 2050 | 469 | 830 | 539 | 1088 | 605 | 1191(1451$km^3$) |

Other than three requirements, water also required for energy development, inland Navigation, Flood Moderation, ecological considerations, to compensate evaporation losses from reservoirs, environment afforestation, etc. If the country desires to irrigate all cultivable land at 100% intensity of irrigation by the year 2050, the water requirement would be 1451 $km^3$ and the total requirement will be 1824 $km^3$. This is a hypothetical and rather unrealistic scenario.

## HYDROLOGICAL CYCLE (WATER CYCLE)

In this solar-driven water cycle, (see figure on next page) water evaporates from the earth's surface into the atmosphere and is returned back as rain or snow. Part of this precipitation evaporates back into the atmosphere. Another part flows into streams, rivers, and lakes, commencing its journey back to the sea. Still another part percolates into the soil and becomes soil moisture or groundwater. Plants extract soil moisture into their tissues and release it into the atmosphere in the process of evapotranspiration. Much of the groundwater eventually works its way back into the flow of surface waters.

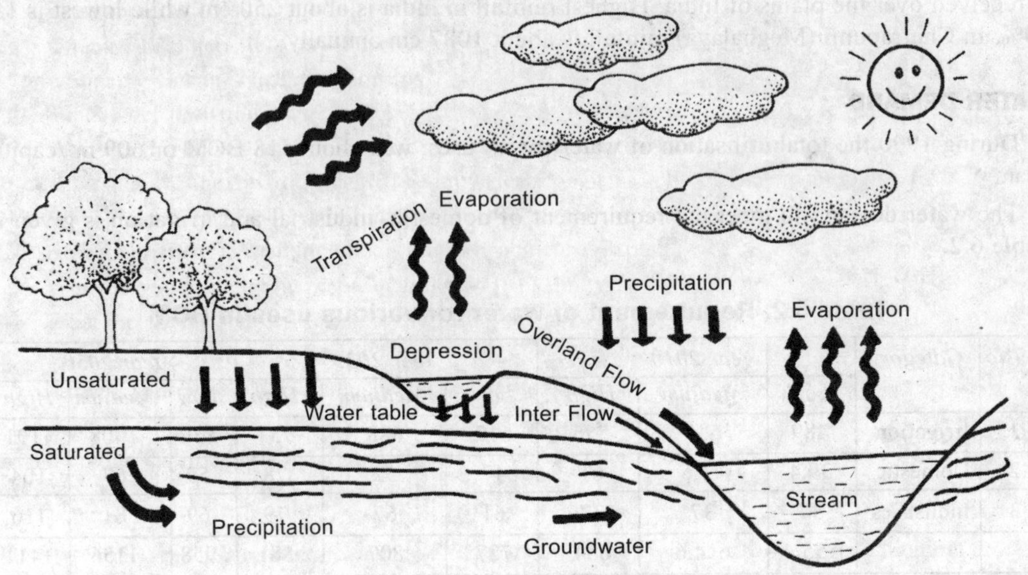


**Fig. 6.3.** Hydrological Cycle (Water Cycle)

## FLOODS

Floods are the most frequent natural calamities faced by India. Floods are more regular and severe in the Brahmaputra Barak basin and Ganga basin in the country. On an average floods are affecting about 33 million persons per year. Total flood prone area in the country is about 40 m.ha. Out of above about 14 m.ha of flood prone area has been provided with reasonable degree of protection.

**Floods.** Many parts of the country, particularly the areas drained by large river systems, suffer from devastating floods. Widespread floods occurred in 1878, 1892, and 1917 in the then many provinces of India, and again in many states of the country in 1954. Rainfall, exceeding 40–50 cm in 24 hours, occurs occasionally along the west coast, in the coastal districts of Tamil Nadu and in southern Andhra Pradesh in Assam, West Bengal and in the foot-hills of the Himalayas.

Some of the long-term measures to minimize the damage because of floods consist in the afforestation of the upper catchment areas of the rivers, the construction of river embankments, and the execution of multi-purpose river valley projects. Individual farmers can lessen the damage to their crops owing to heavy rains or floods by making the necessary drainage channels in their fields. Plans are in hand for forecasting floods and for arranging expeditious facilities to take timely action to save the standing crops from devastation by floods. A flood meteorological organization has been drawn up for the country and is under implementation during the current and the subsequent plan periods.

## DROUGHT

Due to deficient rainfall 150 districts are drought prone in India. Out of above 7 districts of Punjab, Haryana, Rajasthan, Gujarat, Tamil Nadu, Uttar Pradesh, Maharashtra, Karnataka and

Andhra Pradesh are provided with irrigation. Among states West Bengal, Madhya Pradesh, Andhra Pradesh, Bihar, Jharkhand and Orissa are suffering from high frequency of droughts.

Food production in India being marginal, drought poses many problems. Irrigation facilities available in this country are also limited and, therefore, when droughts occur, they cause partial or complete crop failure. If failures occur in consecutive years, it becomes a national calamity, putting great strain on the economy of the country.

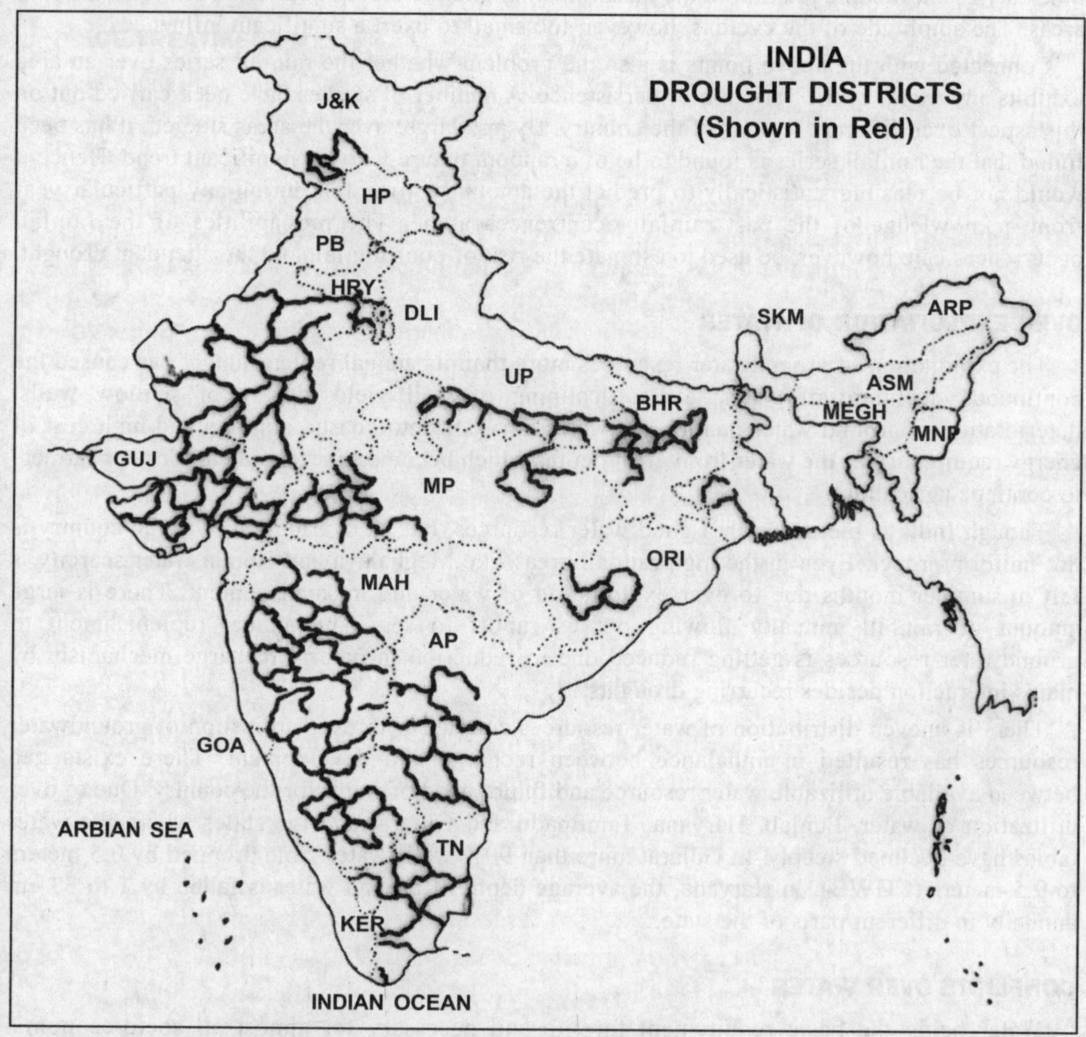


According to the practice followed by the Indian Meteorological Department, drought is taken to have occurred over an area where the annual rainfall is less than 75 per cent of the normal. When the annual rainfall is less than 50 per cent of the normal, it is called severe drought. Drought can occur anywhere in India, depending upon the distribution of rainfall.

Attempts to quantify drought in the form of an index have been made by using various techniques. Pilot studies have been conducted in India, with special reference to Bihar State, in order to arrive at a rational definition. The Palmer drought index, which takes into account rainfall, evapo-transpiration and soil moisture is considered a comprehensive approach to this problem. The computation of the Palmer drought index have been carried out for the different sub-divisions of the country. The computations show that, on the average, drought is experienced on 20-25 per cent of the days in each of the months of the *kharif* season over large areas of the country.

Another aspect of great interest in respect of drought is whether there is any periodicity in its occurrence. The power-spectrum analysis of the rainfall series and that of the Palmer drought index series show some relation to the quasi-biennial and the eleven-year sunspot cycles in some areas. The amplitude of the cycle is, however, too small to exert a significant influence.

Connected with the above points is also the problem whether the rainfall series over an area exhibits any trend, cyclic variation or persistence. A number of studies have been carried out on this aspect over different regions of the country. By and large, over the areas studied, it has been found that the rainfall series is found to be of a random nature with no significant trend. Hence it would not be feasible statistically to predict the amounts of rainfall during any particular year from a knowledge of the past rainfall occurrences alone. The probabilities of the rainfall occurrences can, however, be used to estimate the risk of poor rainfall and the attendant drought.

## OVER EXPLOITATION OF WATER

The exploitation of groundwater resources more than its annual replenishment has caused the continuous declining of water levels, declining of well yield, drying of shallow wells, deterioration of ground water quality, sea water intrusion into coastal aquifers and high cost of energy required to lift the water from great depths which becomes uneconomical for poor farmers to continue agriculture.

Though India is blessed with a good water resources, but its distribution over the country is not uniform/proper. Even in the high rainfall areas like Meghalaya and Kerala water scarcity is felt in summer months due to over exploitation of water and mismanagement. There is large amount of rainfall annually flowing out as runoff to sea. The annual replenishment to groundwater resources is getting reduced due to reduction in natural recharge mechanism by man's interaction desides recurring droughts.

There is uneven distribution of water resources coupled with over utilization of groundwater resources has resulted in imbalance between recharge and development. There exists gap between available utilizable water resource and future need of water for the country. Due to over utilization of water, Punjab, Haryana, Tamilnadu and Gujarat and the states where the water tables have declined steeply. In Gujarat more than 90% wells water table dropped by 0.5 meters to 9.5 meters (CGWB). In Haryana, the average depth of ground water is fallen by 1 to 33 cm annually in different parts of the state.

## CONFLICTS OVER WATER

Water being the basic requirement for life and necessary for almost all socio-economic activities is facing ever greater demand. Its relative demand increases with degree of scarcity. As we have noted above, a large part of the country already faces water scarcity conditions and it is expected that by the middle of the next century most regions of the country would face some degree of scarcity. These conditions have already created a number of inter state water dispute. If the such conditions continue, we can also expect the next world war will be on water. Bitterness over these disputes is increasing with passage of time. These water disputes have multiple facets, examples of two such instances are as :

1. Urban water demands are concentrated in space, therefore, pose serious problems at local levels. Water demands in mega cities are growing much faster than envisaged and are putting

heavy strain on water resources. It is creating difficult problem for the surrounding rural areas, leading to serious conflicts.

2. Since the urban water supply are met from surface (river) flows, there will be conflict with upstream users, specially farmers, over the quantum of withdrawls, while the down stream users will be affected by the polluted waste waters released by urban areas. Such conflicts already exists : between Delhi and Haryana, and between Chennai and the farmers in drought prone districts of Andhra Pradesh.

In future such conflicts are likely to increase in number and escalate in magnitude unless an effective mechanism is evolved to resolve than expeditiously and judiciously.

## MINERAL RESOURCES

Minerals, being the vital raw material for many basic industries, play an important role in the industrialisation and overall development of nation. Minerals are generally called the "stock" as they are the non-renewable resources. Minerals are the definite chemically bonded substances, created through chemical processes between organic and inorganic matters present in the earth's crust. They may be solid or liquid. Since the prosperity of a nation depend upon the proper use of minerals, hence they should be conserved and should not be misused. Govt. should promote the research in this field of mining minerals.

The history is hundreds year old. Iron, steel, copper, zinc, lead, gold, silver, cobalt etc. metals were extracted from minerals in India. But now, building materials coal, iron ore, mangenese ore, gold, petroleum, natural gas, copper ore, ilmenite, glass sand, lead and zinc ores, chromite, raynite, silmenite, magnesite, gypsom, monazite, beryl, dolomite, bauxite etc. are produced from minerals in India. The minerals from metals like bismuth, cadmium, graphite, platinum, tungsten tin, silver, gold are extracted, are in least quantity.

### Types of minerals

Minerals available in earth crust can be divided into three types

1. Metallic minerals
2. Non-metallic minerals
3. Mineral fuels

Some other classifications of minerals are also given by scientists. They are classified as strategic and critical depending on the use and importance.

**1. Metallic Minerals:** We cannot extract metal directly from minerals. There is difference between minerals and ores. Therefore, for extracting metals, minerals are treated by different processes before extraction. Metallic minerals are generally found in combined state. According to availability of metals, metallic minerals are further divided into following :—

(*a*) Ferrous alloys: Most common metal (which is used largely) is iron. Other than iron are aluminium, lead, zinc, copper etc. All are found in rich quantities, found in native as well as in combined state. Iron pyrite, Lynonite, Haematite, Magnetite are examples of ferro alloys. Certain other metals, non-metals are contaminated with these as impurities.

(*b*) Non-ferrous alloys: The minerals/alloys of this type contain the metals like titanium, antimony, arsenic, beryllium, copper, zirconium, cerium, lithium etc. These metals are costlier than proceeding metals. Here the iron found as an impurities.

(*c*) The minerals/alloys containing very least quantity of metals whose extraction is costlier. These metals are generally used in jewellary eg. gold, platinum, silver, irridium etc.

**2. Non-metallic minerals :** Minerals, whose yield products are other than metals comes in this head. They are called the non-metals. They are further divided on the basis of physical and chemical properties. Graphite, pyrolusite, dolomite quartz, kaoline, fire clay, felspar, mica, asbestos, gypsom fluorite, chrome/red ochre, lime stone, borax, phosphorite, ilmanite, flint, dymond, calcite, sand stone, stones like phylite, cyanite, lime stone, ruby, sapphire. Emarald, amber, spodumene etc. are the examples of non-metallic minerals.

**3. Mineral fuels :** These include the materials used to provide energy, for example coal, natural gas, fossil fuels and petroleum etc. These are the important source of energy, hence they have tremendous importance for mankind.

Coal is the most commonly available fuel which is used as domestic as well as industrial fuel. It is of different type i.e. Anthracite, Bituminous, Lignite etc. The type and quality of the coal depend upon the percentage of carbon present in them. It is the principal source of energy in world. It is used in various ways in different industries like cement, glass, railways, textile, sugar, paper, steel etc. It is also largely used in domestic way. USA, China, Britain, Germany, South Africa, Australia are richest coal containing countries in world.

Petroleum is used in the manufacture of large number of petro-chemicals. It is drilled out from the sources as crude oil. Crude oil is refined before use as petrol, diesel, kerosine etc.

**Minerals in nature :** The man is using minerals since long. From lacs of year back primitive man was using flint, quartiz etc. for preparation of their tools. This was called "stone age". Later they use metals therefore, the period was named after as "copper age", "bronze age", and "iron age". Now present age is "machine age" because machines are prepared from minerals and they run by mineral fuels.

The formation of mineral deposits is a very slow geo-chemical or biological process, which takes millions of years to develop mineral deposits. Most of the minerals are widely distributed in earth's crust. Studies shows that, there are number of ways by which mineral deposits are formed. They are

1. Molten rock materials, which is a complex collection of a number of substances, when cooled, the crystallization of different minerals takes at different temperatures. These are settled in different bands, giving the mineral deposits.

2. Sodium chloride, gypsom, salt peter etc. Water soluble minerals are obtained by evaporation of lake/sea water. The compounds of iron and manganese as chemical sediments are also formed by precipitation from lake or sea water.

3. Deposits of minerals like asbestos, talc, graphite etc. are formed intense heat and pressures inside earth's crust.

4. When the pH, temperature, solubilities are changes, the rock materials in solution/ suspension are deposited in sufficient amounts to form mineral deposits as water current slow down.

5. Mineral deposits are also formed by oxidation and reduction reactions.

6. Formation of mineral deposits are also take place by micro-organisms. It is mainly autotrophic bacteria which are involved in mineralization reactions.

There are also other views for formation of mineral deposits. When the plants, dead animals, wild life & other ecosystems are accumulated below in earth. Biological process convert them in to mineral deposits.

**Mineral resources of India :** India has sufficient quantities of iron, aluminium, titanium, copper, lead, zinc ores. India is fairly rich in mineral resources. We possess good deposits of most of mineral elements which we needed in large quantities. However, other economically important minerals are not present in sufficient quantities. Iron minerals, which are most important ingredient of today's economy are found in sufficient quantity in our country. We are presently exporting it to other countries. The similar case is for aluminium also. Our country ranks Vth among the aluminium rich countries of the world. At present we have sufficient aluminium stock for the domestic market. We are exporting to other countries as well.

Zinc, lead ore reserves in India are estimated to be about 390 million tons. Good quality of these elements minerals/ores are being depleted at a fast rate.

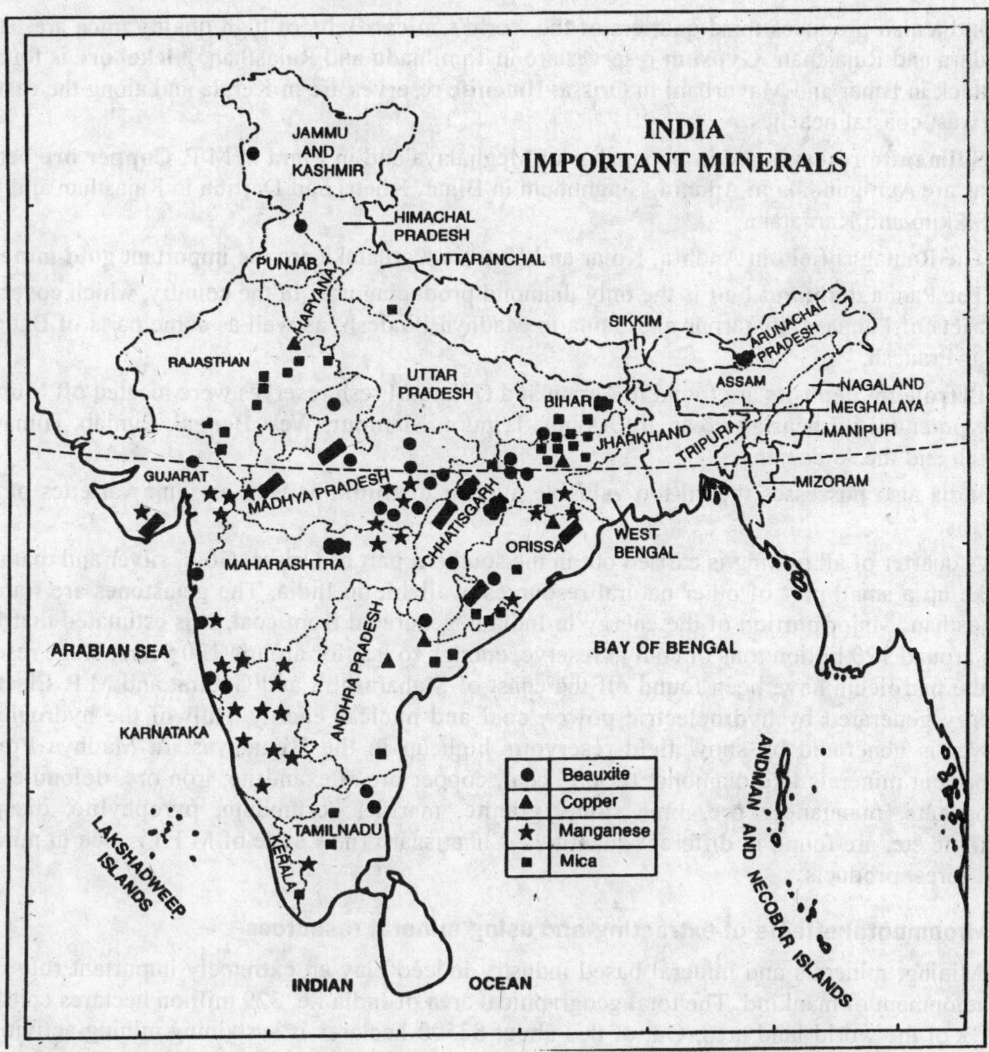


India has a large number of economically useful minerals and they constitute one-quarter of the world's known mineral resources. About two-thirds of its **Iron deposits** lies in a belt along Orissa and Bihar border. Other haemaite deposits are found in Madhya Pradesh, Karnataka, Maharashtra and Goa. Magnetite Iron-ore is found in Tamilnadu, Bihar and Himachal.

India has the world's largest **deposits of coal.** Bituminous coal is found in Jharia and Bokaro in Bihar and Raniganj in West Bengal. Lignite coals are found in Neyveli in Tamilnadu.

Next to Russia, India has the largest supply of **Manganese.** The manganese mining areas are Madhya Pradesh, Maharashtra and Bihar-Orissa area. **Chromite deposits** are found in Bihar, Cuttack district in Orissa, Krishna district in Andhra and Mysore and Hassan in Karnataka. **Bauxite deposits** are found in western Bihar, southwest Kashmir, Central Tamilnadu, and parts of Kerala, U.P. Maharashtra and Karnataka.

India also produces third quarters of the world's **mica.** Belts of high quality mica are, Bihar, Andhra and Rajasthan. **Gypsum** reserves are in Tamilnadu and Rajasthan. **Nickel ore** is found in Cuttack in Bihar and Mayurbanj in Orissa. **Ilmenite** reserves are in Kerala and along the east and the west coastal beaches.

**Silimanite reserves** are in Sonapahar of Meghalaya and in Pipra in M.P. **Copper ore** bearing areas are Agnigundala in Andhra, Singhbhum in Bihar, Khetri and Dartiba in Rajasthan and parts of Sikkim and Karnataka.

The Ramagiri field in Andhra, Kolar and Hutti in Karnataka are the important **gold mines.**

The Panna **diamond belt** is the only diamond producing area in the country, which covers the districts of Panna, Chhatarpur and Satna in Madhya Pradesh, as well as some parts of Banda in Uttar Pradesh.

**Petroleum deposits** are found in Assam and Gujarat. Fresh reserves were located off Mumbai. The potential oil bearing areas are Assam, Tripura, Manipur, West Bengal, Punjab, Himachal, Kutch and the Andamans.

India also possesses the all-too valuable nuclear **uranium** as well as some varieties of **rare earths.**

A quarter of all mining is carried out in the southern part of Orissa. Gold, silver and diamonds make up a small part of other natural resources available in India. The gemstones are found in Rajasthan. Major portion of the energy in India is generated from coal. It is estimated that India has around 120 billion tons of coal in reserve, enough to last for around 120 years. Huge reserves of the petroleum have been found off the coast of Maharashtra and Gujarat and M.P. Electrical energy generated by hydroelectric power, coal and nuclear energy. Half of the hydroelectric power is generated by snow field reservoirs high up in the Himalayas. In Madhya Pradesh important minerals like diamond, tin ore, coal, copper ore, alexandrite, iron ore, dolomite, rock phosphate, manganese ore, lime stone, granite, marble, corumdum, pyrophylite, diaspore, Bauxite etc. are found in different quantities. Chhatisgarh (new state of M.P.) is rich in minerals and forest products.

### Environmental effects of extracting and using mineral resources

Mining, minerals and mineral based industry indeed play an extremely important role in the development of mankind. The total geographical area of India i.e. 329 million hectares constitute 2.4% of the world land area. Out of this about 82500 hectares is sustaining mining activities of some kind or the other. As the mining activity grows, the per capita availability of land is declining at a very high rate. The extra emphasis on mining and minerals is directly related to growing population and better standard of living.

The environment means the surroundings. The components of environment include soil, water, air, land, landscape, and living creatures. The environment is more damaging by open cast mining than underground mining. Not only environment, mining also effects human health. Over exploitation causes the wastage of mineral wealth and derelict of land. Mineral deposits should not be over exploited because they are non-renewable. Derelict land is that land which has been abandoned as useless. Dereliction is the result of thoughtless, uncontrolled ruthless exploitation of natural resources. This land is the permanent damage not usable for agriculture. There are following environmental effects of mining—

(*i*) Land degradation due to lowering of the surface levels at some places and creation of large mounds at other places;

(*ii*) Deforestation in the mining areas, i.e., the loss of valuable cover resulting in the possibility of enhancement of soil-erosion;

(*iii*) The loss of top and sub-soil;

(*iv*) Adverse effect on ground water table. The local water table is lowered as a result of opencast mining. The replenishment of acquifers is adversely affected as the mined out terrains are completely dismembered of acquiducts leading to acquifers. As a result the affluent discharge of rain water is increased leaving the water-table completely or partly unrecharged. This also increases the salinity of remaining ground water;

(*v*) Due to increased discharge of rain water passing through the terrains, disturbed by surface mining, the local drainage system is polluted, which on joining the main drainage feature, affects it also;

(*vi*) The frequency of land slides increases substantially as a combined result of factors as stated above;

(*vii*) The erosion of soil is enhanced;

(*viii*) The agricultural lands are affected by silt and the fine material mined but not recovered. It also clogs the surface water channels;

(*ix*) The disturbance caused adversely affects the well-balanced pH and diminishes the regenerative qualities of soil, etc.;

(*x*) The disturbance caused to the floral and faunal population is immense;

(*xi*) The heavy earth-moving machinery and blasting cause problems of noise, vibration and the release of noxious gases in the atmosphere;

(*xii*) The aesthetic damage caused to the landscape reduces its recreational value;

(*xiii*) Mine drainage has polluted streams, rivers, lakes even seas;

(*xiv*) Fumes from smelters damage forests and spread pollution over large area (air pollution);

(*xv*) Mining and mineral based industries with their effluents create pollution problems. Asbestos, cement and other chemical industries are very hazardous. People are not supposed to live in surrounding areas;

(*xvi*) Mining causes the reduction of forests i.e. deforestation. Thus flora and fauna are also destroyed. Wild life also effected. Land becomes barren and this results in increased incidents of land slides;

(*xvii*) The people related with mining and extraction effected by polluted environment (Dust & poisonous gases) lead to skin and lung diseases;

(*xviii*) Mining affects the sub segments of the environment like forests, vegetation, soil cover, humus and ground water. Dust and toxic gases indirectly affects air, humidity, temperature;

(*xix*) Deforestation and climatic change results poor rainfall and affects flora & fauna.

## WORLD FOOD PROBLEM

Before the 21st century, it was felt that world food production is not sufficient for the present population. Food production was less because people were using the old techniques, seed etc. Later on when population pressure starts, the new ways of food production, using fertilizers, pesticides, insecticides etc. are discovered to increase the yield. In 1999 International Food Policy Research Institute (IFPRI) reported the increase in world food consumption by 2020, discussing the impact of this on both developed and developing countries. The report considers the six emerging issues, nutrition, grain prices, WTO, agroecological approaches to small scale farming, biotechnology, information technology and precision farming. In world food submit 1996 in Rome the following points were discussed—

(*i*) Reduce world hunger

(*ii*) Agricultural supply and demand

(*iii*) Population growth

With respect to crop production and agricultural growth the similar discussion on "goals, solutions and actions necessary to end hunger" were also held in IFPRI conference held in Bonn, Germany 2001. All this shows the awareness of we people to increase the food production in view of increasing population growth. World leaders also agreed that the contribution of irrigation to incremental food production should be substantial.

Different scenarios have been examined to explore a number of issues such as the expansion of irrigated agriculture, the increase in food production in rainfed areas, and the public acceptance of genetically modified crops. Some opinion were that the world may face urgent food and agriculture problems. Analysis believe that what is needed is a new and greener revolution to once again increase productivity and boost production.

Severe droughts and sharply rising food prices spurred national governments and international agencies to address the food crisis of the 1960s and 1970s. The 'Green Revolution', consisting of crop variety improvements, increased use of fertilizers and expansion of irrigation, averted the projected shortages in food production. According to some experts, another food crisis predicted by advocates of a new boom in investment for irrigation is not yet in view. Food grain prices have remained stable for the last 15 years. There is hunger in the world, but that is because the hungry cannot translate their needs into demand or civil disorders disrupt food flows. However, according to the authoritative Consultative Group on International Agricultural Research (CGIAR), the world is entering the 21st century on the brink of a new world food crisis that is as dangerous, but far more complicated than the threats it faced in the 1960s (Shah and Strong, 2000).

Much could be said on the role of demographic and economic factors, such as world trade, price commodities and agricultural subsidies to farmers in meeting the challenge. However, the purpose of this paper is not to contribute more to the dabate between experts on food security. It is to examine the probable consequences of the business-as-usual scenario that has been the prevailing model for the development of irrigated agriculture, particularly of the large-scale irrigation systems, in many countries. It also projects the likely benefits of increased investment in irrigation and advocates a new approach to design and management of irrigation systems in association with institutional and policy reforms.

## CHANGES CAUSED BY AGRICULTURE

Agriculture in US and India was food raising activities, and over half of the current crops comes from plants such as corn, cotton, potatoes, tobacco etc. The agricultural methods in most parts of the world were primitive. Fields were dug by oxen pulling wooden plows, seeds were broadcast by lands and grains were harvested with scythes. From the Indians the first American settlers learned how to clear land till the fields and grow the corn, that was crucial to their initial survival. The mid-1800s began an era of great change which brought advances in cultivation method, breeding of improved crop varieties and use of fertilizers and crop rotations to maintain soil productivity. A vigorous market soon developed for soil amendments such as guano, manure, crushed bone and lime. By 1860 seven factories had been established in US to manufacture mixed chemical fertilizers. The use of pesticides also began. By 1893 there were 42 patented insecticides offered by several manufacturers. In 1840 the benefits of irrigation were discovered. Govt. passed several laws to assist western states in developing extensive and costly irrigation systems. Farm labour requirements diminished with the introduction of mechanised invention of machines for tilling, planting, reaping and threshing vastly increased farm efficiency. In 1892 the first successful gasoline powered tractor was introduced in Iowa. Later on tractors gradually became popular.

Ever since colonial days, agricultural leaders have been interested in increasing the productivity. As land became less available, people became more interested in maintaining soil fertility and increasing crop yields. In 1914 Govt. responded to this need by providing funds for state agricultural extension programme assist farmers in adopting improved farming methods. In 1930s national attention was focussed on the need for soil and water conservation measures to maintain farm productivity. The rich soils produced bountiful crops and between 1870 to 1910 the population of plains states increased by a factor of 10. In response to soil and water conservation programmes, the soil conservation service was established in 1935. In America untill world war II, productivity increased in view of additional demand for food. Conversion of animal power to mechanical power resulted in increased output. Use of fertilizers and pesticides including DDT increased by 50% in between 1940–1944. But in India the productivity was not so, because England ruled India till 1947. The production was not sufficient. Some time we import the food grains and other food materials. After the independance we progressed by giving the facilities to farmers. Farmers didn't have high quality mechanical power, seeds, fertilizers, pesticides etc. For all people were dependent on Govt.

## FOOD RESOURCES

We know that only plants can produce food, hence they are called producers. Food, which is necessary for all living organisms consists of proteins, enzymes, carbohydrates, minerals etc. There are various types of animals depend upon the types of food. Herbivores take plants directly as food. Meat eaters are called Carnivores. Animals eat every thing i.e. vegetables or meat called Omnivores. Some animals eat flesh of dead animals are called Scavengers. Thus agriculture and domestic animals are the principal sources of food.

Cereals, pulses, grains, vegetables, fruits etc. we get from agriculture. Domestification of cattle and poultry are necessary for food production from animals. Fish is the another animal source of food. For the last so many years, our food production increased by 50%, at the same time population growth is increased. Still we are not getting production at par with other countries like U.S.A. The 1996 world food summit warmed the countries to reduce the number of undernourished and malnutrition.

The land and water resources are limited, therefore, it is needed to increase the productivity, urgent steps are also needed to reduce the population. Fishes also play an important role to fulfil the world food demand to some an extent. In India, open ocean is under utilised and coastal areas are overexploited for fish industry. Thus fish production from aquatic resources goes high, solving the food problem.

## OVERGRAZING

Grazing management is the foundation of grassland based live-stock production since it affects both animal and plant health and productivity. Overgrazing can occur under continuous or rotational grazing. It can be caused by having too many animals on the farm or by not properly controlling their grazing activity. Overgrazing reduces plant leaf areas which reduces interception of sunlight and plant growth. Plants become weakened and have reduced root length and pasture sod weakens. The reduced root length makes the plants more susceptible to death during dry weather. The weakened sod allows weed seeds to germinate and grow. If the weeds are unpalatable or poisonous, major problems can result.

One indicator of overgrazing is that the animals run short of pasture. Under continuous grazing, overgrazed pastures are predominated by short grass species such as blue grass and will be less than 2-3 inches tall in the grazed areas. Soil may be visible between plants in the stand, allowing erosion to occur. Under rational grazing overgrazed plants do not have enough time to grow to the proper height between grazing events. The animals are turned in to a paddock before the plants have restored carbohydrate reserves and grown back roots lost after the last defoliation. As the sod thins, weeds encroach in to the pasture. Another indicator is that the livestock run out of pasture, and hay needs to be fed early in the fall. Overgrazing is also indicated in livestock performance and condition.

Overgrazing can increase soil erosion. Reduced soil depth, soil organic matter and soil fertility hurt the land's future productivity. Soil fertility can be corrected by applying the appropriate lime and fertilizers. However the loss of soil depth and organic matter takes years to correct. Their loss is critical in determining the soil's water holding capacity and how well pasture plants do during dry weather.

To prevent overgrazing, match the forage supplement to the herd's requirement. This means that a buffer needs to be in the system to adjust for the last spring growth of cool season forages. One buffer may state producers use is to harvest hay in May and June and allow the cattle to graze the after math in August and September.

Another potential buffer is to plant warm season perennial grasses such as switchgrass, which do not grow early in the session. This reduces the acreage that the live stock can use early in the session, making it easier for them to keep up with the cool season grasses. The animals then use the warm-season grasses during the heat of the summer, and the cool season grasses recover for all grazing.

## EFFECTS OF MODERN AGRICULTURE

Between 1950 and 1975 agricultural productivity in American history changed more rapidly. Total farm output increased more than half. This change was due to technological innovations, development of hybrid strains and other genetic improvements and a fourfold increase in the use of pesticides and fertilizers. Not only in America, all over world productivity and means of farming were changed. Thus the agriculture has become more intensive producing higher yields per acre. It also has become more expensive, relying on purchase of machinery and chemicals to replace heavy labour requirements of the past. To remain competitive farmers have been forced to become more efficient.

Although the intensification of agriculture has vastly increased productivity, it also has had a number of potentially detrimental environmental consequences, ranging from rapid erosion of fertile topsoils to contamination of drinking water suppliers by the chemicals used to enhance farmland productivity.

## IMPACTS

**Damage to Soil.** Soil erosion from farmland threatens the productivity of agriculture fields and causes a number of problems elsewhere in the environment. An average of 10 times as much soil erodes from American agricultural fields as is replaced by natural soil formation processes. Because it takes up to 300 years for 1 inch of agricultural topsoil to form, soil that is lost is essentially irreplaceable. The consequences for long-term crop yields have not been adequately quantified. The amount of erosion varies considerably from one field to another, depending on soil type, slope of the field, drainage patterns, and crop management practices; and the effects of the erosion vary also. Areas with deep organic loams are better able to sustain erosion without loss of productivity than are areas where topsoils are shallower.

Erosion affects productivity because it removes the surface soils, containing most of the organic matter, plant nutrients, and fine soil particles, which help to retain water and nutrients in the root zone where they are available to plants. The subsoils that remain tend to be less fertile, less absorbent, and less able to retain pesticides, fertilizers, and other plant nutrients. Why then is erosion allowed to continue at excessive levels on many U.S. farms? Often the short-term costs of implementing erosion control measures far exceed the immediate economic benefit to the farmer, but such cost-benefit analyses fail to take into account the long-term losses of fertility and water-holding capacity of the soil. Up to a certain point, increased fertilization and irrigation will compensate for the lower soil fertility. Long-term loss of farmland productivity and damage to the environment from eroded sediments, therefore, often are overlooked in the need for short-term economic gains.

Over the past 50 years, the negative effects of soil erosion on farm productivity have been masked by improved technology and increasing use of fertilizers and pesticides. Ironically, many of these measures used to increase the short-term productivity of American farms are also causing excessive erosion, which threatens productivity over the long term. For example, diminished use of cover crops leaves soils unprotected from wind and rain during much of the year, and increased mechanization has led to use of larger fields without windbreaks or drainage contours.

The effects of erosion are also felt elsewhere in the environment. A recent study estimated the off-site cost of cropland erosion in the United States to be in the range of a billion dollars per year (Clark, Haverkamp, and Chapman 1985). Eroded soil clogs streams, rivers, lakes, and reservoirs, resulting in increased flooding, decreased reservoir capacity, and destruction of habitats for many species of fish and other aquatic life. The eroded soils contain nutrients and other chemicals that are beneficial on farm fields, but can impair water quality when carried away by erosion. As a result, drinking water supplies may contain nitrate or organic chemicals in concentrations that exceed public health standards, or surface waters may become clogged with excessive plant growth from the added nutrients.

Even when soil erosion is not excessive, intensive agriculture can impair soil quality by depleting the natural supplies of trace elements and organic matter. In natural ecosystems, soil fertility is maintained by the diverse contributions and recycling of nutrients by a wide range of plant and animal species. When this diversity is replaced by a single species grown year after year, some trace elements are depleted if not replaced by fertilization. The organic content of the soil also diminishes unless crop residues or other organic materials are supplied in sufficient quantities to replace that consumed over time.

**Contamination of Water.** In the Northeast water supplies are generally plentiful, but are increasingly becoming threatened by contamination. Farming is one potential source of such contamination. Surface runoff carries manure, fertilizers, and pesticides into streams, lakes, and reservoirs, in some cases causing unacceptable levels of bacteria, nutrients, or synthetic organic compounds. Similarly, water percolating downward through farm fields carries with it dissolved chemicals, which can include nitrate fertilizers and soluble pesticides. In sufficient quantities these can contaminate groundwater supplies.

**Fertilizers.** Nutrients are lost from agricultural fields through runoff, drainage, or attachment to eroded soil particles. The amounts lost depend on the soil type and organic matter content, the climate, slope of the land, and depth to groundwater, as well as ~~on~~ the amount and type of fertilizer and irrigation used.

The three major nutrients in fertilizers are nitrogen, phosphorus, and potassium. Of these, nitrogen is the most readily lost because of its high solubility in the nitrate form. Leaching of nitrate from agricultural fields can elevate concentrations in underlying groundwater to levels unacceptable for drinking water quality. In the Suffolk County area of Long Island, for example, almost 10 per cent of private wells tested for nitrate exceed the 10 mg/l drinking water standard.

Phosphorus does not leach as readily as nitrate because it is more tightly bound to soil particles. However, it is carried with eroded soils into surface water bodies, where it may cause excessive growth of aquatic plants. If this process proceeds far enough, lakes and reservoirs become choked with decaying mats of algae, which have offensive odour and can cause fish kills from the resulting lack of dissolved oxygen.

Potassium, the third major nutrient in fertilizers, does not cause water quality problems because it is not hazardous in drinking water and is not a limiting nutrient for growth of aquatic plants. It is tightly held by soil particles and so can be removed from fields by erosion, but generally not by leaching.

**Pesticides.** The trend toward intensive crop production in modern farming has led to increased potential for damage by pests and diseases. Predators that would be present in a mixed biological community are not supported by large fields of a single crop; so farmers, instead, rely on chemical measures for crop protection. Use of pesticides on U.S. farms has risen 10-fold over the past 40 years as agriculture has become more intensive. One drawback to this is that pesticides generally kill not only the pest of concern, but also a wide range of other organisms, including beneficial insects and other pest predators. Once the effect of the pesticide wears off, the pest species is likely to recover more rapidly than its predators because of differences in the available food supply. Previously unimportant species may also become significant crop pests when their natural predators are killed by pesticide applications.

Another drawback to the increasing pesticide use is the development of resistance in pest species. The individual pests that survive pesticide applications continue to breed, gradually producing a population with greater tolerance to the chemicals control pest populations.

Following World War II, DDT and related chlorinated hydrocarbons were introduced as potent new pesticides and were used throughout the world for protection of agricultural crops, as well as control of mosquitoes, lice and other human pests. In 1962 Rachel Carson's book *Silent Spring* brought public attention to the fact that these organic compounds are highly persistent in the environment and accumulate in animal tissues, causing water contamination, fish kills, and decline of some bird populations. DDT was banned for agricultural use in the United States in 1973, and since that time it and similar chlorinated hydrocarbons have been replaced by less persistent, but more acutely toxic, compounds.

In the past couple of decades, awareness has been growing of the many potential problems caused by the heavy use of chemicals in modern agriculture. This, combined with the rapid rise in the cost of fertilizers and pesticides, has led many farmers to seek ways of reducing their reliance on chemical intensive methods of farming. A small but growing percentage of farmers are farming with no synthetic chemicals, and many others are reducing their overall chemical use. Agriculture research has begun to focus on ways of maintaining environmental quality while producing acceptable crop yields. One example is integrated pest management, aimed at controlling pests through a combination of methods that minimize undesirable ecological effects. Continuing research and education need to be conducted on farming practices that produce profitable yields while maintaining environmental quality and the long-term productivity of the land.

## WATERLOGGING

Another problem associated with excessive irrigation on poorly demand soils is waterlogging. This occurs (as is common for salinization) in poorly drained soils where water can't penetrate deeply. For example, there may be an impermeable clay layer below the soil. It also occurs on areas that are poorly drained topographically. What happens is that the irrigation water (and/or seepage from canals) eventually raises the water table in the ground – the upper level of the groundwater – from beneath. Growers don't generally realize that waterlogging is happening until it is too late – tests for water in soil are apparently very expensive.

The raised water table results in the soils becoming waterlogged. When soils are water logged, air spaces in the soil are filled with water, and plant roots essentially suffocate – lack oxygen. Waterlogging also damages soil structure.

Worldwide, about 10% of all irrigated land suffers from water logging. This is an area about the size of Idaho. As a result, productivity has fallen about 20% in this area of cropland.

Both waterlogging and salinization could be reduced if the efficiency of irrigation systems could be improved, and more appropriate crops (less water hungry) could be grown in arid and semi-arid regions. In addition, increasing the cost of water to more closely reflect its true value would encourage its conservation, rather than using incentives that essentially encourage wasting water, as some water rights laws in the US do (e.g., the "use it or lose it" approaches discussed above). We can consider excessive irrigation to be another example of humanities' attempt to increase "K" in ways that are unsustainable. For example, one could argue that the Central Valley of CA shouldn't be lush and green; it is one of the most agriculturally productive regions of the world only because of irrigation, which may not be sustainable there.

## SALINITY

In many areas of India, crop production is limited because of salinity or alkalinity or both. It is estimated that about 7 million hectares in the country have either gone out of cultivation or this area produces low yield of crops. Three class of saline and alkali soils are recognized. They are—

**1. Saline Soils.** The soils containing toxic concentrations of soluble salts in the root-zone are called saline soils. Electrical conductivity in the saturation extract of such soils taken as a measure of salts is greater than 4.0 mmhos/cm. Exchangeable sodium percentage is less than 15 and the pH is less than 8.5. The soluble salts mainly consist of chlorides and sulphates of sodium, calcium and magnesium. Because of the white encrustation due to salts, the saline soil is also called white alkali.

**2. Non-saline Alkali Or Sodic Soils.** These soils do not contain any large amount of neutral salts and, as such, the electrical conductivity is less than 4 mmhos/cm. The detrimental effect of alkali soil on plants is largely due to toxicity of a high amount of exchangeable sodium and the pH. Alkali soils have an exchangeable sodium percentage of more than 15 and a pH greater than 8.5. Such soils have low infiltration rate and the physical condition is unfavourable. Because of high alkalinity, resulting from sodium carbonate, the surface soil is discoloured and black, and, hence the term black alkali is frequently used to designate the non-saline alkali soil.

**3. Saline-alkali Soils.** This group of soils is both saline and alkali. They have appreciable amounts of soluble salts, as indicated by the electrical conductivity values of more than 4 mmhos/cm. Also, the exchangeable sodium percentage is greater than 15. The pH, however, is likely to be less than 8.5.

The soil salinity or alkalinity or both have many adverse effects, which are summarized below :

1. Causing low yields of crops or crop failure in extreme cases.

2. The limiting of the choice of crops, because some crops are sensitive to salinity or alkalinity or to both.
3. Rendering the quality of fodder poor as, at times, the fodder grown on alkali soils may contain a high amount of molybdenum and a low amount of zinc, causing nutritional imbalance and diseases among live-stock.
4. Creating difficulties in the construction of buildings and roads and their maintenance.
5. Causing excessive run-off and floods owing to low infiltration, resulting in damage to crops in the adjoining areas.

**Causes of salinity.** In arid and semi-arid areas, salts formed during the weathering of soil minerals are not fully leached. During the periods of higher-than-average rain-fall, the soluble salts are leached from the more permeable high-lying areas to the low-lying areas, where, if the drainage is restricted, salts accumulate on the surface as water evaporates. The excessive irrigation of the uplands containing salts thus results in the accumulation of salts in the valleys. In areas having a salt layer at lower depths in the profile, seasonal irrigation may favour the upward movement of the salts. Salinization is also caused owing to the irrigation of soils with saline water. In all these cases, restricted drainage is usually the main reason. Rise in the water-table within 2 metres of the surface due to irrigation, the obstruction of natural drainage because of developmental activities, e.g. roads and canals, and the siltation of natural drainage may also cause soil salinity. In the coastal areas, the ingress of sea-water induces salinity in the soil. When sodium ions predominate in the soil solution, and carbonates are present, alkali soils are formed.

**Reclamation.** The salinity or alkalinity depend upon their mode of formation and physiographic position. Since the degree of salinity or alkalinity may vary as such methods of reclamation also differ.

(*i*) If the problem is only of salinity, the salts need to be leached below the root-zone and not allowed to come up. In practice, however, this might be difficult to accomplish, especially in deep and fine-textured soils containing more salts in the lower layers. Under these conditions, a provision of some kind of subsurface drains becomes important. If the soil contains a sandy layer at a lower depth, the leaching of the salts below this layer will check the rise of salts.

(*ii*) The reclamation of alkali soils needs the addition of a soil amendment, containing soluble calcium salts. The commonly used amendment is gypsum. In the course of reclamation, sodium on the exchange complex is replaced by calcium. The sodium salts, thus formed, are leached down.

(*iii*) The number and frequency of leaching, the quantity of gypsum to be added and the techniques involved vary from region to region, depending upon the clay mineralogy of the soils, the intensity of the problem, the subsequent use of the soils, the availability and quality of irrigation water and the economics of these operations. Hence, the state authorities engaged in this work should be consulted to draw a schedule of operations for reclamation.

## ENERGY RESOURCES

Energy is needed by all living organisms and vegetations for biochemical reactions of their cells. It is a power which is needed in one form or other for work done. Long before most of the power available to human society was limited to solar energy trapped by green plants which produced organic matter. Biological oxidation of the organic matter provided fuel to muscle power. The fire was the first form of known energy used for cooking, heating purposes. The formation of fossil fuels (coal, oil & natural gas) is also due to photosynthesis carried on by plants which occurred millions of years ago. Now the things are changed drastically. For the developmental activities, energy sources have their own importance. Energy consumption of a nation is usually considered as an index of its development.

## GROWING ENERGY NEEDS

Energy is the prime input of a country. It is converted into heat & electricity. For every activity to be performed required energy in the form of heat, light, electricity and even food (in the form of energy) for our body. Food energy is measured in calories. India has fast growing developing economy, with the GDP growth rate exceeding 6% in recent years. Xth plan projected 8% growth rate economy. This growth has been accompanied by a steady increase in energy consumption. Primary commercial energy demand grew at annual rate of 6% upto 2001. It will go more rapidly than in the past as country's reforms process accelerates.

As the economy grows, energy intensity rises following corresponding increase in energy consumption. However, beyond a certain level of per capita income, energy intensity begins to decline. These linkages between energy and economic factors, manifested in energy elasticity and energy intensity are broadly related to—

(*i*) Demographic changes, including a relatively faster growth in urban areas, higher per capita GDP and per Capita gross saving.

(*ii*) Efficient end-use devices.

(*iii*) Technological improvements in conversion equipments.

(*iv*) Inter fuel substitution with more efficient alternatives.

As the world population is growing (@ 6 billion) we need energy much faster rate to meet the industrial, food, residental, transport, agriculture etc. requirements. So it was realised that, energy sources will not meet the requirement last long. In early 70s interest was shown world over to look for other alternatives sources of energy. Therefore, thrust was given for renewable sources of energy like solar, wind, hydro, biomass, tidal, hydrothermal, hydrogen energy etc.

## ENERGY CONSUMPTION

India's commercial energy consumption is about one-fourth that of the world average. Therefore, India's consumption of coal, oil, and natural gas is only a small part of total world consumption. The consumption of commercial fuels is steadily rising throughout the country, with coal continuing to be the most prominent energy source. The fuel composition of commercial energy consumption varies significantly from sector to sector. Coal continues to meet over 70 per cent of the industry's energy needs, but petroleum products. especially natural gas and power, are steadily replacing coal. In the field of agriculture, draught animal power continues to be predominantly used in various farming activities, but this use is also not well documented. In transportation, the direct use of coal is virtually nil, following the substitution of steam traction by electric traction. Consequently, the use of electric power has gone up in the Railways sector.

The consumption of petroleum products has significantly increased following a seven-fold increase in registered motor vehicles between 1981 and 1997 (MoST, 1999). While a large part of the household sector continues to depend on traditional fuels for cooking, this data has not been documented formally. There appears to be some substitution of petroleum products by power in the household sector, probably resulting from extensive electrification, whereby kerosene gets replaced by electricity for lighting purposes. But in overall terms, petroleum consumption has grown manyfold and the oil intensity of the Indian economy has been sharply increasing.

## LAND RESOURCES

In India, land is generally called as "*MOTHER LAND*". It is because of our life depend on it for food, fibre, fuel and other basic amenities. Therefore, it is the valuable gift of nature to human beings. Top layer of the land is called soil, which is renewable resource and essentail for survival of life. Though it is life support system, but it is overused which causes environmental problems. The existing land use pattern in different regions in India has been evolved as the result of the action and interaction of the various factors i.e. physical characteristics of land, the institutional frame work, the structure of other resources available and the location of the region in relation to other aspects of the economic development *e.g.*, those relating to transport as well as to industry and trade.

Out of the total geographical area of 328 million hectares, the land use statistics are available for roughly 306 million hectares, constituting 93% of the total (1970-71) land available for cultivation is approximately 14 million hectares. But it is reducing day by day. It is due to mismanagement. The earth is made up of three principal layers cores, mantle and crust. Cores are inner most fluid layers. Mantle is the middle layer and crust is the upper most layer.

Till 1949-50, the land area in India was classified into five categories—(*i*) forests (*ii*) area not available for cultivation (*iii*) other uncultivated land (*iv*) fallow land (*v*) the net area sown. But after 1950-51 the new classification was adopted, where, classified in to 9 categories instead of 5. They are (*i*) forests, (*ii*) land put to non-agricultural use (*iii*) barren land unculturable land (*iv*) permanent pastures and other grazing lands (*v*) misc. tree crops and groves (*vi*) culturable waste (*vii*) fallow land (*viii*) current fallow (*ix*) net area sown.

**Land-capability classification.** Any soil and water-conservation project includes two distinct sets of operations, viz. (1) the mapping of land for classification according to its capability, and (2) planning and executing measures to check erosion, improve land productivity and reclaim wasteland. The farm plans for effective soil and water conservation are based largely on the capability of the land. The land-capability classification map is normally prepared by interpreting a standard soil-survey map.

Land-capability classification is a systematic arrangement of different kinds of land according to those properties that determine the ability of the land to produce crops on a virtually permanent basis.

**The factors determining land-capability.** These are the major soil characteristics of the land, e.g. the texture of the top soil, its effective depth, permeability of the top soil and subsoil, and associated land features, e.g. the slope of the land, the extent of erosion, the degree of wetness and susceptibility to overflowing and flooding.

The grouping of soils into capability classes is done primarily on the basis of their capability to produce common cultivated crops and pasture plants without deterioration over a long period.

**Land-capability classes.** The land-capability classes are based on the intensity of hazards and the limitations of use. The land-capability classes range from the best and most easily farmed land to that which has no value for cultivation, grazing or forestry, but which may be suited to wild-life, recreation or for watershed protection. They all fall into 2 broad groups : one suitable for cultivation and other land uses, and the other not suitable for cultivation, but suitable for other land uses.

## LAND SUITABLE FOR CULTIVATION AND OTHER USES

There are four class of land which are suitable for cultivation and other purposes. Their details and limitations are as

**Class I (Green Colour).** Soils in class I have very few or no limitations that restrict their use. This type of land is nearly level and the erosion hazard is low. The soils are deep, well-drained, easily worked, hold water well and are either fairly well supplied with plant nutrients or are highly responsive to the application of fertilizers. The soils are not subject to damage because of overflow. The local climate must be favourable for growing many of the common field crops. In irrigated areas, the soils may be placed in class I, if the limitation of the arid climate has been removed by relatively permanent irrigation works.

These soils need ordinary management practices to maintain productivity. Such practices may include the use of one or more of the following : fertilizers, lime, cover and green-manure crops, conservation of crop residues and crop rotations.

Soils in this class are suited to a wide range of plants, may be used for cultivated crops, pastures, forests and wild life, food and cover.

**Class II (Yellow Colour).** Soils in class II have some limitations which reduce the choice of plants or require simple conversion practices.

The limitations of soils in class II may result from the effects of one or more of the following factors : (*i*) a gentle slope, (*ii*) a slight susceptibility to erosion, (*iii*) less than ideal soil depth, (*iv*) occasional damaging overflow, (*v*) wetness which can be corrected by drainage, but existing permanently as a moderate limitation, (*vi*) slight to moderate salinity or sodium, easily corrected but likely to recur, and (*vii*) a slight climatic limitation on soil use and management.

They needy terracing, strip-cropping, contour cultivation, water-disposal area, covered with vegetation crop rotation, cover and green-manure crops, stubble mulching, the use of fertilizers, manure and lime. These soils may be used for growing cultivated crops, raising pastures, forests and for wild-life food and cover.

**Class III (Red Colour).** Soils in class III have moderate limitations which reduce the choice of plants and require special conservation practices.

Limitations of soils in class III may result from the effects of one or more of the following factors : (*i*) moderately sloping land, (*ii*) moderately susceptible to water or wind erosion, (*iii*) frequent overflow accompanied with some crop damage, (*iv*) very slow permeability of the sub-soil, (*v*) wetness or continuing water-logging after drainage, (*vi*) shallow soil depth up to the bed-rock, hard-pan or clay-pan which limits the rooting-zone and the water storage, (*vii*) low moisture-holding capacity, (*viii*) moderate salinity or sodium, and (*ix*) moderate climatic limitation.

The soils can be used for raising cultivated crops, pastures, forests and wild-life food and cover.

**Class IV (Blue Colour).** Soils in class IV have severe limitations that restrict the choice of plants and require careful management.

The restrictions in the use of these soils are greater than those in class III and the choice of plants is more limited. When these soils are cultivated, very careful management is required and the conservation practices and more difficult to apply and maintain.

The use of these soils for cultivated crops is limited as a result of the effect of one or more permanent features, such as (*i*) steep slopes, (*ii*) severe susceptibility to water and wind erosion, (*iii*) severe effect of past erosion, (*iv*) frequent over-flow accompanied with severe crop damage, (*vii*) excessive wetness with a continuing hazard of water-logging after drainage, (*viii*) severe salinity or sodium, and (*ix*) moderately adverse climate.

These soils can be used for crops, pastueres forests and wild life food and cover.

## SOIL

The word "*SOIL*" is derived from the latin word "*SOLUM*" meaning ground. Here, the word soil is used in a strictly scientific sense. To the scientist soil is not merely the product of simple rock decay - the result of both chemical decomposition and mechanical disintegration of rocks, whether igneous or sedimentary, but a reworked natural material having special and definite characteristics of its own.

Soil is defined as a thin layer of earth's crust which serves as a natural medium for the growth of plants. Soil is formed by weathering process of parental material rocks. Soil differ from parental material in the morphological, physical, chemical and biological properties. Generally it is said that, it is a mixture of mineral constituents and of organic matter, which later consists, more or less of the decomposed residues of plant materials and to a lesser extent of animal remains and excreta.

There are various types of soils like red, black, yellow - alluvial etc. Depth of soil is variable and it differ in texture like coarse, fine, soils serve as reservoirs of nutrients and soil's flora. The most characteristic constituents of soil are the colloidal organic matter and colloidal clay. Soil organic matter and particles conserve soil moisture for crops, plantation, soil living organisms. Soil provide even mechanical anchorage and favourable tilth. The components of the soil are mineral material, organic material,water and air the proportion of which vary.

Since, the main interest of soils to MAN is for growing his crops, it is necessary to learn also about the physical and chemical properties of the soils. Because for growing with success particular crop is dependent. In soil studies, the following physical properties are important

(*i*) Specific gravity (*ii*) Pore space (*iii*) Plasticity
(*iv*) Cohesion (*v*) Colour (*vi*) Texture
(*vii*) Soil temperature (*viii*) Permeability.

### Soil Forming Materials

Rocks are the chief sources for the parent materials over which soils are developed. There are three main kind of rocks viz (*a*) igneous rocks (*b*) sedimentary rocks (*c*) metamorphic rocks. Soil differs in chemical composition. The main chemicals are

$SiO_2$, $Al_2O_3$, $Fe_2O_3$, MgO, FeO, CaO, $Na_2O$, $K_2O$, $CO''_3$, $P_2O_5$, MnO, $TiO_3$ etc. and water moisture.

Weathering, soil formation and development proceed simultaneously. The weathering may be chemical or physical. Physical weathering comprised of — water, temperature, wind, plants/animals, while chemical weathering comprised of solution, hydration, hydrolysis, carbonation, oxidation-reduction.

### Soil types and Soil groups

Depending upon weather a particular soil had originated from the direct decomposition of the crystalline rocks or it had been formed from deposits which have already passed through a cycle or cycles of weathering, transport or consolidation, we have respectively the *Primary and Secondary soils.*

The important soil groups of the world are the following :

**1. Soils of the Podsolic group :** They are found in cold to temperate regions under humid conditions. The Pod sols and their congeners are completely leached soils in which Calcium carbonate and Calcium sulphate are present only in very subordinate amounts. Soils of this include the Tunda soils, Podsols, Brown forest soils, Prairie soils and Mountain soils.

**2. Tschernosems and their related groups :** Such soils develop best in semi and climate. Hence leaching is rather incomplete and generally there is always the development of horizons of carbonate deposition. The Tschernosems are also known as Black Earths. Owing to the invariable presence of humified organic matter in these soils. It includes the typical Tschernosems, black and cotton soils, chestnut Earths and Desert soils.

**3. Groundwater soils including Peat :** Such soils are found throughout the middle, northern and western parts of Europe and their greatest development is in the Pripet Marsh region of Poland. In these areas, the ground water table is either at or near to the surface.

**4. Brown and grey soils of the semi-desert :** As one moves from regions covered with Chestnut soils, into more arid areas, brown and grey semi-desert soils are noticed. The latter have a lower organic matter content and calcareous and gypseous salts occur in such soils within a few inches of the surface.

**5. Saline, Alkaline and Soloth soils :** Such soils are formed in areas of inland drainage without an outlet to the sea such as the inland seas or salt lakes, there is always a characteristic accumulation of mainly the sodium salts with some amount of potassium salts in them. Alkaline soils are characterised by the presence of the carbonate of sodium.

**6. Soils of the Humid Tropics and the Sub Tropics :** The tropical soils are made up almost entirely of the products of chemical weathering, together with a greater or less admixture of unweathering minerals such as quartz, magnetite etc. It is red in colour but modified by organic matter content. Red loams, red earths and the laterites belong to this group.

**7. Soil associated with Calcareous parent rocks :** Two types of this soils are

(*a*) Soils of a brownish or brownish grey colour, it belongs to Rendzimes.

(*b*) Soils predominantly red or reddish brown in colour, it belongs to Terra Rosa soils.

### Classification of soils in India

Soils widely vary in their characteristics and properties. Understanding the properties of the soils is important in respect of the optimum use they can be put to and the best management requirements for their efficient and productive use. Classification helps to reduce the study of the number of individuals to a few well defined units. Different systems of soil classification have been used from time to time, but brief descriptions of major soil groups and given below—

**1. Alluvial soils :** This include the deltaic alluvium calcareous alluvial soils, Coastal alluvium and Coastal sands. It is the most important soil group of India contributing the largest share to its agricultural wealth. This soil are derived from the deposition laid by the numerous tributaries of the Indus, Ganges and the Brahmputra systems. These streams draining the Himalayas. Geologically, alluvium is divided in to Khadar generally light coloured sandy and Bhanger generally dark and full of "Kankar".

**2. Black soil :** These soils vary in depth from shallow to deep. The typical soil derived from Deccan trap is the "Regur" or black cotton soil. It is common in Maharashtra, Western parts of Madhya Pradesh, A.P., Gujarat, Tamilnadu. It is comparable with the "Chernozems of Russia and Spairie soil of U.S.

**3. Red soils :** The ancient crystalline and metamorphic rocks on meteoric weathering have given rise to the red soils. The red colour is due to the wide diffusion of iron rather then to a high proportion of it. The soils grade from poor thin gravelly and light coloured varieties of the plains and valleys. They are poor in nitrogen, phosphorus, lime, potash, iron oxide. The soils comprise vast areas of Tamil Nadu, Karnataka, Goa, Daman Diu, M.P. Orissa, Chhotanagpur, Bihar, West Bengal, U.P. etc.

**4. Laterites and Lateritic soils :** It is a formation peculiar to India and some other countries. It is a mixture of the hydrated oxides of aluminium and iron with small amounts of manganese oxide and tatania, etc. These soils are poor in lime, magnesia nitrogen, potassium oxide but humus and $P_2O_5$ is high.

**5. Desert Soils :** The most predominant component of this soil is quartz but feldspar and hornblende grains also occur with a fair proportion of calcareous grains. Some of these soils contain high percentage of soluble salts, possess high pH, a very high percentage of calcium carbonate and are poor in organic matter.

**6. Problem soils :** These are the soils which owing to land and soil characteristics, cannot be economically used for the cultivation of crops without adopting proper reclamation measures. High eroded soils, revive lands, soils on steeply sloping lands etc. constitute one set of problem soils.

**7. Acid soils :** Soils having pH below 7 are considered to be the acidic, but those which have pH less than 5.5 and which respond to liming may be considered to qualify to be designated as acid soils. It occurs widely in Himalayan region, Peninsula, Gangetic delta.

## SOIL EROSION

Soil erosion means the removal of material from the surface of the soil by the agency of running water, wind or even by gravity. Since the superficial layers of the soil are the richest in plant food and thus the feeding ground of plant roots, the process of soil erosion involves a definite loss of valuable plant nutrients and if it becomes sufficiently intense, may lead to the complete destruction of the soil as the seat of plant growth. Where soil erosion is intense, the natural soil profile is destroyed or truncated and indeed, may never attain full development.

In India soil erosion is a serious problem in certain parts as in Chhota Nagpur. Gully erosion is welmarked in Gwalior, Mandsaur, Shivpuri of Madhya Pradesh. The following types of erosion are observed in India.

**Normal or geologic erosion.** This is a normal feature of any landscape. Geologic erosion takes place steadily but so slowly that ages are required for it to make any marked alteration in the major features of the earth's surface. There is always an equilibrium between the removal and formation of soil, so that unless the equilibrium is disturbed by some outside agency, the mature soil preserves, more or less, a constant depth and character indefinitely.

**Accelerated soil erosion.** The removal of the surface soil from areas denuded of their natural protective cover as a result of human and animal interference takes place at a much faster rate than that at which it is built up by the soil-forming processes. This accelerated detachment rapidly ravages the land and it is with this type of soil erosion that we are so seriously concerned. Nature requires, on an average, about 1,000 years to build up 2.5 cm of top soil, but wrong farming methods may take only a few years to erode it from lands of average slope.

**Wind erosion.** Wind erosion takes place normally in arid and semi-arid areas devoid of vegetation, where the wind velocity is high. The soil particles on the land surface are lifted and blown off as dust-storms. When the velocity of the dust-bearing winds is retarded, coarser soil particles are deposited in the form of dunes and thus fertile lands are rendered unfit for cultivation. In other places, fertile soil is blown away by winds and the subsoil is exposed, as a result the productive capacity of the soil is considerably reduced.

**Water erosion.** Soil erosion caused by water can be distinguished in three forms, viz. (1) sheet erosion, (2) rill erosion, and (3) gully erosion.

**Sheet erosion.** Sheet erosion removes a thin covering of soil from large areas, often from entire fields, more or less, uniformly during every rain which produces a run-off. This type of erosion is very insidious, since it keeps the cultivator almost ignorant of its ill-effects. It is generally neglected, although the soil deteriorates slowly and imperceptibly. Its existence, however, can be detected by the muddy colour of the run-off from the fields.

**Rill erosion.** When sheet erosion is allowed to continue unchecked, the silt-laden run-off forms a well-defined, but minute finger-shaped grooves over the entire field. Such thin channelling is known as rill erosion.

**Gully erosion.** When rill erosion is neglected, the tiny grooves develop into wider and deeper channels, which may assume a huge size. This is called 'gully' erosion. Gullies are the most spectacular evidence of the destruction of soil. The gullies tend to deepen and widen with every heavy rainfall. They cut up large fields into small fragments and in course of time, make them unfit for cultivation.

**Landslides or slip erosion.** A landslide is defined as an outward and downward movement of the slope-forming material, composed of natural rocks, soil, artificial fills, etc. The fundamental causes of landslides are topography of the region and geological structure, the kinds of rocks and their physical characteristics. The immediate cause of a slide may be an earthquake, or a heavy rainfall, which unduly saturates the ground or a part of a road. However, these are accidents rather than fundamental causes.

**Stream bank erosion.** Torrents are defined as hill streams characterized by wide-spreading beds on emergence from the hills with ill-defined banks, flashy flows and swift currents. Usually, they are dry water-courses, except during the rainyseason when with every heavy downpour in their catchment, they get very much swollen with flood and subside almost to its normal tiny size immediately after the storm is over.

These sudden and violent flows are responsible for moving immense quantities of detritus, comprising boulders, shingle, sand and silt, depending upon the geology of the terrain. This debris gets deposited in the torrent bed in the form of scattered islands owing to the sudden widening of the torrent channel after it emerges from the hills, or owing to the flattening of the gradient in the lower reaches, or because of obstructions caused by wild vegetation and uprooted trees. The bed level of the torrent is raised by these deposites. These deposits, in turn, reduce the transporting capacity of the torrent, resulting in overflows and the meandering of the course and in the erosion of the banks.

## MECHANISM OF EROSION

### Water erosion

Soil erosion caused by rainfall is the result of the application of energy from two distinct sources, namely (*i*) the falling rain drops, and (*ii*) the surface flow. The energy of a falling rain drop is applied slantingly or vertically from above, whereas that of surface flow is applied more or less horizontally along the surface of the ground. The chief role of the falling rain drop is to detach soil particles, whereas that of the surface flow (outside the rills and gullies) is to transport the soil. The falling rain drop also makes a major contribution to the movement of the soil on unprotected sloping lands during the periods of heavy impact storms, by splashing large quantities downslope and by imparting transporting capacity to the surface water by keeping it turbid. More than 100 tonnes of soil per hectare can sometimes be lost yearly in this fashion from a bare and highly detachable soil on slopping land.

### Wind erosion

Wind is responsible for three types of soil movement in the process of wind erosion. They are known as : (*i*) saltation, (*ii*) suspension, and (*iii*) surface creep.

(*i*) *Saltation.* The major portion of the soil carried by the wind is moved in a series of short bounces called "saltations". The soil carried in a saltation consits of fine particles ranging from 0.1 to 0.5 mm in diameter. Saltation is caused by the direct pressure of wind on soil particles and their collision with other particles. After being pushed along the ground surface by the wind, the particles leap atmost vertically in the first stage of saltation. Some grains rise only a short distance; others leap 30 cm or higher, depending directly on the velocity of the rise from the ground.

(*ii*) *Suspension.* Very fine soil particles, less than 0.1 mm in diameter, are carried into suspension, being kicked up into the air by the action of particles in saltation. The movement of fine dust in suspension is completely governed by the characteristics movement of the wind. Suspended material is carried long distances from its original location and is thus a complete loss to the eroded area, whereas the soil moved in saltation and surface creep usually remains within the eroded areas, especially when erosive winds are from different directions.

(*iii*) *Surface creep.* Soil particles, larger than about 0.5 mm in diameter but smaller than 1.0 mm, are too heavy to be moved in saltation but are pushed or spread along the surface by the impact of particles in saltation to form a surface creep.

About 90 per cent of the total soil movement in wind erosion is below the height of 30 cm, and about 50 per cent of it is within 5 cm of the ground level. The control of wind erosion is mainly based on the reduction or eliminaton of movement in saltation.

The remedial measures suggested to arrest soil erosion are

(*a*) Deep tillage, in order to facilitate the absorption of the excessive water in to the soil.

(*b*) Direct control processes such as "Contour-bunding".

(*c*) Under-draining of the lands – this consists of laying down drains in gullies to enable the natural drainage of water.

(*d*) Re-forestation.

Attention has already been drawn to the important role played by organic matter *e.g.*, humus in plant growth.

### Land Degradation

The total land under agricultural use is around 58.4% i.e. grossed cropped area is 167.41 million hectares. The land not fit for cultivation i.e. barren land is around 9.9%. The area under forest is 21.6%, but it needs to be raised.

In addition to water, land resources are the precious resources. Food security depend on conservation and proper utilization of all resources. Due to use and over exploitation land resources are degraded. It is due to the more & more pressure with increasing population. Land degradation is a real alarm. Because soil formation is a very slow process. In millions of years we have a layer crust of fertile soil. In general, formation of 1.0 cm soil crust from parent material take 300–400 years. Fertile soil have high percentage of organic matter vis-a-vis micro-organisms. Each gram of fertile soil have 30 billion micro-organisms.

Some 1.9 billion ha of agricultural land have been degraded to some an extant and 8 million ha are converted to non agricultural use such as homes, highways, shopping centres, factories, reservoirs etc. In India about 175 million land is effected by degradation problem. Land degradation is caused by impoverishment or erosion of soil, water runs off, vegetation by shrinking of lakes etc.

**Significance of the problem**

Agricultural production has witnessed dramatic rise in the last three decades or so in the countries world over. In India, green revolution brought about technological break through, which led to the use of short duration high yielding varieties helping intense use of land in a year, increasing area brought under irrigation and prolific use of Chemicals such as fertilizers and pesticides. India, being vastly agriculture oriented, historically has had policies in various phases for the development of agriculture with the expectation that development of agriculture would lead to overall development of the nation and help eradication of poverty. It has been of late recognised that the increasing efforts to raise agricultural growth has cost us clearly in the form of land & water degradation.

Large scale ecological losses were reported in cropland grassland and forest land, such as soil erosion, soil alkalinity and salinity, micronutrient deficiency, water logging and fast depletion and contamination of ground water. These factors limit future gains from the land & water resources.

Department of land resources, ministry of rural development, Govt. of India has identified different types of degraded land 6.4.

It is clear that soil erosion is the major problem of land degradation.

**Table 6.4. Categorywise percentage of degraded land as on 2000**

| | | *(Percentage of total degraded land)* |
|---|---|---|
| 1. | Gullied/ravinous land | 3.22 |
| 2. | Upland with or without scrub | 30.40 |
| 3 | Water logged & marshy land | 2.58 |
| 4. | Land affected by alkalinity/salinity | 3.22 |
| 5. | Shifting cultivation area | 5.50 |
| 6. | Degraded notified forest | 22.02 |
| 7. | Grazing land | 4.07 |
| 8. | Degraded under plantation crops | 0.90 |
| 9. | Sands - Inland/Coastal | 7.84 |
| 10. | Mining/Industrial | 0.20 |
| 11. | Barren Rocky/Stony | 0.12 |
| 12. | Steep slopping area | 1.20 |
| 13. | Snow covered/glacial | 8.73 |

## LANDSLIDES

A landslide is a sudden collapse of a large mass of hillside. There are many different types of landslides where not only earth, but rock, mud, and debris flow down the side of a slope.

Since the beginning of the monsoon season in June India has been hit by heavy rains and landslides affecting in particular, Arunachal Pradesh, Assam, and Bihar states. According to the latest information on the impacts of the landslide, more than 12 million people are affected and more than 270 people were killed in these three states.

**Landslides mostly occur**

1. Where landslides have occurred before.
2. On steep slopes.
3. On benches.

4. Where drainage is causing a problem.
5. Where certain geologic conditions exists.

### Types of landslides

1. Shallow, disrupted landslide—Example of this type is the Santa Susana Mountains and the mountains north of the Santa Clara River Valley. Here more than 75% of the slope area was denuded by landslides triggered by strong shaking.

2. Deep, Coherent Landslides—These triggered by the earthquake were far less numerous than disrupted slides, they contributed significantly to the total volume of landslide material because they tended to be much larger. Some of these landslides are

(*a*) San Martinez Graude Landslide

(*b*) Rancho Camulos Landslide

### Factors causing landslides

There are following main causes of landslides—

1. Landslides are the sudden downhill movements on earth or other solid materials and are usually caused by rain thaws or forces either increasing the top material layers or making the slope too steep. They can be triggered by earthquakes, saturation with heavy rain or crashing waves.

2. Excessive rainfall or snowmelt, however is also known to saturate and lubricate soil on steep angles. Rapid temperature changes can also cause land to slide by alternatively shrinking and expanding soil formations, or forming ice heaves between layers of rock.

3. Forest fires are indirectly responsible for landslides because they take away slope vegetation making erosion easier.

4. Man can also cause slides by mining the earth, underground excavation, pumping and draining groundwater levels or overdeveloping hillsides. Man induced landslides are generally done for the development purposes i.e. industrial, forming roads, agricultural use, homes, etc. They use heavy explosives for that. In this case no serious casualties or damage occur because proper warned earlier to shift in safer places.

### Effects

No heavy damage occur in man induced landslides but thousands of people affected and killed due to landslides. Many houses can be damaged and the loss of public properties is also noticed. Roads and rail communication may remain cut off from rest of the regions. Thunder storms cause debris flows on hill slopes leading to deposits of mud. Heavy rains at the same time may worsen the situation.

## DESERTIFICATION

Desertification is a process by which productive potential of arid or semiarid land falls. The decrease in productivity is varies from 10%—50%. Thus desertification leads to the conversion of irrigated crop land to desert (where productivity is minimum). It is characterized by devegetation loss of vegetal cover, depletion of ground water, salinization and soil erosion.

Water have a binding capacity as well as it provide vegetal cover. Draught in three consecutive years in Central India in recent years has accelerated the process of wind erosion and desertification. In Rajasthan and Kutch-Saurashtra area, extension in the area of desert is assuming serious proportion. Climatic change and anthropogenic activities are also responsible for desertification.

During last so many years large area has destroyed (agriculture land) by Sahara desert. In India, also, so many places which affected by desertification. An extreme example of sand movement in the Coastal area of Saurashtra has hampered activities at the ports.

Deforestation is also one of the cause of desertification. Because, after forests grasslands are used by human. So human activities are also responsible for desertification. Govt. and worldwide ecologists are seriously thinking about this problem. United Nations Environmental Programme (UNEP) organised a conference in 1977 at Nairobi to discuss about the change of desert land in to productive agriculture land. The increasing cattle population heavily graze in grasslands or forests and denude the land area, which is not suitable for seed germinations. The dry barren land becomes loose & more proper to soil erosion. Thus overgrazing is also one of the cause for desertification.

Mining and quarrying activities are also responsible for conversion of productive land in to desertification. It is studied that, the last 50 years about 900 million ha of land have undergone desertification over the world. Due to mining disappearance of vegetation takes place and defacting of land occurs. Therefore, it is advised to use the eco-friendly mining technology.

Salinization is also one of the cause for conversion of agricultural land to desert. Govt. planning to take such projects, through which conservation of land and desertification can be checked. For salinization farmers are advised to use gypsom, organic fertilizers, manures with proper irrigation. Then sodium will go deeper in the soil. Awareness programmes should be conducted for affected people. It will supplement in the check of desertification.

## ROLE OF AN INDIVIDUAL IN CONSERVATION OF NATURAL RESOURCES

Planning of a suitable strategy for the conservation of our natural resources and most judicious execution of planned strategies is called as Conservation Management. Environmental planning, evaluation, monitoring, and impact assessment are methods of conservation management. The Indian philosophy of conservation is to keep "*Harmony with Nature*". Therefore, we have to learn to live with nature. For this every individual has to play his role to conserve the nature & natural resources.

It is well known that people destroying, over utilising natural resources for their own interest. Resources are limited, if they will not be properly used, they will exhaust. Therefore, before doing anything, awareness should be aroused by various methods. People should understand the importance of resources i.e. land water, air, forest, minerals, energy etc. that these are precious and should be used with great care. Every individual must be made to realise that he has to make a personal contribution to the programme to conserve the nature. For this we can take help from media, education and Govt. institutions by organising small programmes, documentary, seminars, conferences etc.

It should not be optional but for all i.e. old, young, rich, poor, industrialist, common man, consumer, businessman, resident of a posh colony, slum dweller. Every body should take part in this work. Voluntary organisations are doing some work in this regard but it is not sufficient. Some important roles of individuals in maintaining peace, harmony and equity in nature are as

1. People should at once stop the over utilization of natural resources instead they must be properly used.
2. Instead of deforestation, representation should keep in mind. We should take help from the Govt. for plantation programmes. Everybody should take part in plantation and care the plants.
3. We should protect wildlife. Though hunting is not allowed even then the persons are doing so. For this educated young should teach the lesson of wild life act.

4. Mixed croping, crop rotation, and proper use of fertilizer insecticide, pesticides should be taught to farmers. Encourage the use of manures, biofertilizers organic fertilizers.
5. We should make habit for waste disposal, compose and to restore biodiversity.
6. Try to educate local people for the protection and judicious use of natural resources.
7. We should use light, fans and other domestic appliances when it is needed.
8. Maintain a balance between resources and human needs.
9. Maintain the essential ecological processes and the life support systems.
10. Install rain water harvesting system in houses, colonies.
11. We should recycle the waste and waste water for agriculture purposes.
12. The fossil fuel should be used only when no other alternative source is available.
13. We must develop energy saving methods to avoid wastage of energy. We should remember *energy saved is energy produced.*
14. Prevent soil erosion.
15. Use drip irrigation and sprinkling irrigation to improve irrigation efficiency and reduce evaporation.
16. Utilize renewable energy sources as much as possible. Encourage use of solar cooker, pump etc.
17. Discourage the frequent use of car, bike, encourage the walk and bicycle.

We know "collecting drop-drop-drops form a big ocean", similarly if each of us will aware about the judicious use of natural resources, all of us will conserve the nature.

**Equitable use of resources for sustainable life Style:** The equal distribution of natural resources should be for all irrespective of rich or poor. There must be balance between the need and consumption particularly for drinking water, food, fuel etc. The developed countries are utilizing more resources as compared to developing countries. This imbalance is responsible for rich become richer and poor gone poorer. This is due to sharp increase in population in developing countries. But it does not mean that people of developed countries are rich and having good life style, and less developed countries people are poor. Less developed countries also have rich and poor both but facing the problem of population and available natural resources.

Developed countries like USA, Canada, Japan, Australia etc. have 22% of world's population utilising 86% of natural resources. Thus it is needed to divert the resources to poor countries to narrow down the gap between the two. To achieve sustainable life style, there should be equal distribution of global resources and income to meet everyone's need. But in the long process of economic development only the powerful and strong people exploited most of the environmental resources even at the cost of migration of poor people already using those resources. Now stress is put on our natural resources and on the poor and inefficient people.

Owing to these conditions our Govt. inclined to change the existing model of economic development which should be based on the principles of peace, harmony and equity i.e. sustainable development.

## BIOGEOCHEMICAL OR MATERIAL CYCLES

As we know Biosphere is earth's that part where living organisms are. Biosphere is divided into hydrosphere, Lithosphere and atmosphere. The maximum spread of living organisms i.e, the extent of biosphere is about 13 kms. The earth is like an energy or entropy pump that receives energy from sun and uses it to power and drive the cycles of water nutrients and other elements known as biogeochemical cycles.

The biogeochemical cycles and the ways/paths where materials or chemical elements of protoplasm like water, carbon, nitrogen etc are circulate-solar energy continue on earth, about 1% of that converted in to energy by photosynthesis or so. It is the energy which is responsible for driving the mineral and hydrological cycles.

The word *biogeochemical cycles* derived from the followings: *Bio* means *living organism*, *geo* denotes *soil environment*, while *chemical* means *elements such as carbon, phosphorus, nitrogen & sulphur etc.* Which undergo transition regularly.

Biogeochemical cycles are generally of two types.

***(a) Sedimentany Cycles.*** In these cyles, the main reservoir is the soil and the sedimentary and other rocks of the earth's crust. Examples are sulphur & phosphorus cycles.

***(b) Gaseous Cyles.*** In these cycles, the main reservoir of the nutrients is the atmosphere and the oceans. Examples oxygen, carbon & nitrogen cycles etc.

*Both types of the cycles involve biotic and abiotic agents. Both are driven by the flow of energy and both are tied to the water energy cycle or hydrological cycles.*

## NITROGEN CYCLE

The chief source of Nitrogen and its compounds is atmosphere. Nitrogen is essential for plants and animal growth. Nitrogen is thus a constituents of various organic compounds which are formed in plants. Green plants combine with nitrogen & carbohydrates to make proteins and nucleic acids. Neither the animals nor the plants are capable of absorbing nitrogen directly from the atmosphere.

From the atmosphere nitrogen is made available to plants and animals on earth and from earth it again reaches to atmosphere. This whole phenomenon is known as "Nitrogen Cycle". This has been nicely described by Mellor as : "Thoday a nitrogen atom may be throbbing in the cells of the meadow grass, tomorrow it may be pulsating through the tissues of a living animal. The nitrogen from afterward's may rise from decaying animal refuse and stream to the upper regions of the atmosphere where it may be yoked with oxygen in a flash of lightning and return as plant food to the soil in a torrent of rains or it may be directly absorbed from the atmosphere by the soil and there rendered available for plant food by the action of symbiotic bacteria. Thus nitrogen atom has doubtless undergone a never ceasing cycle of change through the countless aeons of time".

Nitrogen cycle is the cycle in which atmospheric nitrogen is converted into its compounds such as nitrates and the combined nitrogen is again constantly passing back to the atmosphere. The to & fro flow of nitrogen is achieved through several processes.

**(*i*) Lightning.** Atmosphere also contain nitric and nitrous oxides formed by the photochemical reaction between nitrogen and oxygen under sunlight. These oxides are also formed under electric discharge (as lightning). They are brought down along with rain and the nitrous and nitric acids present are neutralised by the lime in the soil giving the nitrate of calcium which is the food of the plants.

$$N_2 + O_2 \xrightarrow{\text{Lightning}} 2NO$$

$$2NO + O_2 \longrightarrow 2NO_2$$

$$4NO_2 + 2H_2O + O_2 \longrightarrow 4HNO_3$$

$$2HNO_3 + CaCO_3 \longrightarrow Ca(NO_3)_2 + H_2O + CO_2$$

**(*ii*) By Bacteria.** There are also some bacteria, known as *symbiotic bacteria* which thrive in the nodules on the roots of leguminous plants like beans, peas, clover etc. These bacteria have the capacity of converting the atmospheric nitrogen into compounds of nitrogen which the plants can assimilate. Elementary nitrogen is also utilised by bacterium *Azotobacter* present in soil, and algae, fungi and mosses.

***(iii)* Through Synthesis.** Plants give out ammoniacal compounds to the earth on decay which under the action of *nitrosifying* and *nitrifying bacteria* form nitrites and nitrates. Most of the ammonia and ammonium compounds are produced, however, as a result of the decomposition of urea, amino acids, etc. which are secreted by the animal body:

$$NH_2CONH_2 + 2H_2O \longrightarrow (NH_4)_2CO_3$$

These substances are formed in the animal body by the process of peptic and tryptic digestion of the nitrogenious plant proteins taken by the animals as food. This constitutes the half of nitrogen cycle. In the other half cycle, nitrogen so fixed as sent to back atmosphere by following steps.

***(i)* Ammonification.** Nitrogen compounds of plants, animals and their wastes are converted in to Ammonia or its salt by Ammonifying bacteria. This happens during the decomposition of these bodies.

***(ii)* Nitrification.** Ammonia and its salts so formed changed to nitrates by nitrosifying bacteria and then to nitrate by nitrifying bacteria. Nitrates are so formed used by plants. Some nitrogen is thus again returned to plants by this short route.

***(iii)* Denitrification.** Nitrogenous compounds formed by ammonification and nitrification as above and converted into nitrogen (free) by denitrifying bacteria present in the soil. Thus, nitrogen of there nitrogenous compounds after composting into for nitrogen escapes to atmosphere and complete the nitrogen cycles.

The complete idea of the nitrogen cycle can be understood from the following figure.

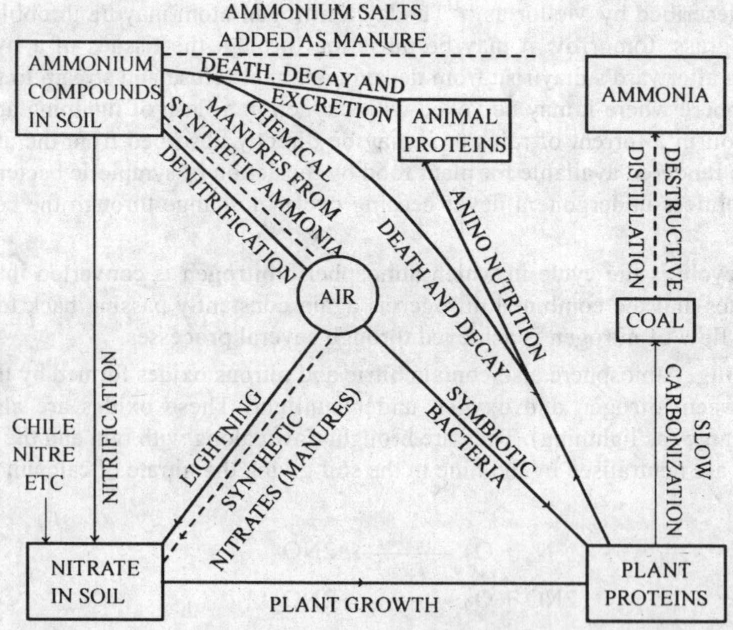


**Fig. 6.4.** The Nitrogen Cycle

## PHOSPHORUS CYCLE

Phosphorus is required the small quantities by plants and animals both. Animals usually derive their phosphorus from plants and plants from soil. It is significant that the soil does not lose its

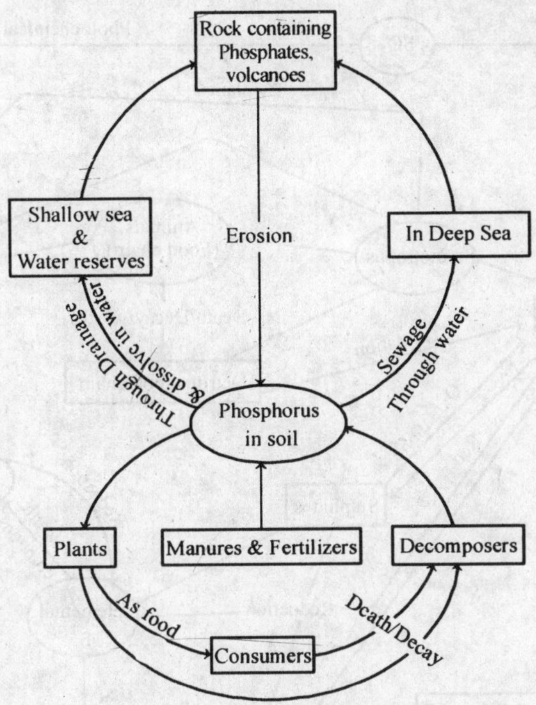


**Fig. 6.5.** Phosphorus cycle

phosphorus as nitrogen is lost by bacteria. Plants and animals return their phosphorus content to the soil on decay. But the cultivated soil loses its phosphorus from year to year as the crops are removed from the field and consumed elsewhere. The phosphorus runs in the river with seawage or buried deep in the cementaries. This necessitates the supply of phosphate manures to our fields.

Phosphates in the soil are also derived from erosion of phosphate rocks. The living organisms are capable of getting phosphorus and their elements from the environment and they also have a phosphate cycle.

Elements like phosphorous, calcium etc reach the sea & other water bodies through the water where they are deposited in the form of rocks. Thus they get separated from the cycle. They again dissolve in water as a result of disintegration of rocks, absorbed by the plants. In this may cyle goes on. The sedimentary phase of phosphorus cycle remains comparatively slower than the organic phase Figure 6.5.

## SULPHUR CYCLE

Sulphur exists in gaseous (as oxides) and solid (compounds) both forms. It is an important constituent of proteins and amino acids. Because it exists in both gas & sold its cycle may be gaseous/sedimentary both. It occur in nature in the form of inorganic and organic deposits from where it releases by weathering and decomposition. The main sources of sulphur in atmosphere are combustion of fossil fuels, volcanic eruptions, surface of ocean and gaseous release.

The main source of sulphur reaching in atmosphere as $H_2S$. Which oxidises to $SO_2$.

$$H_2S + 2O_2 \longrightarrow SO_4^{-2} + 2H^+$$

$$2H_2S + 3O_2 \longrightarrow 2H_2O + 2SO_2$$

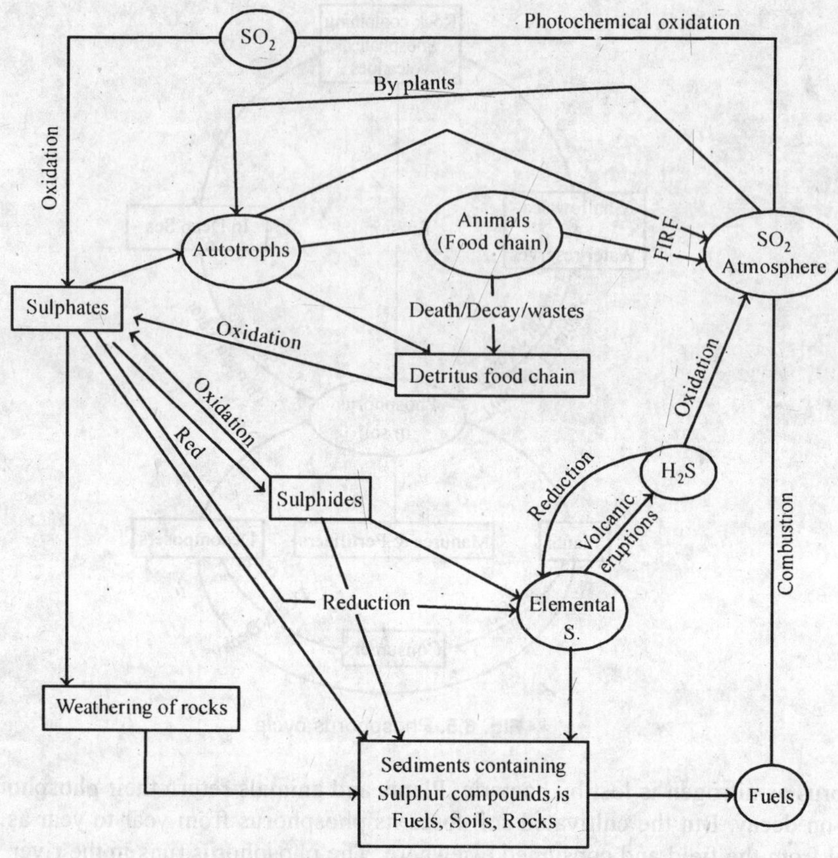


**Fig. 6.6.** Sulphur Cycle

The –SH group of amino acids also gets separated from rest of the molecules as $H_2S$. This again oxidised to $SO_2$.

$SO_2$ present in the atmosphere is soluble in water and in carried back to earth thing rain in the form of $H_2SO_4/H_2SO_3$. Sulphur in the form of $SO_4^{-2}$ absorbed by plants where it is incorporated in to certain organic molecules such as amino acids and proteins. Through decay, excretion & death of livings go back to soil and also in the bottom of ponds & sea.

Bacteria may also cause oxidation of sulphur to $SO_4$.

$$2S + 3O_2 + 2H_2O \longrightarrow 2SO_4^{-2} + 4H^+$$

Sulphates can also be reduced to sulphide by some bacteria under certain conditions. This happens in the organic phase of sulphur cycle.

From the fig. 6.6 we may get the complete idea about the sulphur cycle.

## CARBON CYCLE

The major source of carbon is also atmosphere. In atmosphere it exists in the form of $CO_2$. Carbonate rocks, fossil fuels, and sediments in the form of carbonate or so are also the huge reserves. It is the basic unit required to build the organic matter.

The $CO_2$ enters the living world through photosynthesis and fixed as starch in plants-plants synthesize the organic compounds and the consumers use them along the food chain. The plants

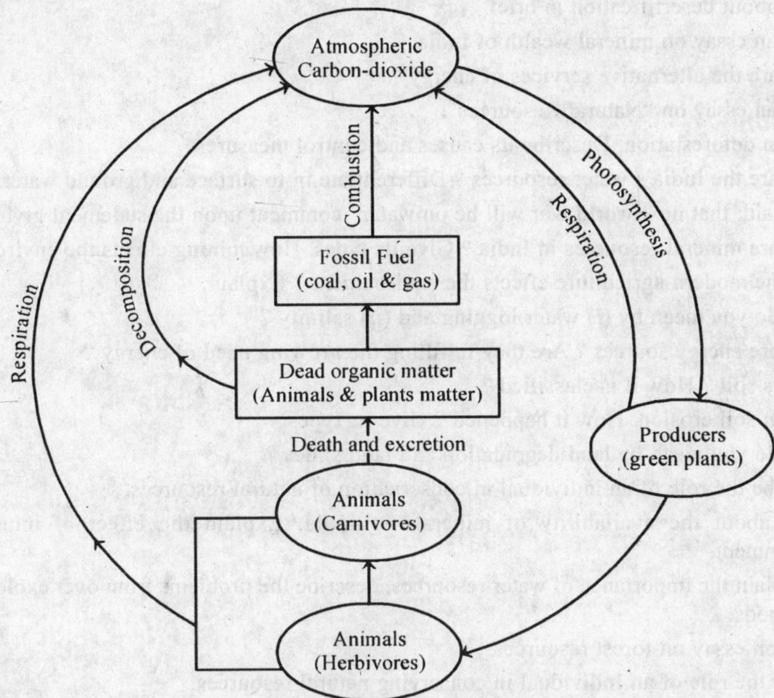


**Fig. 6.7.** Carbon-cycle

and animals release carbon back to the atmosphere in the form of $CO_2$. The dead bodies, rotten plants & wastes also release $CO_2$ by aerobic decomposition. $CO_2$ is also go back to atmosphere through oxidation by bacteria & fungi. During respiration of terrestrial plants, animals & human take oxygen & release $CO_2$ in atmosphere. The sinking of carbon into the geological strata takes place when plant materials get burried and undergo slow degradation & compaction. This forms coal & oil having huge amount of carbon. Animals also have carbon in their parts. Fossil fuel on burning release $CO_2$ to atmosphere. Thus cycle completes. The diagrammatic representation of carbon cylce is represented in fig. 6.7.

## QUESTIONS

1. What do you understand by "Natural Resources" ? Give its types
2. State and explain the renewable and non renewable resources.
3. What are Fossil fuels ? Give examples.
4. Write note on the "problems of natural resources".
5. Discuss the environmental effects of deforestation.
6. Describe the benefits of dams.
7. What are effects of mining ?
8. Why is water regarded as a precious resources ?
9. What is Chipko movement ?
10. Differentiate between Surface and Ground Water.
11. What are the consequences of over exploitation of water ?
12. Write note on world food problem in India.
13. Explain overgrazing.

14. Write about desertification in brief.
15. Write an essay on mineral wealth of India.
16. What are the alternative services of energy ?
17. Write an essay on "Natural Resources".
18. Explain deforestation. Describe its causes and control measures.
19. What are the India's water resources ? Differentiate in to surface and ground water.
20. "It is said, that next world war will be on water" comment upon the statement giving facts.
21. What are mineral resources in India ? Give its types. How mining effects the environment ?
22. How the modern agriculture effects the environment ? Explain.
23. What do you mean by (*i*) water logging and (*ii*) salinity ?
24. What are energy sources ? Are they fulfilling the growing need of energy ?
25. What is soil ? How it is classified ?
26. Explain soil erosion. How it happened ? Give its types.
27. What do you mean by land degradation and land slides ?
28. Describe the role of an individual in conservation of natural resources.
29. Write about the availability of minerals in India. Explain the effect of mineral mining on environment.
30. To explain the importance of water resources, describe the problems from over exploitation, drought and floods.
31. Write on essay on forest resources.
32. Define the role of an Individual in conserving natural resources.
33. Describe the impact of modern agriculture practices on environment.
34. What do you mean by Biogeochemical or material cycles? Discuss Nitrogen Cycle in detail
35. Write note on (i) Nitrogen cycle (ii) Sulphur cycle (iii) Phosphorus cycle.

# Energy and Environment

Increasing industrialization and unsustainable consumption patterns are escalating the environmental problems due to depletion of resources and energy. The unsustainable use of renewable resources and the generation of toxic materials during industrial operations are creating problems to biodiversity, environment and human health. Industrial production by using environmentally sound technologies is considered to be one of the best strategies to provide the paradigm to put our society on the path of sustainability.

Energy is a primary input in any industrial operation. Energy is also a major input in sectors such as commerce, transport, telecommunications, etc. besides the wide range of services required in the household and industrial sectors. Owing to the far-reaching changes in the forms of energy and their respective roles in supporting human activities, research and training on various aspects of energy and environment have assumed great significance.

## SOURCES OF ENERGY

Energy is the capacity for doing useful work. Energy resources are broadly classified as *primary* and *secondary.*

(*i*) **Primary energy resources** are those which are mined or otherwise obtained from the environment.

**Ex.** (*a*) *Fossil Fuels* : Coal, Lignite, Crude oil, Natural gas, etc.

(*b*) *Nuclear Fuels* : Uranium, Thorium, Deuterium, other Nuclei used in fission or fusion reactions.

(*c*) *Hydroenergy* : The energy of falling water used to turn turbines or mill wheels.

(*d*) *Geothermal* : The heat from the underground streams or the heat stored in the hot rocks beneath the earth's surface.

(*e*) *Solar Energy* : Electromagnetic radiation from the sun.

(*f*) *Wind energy* : The energy from moving air used by wind mills.

(*g*) *Tidal energy* : The energy associated with the rise and fall of the tidal waters.

However, the sun is the source for most of our energy resources; circulation of the atmosphere due to the sun's heat is the basis of solar energy and wind energy. The evaporation of water thereby causing rain and water flow is the basis of hydroenergy. Photosynthesis because of solar radiation forms the basis for chemical energy in the plants. The chemical energy stored in fossil fuels is also due to the solar heat.

The primary energy resources can be further classified as *renewable resources* and *non-renewable resources. Renewable resources* are those which are not exhaustible and which can hence provide continuous supply *e.g.* wood, tidal energy, solar energy, wind energy, geothermal energy etc. *Non-renewable resources* are those which are finite and exhaustible *e.g.*, fossil fuels (*viz.*, coal, petroleum, natural gas, nuclear fuels etc.

(*ii*) **Secondary energy resources** are those which do not occur in nature but are derived from primary energy resources.

**Ex.** : Petrol or gasoline, electrical energy from coal burning, hydrogen obtained by electrolysis of water etc.

## FOSSIL FUELS

During the early parts of the biological history of the earth, photosynthesis outpaced the activity of the consumers and docomposers. Therefore, large amount of organic matter accumulated, particularly on the bottoms of swamps and shallow seas. Gradually, this material was buried under sediments eroding from the land, and, over millions of years, was converted to coal, crude oil and natural gas. These three are called Fossil fuels because all of them once were living matter.

Formation of additional quantities of fossil fuels by natural processes may be still continuing even today. However, we are using fossil fuels outrageously faster than their formation. Hence, there is no hope of replenishment of the fossil fuels. Thus, sooner or later, we shall run out of the fossil fuels.

Almost the entire transportation in the world is dependent on crude oil. Nuclear energy, coal and hydroenergy are exclusively used for the production of electrical power. Natural gas and oil are more versatile energy sources but they should be used as efficiently as possible. For instance, using natural gas to generate electrical power and then using the electricity for space heating is extremely inefficient because of the 60–70% thermodynamic loss that occurs in the production of the electricity. The natural gas can be used more efficiently in a modern furnace, in which more than 90% of the energy content of the gas goes into heating the home and the rest 10% goes up to chimney.

As the supplies of fossil fuels in nature are finite, we shall be running out of them in near future. Therefore we should make every effort to accommodate ourselves to the decreasing supplies of fossil fuels and we cannot afford to wait until the crisis develops. There are only two options :

(1) Conservation

(2) Developing non-fossil fuel energy alternatives.

While exploratory drilling in recent years has not proved any appreciable reserves of crude oil, it has shown considerable reserves of natural gas. With the installation of a tank for compressed gas in the trunk and some modifications of the engine fuel intake system, cars can run perfectly well on natural gas. Such cars are widely used in Buenos Aires, where service stations are equipped with compressed gas to refill the tank. Natural gas is a clean-burning fuel, free from hydrocarbon emissions. However, using natural gas as a substitute for total oil demand is not a permanent solution. However, natural gas can provide an interim relief until non-fossil fuel energy alternatives are developed.

As against the limited crude oil reserves, several countries are endowed with coal reserves which can meet the energy needs for over 150 years. However, coal cannot be used directly as fuel in vehicles. However, coal can be converted to a liquid or gaseous fuel by chemical processes. However, these coal-based synthetic fuels are too expensive atleast for the present.

These processes may also generate some pollutants. Further, mining and using coal as fuel are accompanied by adverse environmental impacts. Many coal deposits can be exploited practically only by strip-mining. In underground mining, atleast 50% of the coal must stay in place to support the mine roof. In strip mining, gigantic power shovels turn aside the rock and soil above the coal seam and then remove the coal. Obviously, this procedure results in total destruction of the ecosystem. Strip-mined areas may be turned into permanent deserts. Furthermore, erosion and acid reaching from the disturbed earth may have adverse effects on surface and ground water of the area. Burning of coal produces considerable quantity of $CO_2$, $SO_2$ and other pollutants.

An important problem with all fossil fuels is that their combustion yields carbon dioxide which can cause global warming through the greenhouse effect.

There are only two alternatives to fossil fuels. These are nuclear power and solar energy.

## NUCLEAR FISSION AND NUCLEAR FUSION

Nuclear fission reactors are based on the fission of U–235 nuclei by thermal neutrons :

$$_{92}U^{235} + {}_0n^1 \longrightarrow {}_{51}sb^{133} + {}_{41}Nb^{99} + 2.5\ {}_0n^1 + 200 \text{ meV}$$

The energy from these nuclear reactions is used to heat water in the reactor and generate steam to drive a steam turbine.

High Temperature Gas-cooled reactors and Fast-Breeder reactors convert non-fissionable $pu^{238}$ and $Th^{232}$ to fissionable $pu^{239}$ and $U^{233}$

A major problem in the widespread use of nuclear fission power reactors is the generation of large quantity of radioactive fission-waste products, some of which may remain lethal for thousands of years. The safe disposal of the radioactive wastes also poses some problems.

Nuclear fusion power reactions are based on deuterium-deuterium reactions and deuterium-tritium reactions :

$$_1H^2 + {}_1H^2 \longrightarrow {}_2He^3 + {}_0n^1 + 3.3 \text{ meV}$$

$$_1H^2 + {}_1H^3 \longrightarrow {}_2He^4 + {}_0n^1 + 17.6 \text{ meV}$$

The deuterium-deuterium reactions promise an unlimited source of energy. However, harnessing of fusion energy will take several more years due to the technical problems involved.

## SOLAR ENERGY

Solar energy originates from the thermonuclear fusion reactions taking place in the sun. The earth continuously receives energy from the sun, a part of which is absorbed while the remaining is emitted back into space. Out of the solar radiations reaching the earth, 92% consists of radiations in the range 315 nm to 1400 nm. 45% of this radiation is in the visible region (400 nm to 700 nm). The earth absorbs radiations mainly in the visible region and emits radiation in the infra-red region (2 μ to 40 μ, with maximum at 10 μ). The value of the solar flux reaching the earth's upper atmosphere is estimated to be about 1400 watts $m^{-2}$ $min^{-1}$. The heat equivalent of the solar radiation reaching the earth is estimated to be about $2.68 \times 10^{24}$ Joules per year.

Light, through photosynthesis, is the energy source for all major ecosystems. Solar energy, being non-polluting and non-depletable, is considered, as renewable energy and thus fits into the principle of sustainability. Only 0.2 to 0.5% of the solar energy reaching the earth is trapped by photosynthesis. Thus, only a tiny fraction of the solar energy reaching the earth drives all our ecosystems. Similarly, a small percentage of the solar energy reaching the earth could sustainable supply all the energy needs of human societies without altering the biosphere in any way. In spite of these advantages, large scale use of solar energy is limited by the following :

(*i*) Problem of economically collecting solar energy over large areas and converting it to other forms that can be conveniently transported, stored and used in existing equipment.

(*ii*) Problems of designing facilities that can utilize diffused sunlight. Recent technological advances have tackled these problems to a considerable extent. Solar energy is now beginning to be used as an economically feasible and sustainable energy sources.

## SPACE HEATING AND WATER-HEATING BY SOLAR ENERGY

About 25% of the U.S. energy budget is used for heating buildings in cold weather and for providing hot water for washing purposes. Temperature of about 20°-22°C are used for space heating and 50°-60°C are used for hot water. Using an oil or gas flame to provide such low-temperature heat is considered as thermodynamically inefficient use of energy. Utilizing electrical power for this purpose is even more wasteful. Solar energy is ideally suited for providing low-temperature heat for such purpose because sunlight falling on any black surface is readily absorbed and converted to heat in the desired temperature range. Therefore, complex collection or conversion equipment is not required when solar energy is to be used for obtaining low-temperature heat. A simple *flate-plate* collector is sufficient. There are several designs of flat-plate collectors but all of them basically consist of a black surface convered by a clear plastic or glass "window" (Fig. 7.1). The black surface absorbs sunlight and converts it to heat, and the window prevents the heat from escaping out. Air can be heated by passing it between the window and the black surface whereas water heating can be done by passing it through tubes embedded in the surface. Thus, minimum cost is involved in collecting and converting solar energy to heat.

However, beyond the collector, solar heating system may be "*Active*" or "*Passive*" and may or may not include a means of heat storage. An *active solar heating system* uses either pumps or blowers to circulate the air or water from the collector to the desired location. On the other hand,

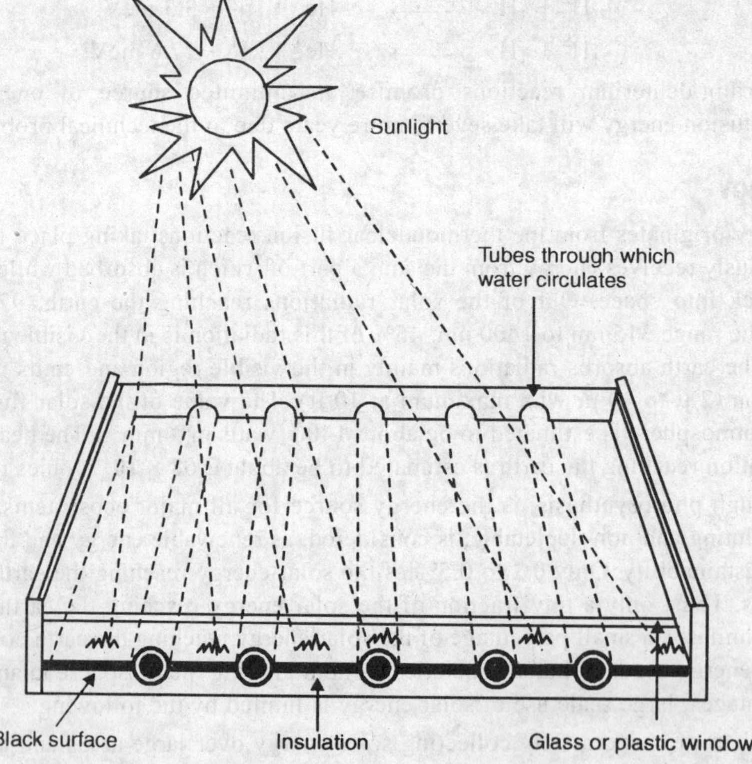


**Fig. 7.1.** Flat-plate solar collector.

a *passive solar heating system* depends upon natural convection currents (created when hot water or air rises due to lower density) to move water or air. Passive systems are relatively inexpensive and practically free from maintenance. Several plants are available for passive solar heating of buildings. The basic concept involves using of large, sun-facing windows due to which the buildings itself acts as the collector. In winter, sun-light beams through the window, and heats the interior while at night, insulated drapes or shades are pulled down to trap the heat inside. In summer, excessive heat load can be avoided by shielding the windows from the high summer sun using suitable awnings or overhangs or deciduous plantings or vines. There is no need of additional heat-storage facilities. Good insulation also serves to keep the buildings cool in summer, reducing the use of air-conditioning. Along with improved insulation, appropriate landscaping can contribute to the heating and cooling efficiency of both solar and non-solar designs. Particularly, deciduous trees or vines on the sunny side of a building will block much of the excessive summer heat while letting the desired winter heat pass through. Thus, suitable architecture can provide heating and cooling virtually free of cost.

A common criticism regarding passive or active solar heating is that a backup heating system is still needed during inclement weather. While this is true, it misses the point that even if solar heating and improved insulation provides only 20% of the overall heating needs, still it reduces the demand for conventional fuels by 20%. The fuel costs are also reduced by 20%. The same holds true for solar water heating.

On a per capita basis, cyprus is the world's largest solar energy user; 90% of the homes and several hotels and apartment buildings have solar water heaters. In Israel, 65% of domestic hot water is provided by solar energy. In the U.S., about 7 lakh solar hot water systems are operating although it represents only 0.5% of the total requirement.

Thus it is clear that there is tremendous opportunity to make use of solar energy in conjunction with improved insulation for space heating and water heating in a cost-effective manner.

## PRODUCTION OF ELECTRICITY USING SOLAR ENERGY

Solar energy can be used to generate electrical power, thereby providing an alternative to thermal and nuclear power. At present, the following two methods seem to be cost-effective:

(*i*) Photovoltaic cells (Solar Cells).

(*ii*) Solar trough collectors.

(*i*) **Photovoltaic cells**—These solar cells were primarily developed for providing power supply for space satellites in 1950s. A typical solar cell consists of two very thin layers of material. The lower layer has atoms with single electrons in the outer orbital, which are easily lost. The upper layer has atoms lacking one electron from their outer orbital and hence can readily gain electrons. The kinetic energy of light striking this "Sandwich" dislodges electrons from the lower layer, which are trapped into the upper layer, thus creating an electric potential between the two layers. This potential provides the electric current through the rest of the circuit, which connects the upper side through a motor or other electrical device back to the lower side. Thus without any moving parts, solar cells convert light energy directly to electrical power. Since there are no moving parts in them. Solar cells do not wear out. However, their current life span is about 20 years because they deteriorate due to exposure to weather.

Semiconductor crystals can be grown in such a way that one region is n-type and an adjacent region, p-type. The boundary between such regions within a single crystal is called a p-n Junction (Fig. 7.2a). In the absence of an external voltage, a hole from the p-side penetrating into the n-side is repelled by the positive ions of the donor in the crystal lattice and similarly, conducting electrons are prevented from entering into the p-side by the negative ions of the acceptor in the lattice. This "Junction-charge" *viz.* positive in the *n*-side and negative in the *p*-side, acts as a barrier for both the holes and the conducting electrons (Fig. 7.2a). However, when light falls on the p-n Junction, electrons from the valence band are promoted to the conduction band, thereby generating electron-hole pairs on both sides of the junction. In other words, an electron-pair bond is "dissociated" into an electron and a hole. Since the p-n Junction is a barrier to both types of charge carriers, concentration of holes builds up on the *p*-side of the junction and a concentration of electrons builds up on the *n*-side of the Junction. When these concentrations exceeds the equilibrium concentration of the carriers in the corresponding parts of the semiconductor, a drift of holes towards A and a drift of electrons towards B takes place. Therefore, if A and B are connected with a conductor, there occurs a flow of current from A to B and flow of electrons from B to A. (Fig. 7.2b). The electrons recombining with the holes ensures an uninterrupted flow of electrons as long as light is falling on the p-n Junction.

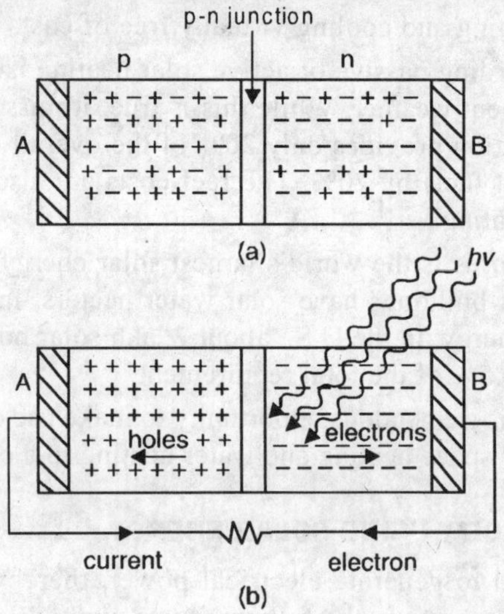


**Fig. 7.2.** Use of Semiconductors as Solar cells.

The solar cells made of silicon have been used as souce of power in space-craft and satellites. Other solar cells developed are Cd S (n-type)/$Cu_2$ S(p-type) gallium arsenide and indium phosphide.

A part of the electrical energy derived from solar energy by using solar cells can be stored by charging Ni-Cd batteries or lead-acid batteries. Alternatively, it can be used to produce hydrogen by electrolysis of water. The liberated hydrogen can be used in hydrogen-oxygen fuel cells.

The huge capital cost is the major limitation for the large–scale use of solar cells.

The largest solar cell manufactured in large scale are about 5 cm in diameter, which in full sunlight can provide power output similar to a standard flash-light battery. However, several cells can be connected together to obtain large amount of power. The present solar cells can achieve a conversion efficiency of light energy to electrical energy in the order of 10 to 20%. This low efficiency is immaterial in one respect because the primary energy source, namely sun-light, is available free of cost. However, the cost of producing the number of solar cells required to obtain a given amount of power is considerable. As is common with any other new technology, the first cell produced were very expensive. In the early 1970's, photovoltaic power was costing about 30 dollars per killowatt-hour. With improvements in production techniques, the cost came down to about 30 cents per killowatt-hour in 1991. However, even this cost is still 4 to 6 times the cost of power from traditional sources. In spite of this, even at this cost photovoltaic cells are economically practical for many applications such as in pocket calculators where they replace batteries. Panels of solar cells are the most economical way of providing power at points that are far from utility lines *e.g.* rural homes, light-houses, irrigation pumps, radio-transmitters, offshore oil drilling plateforms, traffic signals etc.

Recent breakthroughs promise to cut the costs by 50 to 80%. Rural electrification projects based on photovoltaic cells are becoming popular in the Third World Countries. In the near future, photovoltaic power plants may be producing power nearly at the same cost as traditional power plants. The Japanese are developing solar-powered air-conditions to be marketed soon. Experiments with vehicles run on photovoltoic cells are under way using ultra-efficient designs.

## SOLAR TROUGH COLLECTORS

The solar trough concept was invented by Charles Abboutt in the 1930's. Solar trough collection systems represent another economically practical method for producing electrical power from sunlight. Sunlight hitting the solar trough collector is reflected onto a pipe running down the centre of the trough, and heats fluid circulating through the pipe. The heated fluid is then used to boil water, thereby generating steam to run conventional turbogenerator.

Lux International Company is reported to have built a large number of solar trough collection systems in southern California with a total capacity of 350 megawatts, which is about 30% of the capacity of a large nuclear power plant. The recent facility built by Lux International achieved 22% conversion of incoming sunlight to electrical power and produces power at a cost of 8 cents per kilowatt-hour. This is slightly higher than the cost of nuclear or coal–fired power plants. This solar power may be considered cheaper if one considers the hidden costs of air-pollution, strip mining, and nuclear waste disposal in the production of thermal and nuclear power.

The drawback that solar power system require large stretches of land to "harvest" the sunlight may not be treated as serious if we consider the thousands of acres which are being wasted by strip-mining of coal every year or the huge area of land which was rendered uninhabitable due to nuclear accidents. Further, photovoltaic systems can utilise existing rooftops without using additional land.

Since about 65% of the electrical power demand occurs in daytime hours when industries and offices are in operation, the problem of sunlight being not available during the night is not very serious. Atleast, solar power will conserve the traditional fuels to a considerable extent for future use. Further, in the long run, the night time power demands can be met by different forms of indirect solar energy such as water power or wind power discussed later. Also, the excess solar power generated during day time may be stored by using it to pump water from a low-elevation reservoir to a high elevation reservoir. When additional power is required, the water from the high elevation reservoir can be allowed to run down through turbo-generators. Alternatively, the excess solar power can be used to pump air into underground caverns and let this compressed air to drive turbines whenever the power is needed.

Two more approaches of producing electrical power from sunlight are receiving attention :

(*i*) **"Power Tower"** : In this method, an array of sun tracking mirrors is used to focus the sunlight falling on a large area of land (several acres) onto a boiler mounted on a tower in the centre of the land. The intense heat produces steam in the boiler which drives a turbogenerator. This system is expected to generate power more economically than the solar trough system.

(*ii*) **"Solar Pond"** : In this method pioneered in Israel, an artificial pond is partially filled with brine and fresh water is placed over the brine. Brine, being denser than fresh water, remains at the bottom without any appreciable mixing. Sunlight passes through the fresh water but is absorbed and converted to heat in the brine. This fresh water then acts as an insulating blanket and holds the heat in. The hot brine solution can be circulated through buildings for heating. Alternatively, it can be converted to electrical power by vaporizing the fluids with low boiling points and using the Vapours to drive a low–power turbogenerators. Since the solar ponds also acts as an efficient heat storage unit, it supplies power continuously.

## SOLAR ENERGY FOR DRIVING VEHICLES

While considering energy sources to sustain our transportation system, conservation by increasing vehicle mileage through better engine design, proper use and maintenance still remains the choicest option for reducing demand for crude oil in the short term. Use of solar heating facilities reduces the demand for natural gas in commercial and residential sectors thereby transferring the conserved natural gas to the transportation sector.

Use of solar energy for running electric vehicles, though feasible, still suffers from the following limitations :

(*a*) Storage batteries are expensive and heavy relative to the amount of power stored. For instance, General Motors' newest prototype electric car, which is likely to be marketed shortly, carries 360 Kg of batteries to store the equivalent of about 4 litres of petrol.

(*b*) The range of even an ultra-efficient car is less than 140 km.

(*c*) Lack of a breakthrough in battery technology.

Solar energy can be used to produce hydrogen which in turn, can be used as fuel for existing vehicles. Hydrogen is a clean-burning gas without any pollution because the product of combustion is only water vapour.

$$2H_2 + O_2 \longrightarrow 2H_2O + \text{energy}$$

Present vehicles could easily be adapted to run on hydrogen with some modification in the fuel tank, fuel intake system and the type of service stations. The service stations should dispense hydrogen gas into vehicle "fuel tanks", which would consist of materials that would adsorb large amounts of hydrogen and release it only slowly, on being heated. This slow release would ensure safety against the hazard of explosion.

However, hydrogen does not exist in free state on earth and it should be produced by electrolysis of water. Solar energy can be used to produce hydrogen by electrolysis of water at places where land is cheap and sunlight is plenty.

## POWER FROM INDIRECT SOLAR ENERGY

Hydropower, wind and biomass are considered as indirect forms of solar energy because water cycle, atmospheric circulation and photosynthesis are driven by solar energy. These indirect forms of solar energy also provide substantial sources of energy, just like direct solar energy. Mankind has been utilizing these indirect forms of energy from ancient times.

### (*a*) Hydropower

Water power has been used since ancient times by diverting water from natural streams or rivers over various kinds of paddle wheels or turbines. The power output from water wheels being low, people started building high dams from the last century to obtain a substantial head of hydrostatic pressure. Thus, the water under high pressure, flows through the base of the dam and drives turbo-generators producing hydroelectric power. In U.S., about 300 large dams generate 9.5% of its total electrical power production.

Although hydroelectric power is basically a nonpolluting renewable energy source, it is still associated with serious problems :-

(1) Dams have drowned out beautiful stretch of rivers, wildlife habitat, forests, productive farmlands, and areas of historic, archeological, and geological significance. The construction of big dams have also rendered several farmers and tribals homeless and without any livelihood.

(2) The reservoir behind the Aswan High Dam in Egypt has caused the spread of a parasitic worm which caused a debilitating disease. Further, the increase in humidity over a large area because of the reservoir is causing rapid deterioration of ancient monuments and artifacts which were existing over many centuries.

(3) Since water flow from the dam is regulated as per the requirement of power, dams play havoc down-stream because water levels may changes from extremes of near flood levels to virtual dryness and back to flood even in a single day. Other ecological factors are also affected because sediments rich in nutrients settle in the reservoir and only small amounts reach the river's mouth.

(4) Devastating earthquakes, observed near Koyana in India, are attributed to the Koyana dam (Maharashtra) by some Scientists.

Many developing countries have great potential for large hydel power projects but due to the above problems, there is lot of opposition from people as well as from Environmental protection organisations.

### (b) Wind Power

Wind is essentially a mass of air in motion, which carries with it kinetic energy. The amount of energy contained in the wind at any instant is proportional to the wind speed at that instant.

Wind was used in ancient times by Egyptians, Persians, Chinese, Europeans and others for propelling ships, and to provide rotary windmill power for pulling up water from wells, grinding of agricultural produce (*e.g.*, grains), and later on for turning machines of factories. The use of wind as source of producing electric power began in the 19th century.

Wind is a dispersed source of energy and this is the limiting factor for the large scale utilization of wind as a source of power. For obtaining substantial amounts of power from wind, massive wind machines must be used. However, it is more practical to use several smaller wind machines than using a single giant machine. It is from this realisation that the concept of ***"wind farms"*** has emerged.

Wind mills have been in extensive use in the Peoples Republic of China, Portugal, USA, Dutch, England, Russia, etc. using a variety of designs suitable for their own uses.

Many wind turbines are in operation today throughout the world.

Most of the wind turbines built in USA during the 1970s used a horizontal axis, and mostly with 2 blades, although some very large machines had 3 or even 4 blades. These turbines produced alternating current and successfully synchronized their input to commercial electrical grids.

The limitations on the extraction of energy from the wind include practical size of wind machines, their density, frictional losses in the rotating machinery, and efficiencies of conversion from rotational energy to electrical energy.

A wind mill works on the principle of converting kinetic energy of the wind to mechanical energy. No windmill device, however well designed, can extract all of the wind's energy because the propellers of the wind mill oppose the movement of the wind; this would prevent the passage of as much air through the rotor as theoretically possible on the basis of wind speed. Even a 100% efficient aero-generator would only be able to convert upto a maximum of about 60% of the available energy in wind to mechanical energy.

## WIND ELECTRICAL ENERGY SYSTEMS

Wind electric energy systems are classified into the following categories of utilization :

| *S.No.* | *System type and Range* | *Utility* |
|---|---|---|
| 1. | Wind Electric Energy Systems connected to grid without need for energy storage facility. (wind farms–0.5 MW to 2 MW; Large scale units 1 MW to 14 MW) | Used where energy conservation is the main objective. |
| 2. | Stand alone (i.e. Isolated) wind electric Energy Systems with need for energy storage facility (5 KW to 750 KW for wind turbine generators). | Economically viable choice for people living in remote areas, which are far away from power supply grid. |
| 3. | Non-critical Wind Electric or Wind Mechanical Energy Systems, without need for energy storage. (5 KW to 250 KW unit size) | Pumping water, grinding grains, farm power supply, electric heating, and energy conservation. |
| 4. | Wind Electric (5 KW to 2 MW) (Energy storage not necessary) | During favourable winds, a wind turbine generator supplies power. |
| | Diesel Electric (Batteries provided for uninterrupted power supply) Hybrid | During non-favourable wind, batteries supply energy. |
| 5. | Wind Electric and Solar PV Hybrid. (350 KW to 750 KW) Storage batteries needed. | For remote stand alone power plants away from grid. |

## SITING OF WIND FARMS

For deciding the location of wind farms, siting studies are carried out in areas of medium winds of average speed of 7 m/s. The sites can be *"onshore"* or *"off-shore"* shallow water. ***"On-shore"*** sites preferred. "Off-shore" sites on shallow sea-bed are chosen by countries like U.K., Sweden and Denmark, because these sites offer wind conditions. The following criteria are generally used for selecting the site for a wind farm :

(1) An average and uniform wind speed of 8 m/s throughout the year.

(2) Minimum chances of cyclones, floods and lighting strikes.

(3) Suitable terrain and soil for installing the wind turbine, tower movement of crane, etc.

(4) Availability of adequate vacant land.

(5) Availability of a distribution sub-station connected to the electrical grid within 10 km.

(6) Availability of approach roads to the selected site for carrying the necessary equipment.

(7) Clearance from the environment protection authorities.

### Wind Turbine Sizes

Wind machines used for generating electricity may be classified into three categories :

(*i*) Mini-converters having an average output of about 1 KWe. These are used to run small irrigation units, lighthouses and ranger stations.

(*ii*) Machines with an average output of 50 KWe or more. In developing countries, these machines are used in village farms, rural industries and isolated houses. In rural areas of North America, Scandinavia, etc., such machines are used for energy required for space heating, pumping of irrigation water and hot water generators.

(*iii*) Wind farms with average outputs in the range from 500 KWe to many megawatts. They supply power to electrical power grids.

## DESIGN OF WIND TURBINE

A wind turbine is a device that turns kinetic energy from wind into electrical energy. The flow of air against the windward side of the blades creates suction on the reverse side, which turns the rotor shaft on which the blades are mounted. The rotor turns the generator via a gear. The generator is linked directly to the electricity grid. Besides these main components, the turbine has a yaw system that ensures that the turbine is always facing the wind. When the wind speed is greater than about 25 m/s, a disc brake stops the turbine from operation. There are also blade tip brakes which are controlled by a hydraulic system in the hub, and these also assist in the breaking function. The operation of the turbine is monitored and controlled by a computer in the bottom of the tower. The computer receives information regarding wind speed and wind direction via an anemometer and a wind wane mounted on top of the turbine housing.

Current state of art wind turbines are of the horizontal axis design with high hub heights (~328 ft) and large rotor diameters (~ 180 ft). They capture the wind energy with 2 or 3 propeller-like blades, which are mounted on a rotor to generate electricity. The turbines sit on the top of tall towers, taking advantage of increased wind speed and decreased turbulence with altitude. Strong winds are critical since power generation increases with the cube of wind speed. It increases linearly with the area swept by the propellers, which in turn is proportional to the square of their length. Hence there is a premium on using long propellers mounted on tall towers. The limiting factor on size is the possibility of structural failure in high winds. Innovative tower designs utilizing stronger light-weight materials now allow taller towers to be built at reduced cost. Currently used hub-heights are around 30 meters and they may reach 70 meters or higher by 2005.

A prototype electrical generator is developed at Rensselaur polytechnic Institute in Toy, New York in order to make wind power more useful for small consumers, such as farms, isolated homes and remove third world settlements. A limitation with most of the current wind turbines is that they operate most efficiently with strong steady winds. At high speed gusts, they induce stress rather than producing more power. Using variable speed turbines is expensive because of high costs of electronic circuity required for variable speed systems. The solution developed by the Institute is to employ a generator that could generate power from variable-speed direct-drive wind turbines. A variable reluctance generator that could develop power at a maximum of 120 rpm, which is about 15 times *slower* than commercial wind turbine generator. In this case, a gear box is not required because of the slow speed. The resulting wind turbine is less expensive than commercial models. The prototype generator developed 20 KW power.

## MERITS OF WIND ENERGY

(1) It is a renewable energy source available free of cost.

(2) Wind machines can be built *"on shore"* or *"off shore"*.

(3) Wind machines are useful in supplying electric power to remote and rural areas.

(4) Cost effective and reliable wind power generators are now being produced.

(5) It is economically competitive for remote areas.

(6) It does not cause pollution during energy generation.

(7) Dispersed wind energy systems are more environmentally being than any other alternative source of energy.

## DEMERITS/LIMITATIONS OF WIND ENERGY

(1) Low energy density.

(2) Economics are favourable only for remote areas

(3) Variable, irregular and intermittent.

(4) Small units are more reliable but have higher capital cost per KWh. Large units require high technology although they have lesser capital cost per KWh.

(5) Design, manufacture and installation of wind turbines is complex due to varying atmospheric conditions where they have to operate.

(6) Require energy storage batteries which indirectly contribute to environment pollution.

(7) Require vast open areas which are generally far away from load centres.

(8) Wind generators may interfere with habitats, can cause noise pollution, aesthetic degradation and may interfere with bird flights.

## INDIAN SCENARIO

Wind energy has proved to be economically competitive renewable source of energy in the following locations in India :

(*i*) Coastal regions in Gujarat, Kachchh and Saurashtra.

(*ii*) Plains in UP and Rajasthan.

(*iii*) Hill tops in MP.

(*iv*) Gaps in the Eastern Ghats in Tamil Nadu, Tuticorin and Kayathar.

Total estimated wind energy resource potential in India is about 25000 MW. The installed capacity is 1167 MW. The production of wind energy is 2.33 million MWh, which represents 0.6% of the total electricity produced.

As of today, the countries with largest wind energy production include Germany, U.S., Spain, Denmark followed by India, Netherlands, Italy, U.K., China and Sweden. As of the year 2000, the installed wind capacity of all the above 10 countries together is about 16,000 MW and the total wind energy produced is 31.2 million MWh. This works out to be 0.5% of the total electricity produced in these 10 countries, viz. 6300 million MWh.

### (C) Biomass Energy

Biomass energy or Bioconversion refers to the direct burning of wood, waste paper, manure, agriculture or any form of biomass or converting them to a fuel. Certain micro-organisms when they digest biomass in the absence of air, produce either alcohol or methane gas, which themselves give energy on combustion. Since biomass is obtained through the process of photosynthesis, biomass energy is considered to be another form of indirect use of solar energy.

Firewood can be considered as a sustainable energy resource where forests are plenty relative to population, which was the case in ancient times. However, any idea of using firewood as a large scale energy resource must be considered in the context of maximum sustainable yields and preservation of the bio-diversity of eco-systems. Many Third World countries where millions of people still depend on firewood as fuel for cooking food and getting warmth, are already suffering from the severe ecological effects of deforestation and human deprivation caused by non-availability of wood. In this backdrop, any blanket recommendation to burn wood because it is renewable, is environmentally irresponsible.

Biomass energy should be preferred wherever energy can be produced as a by-product of waste, disposal, (e.g., forest waste, farm waste, saw mill waste, sugar refinery waste, municipal refuse etc.). The wood products industry already uses most of its mill residues to generate steam and electricity for use in the mills themselves. Wheat straw, corn stalks, and other plant residues on the forms are usually left there itself to decompose in the fields. In fact, much of this is needed to replenish the soil organic matter and to limit erosion. However, 40% of this could be harvested for energy production wherever economics permit. Municipal waste also contains a considerable fraction of combustible matter. Much of this is land-filled, but there are many municipalities which burn the trash in incinerators. In some cases, the heat of combustion is used to generate electricity. However, the cost of trash collection offsets the income due to the generation of electricity in most cases. Further, incinerators are contentious because of concerns due to toxic emissions such as dioxins and mercury. Emissions are minimal in properly run, modern state-of-art incinerators. However, operating conditions may not be optimal always. In all the cases where energy is produced as a by-product of waste disposal, advantages of recycling, reuse or composting should always be kept in mind.

Quite apart from utilization of wastes, there exists a distinct possibility of planting fast-growing trees and grasses solely as energy crops. Switch grass seems to be a good example. It can grow profusely for 10 years, resist drought, and thrive even on low fertilizer application. Fast-growing trees, such as poplar, cottonwood, willow and green ash are already serving other environmental purposes as well. They are planted next to crops in buffer strips along stream banks to mitigate erosion and water pollution. They can also be planted around live-stock feedlots to control odours and intercept runoff. Further, when these trees are harvested, the stumps grow fresh sprouts, thus enabling the trees to re-grow for several rotations before replacing the rootstocks. Thus, harvesting of switch-grass or woody crops for energy production offers an excellent opportunity for farmers with marginal lands that are unsuitable for row-crops.

However, proposals for growing trees or other crops solely from the point of view of the biomass energy, either for direct burning or for conversion to alcohol or methane, must be considered in the light of possible soil erosion and requirement of fertilizers, pesticides, etc. for growing such crops and their possible environmental impact. Further, the possibility of depriving the land for growing food crops in favour of energy crops should also be kept in view because both may be grown on the same land.

In spite of the fact that industrialization is responsible for switching from traditional biomass fuels to fossil fuels, biomass still represents a large energy resource. Since the carbon present in biomass is continuously converted to $CO_2$ by respiration, burning it to extract energy does not add to the net $CO_2$ content of the atmosphere. The energy density of biomass is variable, and is substantially lower than that of fossil fuels. Further, collection and processing of the biomass is rather difficult. Because of these reasons, biomass is not economically competitive with fossil fuels. However, if emission of $CO_2$ comes under regulation or taxation, then the prospects of biomass fuels to compete with fossil fuels will become brighter. It is interesting to note that $CO_2$ emission is already under regulation or tax net in several countries.

## PRODUCTION OF ETHANOL FROM BIOMASS

Although there are several schemes for chemical conversion of biomass to liquid fuels, biological fermentation of sugars to ethanol is the one which is used in relatively large scale. Brazil has been the leader in producing ethanol from its fast-growing sugar cane. At one point of time, 50% of its automotive fleet was running on ethanol. In the U.S., ethanol is produced from corn. Ethanol so produced is blended with gasoline to a level of 10% (E 10) or 85% (E 85).

The U.S. EPA, in the year 1990, has mandated oxygen-containing additives (called "oxygenates") for gasoline to reduce emissions of carbon monoxide and other smog-forming pollutants. The two oxygenate additives vying for the gasoline market have been ethanol and methyl t-butyl ether (MTBE). However, MTBE has now fallen from favour because it was found to cause ground-water contamination. Ethanol currently accounts for about 1.2% of the U.S. gasoline supply and it may increase further.

In the US, the use of corn for producing ethanol is found to be uneconomical but for the farm subsidies and tax concessions. In fact, the energy available from a gallon of ethanol is only slightly higher than the energy required to produce, it, if the inputs for fertilizer, farm machinery, and the energy for fermentation and distillation are taken into account. The ratio of "energy output" to "energy input", even with the recent technological developments, stands at 1.24. However, the economics can be improved if the entire corn plant can be converted to ethanol, instead of just the Kernel as is done now. The reason for this is that the Kernel mostly comprises of starch, whereas the rest of the plant (consisting of cob, leaves and stalk) is mostly made up of cellulose. The enzyme required for degradation of starch is amylase, while that required for cellulose is cellulase. Many micro-organisms contain amylase whereas only a few, viz. certain fungi and bacteria contain cellulase. Chemical conversion of cellulose to ethanol is rather difficult because of the following reasons :

(*i*) cellulose conversion requires the enzyme "cellulase" which is difficult to obtain in industrial quantities. The fungi or bacteria that produce cellulase are difficult to grow. (Termites and ruminants can digest cellulose, because their digestive systems harbor cellulase-producing bacteria).

(*ii*) cellulose in plants is protected by a protective sheath of hemicellulose (a polymer of five-carbon sugars) and lignin. Therefore, for chemical conversion to take place, this sheath must be first disrupted. This can be done by treatment with dilute sulfuric acid at elevated temperature and pressure.

In the industrial production of ethanol from corn, the starch is first degraded to glucose with the readily available amylase, and the resulting glucose is then fed to yeasts, which metabolize the glucose and release ethanol (fermentation). However, for conversion of cellulose (from biomass) to ethanol, the cellulase needed is difficult to obtain in industrial quantities. Moreover, the lignin and hemicellulose sheath which protects the cellulose in plants, has to be first disrupted with dilute sulfuric acid at elevated temperature and pressure. This treatment breaks down hemicellulose to xylose and other five-carbon sugar moments. Thus, the glucose produced by cellulase gets mixed with xylase and other five-carbon sugars. The common types of yeast which can ferment glucose are not effective to ferment the five-carbon sugars. This has to be done with the help of other organisms but a separate fermentation unit is required. Research is currently underway for developing new strains of bacteria which are capable of metabolizing all the sugars in a single fermentation process. The biocatalytic routes leading to the formation of ethanol from corn starch and cellulosic biomass are given below :

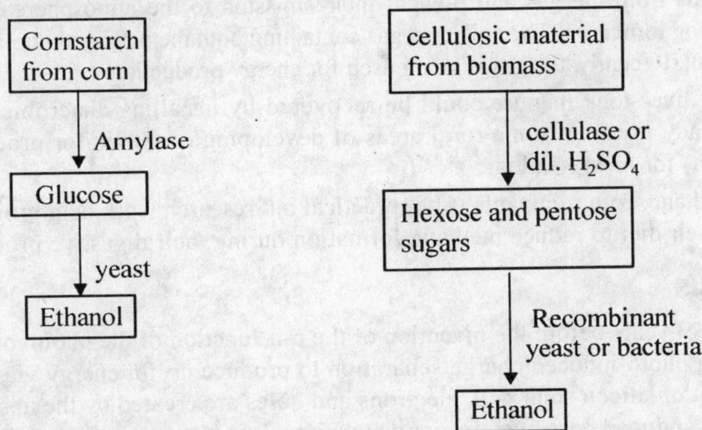


(*Note* : Starch contains polymerized α-glucose units, while cellulose contains polymerised β-glucose units. Owing to the different orientation of the connecting bonds in starch and cellulose, the enzymes must have different placements of the catalytic groups in their active sites. The enzyme required for conversion of starch is amylase, whereas the enzyme required for conversion of cellulose is cellulase).

Use of alcohol as fuel is being vehemently promoted in some grain growing regions of developed countries such as U.S. Alcohol is produced by fermentation of grains, starches, sugar or similar food products. However, developing and under developed countries should be very cautious in entertaining such ideas. Since many parts of the world suffer from shortage of grains, production of alcohol from grains may lead to shortage of food for their rapidly growing population thereby leading to widespread malnutrition. Further, while alcohol by itself is clean burning, we should remember that producing fuel grade alcohol requires distillation for which cheap, dirty burning fuels such as soft coal are used as fuels. The net energy yield of alcohol is therefore modest, because fuel equivalent of about 0.5 gallon of alcohol is used for every gallon of alcohol produced.

Anaerobic digestion of sewage sludge and animal manure is a biomass utilizing method which creates a valuable synergism between recycling and energy production. Such a method yields biogas and a nutrient-rich compost which is a good organic fertilizer which can be recycled back to the land in order to maintain the fertility of the fields growing forage for the animals. Further, we can thus get cheap energy which is sustainable and is free from adverse environmental consequences. In China, millions of small farmers maintain a simple digester in the form of a sealed pit into which they put agricultural wastes. The biogas generated is used as domestic fuels. This is certainly a better alternative than using wood as fuel.

## PRODUCTION OF METHANE FROM BIOMASS

The carbon present in dead biomass from landfills, manure piles etc. under anaerobic conditions is converted to methane by anaerobic bacteria. This happens also in the rumens of ruminants, where the bacteria which breaks down cellulose also release methane. This methane generally escapes as wasted energy source. It also contributes significantly for global warming. The global warming potential of $CH_4$ is 23 times greater than that of $CO_2$. This fact is recognised in the Acts and Schemes formulated to regulate greenhouse gases, including Kyoto protocols. If these come into effect, the incentive to collect $CH_4$ from landfills and manure piles is expected to increase further.

There were several cases of landfill fires in U.S.A., U.K. and other countries. Subsequently, the gas from land fills was collected and vented to the atmosphere. But, with the advent of promulgation of the 1996 clean Air Act in USA, it became mandatory for large landfills in USA to collect non-methane organic compounds from the gas and prevent their emission to the atmosphere, because of their contribution to smog formation. The collected gas containing both methane and non-methane organics was generally burnt. Recently, the gas is being used for energy production.

Methane from live-stock manure could be recovered by installing anaerobic digesters. Such digesters are already in use in some rural areas of developing countries for producing gas (e.g., gobar gas in India) for local heating.

Collecting methane from ruminants is not practical but research work is in progress to alter the composition of their diet to reduce methane formation during their digestive process.

## PHOTOSYNTHESIS

Since millions of years before the invention of the p-n Junction of the photovoltaic cell, nature started harnessing photo-induced charge separation to produce useful energy via photosynthesis. Just similar to that of silicon solar cell, electrons and holes are created by the absorption of solar photons, and are induced to travel in opposite directions. Instead of a p-n junction, nature evolved a photo-reaction center at which charge is separated across a biological membrane (Fig. 7.3). The important components of the photo-reaction centre are nothing but a collection of chlorophyll molecules (Fig. 7.4) capable of absorbing sunlight (strongly in the region of 400-700 nm) and generating electrons and holes in their photo-excited states. The electrons jump from one molecule to another along an energy gradient of empty energy levels. This gradient serves the same function as the charge asymmetry in a p-n junction, namely, to physically separate electrons and holes. Instead of leading to the generation of electric current, the charge separation leads to the production of energy-storing chemicals. Since the photo-reaction center is held in a membrane, the electron and hole are able to react separately with molecules that are situated on opposite sides of the membrane. The electron is taken up in a series of biochemical steps that lead to the reduction of $CO_2$ to carbohydrates, while the hole is involved in another series of steps that oxidize water to $O_2$.

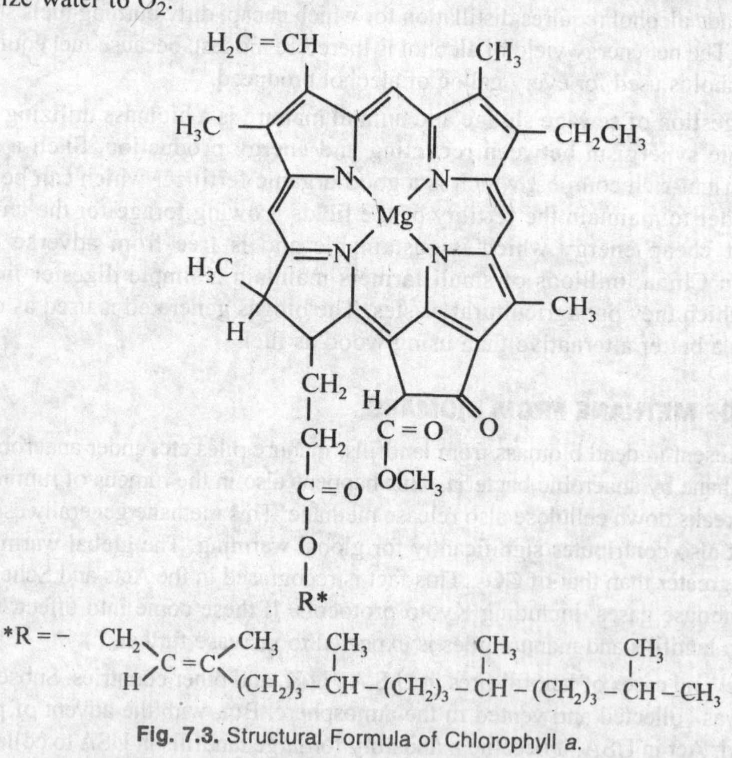

$$^*R = -CH_2-CH=C(CH_3)-(CH_2)_3-CH(CH_3)-(CH_2)_3-CH(CH_3)-(CH_2)_3-CH(CH_3)-CH_3$$

**Fig. 7.3.** Structural Formula of Chlorophyll *a*.

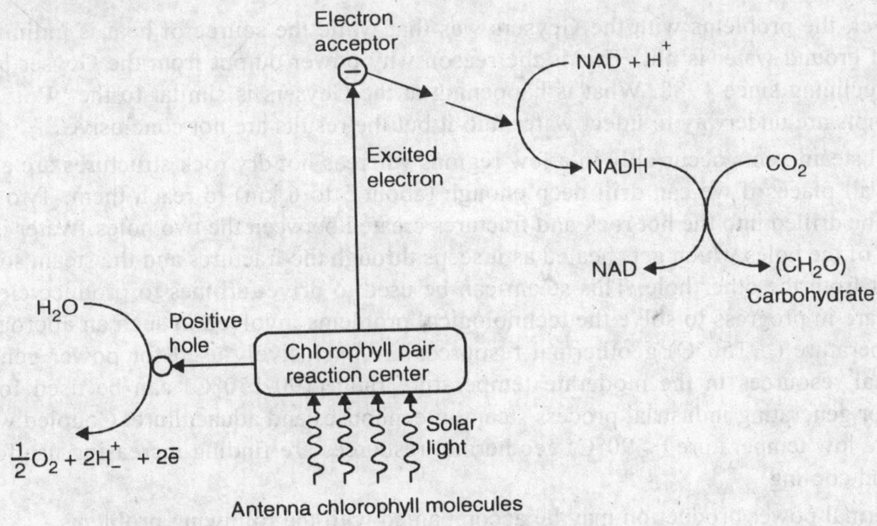


**Fig. 7.4.** The photo-reaction center and energy storage in photosynthesis.

## PHOTOELECTROCHEMISTRY

With the help of a photoelectrochemical cell, we can carry out chemical processes analogous to photosynthesis. In such a type of cell, a *photosensitizer*, usually an organic dye, analogous to chlorophyll, captures sunlight and transfers the photoexcited electrons and holes to a pair of electrodes, at which they react separately with molecules. If the electrons and holes can be induced to react with water, then hydrogen and oxygen are generated, exactly in a reversal of a hydrogen-oxygen fuel cell reaction (discussed elsewhere in this book). Although the same result can be obtained by using the electricity from a solar cell to electrolyze water, the photoelectrochemical cell eliminates the necessity of an electrolyzer and thus could be more efficient. However, the basic problem in this case is that the electrodes carrying out these reactions have a tendency to degrade with time. Research is in progress to develop improved electrode materials and catalysts to enhance the life and efficiency of the electrodes and to accelerate the electrode reactions.

## OTHER RENEWABLE ENERGY SOURCES

### (1) Geothermal Energy

Heat from the molten core of the earth offers a huge and sustainable energy source. This geothermal energy can be used for space heating or for boiling water and generate steam to drive turbogenerators. In volcanic regions of the earth, the hot rock is relatively close to the surface. Also, in such regions sometimes, the natural ground water comes in contact with the hot rock and the heated water or steam may find its way to the surface through natural steam vents. In such regions, holes can be drilled into the hot-rock ground-water structure and make the rising steam to drive turbogenerators to produce electric power. The world's largest Geothermal energy production facility exists at a location known as **"The Geysors"** near San Fransico in U.S. The electrical output from this facility in the year 1988 was about 2000 megawatts, which is equivalent to the power produced by two large nuclear power plants. Similar geothermal facilities exist in the Phillippines, Mexico, Japan, Italy and Iceland, generating a total power of about 3,000 megawatts.

However, the problems with the Geysers was that while the source of heat is unlimited, the amount of ground water is not. This is the reason why power output from the Geyser has been steadily declining since 1988. What is happening at the Geysers is similar to the **"Pot"** boiling dry. Attempts are underway to inject water into it but the results are not conclusive.

Natural steam vents occur only in a few regions whereas hot dry rock structures are available in almost all places if we can drill deep enough (about 5 to 6 km) to reach them. Two parallel holes can be drilled into the hot rock and fractures created between the two holes. Water is forced down one of the holes which gets heated as it seeps through the fractures and the steam so formed comes out from the other hole. This steam can be used to drive turbines to produce electricity. Attempts are in progress to solve the technological problems involved in such an approach. The high temperature (> 150°C) geothermal resources are exclusively used for power generation. Geothermal resources in the moderate temperature range (90-150°C) can be used for space heating, for generating industrial process steam, greenhouses and aquaculture. Coupled with heat pumps, the low temperature (< 90°C) geothermal resources are finding increasing use for home heating and cooling.

Geothermal power production may be accompanied with the following problems :

(*i*) Hot steam and water coming to the surface are usually contaminated with salts and other pollutants such a sulphur compounds. Some of these contaminants are highly corrosive to turbines and other equipment. They may also cause air pollution if the steam escapes into the atmosphere. $SO_2$ pollution from a geothermal plant may be as much as that of a high sulphur coal based thermal power plant.

(*ii*) Hot brine released into surface waters may be ecologically hazardous.

**(2) Ocean Thermal Energy Conversion (OTEC)**

In oceans, a thermal gradient (*i.e.* temperature difference) of about 20°C exists between surface water heated by the sun and colder deep water. (In tropical regions, the temperature of the surface layers of the ocean is as high as 26°C, whereas at a depth of 1,000 meters, the temperature is 5 to 6°C). This temperature difference can be harnessed to produce power. This concept is known as Ocean Thermal Energy Conversion (OTEC). An OTEC power plant can be built on a brage (*i.e.*, sailing vessel) that could travel anywhere in the ocean. It uses the warm surface water to heat and vaporise a low boiling liquid such as ammonia. The increased pressure of the vaporized liquid would drive turbo generators. The ammonia vapour leaving the turbines would then be condensed by cold deep water which is about 100 m below the surface and is returned back to start the cycle again. The electrical power so generated could be used to produce hydrogen and shipped to the shore. Alternatively, an energy intensive industry can be located on factory ships that would anchor alongside the OTEC Plant. A few OTEC Plants have been tested.

Owing to the small temperature difference between the surface water and deep water, the conversion efficiency is as low as 2-3%. This low efficiency by itself is immaterial since the primary energy source viz., the temperature difference between the surface and deep water over most of the ocean, is freely available. However, this low efficiency coupled with other drawbacks such as high capital costs, persistent maintenance problems and fouling of pipes and pumps due to marine organisms, results in meager energy yields thereby rendering OTEC power uneconomical at the present state of the OTEC technology.

In addition to electricity, an OTEC plant delivers cold sea water to the surface, which can be used to air-condition nearby buildings. The water from the deep ocean, which is rich in nutients, can support agriculture and mariculture enterprises. An experimental station in Hawaii is exploring the possibilities for integrated power, water, air-conditioning and farming activities on the shore.

Further, the difference in salt concentration between the oceans and fresh water represents a large osmotic pressure which is equivalent to a 240-meter water head. A practical application of using this mixing energy is worth exploring.

**(3) Tidal Power**

A lot of energy is inherent in the twice-a-day rise and fall of the tides. Scientists have tried to utilize this tidal energy which is eternal and pollution free. One of the simple schemes is to construct a dam across the mouth of a bay and mount turbines in the structure. The incoming tide flowing through the turbines generates power. As the tide shifts, the blades may be reversed so that outflowing water continues to generate power. At present, tidal power plants are in operation in Russia, France and Nova Scotia.

For any tidal power project to be of practical use, a fluctuation of atleast 6 meters is needed between the high tide and low tide. Such a situation exists only at about 15 locations in the world. The Bay of Fundy in North America is one of such locations and a large tidal power plant is being developed.

Tidal power plants are accompanied by the adverse environmental effects because of the dams which may trap sediments, impede the migration of marine organisms, change water circulation and cause mixing of fresh water with salt water.

**(4) Air Energy (Generation of power using air)**

Air energy is distinctly different from wind energy. The development of this form of energy is relatively recent. It is not yet commercialised.

A concept for developing air power was conceived by Philip Carlsson, which has been patented by Lockheed. The power unit is a tall tower, which has water pumped to the top of the tower and then dropped in the form of a fine mist. The mist cools the air, thereby making it heavier and to descend. The air at the bottom of the tower moves at around 50 mph. It turns the blades of the turbines stationed at the bottom of the tower (in a similar fashion as a waterfall turns a hydro turbine) and develops power. As hot air descends, cool air would be drawn in at the top, thereby making the system self-sustaining.

The original calculated cost of test unit which is 8000 ft high was \$ 2.7 billion which is supposed to produce 7500 MW. Less ambitious schemes had lower cost estimates. A 3000 ft high version made of aluminium and fabric was estimated to cost \$ 3 lakh.

In 1991, Melvin Pruiett suggested that the principles of the air energy design might be used to clean up the smoggy air over highly air-polluted cities like Los Angeles, Mexico, etc. In Pruiett's scheme, water droplets descending through the tower would pull along particles of soot and other impurities and end up as a sludge of the bottom of the tower, which could then be pumped away. An investment cost of \$ 10 million per tower or \$ 1 billion for Los Angeles, would pull in half the air above the city in 1 day.

Such a system might be useful to clean up residual particulates that makes the haze in the air, in citites like Mexico which suffer from temperature inversion.

The best time to clean air is just before the pollutants are emitted or just afterwards when the rate of condensation of the pollutant gases is the fastest.

Another approach suggested was to use the land around a 3000 feet high tower as farmland. The moisture – laden air released at night would spread out over an area estimated at 6 miles in diameter, and condense on farm plants.

## QUESTIONS

1. Discuss the use of solar energy for space heating, water heating and for production of electricity.
2. Write informative notes on any two of the following :
   (*a*) Flat plate solar collectors.
   (*b*) Photovoltaic cells.
   (c) Solar trough collectors.
3. Write short notes on any three of the following :
   (*a*) Solar energy for producing fuel for vehicles.
   (*b*) Solar production of electricity.
4. Write informative notes on any two of the following :
   (*a*) Wind power (*b*) Biomass energy (*c*) Tidal power (*d*) Air energy
5. Discuss the principles involved in Ocean Thermal Energy Conversion (OTEC) Discuss its merits and demerits.
6. How is Geothermal energy used for generation of electrical power ? Discuss its advantages and disadvantages.
7. Write informative notes on how the power from indirect solar energy is harnessed.

# Ecosystem

## INTRODUCTION

Every living organism is surrounded by materials and forces which constitutes its environment and from which it must derive its needs. Thus, for its survival, a plant, an animal or a microbe can not live completely sealed in an impervious skin or shell, but requires from its environment a supply of energy, a supply of materials and a removal of waste products. For these basic requirements each living organism has to depend and also to interact with different non-living or abiotic and living or biotic components of the environment. The scientific study of the interactions of organisms with their physical environment and with each other is called ECOLOGY (Helena Curtis 1975). It mainly concerns with the directive influences of abiotic and biotic environmental factors over the growth, distribution, behaviour and survival of organisms (Herreid II 1977). The credit goes to German Zeologist E. Haeckel in 1966, who used the term as *Oekologie* to the inter-relationships of living organisms and their environment.

The word ecology comes from two Greek words "*oikos*" meaning 'household' or 'home' or 'place to live or habitation' and "*logos*" meaning "discourse" or 'study'. Thus ecology deals with the organism and its place to live i.e. environmental Biology.

## CONCEPT OF ECOSYSTEM

Now, we can say *Ecology* deals with interrelationships between the biotic and abiotic compounds of an *Ecosystem*. Ecology is both Science and Art. The term ecosystem was first coined by *A.G. Tausley 1935.* It is derived by two words '*eco*' means environment and '*system*' implies a complex of co-ordinated units.

Ecosystem is defined as a community of organisms interacting with one another and the environment in which they live i.e. **study of home.** A home can be a drop of water for an amoeba or so, a home for a lion may be many miles of land over which it searches for its food. A pond, forest, lake a river, an ocean, a dam, a cropland, a garden, an orchard, a city, aquarium, wood lots and fields may be an example of ecosystem. Further an ecosystem may be natural, man-made or artificial like an aquarium. There are two types of systems open and cybernetic. The ecosystem is an open system for it receives energy from an outside source, the sun, fixes and utilizes it, and ultimately dissipates heat to space i.e. flow into and out of it.

But the ecosystem also an abstraction and has no particular size. In a sense, asking how big an ecosystem is would be, like asking how big a city, an engine or a triangle is. All of these things are abstractions. No two ecosystems are exactly alike. **An ecosystem concept is that the living organisms of a community not only interact among themselves but also have functional relationship with their nonliving environment. This structural and functional system of communities and their environment is called an ecosystem.**

## FUNCTIONING AND TYPES OF ECOSYSTEM

Functioning of the ecosystem is self-regulating and self-sustaining. This depend upon flow of energy, cycling of materials and perturbations both intrinsic and extrinsic. It is recognised as a dynamic concept with structural hetergeneity based on at least four functional phases. A rapid release phase consisting of tightly bound resources is replaced by a reorganisation phase followed by a exploitative phase, which gradually transformed into conservation phase or climax phase. The control of ecosystem function by nutrient flux and the condition of the physical environment is called Bottom-up-control but function via tropic interactions is called Top-down control.

Depending upon the species, diversity and the manner in which they are organised, Ecosystems are of following types —

**1. Permanent and Natural ecosystem.** These operate under natural conditions without any interference (even by human beings). These can be further classified in to—

(*i*) Terrestrial ecosystem

(*ii*) Aquatic ecosystem

Terrestrial ecosystems operate on land hence Forest, Desert and grassland and Agro-ecosystems included in this type. While Aquatic ecosystem operates in water. It can be divided in two

(*a*) Fresh water ecosystem

(*b*) Marine ecosystem

Freshwater ecosystems are usually named after the size and nature of the fresh water body such as pond, lake & river.

Marine ecosystem is largest ecosystem on earth, which consists of several sub-divisions each having its physico-chemical and Biological characteristics. For example, in the deepest ocean producers are absent but in many other organisms survive which dependent for food on the dead organic matter coming from the upper layers of the ocean.

**2. Temporary and Natural ecosystems.** These are short lived but operate under natural conditions.

**3. Artificial or Anthropogenic ecosystems.** These are man-made like fishery tanks dams, croplands and space ecosystems also. Fish aquarium is also come under this head.

These typologies are determined not only by the species composition but also by the physiognomic characteristics and soil and climatic conditions.

## STRUCTURE OF AN ECOSYSTEM

It is a description of the species of organisms that are present (including information on their life histories, populations and distribution in space). The structure of ecosystem provides information about the range of climatic conditions that prevail in the area, composition and organization of Biological Communities and Abiotic compounds constitute the structure of an ecosystem. According to *Odum,* from the trophic (Food) point of view, an ecosystem has the following components:

**1. Biotic Structure.** Producers, consumers and decomposers are components of biotic ecosystem. Living things are made of control and other chemicals with a lot of water added. Living organisms exchange expel convert, assemble, disassembles, organise and otherwise manipulate the constituents of earth, air and water. Biotic structure includes plants, animals and micro organisms present in an ecosystem. We have identified producers, decomposers and consumers are the basic components of biotic ecosystem. These can be distinguished on the basis of their source of energy and material.

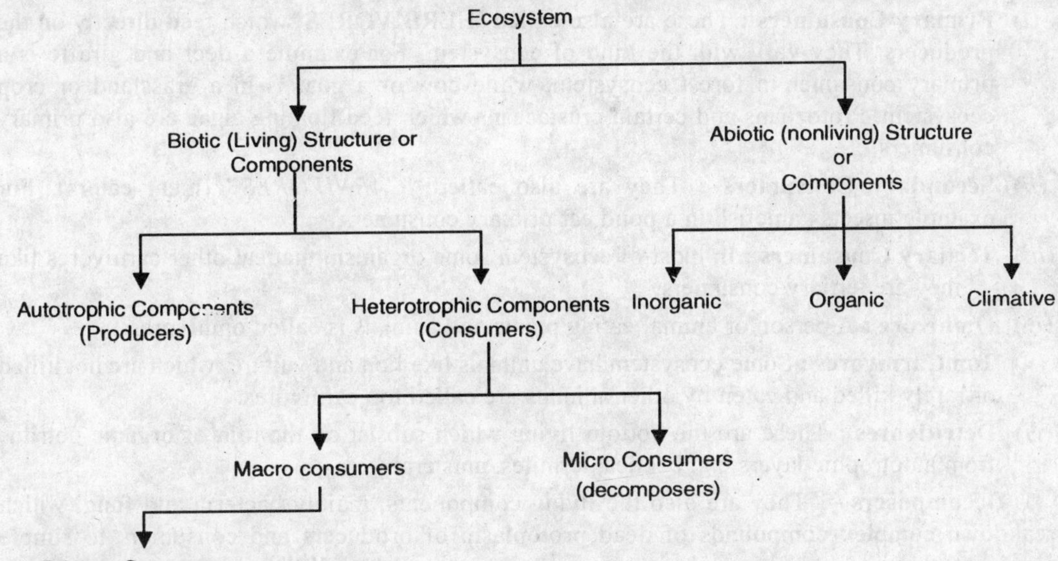


(*a*) **Autotrophic components** (Autotrophic = self nourishing)

In which the fixation of light, energy, the use of simple inorganic substances and manufacture of complex material predominates. These are also called **producers.**

(*b*) **Heterotrophic Components** (Heterotrophic = other nourishing)

These utilizes, rearranges and decomposes the complex materials synthesized by the autotrophs. The most intense heterotrophic activity takes place where the organic matter accumulates in the soils and sediments. These are also called **consumers**.

## PRODUCERS

All green plants are producers. They are also called **"converters"** or **"transformers"**. They are living members of the ecosystem that utilize sunlight as their energy source and single inorganic materials from soil, air and water to transform them by photosynthesis in to more complex energy rich chemicals as their own food. Producers are largely photosynthetic plants and their kind varies with the kind of ecosystem. In dense forest the trees are the most important producers. In lakes and ponds, the producers are rooted or large floating and microscopic plants (phytoplankton) usually the algae. They are also known as **"photo-autotrophs"** (*Photo = light, auto = self, troph = food.*)

Recently, scientists have found ecosystems based on chemical energy at great ocean depth (more than a km), where there is no light. The producers in these systems are bacteria that are able to gain energy from the oxidation of $H_2S$ that seeps from volcanic rents in the ocean. Since these organisms get their energy from chemical reactions rather than sunlight, they are called **"Chemotrophs"**.

**Consumers** — As we have seen earlier, consumers are heterotrophs, the living organisms which ingest other organism. They derive their food directly or indirectly from the producers. The food is then digested i.e. broken down to simple substances which are metabolized in the consumers body and released the waste product to the environment. Consumers are of following types —

(*i*) **Primary Consumers :** These are also called 'HERBIVORES' which feed directly on the producers. They vary with the kind of ecosystem. For example a deer and giraffe is a primary consumer in forest ecosystem, while cow or a goat is in a grassland or crop ecosystem. Protozoans and certain crustaceans which feed floating algae are also primary consumers.

(*ii*) **Secondary Consumers :** They are also called "*CARNIVORES*" (meat eaters). For example insects gamefish in a pond eat primary consumers.

(*iii*) **Tertiary Consumers :** In most of ecosystem some organism that eat other carnivores like — they are tertiary consumers.

(*iv*) **Omnivore :** A person or animal eating plants and animals is called omnivores.

(*v*) **Top Carnivores :** Some ecosystem have animals like lion and vulture, which are not killed or rarely killed and eaten by other animals are called top carnivores.

(*vi*) **Detritivores :** These are the bottom living which subsist on the rain of organic detritus from autotrophic layers e.g. beetles, termites, ants crabs etc.

3. **Decomposers** — They are also the living components, mainly bacteria and fungi which breakdown complex compounds of dead protoplasm of producers and consumers to simple organic compounds and ultimately in to inorganic nutrients. In all the ecosystems, this biotic structure prevails. Molds and mushrooms of the forest are the largest of the decomposers that are visible. The role of decomposers in ecosystem is very important. They are responsible for the completion of ecosystem mineral cycles. They are also called *microconsumers* or *saprobes* or saprophytes or saprotrophs. (Sapros = rotten, trophs = feeder) other examples are bacteria and fungi.

2. **Abiotic Structures Or Components:** The physical and chemical components of an ecosystem constitute its abiotic structure. It includes two things—

(*i*) **Materials Or Chemical Factor:** The materials are like water, minerals, atmospheric gases and other inorganic salts. They also include some organic matter such as amino acids, decay products, lipids, carbohydrates, proteins etc. The quantity of abiotic materials like the minerals present at any given time in an ecosystem is termed as the '*standing state*' or '*standing crop.*'

(*ii*) **Energy Or Physical Factor:** This is in the form of light, heat and stored energy in chemical bonds. Annual rainfall, wind latitude and altitude etc. are also some physical factors, which have a strong influence on ecosystem. For proper functioning of an ecosystem there must be a continuous 'flow of energy' and 'cycling of minerals' among the organisms of the ecosystem.

## ENERGY FLOW IN THE ECOSYSTEM

Energy is needed for every biological activity. Solar energy is transformed in to chemical energy by a process of photosynthesis. This energy is stored in plant tissue and then transformed in to mechanical and in the form of heat during metabolic activities. In the biological world the energy flows from sun to plants and then to all heterotrophic organisms like nitroorganisms, animals and man i.e. from producers to consumers. 1% of the total sunlight falling on the green plants is utilized in photosynthesis. This is sufficient to maintain all life on this earth. There is no 100% flow of energy from producers to consumers. Some is always lost to environment. Because of this, energy can not be recycled in an ecosystem '*it can only flow one way.*'

The flow of energy follows the two laws of thermodynamics.

**Ist law of thermodynamics.** The law states that energy can neither be created nor be destroyed but it can be transformed from one form to another. Similarly, as we have read earlier, solar energy utilized by green plants (Producers) in photosynthesis converted into biochemical energy of plants and later into that of consumers.

**IInd law of thermodynamics.** The law states that energy transformation involves degradation or dissipation of energy from a concentrated to a dispersed form. We have seen dissipation of energy occurs at every trophic level. There is loss of 90% energy, only 10% is transferred from one trophic level to the other.

**Sun as the Source of Energy :** Sun is the source of energy which extends radiations from high frequency to low frequency. Approximately 99% of total energy is in the region between UV and IR. The visible spectrum spreads over 0.38 μ to 0.77 μ involving about 50% of solar radiations. Some autotrophs however utilize energy released from oxidation processes for the synthesis of organic food.

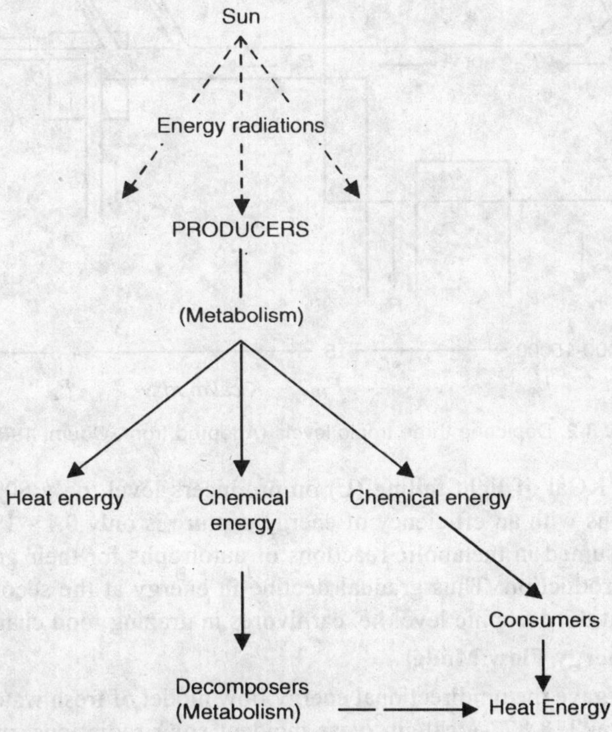


**Fig 8.1.** Sun as the source of energy

## ENERGY FLOW MODELS

As we have seen, that there is unidirectional flow of energy from sun to the producers and then various types of consumers. Therefore, behaviour of energy in ecosystem can be termed **Energy flow**. About 34% of the sunlight reaching the atmosphere is reflected back in to its atmosphere. 10% is held by ozone layer, water vapours and other atmospheric gases. Rest 56% reaches the earth surface. Out of this 1-5% is used by green plants for photosynthesis.

$$6CO_2 + 6H_2O \xrightarrow[\text{Chlorophyll}]{\text{Sunlight}} C_6H_{12}O_6 + 6O_2$$

Rest is absorbed as heat by ground vegetation or water. The flow of energy in an ecosystem can be explained with the help of various energy flow models—

**1. Odum's Energy Flow Model**

E.P. Odum 1963 explained flow of energy involving three trophic levels with the help of his universal energy flow model. (Fig.8.2) As the flow of energy takes place, there is gradual loss of energy at every level thereby resulting in less energy available at next trophic level.

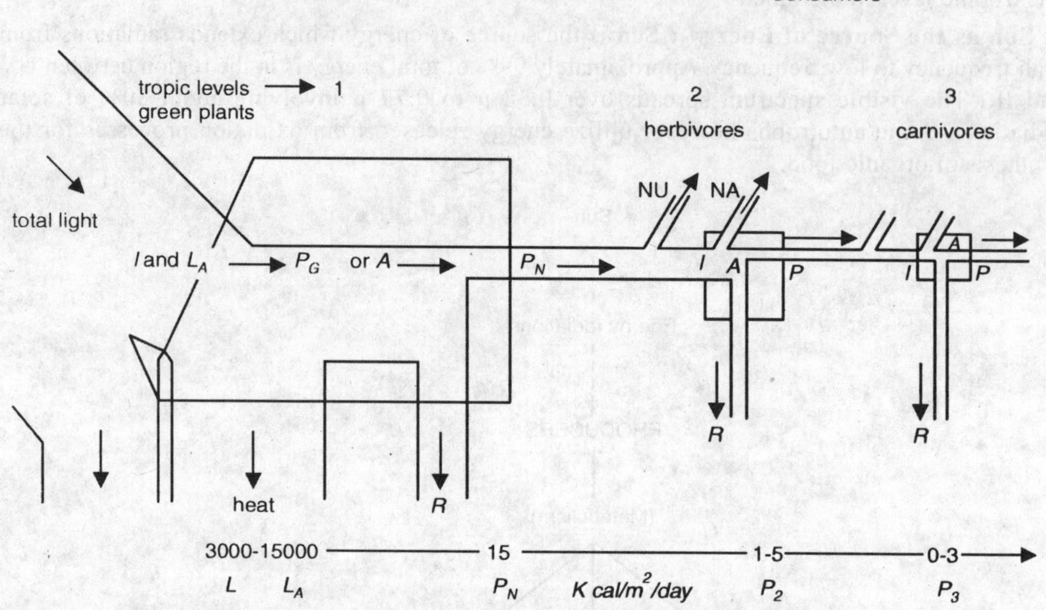


**Fig. 8.2.** Depicting three tropic levels (Adopted from, Odum, 1963)

Out of total 3000 KCal of light falling (L) on producers level, only 50% i.e. 1500 KCal is absorbed by autotrophs with an efficiency of energy capture is only 0.1 - 1%. As reported 21% of this energy is consumed in metabolic reactions of autotrophs for their growth, development, maintenance and reproduction. Thus gradual decline in energy at the second trophic level i.e. herbivores and then at third trophic level i.e. carnivores in grazing food chain is observed.

**2. Lindeman's Energy Flow Model**

Lindeman in 1942 gave the unidirectional energy flow model of fresh water ecosystem. Model shows that out of total 118,872 g.cal/cm$^2$/year incident solar radiations, producers can utilize only 1% (111.0 g cal/cm$^2$/year) in their photosynthesis (Fig. 8.3)

About 21% of this gross production (GP) is utilized in metabolic functions of producers, 3% is utilized in decomposition and 63% remains unutilized in decomposition and 63% remains unutilized. Thus only 13-14% i.e. 15 g.cal/cm$^2$/year of GP is available to herbivores. At this level about 30% of it (i.e.15 g.cal/cm$^2$/year) is utilized in metabolic functions i.e. respiration, growth and reproduction etc. This is more than the autotrophs consumed *i.e.,* 21%. Again 3% of it is utilized in decomposition while 47% is remains un-utilized. Thus only 20% energy of the autotrophs is available (i.e. 3 g.cal/cm$^2$/year) to carnivorous. It is also reported that about 70% energy is available for carnivores, which is not utilized and only 28-6% of net production passes to carnivorous. Carnivores utilized 60% energy at this level in metabolic activities and rest is remains as un-utilized.

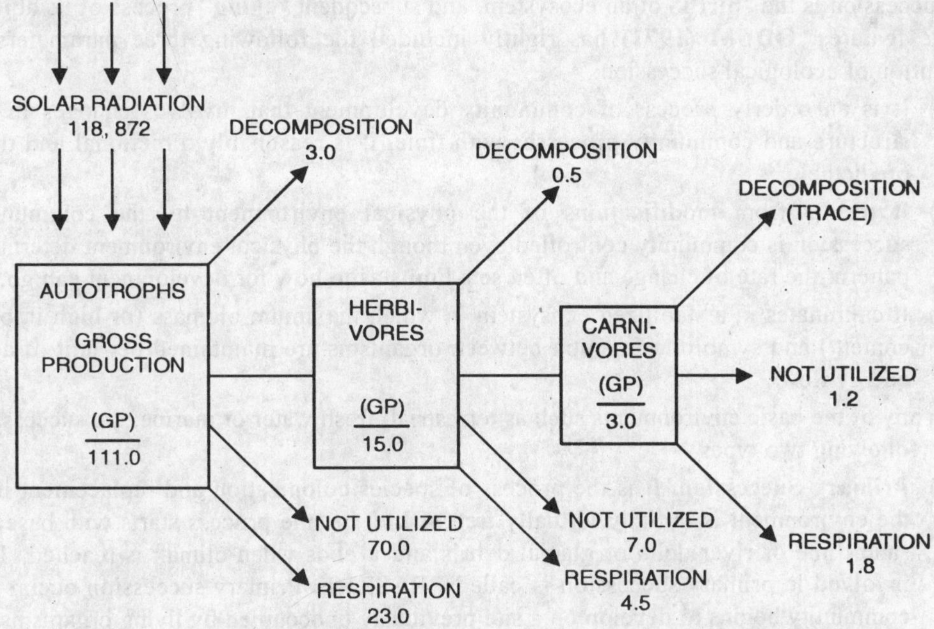


**Fig. 8.3.** Energy flow diagram for a lake (Freshwater ecosystem) in g.cal./cm²/yr (modified from Lindeman, 1942)

From both the figures it is clear that —

1. There is unidirectional flow of energy i.e. the system would collapse if the primary source, the sun were cut off.
2. Progressive decrease in energy at each trophic level.
3. There is also a corresponding decrease in biomass. But there is no correlation between biomass and energy. This relationship may differ according to situations.

## ECOLOGICAL SUCCESSION

Biotic communities are not static, they change with time. This change can be understood on several levels. Changes take place continuously in the community structure, organization, physiognomy, the associated animals and the environment at a place in the course of time, this phenomenon is called **ecological succession.** The rate of successional changes is rapid initially and gradually it slows until a point of dynamic equilibrium is reached, and the community is more or less stable. A complete succession is called a **SERE.** A sere is made up of a number of seral stages.

Seral communities are large in number depending on minor differences in edaphic, topographic or other local conditions. These communities are open to a large scale invasion of other species. Seral communities react to the prevailing environment in such a way that the environment is modified sufficiently to become less favourable to the existing sere. Seral communities are therefore, short lived and quick in development. As against these the climax communities are only a few and similar in identical climatic conditions. These are more or less closed to invasion of outside floristic elements. Climax communities attain a dynamic equilibrium with their environment. They cannot change their environment to such an extent as to make it less suited to the existing communities. Climax communities are formed after a long period of time and once formed they maintain their structure and composition. It is the final or the last seral stage.

Succession is the "**birth**" of an ecosystem, and subsequent "**aging**" process of its abiotic and biotic features. **ODUM (1971)** has rightly included the following three parameters in his definition of ecological succession.

(1) It is an orderly process of community development that involves changes in species structure and community processes with time, it is reasonably directional and therefore predictable.

(2) It results from modifications of the physical environment by the community, i.e. succession is community controlled even though the physical environment determines the pattern, the rate of change and often sets limits as to how for development can go.

(3) It culminates in a stabilized ecosystem in which maximum biomass (or high information content) and symbiotic function between organisms are maintained per unit of available energy flow.

In any of the basic environments such as terrestrial, fresh water or marine, the succession may be of following two types—

(*i*) **Primary Succession.** It is the process of species colonization and replacement in which the environment is initially virtually free of life, i.e. the process starts with base rock or sand dime or river delta or glacial debris and it ends when climax is reached. The sere involved in primary succession is called **PRESERE.** Primary succession occurs when a community begins to develop on a site previously unoccupied by living organisms.

(*ii*) **Secondary Succession.** The term secondary succession refers to community development on locations or sites previously occupied by well developed communities. It occurs where a community has been disrupted and the surface is completely or largely devoid of vegetation. It may be due to earthquake, fire or even clearing of forests by man. In each case organism modify the environment in a way that allow one species to replace another. The sere involved in secondary succession is called **SUBSERE.**

Depending on the moisture contents, the primary and secondary successions may be of the following types—

**(*a*) Hydrach Or Hydrosere:** The succession when starts in the aquatic environment such as ponds, lake, streams, swamps, bogs etc.

**(*b*) Mesarch Or Mesosere:** It is an intermediate type with adequate moisture. The succession when begin in such an area is called mesarch.

**(*c*) Xerach Or Xerosere:** The succession when starts in Xeric or dry habitat having minimum amounts of moisture, such as rocks, dry deserts etc. is called xerach. A temporary community in an ecological succession on dry and sterile habitate is called Xerosere. It may be of three types—

**(*i*) Lithosere :** i.e. succession initiating on rocks.

**(*ii*) Psammosere :** i.e. succession initiating on sand.

**(*iii*) Halosere :** i.e. succession initiating on saline water or soil.

Some times succession is also classified into two on the basis of community metabolism.

**(a) Autotrophic Succession :** It is characterised by early and continued dominance of autotrophic organisms like green plants. It begins in a predominantly inorganic environment and the energy flow is maintained indefinitely.

**(b) Heterotrophic Succession :** It is characterised by early dominance of heterotrops such as bacteria, actinomycetes, fungi and animals. It begins in a organic environment and there is a progressive decline in energy content.

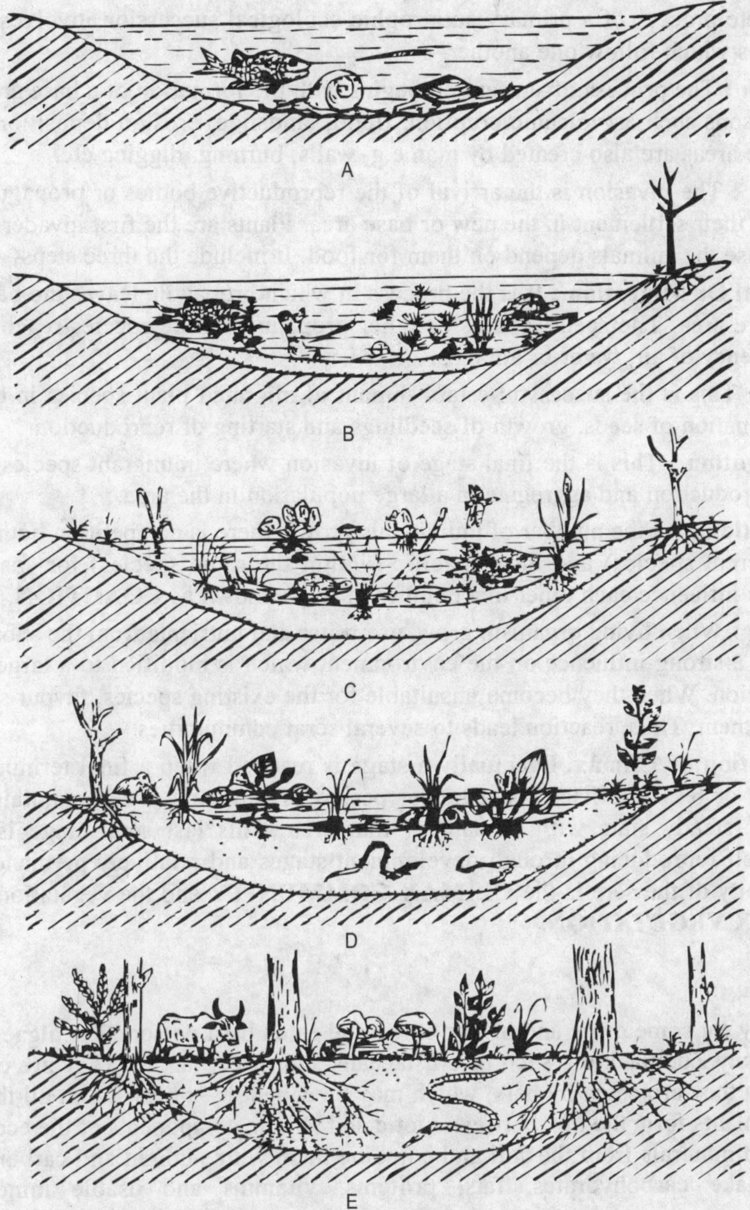


**Fig. 8.4** Hydroseric succession showing through stages A to E formation of different communities at the same place. Gradual silting of pond bed reduces the depth. Finally a terrestrial community has formed.

In some cases the replacement of one type of community by another is due to modification of the environment by communities themselves. Such a successional process is called **Autogenic succession.** But in some cases, if replacement is largely due to forces other than the effects of communities on the environment, this is called **Allogenic succession.** It occur in a highly disturbed or eroded area or in ponds nutrients and pollutants enter from outside and modify the environment and in turn the communities.

## GENERAL PROCESS OF SUCCESSION

The complete process of a primary autotrophic ecological succession involves the following sequential steps which follow one another—

**1. Nudation :** The process of succession begins with the formation of a base area or nudation by several reasons such as volcanic eruption, flood, landslide, erosion deposition, fire, disease etc. Some base areas are also created by man e.g. walls, burning, digging etc.

**2. Invasion :** The invasion is the arrival of the reproductive bodies or propagules of various organisms and their settlement in the new or base area. Plants are the first invaders (pioneers) in any area because the animals depend on them for food. It include the three steps—

**(i) Dispersal Or Migration :** It is the process in which propagule leaves the parent plant and arrives the bare area. The seeds, spores or other propagule of the species reach the bare area through the agency of air, water or animals.

**(ii) Ecesis :** This is the successful establishment of migrated plant species in to new area. It includes germination of seeds, growth of seedlings and starting of reproduction.

**(iii) Aggregation :** This is the final stage of invasion where immigrant species increase their number by reproduction and aggregate in a large population in the area.

**3. Competition :** As the number of individuals grows, there is competition both inter-specific (between different species) and intra specific (within the same species) for space, water and nutrition. They influence each other in a number of ways known as **COACTION.**

**4. Reaction :** When living organism grow, use water and nutrients from the substratum and in turn they have a strong influence on the environment which is modified to a large extent and is known as reaction. When they become unsuitable for the existing species, favour some new one, which replace them. Thus, reaction leads to several seral communities.

**5. Stabilization or Climax.** Eventually a stage is reached when a final terminal community becomes more or less stabilised for a longer period of time and it can maintain itself in the equilibrium or steady state with climate of that area. This last seral stage is mature, self maintaining, self reproducing through development stages and relatively permanent. This final stable community of the sere is the **CLIMAX COMMUNITY** and the vegetation supporting it is the **CLIMAX VEGETATION.**

## FOOD CHAINS

A sheep may eat some grass and in turn it may be eaten by a person. The algae of a lake will be eaten by many Zooplankton, such as crustaceans and insect larvae. They are eaten by small fish and in turn they are eaten by bars, which may be caught by a bear. From all these examples it is clear that plants form the link between biotic and abiotic components of the ecosystem. They draw water and minerals from the soil and combine them with sunlight and carbondioxide from the air to make carbohydrates, fats, proteins, vitamins and usable minerals through photosynthesis.

Small harbivorous organisms such as Caterpillars field mice etc. Consume this vegetable material and convert it to animal material, which serve as food to meat eating animals. They are eaten by larger carnivores. This sequence of eaten and being eaten, with the resultant transfer of energy is known as **FOOD CHAIN.** Thus in food chains organisms of an ecosystem are linked together. Each step is known as **trophic level** and the study of the energy flow through these steps is called **trophic ecology.** Food chains are not isolated from each other.

Primary producers trap radiant energy of sun and transfer that to chemical or potential energy of organic compounds such as carbohydrates proteins and fats. When herbivore eats a plant and these compounds are oxidised. As we have read earlier the energy liberated is just equal to the amount of energy used in synthesizing the substances. When this animal is eaten by another one, along with transfer of energy from a herbivore to carnivore a further decrease in energy occurs as the carnivore oxidise the organic substances of the first (herbivore) to liberate energy to synthesize its own cellular constituents. Such transfer of energy from organism to organism sustains the ecosystem (Fig. 8.5)

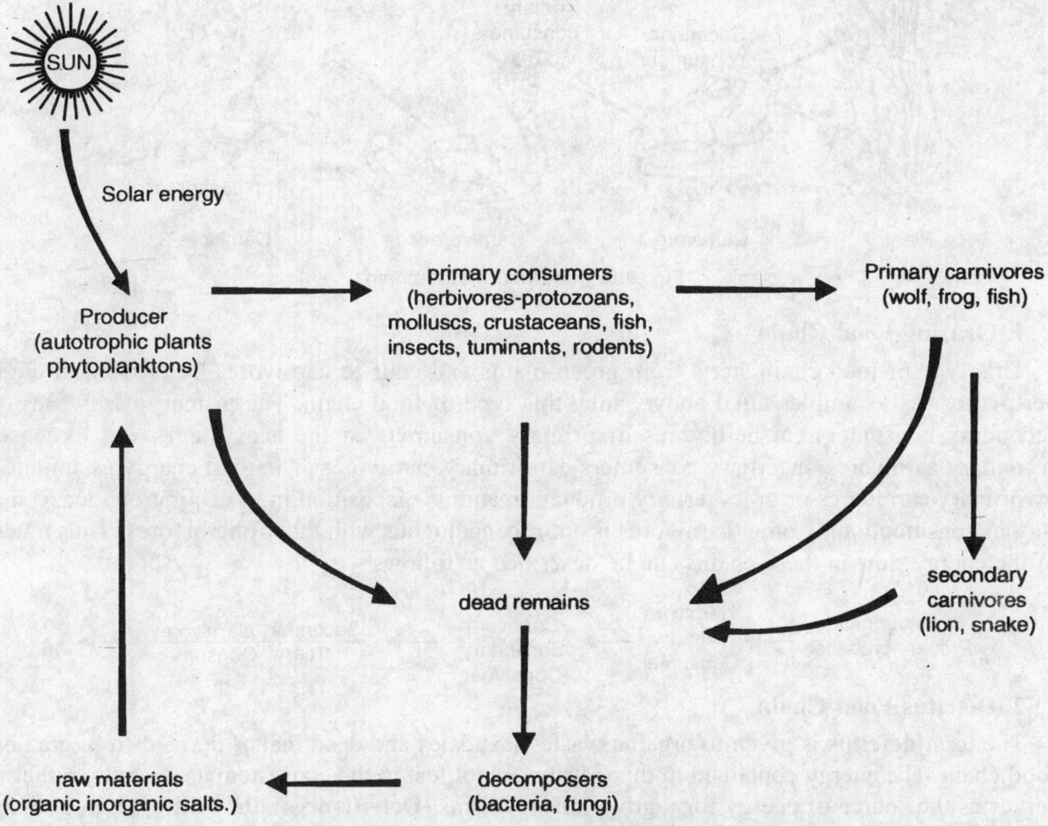


**Fig. 8.5.** A generalized food chain of the ecosystems.

Thus energy flows from primary producers to primary consumers, from primary consumers to secondary consumers and from secondary consumers to tertiary consumers and so on. This simple chain of eating and being eaten away is known as food chains. A food chain in grassland ecosystem starts with grass and forbs and goes through grasshoppers, the frogs, the snake, the hawk in an orderly sequential arrangement based on the food habits. (Fig. 8.6)

Other examples of food chains are :

1. Grass → Rabbit → Fox → Wolf → Lion.
   (Grass land ecosystem)
2. Phytoplanktons → Waterfleas → small fish → Tuna
   (Pond ecosystem)
3. Lichens → reindeer → Man
   (Arctic Tundra)

In nature, we can distinguish two types of food chains.

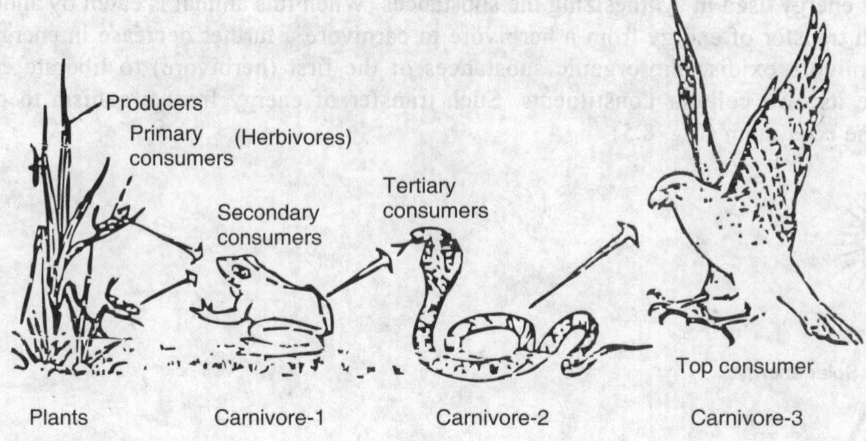


**Fig. 8.6.** Food and a food-chain

### 1. Grazing Food Chain

This type of food chain starts from green plants and ends to carnivores by passing through herbivores. All examples cited above show this type of food chain. The primary carnivores or secondary consumers eat herbivores or primary consumers of the ecosystem. And likewise, secondary carnivores or tertiary consumers eat primary carnivores. The total energy assimilated by primary carnivores or gross tertiary production and its disposition in to respiration, decay and further consumption by other carnivores is entirely analogous with that of herbivores. Thus much of the energy flow in these chains can be described as follows —

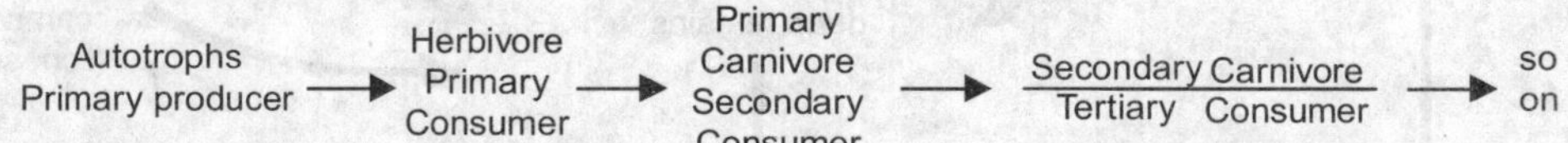


### 2. Detritus Food Chain

The term **detritus** is given to organic wastes, exudates and dead matter derived from grazing food chain. The energy contained in this detritus is not lost to the ecosystem as a whole, rather it serves as the source of energy for a group of organisms (**Detritivores**), they differ from grazing food chain called the detritus food chain. These food chains are less dependent on solar energy, but chiefly depend on the influx of organic matter produced in another system. Such food chains operates in the decomposing accumulated litter in a temperate forest.

In some ecosystems, considerably more energy flows through the detritus food chains than through the grazing food chains. The organisms of the detritus food chains are, algae, bacteria, slime molds, fungi, actinomycetes protozoa, insects, mites, crustaceans, molluses worms, nematodes etc. Some species are highly specific in their food requirements and some can eat almost anything. All these are detritus consumers. They ingest large amounts of the vascular plant detritus. These animals are in turn eaten by some minnows and small game fish i.e. small carnivorus which in turn serve as food for larger game fish and fish eating birds i.e. top carnivores.

**Heald (1969)** and **Odum (1971)** have studied the detritus food chain of mangrove leaves (Rhizophora mangle) of Southern Florida. Conclusively, we can understand detritus food chain as follows —

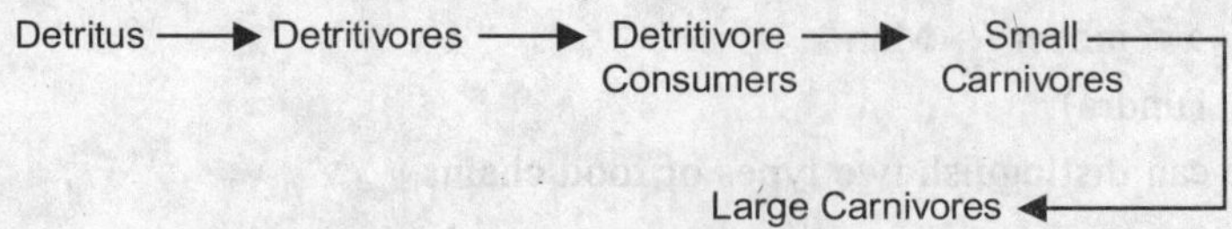

## FOOD WEBS

In nature simple food chain occur rarely. The same organism may operate in the ecosystem at more than one trophic level i.e. it may derive its food from more than one source. Even the same organism may be eaten by several organisms of a higher trophic level or an organism may feed upon several different organisms of lower trophic level. In this way individual food chains interconnect to form a complex network with several linkages and are known as food web. Thus food web is defined as—**"A network of food chains where different types of organisms are connected at different trophic levels, so that there are a number of options of eating and being eaten at each trophic level.**

Fig. (8.7) shows the inter-linking of five food chains in a terrestrial food web.

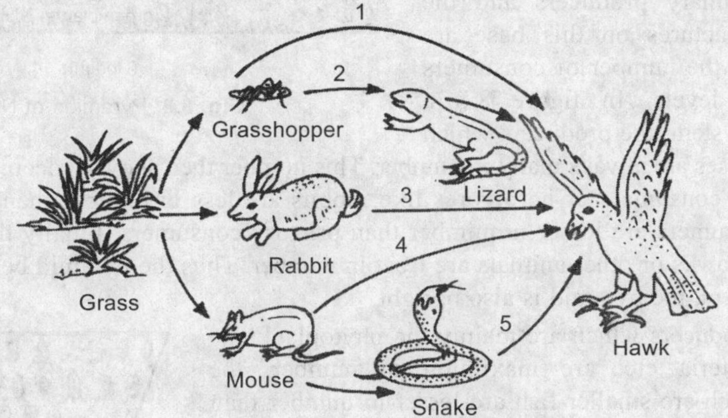


**Fig. 8.7.** A food web in a grassland ecosystem with five possible food chains.

The following five types of food chains are interconnected to form food web in this figure.

(1) Grass → Grasshopper → Predatory bird → (Hawk)

(2) Grass → Grasshopper → Lizzard → Hawk.

(3) Grass → Rabbit → Hawk (or vulture or man)

(4) Grass → Mouse/Rat → Hawk

(5) Grass → Mouse/Rat → Snake → Hawk.

This shows, food chains in natural conditions never operate as isolated sequences but are interconnected with each other forming some sort of interlocking pattern (which is referred to as a food web).

## ECOLOGICAL PYRAMIDS

Charles Elton in 1927, noted that the animals at the base of the food chain are relatively abundant, while those at the end are relatively few in number i.e. there is progressively decrease in between the two extremes. Secondly, there is some sort of relationship between the numbers, biomass and energy content of the primary producers, consumers of the first and second orders and so on to top, Carnivores in any ecosystem. These relationships may be represented in diagrammatic (Graphic) ways and are referred to as **ecological pyramids or Eltonian Pyramids.**

Ecological pyramids are of three general types —

1. Pyramid of numbers - (Based on number of organisms at each level.)
2. Pyramid of Biomass - (Based on biomass of organisms.)
3. Pyramid of energy - (Showing the rate of energy flow and/or productivity at successive trophic levels.

The pyramids of numbers and biomass may be upright or inverted depending upon the nature of the food chain in the particular ecosystem whereas pyramids of energy are always upright.

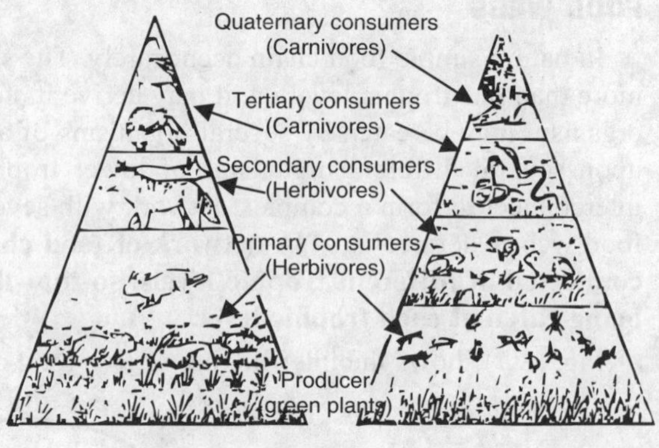


**Fig. 8.8.** Pyramids of Number

**1. Pyramid of numbers.** This deals with the relationship between the number of producers, herbivores and carnivores at successive trophic levels. At the base of such figure (pyramid) is always the number of primary producers and the subsequent structures on this base are represented by the number of consumers at successive levels. In figure 8.8 a grassland ecosystem, the producers which are mainly grasses are always many in number. This number then shows a decrease towards apex, as the primary consumers or herbivores like rabbits are less in number than the grasses. The secondary consumers are lesser in number than primary consumers. Finally the top consumers (tertiary) like hawks or other animals are least in number. Thus the pyramid becomes upright. In a pond ecosystem, the pyramid is also upright.

Here the producers which are mainly the phytoplanktons as algae, bacteria etc. are maximum in number, the herbivores which are smaller fish are lesser in number than producers. The secondary consumers are lesser in number than herbivores. Finally the top consumers (tertiary) are least in number.

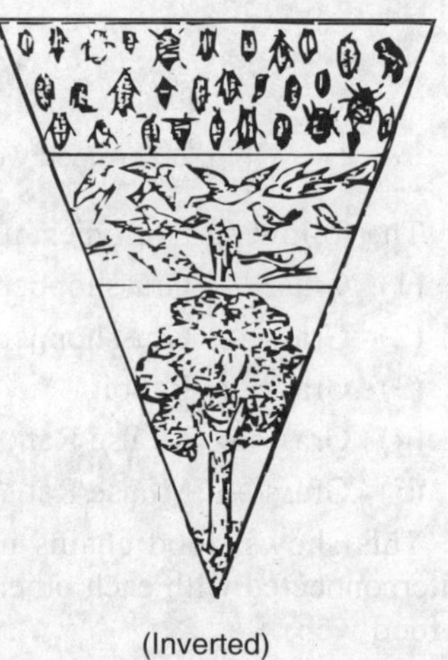


**Fig. 8.9.** Pyramid of Numbers

In a forest ecosystem, however the pyramid is inverted, Fig. 8.9 illustrates an instance where the number of primary producers (a tree) is less than that of herbivore birds feeding upon the tree fruits. The number of parasites like bugs and lice living and feeding upon the birds body is still higher. Thus depending upon the size and biomass the pyramid of numbers may not be always pyramidal, it may even be completely inverted in shape.

Odum (1971) has studied the grassland and forest ecosystems by collecting the data from USA and England respectively.

**2. Pyramid of Biomass.** Pyramids of biomass are comparatively more fundamental, as they instead of geometric factor, show the quantitative relationships. In order to explain the inverted nature of a pyramid of numbers, the idea of pyramid of biomass is given where the weight of primary producers forms the base. In figure 8.10 the ecosystem is shown where the pyramid of biomass is upright. The biomass of one tree is very high. The biomass of a number of birds feeding upon the tree is far less than that of the tree. Similarly, the biomass of even a very large number of parasite in and on the body of the birds is far less. Thus the pyramid of biomass, therefore, becomes upright. But there can be instances where the pyramid of biomass also get inverted as shown in figure 8.11. The biomass of phytoplanktons is quite negligible as compared to the small herbivores i.e. fish that feed on them. The biomass of large carnivores (Fish) feeding on small fishes is still higher. Harvery (1950) studied that in English Channel the biomass of primary producers is only 4 g/m$^2$ whereas that of the consumers is 21 g/m$^2$. This is the case in most aquatic bodies.

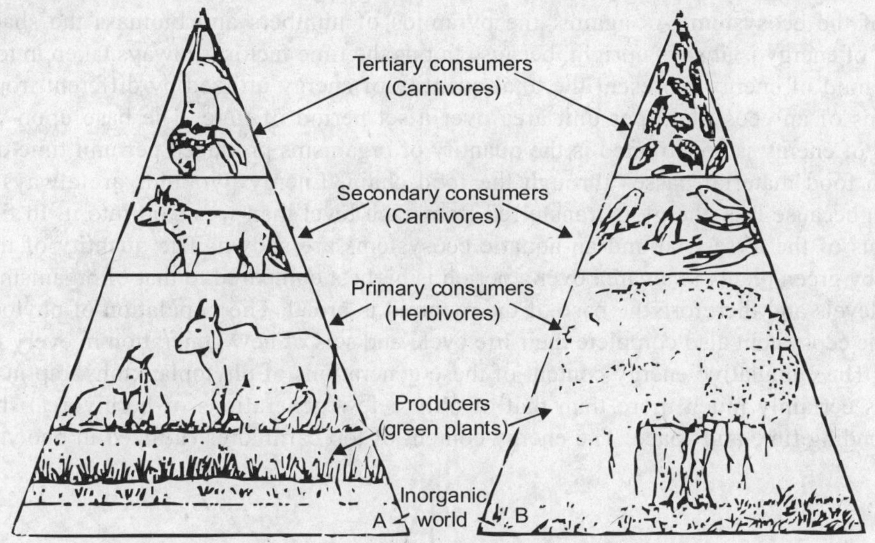


**Fig. 8.10** Pyramids of Biomass

**Fig. 8.11.** Inverted pyramid of biomass in an aquatic ecosystem due to lower biomass of phytoplanktons than of consumers in unit volume of water at any one time.

**3. Pyramid of energy**

Of the three types of ecological pyramids, the energy pyramid give the best picture of overall nature of the ecosystem. As against the pyramids of numbers and biomass the shape of the pyramid of energy is always upright, because in this the time factor is always taken in to account. The pyramid of energy represent the total quantity of energy utilized by different trophic level organisms of an ecosystem per unit area over a set period of time. The base upon which the pyramid of energy is constructed is the quantity of organisms produced per unit time or the rate at which food material passes through the food chain. Energy pyramids are always slopping (upright) because less energy is transferred from each level than was paid into it. In figure 8.12 organisms of the terrestrial and an aquatic ecosystems are shown. The quantity of the energy trapped by green plants in an area over a period is highest compared to that of organisms of other trophic levels and therefore the base of the pyramid is broad. The population of phytoplanktons in aquatic ecosystem also complete their life cycle and sets of new generation in every few hours or days. The cumulative energy content of these generations of phytoplanktons trap in course of a year is certainly much more than that of only a few generations of herbivore fishes in the corresponding time and space. The energy content of top carnivores (utilized in one year) is the

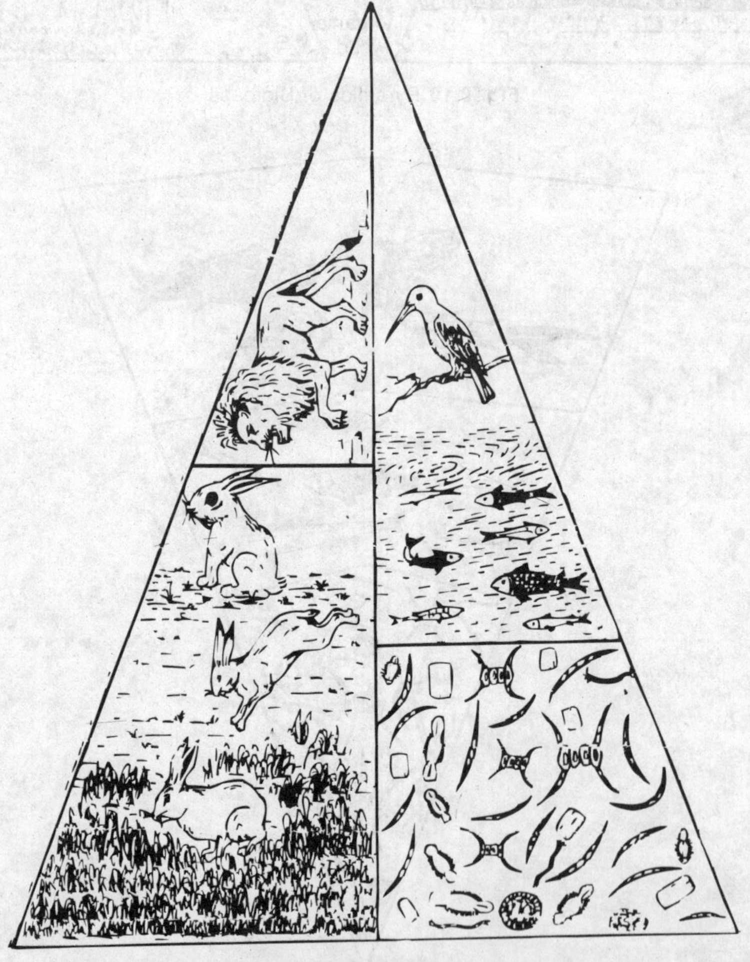

**Fig. 8.12.** Pyramids of energy in grassland and aquatic ecosystems. The cumulative energy contents utilized by primary producers is always higher as compared to energy utilization by successive trophic levels, over a period of time in a given area.

least. Therefore, the pyramid of energy is upright. The ratio of the amount of energy absorbed and the amount of energy which would be retained in biomass is known as ecological efficiency.

## SOME MAJOR ECOSYSTEMS

There are three types of ecosystems in nature—

1. Terrestrial ecosystem
2. Freshwater ecosystem
3. Marine ecosystem

A large geographical are with its specific and complex flora and associated fauna is called a **biome.** The physical factors like, nature of soil, rainfall, temperature, light etc. effects the vegetation of a biome. The ecological characteristics of some major ecosystems are given.

**Terrestrial ecosystems**

The terrestrial ecosystems may be Latitudinal biome or altitudinal. It consists of

(*i*) Forest ecosystem
(*ii*) Grassland ecosystem
(*iii*) Desert ecosystem

## FOREST ECOSYSTEM

Roughly 40% of the land is occupied by forest. But in India it is one-tenth. In India, tropical rain forests are found in Western Ghats, Andamans and North-East Himalayas. So these have maximum bio-diversity. The different components of a forest ecosystem are as:

**Abiotic Component :** These are the inorganic & organic substances present in the soil & atmosphere. In addition to the minerals present in the forests, we find the dead organic debris. The light conditions are different due to complex stratification in the plant communities.

**Biotic Component :** The living organisms present in the food chain occur in the following order—

**1. Producers.** These are mainly trees that show much species diversity and greater degree of stratification specially in tropical moist deciduous forest. In northern coniferous forest needle leaved evergreen tree, specially the spruces, firs and pines are with poor development of shrub and herb layers. Deciduous forest is greatly modified by man and much of it is replaced by cultivated and forest edge communities. Thus trees are of different kinds depending upon the kind of forest. Beside trees, shrubs and ground vegetation are also present. In these forests, producers are tectona grandis, butea frondosa, shorea rubusta and hagerstroemia parviflora. Plant dominants of the evergreen broad leaved forest range from the more northerly live oaks (Ouercus virginiana) magnolias bays and hollies. Palms such as the sabal or cabbage palm are also often prominent vines and epiphytes are characteristic. In temperature deciduous forests the dominent trees are species of Ouercus, Acer, Betula, Thuja, Picea etc.

**2. Consumes.** These are as follows—

**(*a*) Primary Consumers :** These are the herbivores that include the animals feeding on tree leaves as ants, flies, beetles, leafhoppers bugs, spiders etc. Many of the larger herbivorous vertibrates like moose, snowshoe have grouse are found on broad leaved developmental communities. Similarly some animals like elephants, nilgai, deer, moles, flying foxes, fruitbats, mongooses etc. are grazing on shoots and/or fruits.

**(*b*) Secondary Consumers :** These are the carnivores like snakes, birds, lizards, fox etc. feeding on herbivores.

**(c) Tertiary Consumers :** These are the top carnivores like lion, tiger etc. that eat carnivores of secondary consumers level.

**3. Decomposers.** These are wide variety of micro-organisms like actinomycetes (streptomyces), bacteria (Bacillus, clostridium, Pseudomonas etc.), Fungi (species of Aspergillus, Coprinus, Polyporus, Fusarium, Trichoderma etc.) Rate of decomposition in tropical and subtropical forests is more rapid than that in the temperate ones.

**2. Grassland Ecosystem**

This type of terrestrial ecosystem occupy roughly 19% of the earth surface. Grasslands dominated by grass species but some times also allow the growth of a few trees and shrubs. Rainfall is average but erratic. These are three types of grasslands depending upon climatic regions—

**(i) Tropical grassland.** Tropical Biomas (grasslands with scattered trees or clumps of trees) are found in warm regions with 40-60 inches of rainfall but with a prolonged dry season when fires are an important part of the environment. The largest area of this type is in Africa (Fig. 8.13) but sizable tropical biomes or grassland also occur in South America and Australia. In Africa these are known as **Savannas.** Grasses belonging to such genera as Penicum, Pennisetum, Andropogon and Imperata. A view of this African Savanna country, including grass scattered trees and mammalian herbivores is in fig. 8.14.

**(ii) Temperate grassland.** In US and Canada, these grasslands are known as **prairies** in South America as **pumpas**, in Africa as **Velds** and in Central Europe and Asia as **Steppes.** These occur where rainfall is too low (between 10 - 30 inches) to support the forest life form but is higher than that which results in desert life forms. However grasslands also occur in regions of forest climate where edaphic factors favor grass in competition with woody plants. Temperate grasslands generally occur in the interior of continents. In North America tall grass prairies have now been replaced by grain agriculture.

**(iii) Arctic Tundra.** There are two tundra biomes covering large areas of arctic, one in the Palearctic and other in the Nearctic region. In both continents the boundary between tundra and forest lies further north in the west where climate is moderated by warm westerly winds. The ground remains frozen except for the upper few inches during the open season. The permanently frozen deeper soil layer is called **permafrost.** The tundra is a wet arctic grassland consists of lichens, grasses, sedges and dwarf woody plants.

The various components of the grasslands are

**Abiotic Components.** The elements like C, H, O, N, P, S, etc. are supplied by $CO_2$, water, nitrates, phosphates, sulphates etc. present in soil & atmosphere. In addition, some other elements are also present in traces.

Biotic Components

**(i) Primary Consumers.** The herbivores feeding on grasses are grazing animals as cows, buffaloes, deers, sheep, rabbit, mouse etc. Besides them some insects like leptocorisa, dystercus, oxyrhachis, etc. termites and millipeds etc. also feed on the leaves of grasses.

**(ii) Secondary Consumers.** The animals like fox, jackals, snakes, lizards, birds etc. (Carnivores) feed on herbivores.

Sometimes the hawks feed on secondary consumers.

**Decomposers.** The microbes active in the decay of dead organic matter are different species of fungi, some bacteria and actinonycetes. They bring about the minerals back to the soil, thus making them available to the producers.

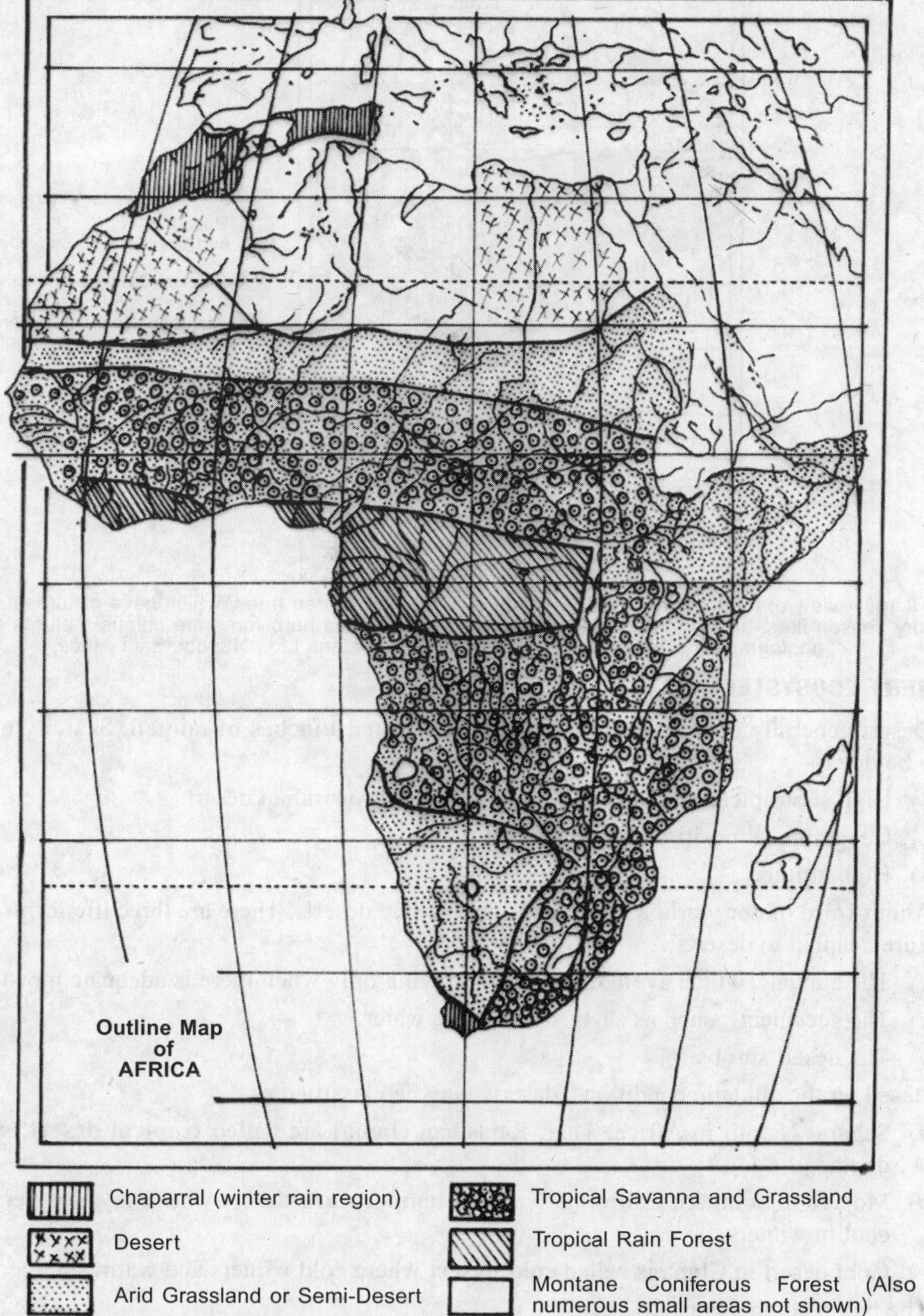


**Fig. 8.13.** The biotic regions of Africa. The six major biomes are indicated by the legend. Note that only a relatively small part of the continent is "jungle" (i.e., tropical rain forest). The very extensive tropical savanna and grassland is the fabulous "big game" country. Some of the larger faunal areas are numbered as follows: 1, Barbary; 2, Sahara; 3, Sudanese Arid; 4, Somali Arid; 5, Northern Savanna; 6, Congo Lowland; 7, Abyssinian Highland; 8, Kenya Highland; 9, Southern Savanna; 10, Southwest Arid; 11, Cape Town "winter rain" region. All of the faunal areas are part of the Ethiopian or African biogeographic region (see Figure 8.14), except the Barbary, which is Palearctic. Map based on sketch-map of Moreau, 1952.)

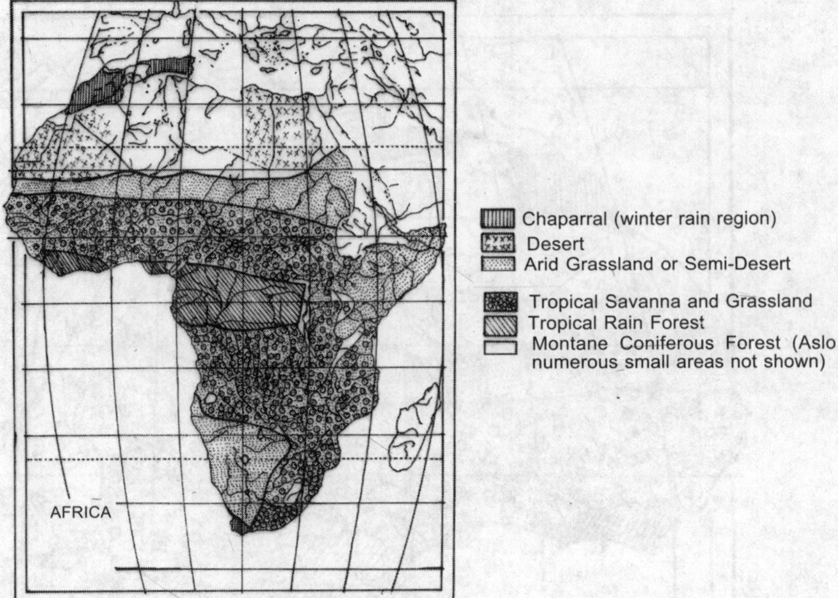


**Fig. 8.14.** A view of the tropical savanna of Africa. Grass, scattered trees, which have picturesque shapes, dry season fires, and numerous species of large mammalian herbivores are unique features of this photograph by Donald I. Ker, Ker & Downey Safaris Ltd., Nairobi, East Africa.)

## DESERT ECOSYSTEM

Desert generally occur in regions having less than 10 inches of rainfall. Scarcity of rainfall may be due to—

(1) High subtropical pressure as in the Sahara and Australian desert.

(2) Geographical position in rain shadow.

(3) High altitude.

About 1/3rd of our world's land area is covered by deserts. There are three life forms of plants that are adapted to deserts :

(*i*) The annuals, which avoid drought by growing only when there is adequate moisture.

(*ii*) The succulents, such as cacti, which store water.

(*iii*) The desert shrubs.

Based on the climatic conditions, deserts may be classified as

(*a*) Sahara, Namib in Africa, Thar, Rajasthan (India) are called **tropical desert,** which are driest.

(*b*) Mojave in Southern California is called **temperature desert** where days are very hot and cool in winters.

(*c*) Gobi desert in China is called **cold desert** where cold winters and warm summers.

The biotic components are

**(1) Producers.** As we have seen, these are shrubs bushes, some grasses and few trees. Some times, Cacti are also present. Some lower plants like lichens and Xerophytic mosses may also be present.

**(2) Consumers.** The most common animals are reptiles and insects. In addition to them, some nocturnal rodents and birds are also found. Camels "the ship of desert" feed on tender shoots of the plants.

**(3) Decomposers.** Due to poor vegetation, these are very few. They are some fungi and bacteria.

## AQUATIC ECOSYSTEMS

More than 70% of the land is covered by water. The important ecosystems are

**Pond Ecosystems.** Ponds are small bodies of water in which the littoral zone is relatively large and the limnetic and profundal regions are small or absent. Stratification is of minor importance. Ponds may be found in most regions of adequate rainfall. They are continually being formed, as a stream shift position, leaving the former bed isolated as a body of standing water where organic materials are accumulated.

**Temporary ponds are dry for part of the year are specially interesting and support a unique community organisms in such ponds which must be able to survive in a dormant stage during dry period**

Ponds created by damming of stream or basin by man or by animals such as the bearer, are among the most numerous. In united states man made ponds were "mill ponds".

Ponds play an important role in the villages where most of the activities like washing clothes, bathing, swimming, cattle bathing etc. are centre around ponds. We may study the pond as an ecosystem by making its convenient division into some basic components as shown in fig. 8.15.

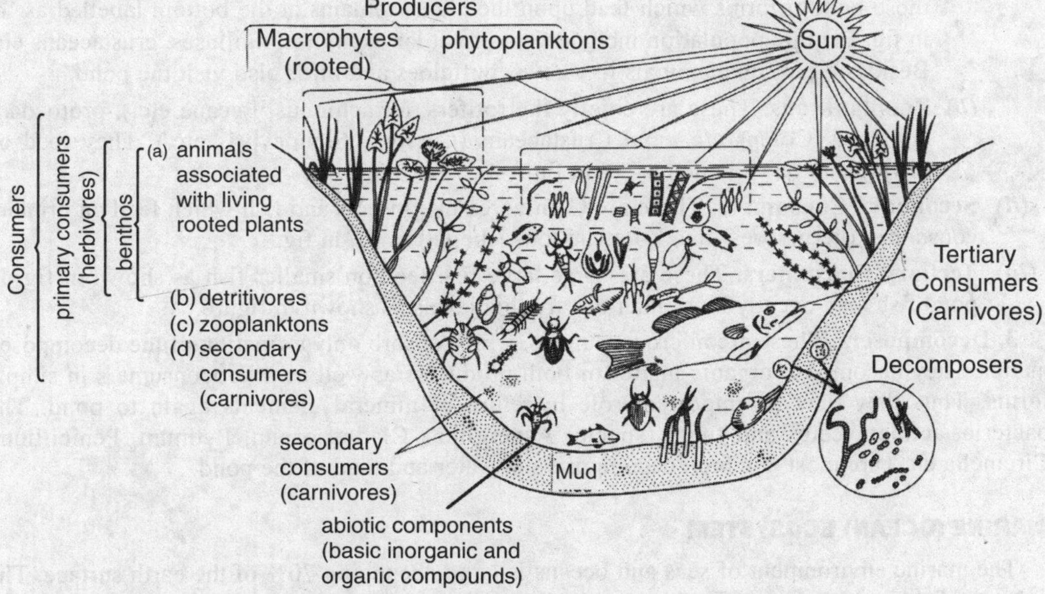


**Fig. 8.15.** Diagram of pond ecosystem, showing its basic structural units—the abiotic (inorganic and organic compounds) and biotic (producers, consumers and decomposers) components.

These components are as

**Abiotic Component.** Apart from heat, light the basic inorganic and organic compounds, elements are water, $CO_2$, oxygen, calcium, nitrogen, phosphorus, amino acids etc. The amount of the minerals present at any time in the physical environment of the pond. "Standing state" may be estimated by appropriate methods. Light intensity and turbidity index of water at different depths can also be measured by lux-photometer and Sechhi disc respectively. pH of water and mud is determined by Electrical pH meter. DO, $CO_2$, phosphate, nitrogen, etc. can also be measured by appropriate methods. Carbohydrates, proteins, lipid etc. are also estimated for biomass determination.

**Biotic Components** They are as follow$—

**1. Producers.** These are autotrophic, green plants and bacteria. They fix radiant energy and with the help of minerals from water & mud form complex organic substances like Carbohydrates, proteins & lipids. Producers are of the following types—

**(*a*) Macrophytes.** These are mainly rooted larger plants which include partly or completely submerged floating and emergent hydrophytes. The common species of the plants are Trapa, Typha, Sagittaria, Nymphaea, Chara, Hydrilla, Utricularia, Marsilea, Azolla, Sylvinia, Spirodella, Lemna etc.

**(*b*) Phytoplankton.** These are minute, floating or suspended lower plants like Ulothrix, Spirogyra, Cladophora, Oedogonium, Cosmarium, Eudorina Pandorina, Volvox, Chlamydomonas etc. and some flagellates. Biomass is estimated as weight of standing crop per unit area or volume. Generally, biomass and energy content of the vegetation decreases from the margin of the pond towards its centre. Energy content is generally expressed in terms of cal/gm dry wt.

**2. Consumers.** Most of the consumers are herbivores except insects and some large fish. But generally are heterotrophs. In pond consumers are distinguished as—

(*i*) **Primary Consumers.** These are herbivores, also known as "primary macro consumers" feeding directly on living plants. They may be large or in small size. They are further differentiated as—

(*a*) **Benthos.** These are the animals associated with living plants labelled as 'a' in fig and those bottom forms which feed upon the plants remains at the bottom labelled as 'b' in fig. Benthic population include fish, insect larvae, mites, molluses, crustaceans etc. Besides there some animals like cows, buffaloes and birds also visit the pond.

(*b*) **Zooplanktons.** These are chiefly the rotifers, (Brachionus, Lecane etc.), protozoans (Euglena, Coleps etc.) and Crustaceans (Cyclops, Stenocypris etc.). They feed on phytoplanktons lebelled as 'c' in fig.

(*ii*) **Secondary Consumers.** These are Carnivores like insects and fish which feed on primary consumers (herbivores) like Zooplanktons lebelled as 'd' in fig.

(*iii*) **Tertiary Consumers.** These are some large fish feed on smaller fish as shown in fig. In pond fish may occupy more than one trophic levels as shown in figure.

**3. Decomposers.** These are microconsumers, which absorb only a fraction of the decomposed matter. They decompose organic matter of both producers as well as microconsumers in simple forms. Thus they play an important role in return of mineral elements again to pond. The bacteria, actinomycetes and fungi (species Aspergillus, Cladosporium, Pythium, Penicillium, Circinella etc.) are most common decomposers in water and mud of the pond.

## MARINE (OCEAN) ECOSYSTEM

The marine environment of seas and oceans is large occupying 70% of the earth surface. The volume of the surface area of marine environment lighted by sun is small in comparison to the total volume of water involved. All the seas are interconnected by currents, dominated by waves, influenced by tides and characterised by saline water. Each ocean indeed represents a very large and stable ecosystem. Oceans play an important role in regulating many biogeochemical and hydrological cycles, thereby regulating the earths climate. They have some major life zones i.e. coastal, Euphotic, Bathyal and Abyssal zones.

The **biotic components** of an ocean are as follows—

**1. Producers**

These are autotrophs, which are mainly the phytoplanktons. They trape radiant energy from sun through their pigments. A number of macroscopic seaweeds (Brown and red algae) are also come in this category. They are in distinct zones at different depths of water.

**2. Consumers**

These are heterotrophic macroconsumers being dependent for their nutrition on the primary producers. These are

(*i*) The herbivores like Crustaceans, molluscs, fishes etc. which feed directly on producers are called primary consumers.

(*ii*) The carnivores fishes like shad, herring etc. feeding on herbivores are called secondary consumers.

(*iii*) The top carnivores fishes like cod, haddock, halibut etc. that feed on secondary consumers are called tertiary consumers.

**3. Decomposers:** The microbes active in the decay of dead organic matter are chiefly bacteria and some fungi.

## ESTUARIES (ESTUARINE ECOLOGY)

Estuarine is derived from the word *aestus means tide.* Pritchard in 1967 defined as a semi-enclosed coastal body of water, which has a free connection with the open sea. It is thus strongly affected by tidal action and within it sea water is mixed with fresh water from land drainage. River mouths, coastal bays, tidal marshes and bodies of water behind barrier beaches are examples. Estuaries could be considered as transition zones or ecotones between the fresh water and marine habitats.

Not all rivers open in to estuaries, some simply discharge their run off in to ocean. Estuaries differ in size, shape and volume of water flow, all influenced of the region in which they occur. *Deltas* are by accumulation of sediments. When silt and mud accumulations becomes high enough to be exposed at low tide, then *tidal flats* are developed.

To illustrate estuaries, the different classifications will be represented based on

(1) Geomorphology (2) Water Circulation and stratification (3) Systems energetics.

According to Pritchard 1967, four subdivisions of estuaries are from zeomorphological point of view—

(*i*) Drowned river valleys

(*ii*) Fjord type estuaries

(*iii*) Bar - built estuaries

(*iv*) Estuaries formed by tectonic processes.

River - delta estuaries found at the mouths of large livers such as Mississippi or the Nile. It is different from formers. On hydrographic basis esturaries can be placed in three broad categories.

(*a*) Highly stratified or salt wedge estuary.

(*b*) The partially mixed or moderately stratified estuary

(*c*) The completely mixed or vertically homogenous estuary

The Hypersaline estuary is a special type.

**Physico Chemical Aspects of Estuaries :**

Current and salinity both are important here. Estuarine currents result from the interaction of a one direction stream flow which varies with the session and rain fall with oscillation ocean tides and with wind. The salinity varies vertically and horizontally and fluctuates amazingly between 0.5 to 0.35%. The water level in the estuary fluctuate regularly unlike that of river. No Echinodermates, Cephalopoda and other molluscs could survive in the estuaries. The temperature in estuaries fluctuate considerably diurnally and sessionably.

The sessional and tidal cycles cause changes in nutrient concentration in the estuary. Any how, all estuaries have high productivity. The concentration of nutrients and fix carbon is very high level of production within the detritus food chain.

**Biotic Communities of Estuaries**

Carrikar in 1967 has classified the regions of estuaries in to upper, middle & lower reaches with increasing range of salinities and the mouth with salinity nearly equal to the sea. He has also classified the animals inhabiting the estuarine region into - oligohaline (.5 to 5%), mesohaline (5-18%). Krishnamoorthy has reported the extent of penetration of the polychacts in Madras from the Bay of Bengal. In Hoogly - Matla estuarine, Gopalkrishnan 1971 has reported an abundance of phytoplanktonic forms, several species of diatous synedra, Naricula etc. and blue green algae like microcystis, oscillatoria, adyar estuary of Madras coast is found rich in invertebrates and vertebrate fauna. In short, the esturine ecosystem is a complex and interesting one. It is also very vulnerable environment, because it has served as condiuts for shipping and as sites for cities.

## LAKE ECOSYSTEM

Lakes are inland depressions containing standing water. They vary in size and depth (few feet to 5000 feet). Some lakes wave outlet streams. In lake there are three to five well recognized horizontal strata namely.

(*i*) **Littoral zone :** Shallow water near the shore forms this zone.

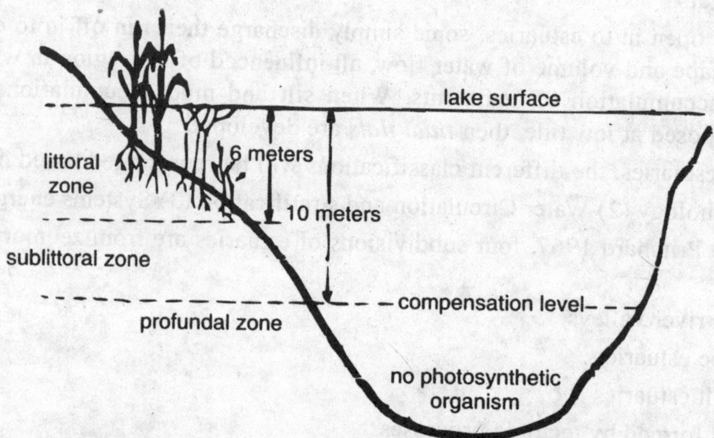


**Fig. 8.16.** Different zone of a deep freshwater lake.

It contains upper warm and oxygen rich circulating water layer, which is called *epilimnion*. It include rooted vegetation.

(*ii*) **Sublittoral zone:** It extends from rooted vegetation to the non circulating cold water with poor oxygen zone i.e. hypolimnion.

(*iii*) **Limnetic zone:** It is the open water zone away from the shore. It is up to the depth of effective light penetration where rate of photosynthesis is equal to the rate of respiration.

(*iv*) **Profundal zone:** It is the deep water area beneath limnetic zone and beyond the depth of effective light penetration.

(*v*) **Abyssal zone:** It is found only in deep lakes since it begins at about 2000 meter from the surface.

**Kinds of lakes:** Based on the physical factors, productivity etc. different classifications of lakes are given. Based on temperation, Hutchinson (1957) classified into *dimictic, monomictic* and *polymictic.* Based on Humic acid contents, the lakes are classified in to clear water lakes and Brown water lakes.

**Physico-chemical properties of lakes**

Lakes have the tendency to become thermally stratified during summar and winter to undergo definite seasonal periodicity in depth. Light too penetrates only to a certain depth, depending on turbidity.

**Biotic Communities of lakes**

Organisms depending on substratum are called *pedonic forms* and that are free from it called limnetic *forms*. The lakes have several type of organisms.

(*i*) **Neuston:** These including floating plants such as duckweeds and many type of animals. Animals are called **epineuston** while others including insects called **hyponeuston.**

(*ii*) **Plankton:** These are small plants and animals whose powers of self locomotion is very limited. Certain zooplanktons are very active some planktons are called as **nektoplanktons.**

(*iii*) **Nekton:** These animals are swimmers.

(*iv*) **Bethos:** These includes the organisms living at the bottom of the water mass. These living above the sediment water interface are termed **benthic epifauna** and those living in sediments itself are termed as **infauna.**

**Stratification in lakes.** The typical sessional cycle illustrated in fig. 8.17.

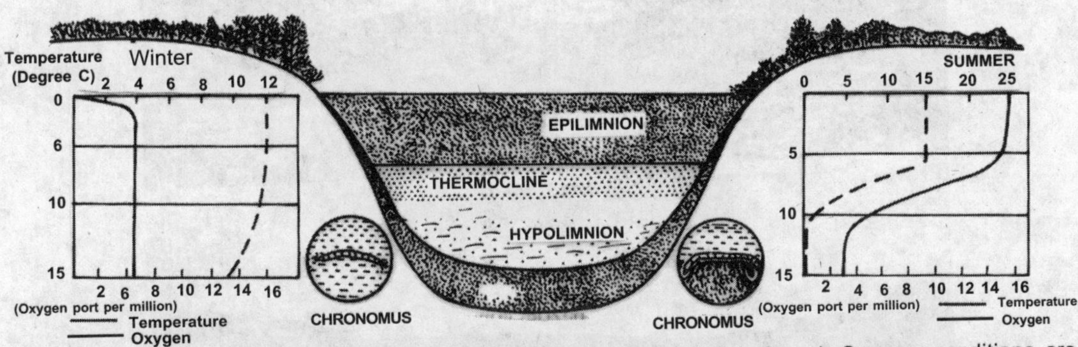


**Fig. 8.17** Thermal stratification in a north temperate lake (Linsley Pond. Conn.). Summer conditions are shown on the right, winter conditions on the left. Note that in summer a warm oxygen-rich circulating layer of water, the epilimnion, is separated from the cold oxygen-poor hypolimnion waters by a broad zone, called the thermocline, which is characterized by a rapid change in temperature and oxygen with increasing depth. Two typical hypolimnion organisms are shown (see also Figure 8.18). (After Deevey, 1951).

During the summer the top water become warmer than the bottom waters, as a result only the warm top layer circulates and it does not mix with the more viscous colder water, called thermocline. The upper water layer is epilomnion. Colder noncirculating water is the hypolimnion. Subtropical lakes having surface temperatures that never fall below 4°C. In terms of water circulation patterns most of the lakes of the world can be conveniently assigned to one of the following categories (Hutchinson 1957).

(a) **Dimictic** (mictic = mixed) Two sessional periods of free circulation.

(b) **Cold monomictic.** Water never above 4°C (polar regions), seasonal overturn in summer.

(c) **Warm monomictic.** Water never below 4°C. One period of circulation in winter.

(d) **Polymictic.** More or less continually circulating with only short, if any, stagnation period.

(e) **Oligomictic.** Rarely mixed.

(f) **Micromictic.** Permanently stratified.

## STREAMS

Biotic community in streams is quite different from that of ponds. Most streams in the vicinity of urban areas are polluted. Streams are fresh water aquatic systems where water current is a measure controlling factor, oxygen and nutrients are in water. Differences between streams and ponds revolve around a triad of conditions.

**(i) Current.** It is a major controlling and limiting factor in streams. Velocity of the current varies greatly in different parts of the same stream and from one time to another. In large streams

**Fig. 8.18** Two streams illustrating the close interdependence of streams and the terrestrial watershed. A small woodland stream whose biota is almost entirely dependent on the import of leaf and other organic detritus from the forest B. A salmon stream in Alaska at a time of a salmon "run" when biomass and mineral nutrients are being "exported" by air-breathing predators (bears and gulls) that are feeding on the fish & soil Conservation Service Photo B. U.S. Dept. Interior, Fish and Wildlife Service Photo.)

the current may be so reduced that virtually standing water conditions result. Current is the most important primary factor which (1) makes for a big difference between stream and pond life and (2) governs differences in various parts of a given stream. The velocity of the current is determined by steepness of the surface gradient, the roughness of depth and width of the stream bed.

***(ii)* Land-water Interchange.** The land-water surface junction is relatively great in proportion to the size of the stream habitat. This means that streams are more intimately associated with the surrounding land. (Fig. 8.18) than are most standing bodies of the water. Most streams depend on land areas and on connected ponds, back waters and lakes for a large portion of their basic energy supply. Streams have producers of their own such as fixed filamentous green algae encrusted diatoms and aquatic mosses. Some time plankton and detritus coming in to stream from quietor water. Streams form an open ecosystem that is interdigitated with terrestrial and leutic systems.

***(iii)* Oxygen.** Stream organisms face more extreme conditions in regard to current and temperature. Because of the small depth large surface exposed to the air and constant motion, streams generally contain an abundant supply of the oxygen even when there is no green plants. For this reason stream animals generally have a narrow tolerance and are specially sensitive to reduced oxygen. Therefore, stream communities are specially susceptible to and quickly modified by any type of organic pollution, which reduces the oxygen supply, fig. 8.19

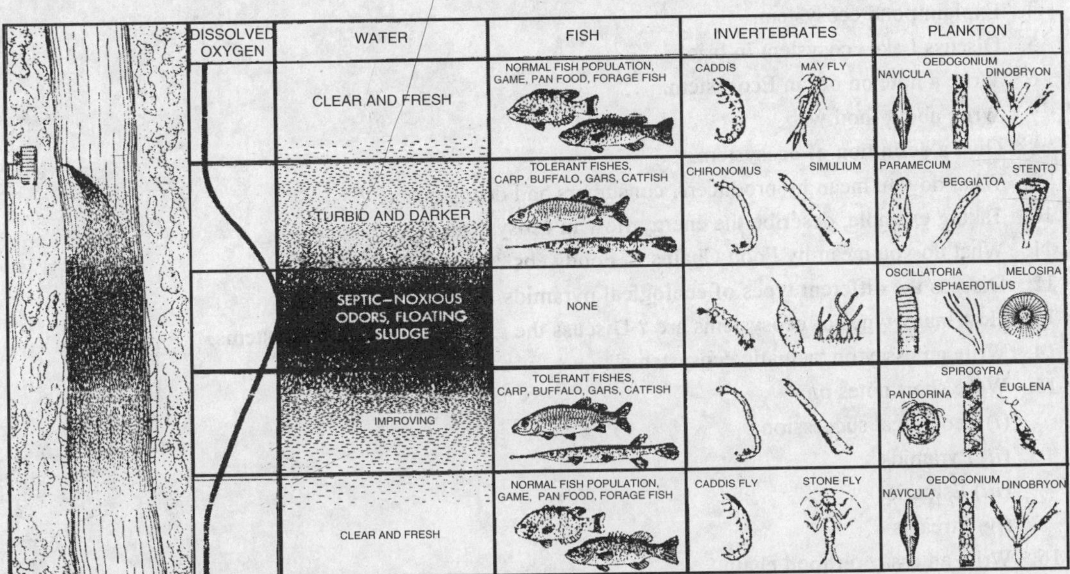


**Fig. 8.19.** Pollution of a stream with untreated sewage and the subsequent recovery as reflected in changes in the biotic community. As the oxygen dissolved in the water decreases (curve to the left), fishes disappear and only organisms able to obtain oxygen from the surface (as in Culex mosquito larvae) or those which are tolerant of low oxygen concentration are found in zone of maximum organic decomposition. When bacteria have reduced all of the discharged material the stream returns to normal. (After Eliassen, Scientific American, Vol. 186, No. 3, March, 1952.)

## ZONATION IN STREAMS

In streams zonation is longitudinal. In streams we find zones increasingly older stages from source to mouth. Charges are more pronounced in the upper part, because of gradient, volumes of flow and chemical composition charges rapidly. The change in composition of communities is likely to be more pronounced in the first mile than in the last fifty miles. The longitudinal distribution of fish in a stream may be selected as a specific example. Thompson & Hunt found that the number of individuals decreased down stream but the size of fish increased so that biomass density remained same.

**STREAM COMMUNITIES**

Streams generally exhibit two major habitats i.e. rapid & pools. Some within these categories the type of the bottom whether sand, pebbels, clay, bedrock or rubble rock is very important in determining the nature of the community and population density. Current is the major factor for rapids but hard bottom may offer favourable surfaces for organisms. Benthic invertebrates have higher density in rapids but claws, burrowing odonata and ephemeroptera are more abundant in pools. Stream fish find refuges in pools and feed in or at the base of rapids. Planktons are at the mercy of the current organism in rapids and pools (in lesser extent) show adaptations for maintaining position in swift water—some of these are Cladophora, Fontinalis, Diptera larvae Simulium, Blepharocera, Snails, flatworms, insect larvae etc.

## QUESTIONS

1. Explain the concept of ecosystem.
2. Describe the structure of an ecosystem.
3. What do you mean by pyramid of energy?
4. Explain pond ecosystem.
5. Discuss Lake ecosystem in brief.
6. Write a note on ocean Ecosystem.
7. Write about food web.
8. Give a definition of ecosystem.
9. What do you mean by producers, consumers and decomposers ?
10. Taking example, describe the energy flow in ecosystems.
11. What do you mean by Food Chains & Foodwebs ?
12. Discuss the different types of ecological pyramids.
13. How many types of ecosystems are ? Discuss the grassland type of ecosystems.
14. Write an essay on "aquatic ecosystems".
15. Write short notes on –
    (*i*) Ecological succession
    (*ii*) Pyramids
    (*iii*) Estuaries
    (*iv*) Streams
16. Write an essay on food chain.
17. Write an essay on structure of ecosystem.
18. Types of ecological succession.
19. Write detailed note on various proplic level of an ecosystem.

# Genetic and Plant Biodiversity

## INTRODUCTION

Biodiversity is the variety of the life forms of earth and its myria of processes. Conservation of biodiversity is one the most pressing environmental issues. The world's biological heritage is threatened by many factors and the most important among them are the anthropongenic (man-made) activities. This challenge, which exists both at local and global levels, has to be faced by the national and international agencies, government agencies, organisation, and individuals, while at the same time catering to the genuine resource requirements by people. If we do not squarely meet this challenge, our posterity will be inheriting a biologically impoverished world which will be incapable of producing the essential resource for their well-being.

Biodiversity being a multi-faceted issue, crosses all resource boundaries. Its conservation requires a truly multi-disciplinary approach. "The conservation of biodiversity encompasses genetic diversity of species populations, richness of species in biological communities, process whereby species interact with one another and with physical attributes within ecological systems, and the abundance of species, communities, and ecosystems at large geographic scales. The demands and expectations placed on these resources are so high and widely varied, that they call for new approaches which go much beyond merely reacting to resource crises and concerns". As the most intelligent species on earth, we the humans have a moral responsibility to preserve all forms of life, as far as possible. The first step in this direction is to determine the amount, variety, distribution of species, ecosystems, landscapes, etc. and prepare inventories followed by monitoring the impacts of management activities. Strategies must then be developed and implemented for the preservation, maintenance and restoration of forest ecosystems, while simultaneously formulating action plan for the sustainable use of forest resources including more efficient utilization, recycling programs and reforestation to meet the human needs.

Biodiversity can be divided into the following four levels: genetic, spices ecosystems, and landscape diversity. Maximum attention is often attracted towards "*Species diversity*", i.e. the number of different kinds of organisms found at a particular locale and how it varies from place and even seasonally at the same place. "*Genetic diversity*", is the variety of building blocks found within individuals of a species. It allows populations of a species to adapt to environmental variation. "*Ecosystem diversity*" is the distinctive assemblage of species that live together in the same area and interact with the physical environment in unique ways. On the regional level, "Landscape diversity" refers to the placement and size of various ecosystems across the land surface and the linkages between them.

## ENVIRONMENT AND ECOLOGY

Every organism is surrounded by materials and forces that constitutes its environment from which its needs are derived. Thus, each organism depends upon its environment and other organisms for its very survival. A microbe or a plant or an animal requires a constant supply of energy for its survival. For this, it depends on intersection with different biotic (i.e., living or abiotic) (*i.e.*, non-living) components of the environment.

The external conditions the affects the life or organisms in their natural habitats together constitute their environment. The environment can be classified as follows:

(*i*) **Biotic environment :** This is composed of all living components such as animals, plant and micro-organisms (e.g., bacteria and fungi)

(*ii*) **Abiotic environment:** This is composed of all external physical factors such as temperature, water, minerals, gases and humidity. All these constituents of the environment are known as environmental factors or ecological factors or simply as factors. The biotic and abiotic components constantly affect each other. Apart from the biotic and abiotic components, the environment is made up of the three major constituents namely, lithosphere, hydrosphere and atmosphere.

The branch of science which deals relationship of organisms or groups of organisms with their environment is called 'ecology". The word ecology was first used in 1869 by Ernst Haeckel and originated from two Greek words " oikos" (which means house, habitation or place of living) and "logos" (which mean 'study'). Ecology deals in general with the effects of environmental factors on all aspects of life such as physiology, morphology, distribution, survival and growth of the various orgnisms. Thus, ecology may be regarded as the science of intersections among individuals, populâtions and communities and their-relations with the biotic and abiotic components of the environment.

Animals, plants, micro-organisms, etc. together with the environment in which they live and interact to form a stable and self-supporting systems, constitute an independent unit, called the "ecosystem". The ecosystem may be of different types as follows:

(*a*) Terrestrial ecosystems, Ex:- grass land, mountain, forest, desert, cropland, engineered ecosystems, etc.

(*b*) Aquatic ecosystem, Ex:- fresh water ecosystem such as ponds, lakes, and marine ecosystems such as seas and estuaries.

The entire zone of water, air or land inhabited by living organisms is called "*biosphere*".

Ecology may be divided into the following two following two major classes or aspects:

(*i*) **Autecology:** This deals with the study on an individual species or organisms in relation to its environment. The studies include the behaviour and adoption of each of the individual species to the environmental conditions at every stage of its life-cycle. That is why autecology is also called as *species ecology.*

(*ii*) **Synecology:** It deals with the study of groups of organism (i.e. communities), their composition, behaviour and relationship with the environment. That is why synecology is also called *community ecology* or ecology of communities. Synecology is further sub-divided as (*a*) population ecology, (*b*) community ecology and (*c*) ecosystemology of ecosystem ecology.

## ORGANISATION OF AN ECOSYSTEM

An ecosystem refers to a biotic community and its abiotic environment taken together. Hence, an ecosystem is made up of biotic and abiotic components.

(*i*) **Biotic Components :** The living organisms of an ecosystem which include animals, plants and micro-organisms (*e.g.*, bacteria, fungi, etc.) constitute the biotic components. From the point of view of their respective role in the ecosystem, the organisms (*i.e.* the biotic components) are generally into the following three types:

(*a*) *Producers* : These are the green plants having chlorophyll, which are capable of synthesizing their own food by trapping energy from the sun and with the help of simple inorganic compounds like carbon dioxide and water. Since these green plants synthesize their own food, they are called "*autotrophs*" (auto = self, trophs = feeder). The chemical energy stored by the producers is partly used by themselves for their growth and the remaining is stored for their further use.

Ex:- Large rooted green plants on land, rootless algae floating or submerged in water, chemosyntheric bacteria etc.

(*b*) *Consumers*: As these cannot systnesize their own food they depends on the producers for their food requirements. Hence they are called " *heterophs*" (hetero = other, trophos = feeder). The consumers are of the following types:

(1) Primary consumers or herbivores: These depends on the green plants for their food.

Ex:- cattle, goats, deers, rabbits, insects.

(2) Secondary consumers or primary carnivores: These feed on the herbivores.

Ex:- cats, wolves, foxes, snakes etc.

(3) Tertiary consumers: These are the large carnivores which feed on the secondary consumers. They are not eaten up by any other animals. Tigers, lions, etc.

(*c*) *Decomposers* : These include micro-organisms such as bacteria and fungi which attack and decompose dead organic matter and dead bodies of producers (plants) and consumers (animals) into simple organic and inorganic susbstances as byproducts of their metabolism. These substances are then utilized by the producers for systhesizing their own food, thus resulting in a cyclic exchange of materials between the biotic community and the abiotic environment of the ecosystem. Hence the decomposers are also called as "Saprotrophs" (Sapros = rotten, trophos = feeder, saprotrophs = those on rotten matter).

In nature, smaller organisms are eaten by large organisms, and all the organisms are inter-linked with each other through food. Thus, the trophic structure of an ecosystem includes primary producers and primary, secondary and tetiary consumers as well as decomposers and all of them together constitute the so-called "*food chain*". Each of the link in the food chain is called as *food level* or *trophic level.* If any link in the food-chain is removed, the other links will be affected. The food chain is aquatic ecosystem may be represented as:

Phytoplankton → Zoo plankton → small fish → big fish. Further, the "*food web*" also provided more than one type of food alternatives to most of the organisms, thereby enhancing the chances of their sustenance and survival.

(*ii*) **Abiotic Components :** These consist of the non-living factors of physical environment which strongly influence the structure, behaviour, distribution and inter-relationship of the organisms. The abiotic components are of the following types:

(*a*) *Climatic factors* : These include light, wind, temperature, humidity and rain.

(*b*) *Edaphic factors* : These include pH, soil, and topography.

Energy is required for maintaining the activities of life in the ecosystem. Energy from the sun is trapped by chlorophyll in the plants and helps in the synthesis of food in the form of starch from the raw materials like water, carbon dioxide, and salts. However, the flow of energy through the ecosystem is non-cyclic and unidirectional, and follow the classical principles of thermodynamics. Within an ecosystem, there will be a gradual loss of energy within each trophic level from producer to consumers. Therefore, shorter the food chain, greater would be the available food energy.

The producers of an ecosystem take up inorganic nutrients from their abiotic environment and transform them into the biomass. These are utilised by the consumer populations and are eventually returned to the environment by intersection with reducers or decomposers. The cycic exchange of nutrients between the living organism and their non-living environment is known as "biogeochemical cycle".

The important biogeochemcial cycles are:

(1) Hydrologic cycle

(2) Carbon cycle

(3) Nitrogen cycle

(4) Oxygen cycle

It is these cyclic process that conserve the valuable and limited nutrients and raw materials in the environment for repeated reuse. (These are discussed in chapter 6).

## BIODIVERSITY

The biosphere comprises of a complex collection of innumerable organisms known as the bio-diversity, which constitutes the vital life support for the survival of human race. As defined in the convention on Biological Diversity signed at Rio de Jenerio (Brazil) in 1992 by 154 countries, the Biological Diversity in defined as "the variability among living organisms from all sources including, inter alia, terrestrial, marine and other aquatic ecosystems and the ecological complexes of which they are part-this include diversity within species, between species and of ecosystems". Biological diversity, abbreviated as Bio-diversity, represents the sum total of various life form such as unicellular fungi, protozoa, bacteria and multicellular organisms such as plants fishes and mammals at various biological levels including genes, species, habitats and of ecosystems.

Nature has developed exceeding by complex spectrum of life forms over 600 million years. It is impossible to accurately assess the number of species of organisms present on earth. It is believed that anywhere between 10 to 80 million species exists on earth. However, only about 1.5 million species could be assessed and enlisted so far.

The existence of human race depends on the well-being of the other life forms present in the biosphere. However, we have been losing this accumulated heritage of millions of years at an alarmingly fast rate over the past 400 years by our own activities, thereby undermining the very basis of our own existence on this planet. Our biological systems are constantly impoverished by human activities. All over the world, human being only have been mostly responsible for the destruction of habitats through intensive agricultural development, over-exploitation of resources, urbanization, industrialisation, over-population, deforestation, pollution, environmental degradation and ethical degradation.

## ASSESSMENT OF BIOLOGICAL DIVERSITY

Biological diversity is analysed or assessed generally at the following four levels:

(*i*) *Diversity of composition of a species within a community* : The biotic component of an ecosystem may be composed of a number of species of plants, microbes and animals which intersect with each other on one hand and interact with the abiotic factors of the environment on the other. The richness of species in an ecosystem is called "*species diversity*".

(*ii*) *Diversity of Genetic Organisation within a species:* Within any given species, we can find several varieties or strains or races which slightly differ from each other in one or more characteristictics such as size, shape, quality of their respectively product, resistance against pests, insects, diseases, etc., and resilience to survive under adverse environmental conditions. Such a diversity in the genetic make-up of a species is called "*genetic diversity*". A species having large number of varieties, strains or races is considered as rich and more diverse in its genetic organization.

(*iii*) *Diversity of Biotic communities and Ecosystems*: An ecosystem develops its own characteristic community of living organisms depending upon the availability of abiotic resources, environmental conditions and other relevant factors. Different types of ponds, lakes, rivers, wetlands, meadow, grasslands, forests etc. represent diverse ecosystems with their own characteristic biotic community. For example, a pond may posses different sets of flora and fauna as compared to another ecosystem such as river.

(*iv*) *Diversity within a landscape* : This refers to the size and distribution of several ecosystems and their interactions across a given land surface.

## IMPORTANCE OF BIO-DIVERSITY

Bio-diversity is of both aesthetic as well as practical importance.

(1) Biodiversity provides us valuable natural resources to satisfy the subtle needs of mankind. Our homes, livestock, fruits, vegetables, grains, grams, etc. are all derived from the products of diverse and healthy ecosystems. Our food, clothings, shelter, and a host of other useful products are derived from a variety of living organisms.

(2) Diverse communities of plants, animals and micro-organisms also provide us valuable and indispensable ecological services. They recycle wastes, maintain the chemical composition of the atmosphere, and play a major role in determining the climate of the different parts of the world. The "ecosystem services" provided to us by nature is "gratis" also include supply of fresh water, generating soils, supply of plant pollinators, and maintaining a huge "genetic library".

(3) In many cultures, maintenance of mountains and other diverse landforms are of the religious significance.

(4) A diversity of abodes of biological communities such as parks, gardens, natural animal habitats, forests, mountains seashores, etc. are useful for picnicking and other recreational activities.

(5) In spite of our intensive scientific efforts, we know only fraction of the multitude of the species thriving on our planet, particularly in tropical ecosystems. Every year, some species are lost even before we had a chance to identify them or study their useful properties.

We have been cultivating hardly about 2500 plant species. This number may be double if we take into account the animals, wild plants microbes and other organisms. However, all these species together account for only a fraction of the total number of plant species recorded, which is about 1.5 million. Even this, in turn, is a minute fraction of the suspected or expected number of species on the earth, which may be 10 to 80 millions. Therefore, even today, we know very little about the biological world in which live. Out of about 2.5 lakh species, hardly about 25,000 species have been broadly assessed and only about 2,500 species have been studied in details. As far as the animal species are concerned, the situation is more inadequate. Thus it is clear that a huge wealth of biological resources are yet to be trapped by the mankind. The price we might have been paying for this neglect may be illustrated by the fact that if "*penicillium*" or "*cinchona*" would have been extinct before their curative properties were discovered, what would have been the plight of mankind.

(6) Biological diversity represents a valuable genetic resource for the mankind . The genetic organisation of wild plant and animals, which has been continually evolving for millions of years survived the trials, tribulations, and vagaries of nature. While we are still cultivating some of those old varieties, we are also using some of those genes to improve the strains we are presently cultivating. Indeed, most of the improved varieties under cultivation today are synthetic of hybrid types by incorporating useful genes to produce better quality of the produce with longer shelf life or heaving better resistance to pests. Such breeding techniques are unlimited in scope. But, however, for getting better strains in future, it is essential to build up a gene-pool containing an assortment of traits

Further, the quality, yield, and resistance to pests, diseases and adverse climatic vagaries mostly depend on genetic factors and combination of genes which may be different strains or varieties of species. Improved varieties containing higher number of useful traits and desirable gene combinations are always preferred which are more productive and have better adoptability. One of the reason for this is that repeated and prolonged cultivation of a particular variety gradually loses its vigour, quality, and productivity, and increase its susceptibility for diseases or attack by pests. There are several examples to illustrate how genetic modification helped in improved quality and pest-resistance of the product.

(*a*) A wild variety of rice grown in U.P. Saved millions of hectares of paddy crop from "Grassy Stunt" virus.

(*b*) The Kans grass known as "Saccharum Spontaneum" from Indonesia provided genes for resistance to red rot disease of sugar cane.

(*c*) The genes from a wild melon grown in U.P. helped in imparting resistance to powdery mildew in Musk melons grown in California.

(*d*) About 20 cultivars of rice grown in several countries were benefited from the useful genes identified from wild varieties from Kerala.

(7) The vast pool of genetic diversity contained within wild populations of plants and animals is of enormous value for the continuing research and development of agriculture, industry and medicine. For instance, the chances of survival of patients suffering from leukaemia have dramatically and appreciably increased in recent years because of synthesis of a drug containing active chemicals derived from a tropical tree from Madagaskar, called "Catharanthus roseus" (rosy periwinkle).

(8) Thousands of species which are edible and demonstrably superior than those which are currently used, are waiting in the wings for their useful exploitation. For example, a plant found in west Africa, called as Katemfe (Thaumatococus danielli), produces proteins that are 1600 times sweeter than Sucrose. Similarly, several plant species containing anti-cancer substances of varying potency, and some others having other potential medicinal applications , are yet to be screened and intensively studied for their applications. It should be remembered that about one third of the modern description drugs contain chemical compounds that plants have evolved to defined themselves from their enemies.

(9) The vast insect fauna contain large number of species that are potentially superior crop pollinators, weed-control agents and are parasites of insect pests.

(10) Biological diversity helps in maintaining a stable and healthy ecosystem. In a simple ecosystem, loss of even one or a few species could be disastrous because of the lack of alternatives. However, in a complicated ecosystem having several trophic levels, each of which, in its turn, is composed of several species, loss of one or more species do not cause

any serious problems because, the alternatives available can maintain the functionality of the system. Thus, biological diversity helps in maintaining a stable ecosystem.

(11) Many species and creatures considered to be useless or undesirable have also their own roles to play in a natural ecosystem.

(*a*) Earthworms and termites help in aerating our soils. But this vital function is stopped by killing them using pesticides.

(*b*) Predators which were facing extinction can no longer control rodents the damage our crops.

(*c*) Mangroves protect our coastline from erosion are being destroyed for use as firewood.

(*d*) Oysters which were present at one time in large numbers in chespeak bay, (USA) could filter the entire water in a few days have declined by 99%, thereby leaving the bay water muddy and oxygen-deficient.

(*e*) An adult from is supposed to consume its own weight of insects every day. But the decreases in frog population in India and Pakistan was associated with increased pest damage and recurrence of Malaria.

Hence it is obvious that each species in nature has its own function in the ecosystem and disappearance of any species may cause serious consequence, sooner or later.

(12) In most tropics of the world having rich biodiversity, most of the nutrients are produced or regenerated at faster rate due to faster rate of decomposition of organic matter and are taken up by the biotic community.

In humid tropics, the luxuriant growth of vegetation hinders rapid flow of water, binds soil particles together and prevents soil erosion. Some of the nutrients that dissolve out are taken up once again as the water is retained for longer time in the system. Reduction in bio-diversity reduces the efficiency of such a phenomenon, thereby resulting in rapid loss of nutrients form the soil and its consequent degradation

(13) Maintaining the bio-diversity helps in preservation of Socio-economic, aesthetic, cultural and religious values of ecosystems.

## FACTORS RESPONSIBLE FOR LOSS OF BIODIVERSITY

The sum total of the all varieties of genes present in a population of species constitutes a gene-pool. Loss of bio-diversity means the genetic erosion of the loss of genes from gene pool. Deforestation and the consequent soil erosion resulted in the categorisation of over 1500 species of animals, plant and microorganisms as "endangered species" in India along. In fact, any disturbance in the eco-system tends to reduce biodiversity.

The estimates of endangered wildlife grossly understates the real magnitude of the loss of biodiversity because it generally deals with higher life-forms commonly known to us. However, all amphibians, reptiles, birds, fishes and mammals put together constitute only about 2-3% of the known species of living organisms. More than a million species of invertebrates and one third of a million species or recorded green plants have been reportedly existing on out planet, quite apart from the enormous number of species in nature which have not yet been assessed and enlisted. The gravity of the situation should be assessed while considering the entire life-spectrum on our planet.

Over the past few decades, the rate of global loss of biodiversity and the consequent biotic impoverishment has been increasing alarmingly. Exponential growth in human population and the consequent growth in consumption of the world's natural resources have led to the accelerated loss of species and habitats. The fragmentation of populations by destruction of habitats has also been contributing to the extinction of species and loss of genetic diversity. In some cases, the introduction of alien species (known as exotic species) has led to the loss of native plant and animal species. For instance, the native flora of "*fynbos biome*" in South Africa (Mediterranian type climate) is under threat, in parts, from the invasion of acacias and pines. Inland communities, which have evolved in isolation, are particularly vulnerable to the introduction of exotic species.

Serious problems are faced in the tropical countries where biodiversity is highest and loss of species as well as ecosystem is relatively rapid. The issues are more serious in developing countries because of huge population, poverty, and lack of awareness.

The destruction of forests in the Third World Countries is estimated to be more than 11 million hectares every year. Recent tropical deforestation is associated with a pervasive cycle of initial timber extraction, followed by shifting cultivation, and acquisition, and subsequent conversion to pasture which leads to loss of forest resources, reduction in biodiversity and impoverishment or rural masses.

Almost about two thirds of the world's plants are present in tropics Over 1.6 lac species known to us occur in the so called mega-diversity countries (i.e. tropical regions of the developing countries) of the world. It is feared that by 2020, about 25% of the tropical plants may be eliminated, if no efforts are made to protect them. In our country, about 15,000 species of flowering plants are known to occur. The "Red data book" complied by the Botanical survey of India enlists about 1000 species as rare and threatened plants. About 16% of the flowering plant population in our country have dwindled to alarmingly low levels, which include about 35% of the endemic plants i.e. plants found nowhere else. Among the non-tropical countries, the arid and semi-arid regions of South Africa contains large collection of rare and endangered plants which include large number of endemic species. About 12% of the 300 plant species of the U.S.A. are reported to be dangerously rare.

It is intersecting to note that some of our ancient civilization (*e.g.*, olmec and Mayo civilizations) existed with high population densities integrated within tropical forest ecosystems. The knowledge of management techniques practiced by such ancient civilization could help in reverting current process of landscape degradation in the tropics.

In some ecosystems, richness of species may be very high, as in the tropical rain forests and coral reefs. Apart from these, the proportion of the endemic species (*i.e.*, those species which are confined to a particular area) is also very significant. The conservation of areas of high endemism is particularly important for the conversation of genetic diversity.

The loss genetic diversity through the mono-culture of crops in another serious cause of concern. The natural ecosystem, with its high number of species is replaced by a much simplified ecosystem, namely the agro-ecosystem, which is typified by only a few species. The world is dependent only on a tiny percentage of the 2.5 lac species of flowering plants for the provision of its staple food crops. The conservation of the genetic diversity found in related wild strains is thus vital to the continuing development of crop plant cultivars, in term of increased productivity and resistance against disease and pests.

Other adverse effects of decrease in biodiversity apart from loss of valuable species are degradation of land, carbon dioxide build-up, acid rain, depletion of stratospheric ozone layer, etc.

The preservation of biodiversity at all levels, is a global concern now. Biodiversity is our most valuable natural resource and it is an insurance for our food and ecological security. Hence its components and genetic resources should be utilized only to support our socio-economic need in a sustainable manner.

The major causes responsible for reduction in Bio-diversity may be summarized as follows:

(1) Destruction of natural ecosystems to meet the requirements of food, space, shelter, etc. for the ever-expanding human settlements.

The species in the Boreal coniferous forests at the northern most belt skirting the Arctic Sea, which is thinly populated, are getting further reduced due to harsh and unfavourable climatic conditions. The biodiversity in several temperature and sub-tropical regions of the world (e.g., the zone between coniferous forests of north and the tropical zone in the south in the Northern Hemisphere, which support extensive agriculture and cattle ranching) is already under threat due to encroachments.

The tropical regions of the world colonized by developing and under developed countries (which are called mega-diversity countries) are endowed with rich and diverse assortment of fauna and flora. These occupy about $1/4^{th}$ total area of the world and contain about $3/4^{th}$ of the world population. In order to feed and sustain millions of poor people, these countries are forced to expand agriculture and industries, which led to over-exploitation of resources and large-scale deforestation. Since tropical forests contain about 50% of the total number of species on the earth, destruction of natural habitats are proving disastrous for the entire biosphere.

Mangroves and other coastal wetlands provide a buffer zone between the sea and land, which protects the coastline from erosion from the oceanic waves and at the same time it clears the silt, sediments, pollutants, etc. before the water flows into the sea. Mangroves also are rich in fauna and flora. Loss of mangrove ecosystems in Asia, West Africa and Latin America are causing extensive damage to biological damage and fish productivity apart from erosion of coastline.

Coral-reefs are marine equivalents of tropical forests and they represent highly productive and richly diverse ecosystem. Loss of mangroves result in destruction of coral-reef ecosystem. The destruction of mangroves in Costa-rica and Philippines is already a matter of grave concern for their rich and diverse biota.

(2) Unfavourable changes in the biotic and abiotic factors result from

(*a*) *Natural Upheavals* : Natural calamities such as earthquakes, floods, volcanic eruptions, forest-fires, epidemics, droughts, etc., cause extensive damage to plant and animals life.

(*b*) *Environmental Pollution* : Pollutants such as pesticides, chlorinated hydrocarbons, and toxic heavy metals destroy the weak and susceptible species, and damage the living organisms. Over -fertilization leads to entrophication which promotes the growth of some species while suppressing others. Industries emitting $SO_2$ and $NO_x$ cause acid rains which destroy plants and fishes.

(*c*) *Invasion by exotic species* : Any species which is not a natural inhabitant of the local habitat but is introduced into the system accidently or deliberately, is called "exotic species". Several instances are known when a natural biotic community of the ecosystem suffered extensive damage because of the invasion by exotic species. A classical example is the contamination of wheat seeds imported by us from USA under PL-480 scheme by "*parthenium hysterophorus*" (known as congress grass) and "*Agrosteamma githago*" (known as corn cockle). Both the species grew and spread all over the country as pernicious weeds in wheat fields Parthernium is an aggressive plant which matures rapidly and generates thousands of seeds. Because of these plants, indigenous herbs and grasses as well as cultivated crops suffered as it depletes the nutrients from the soil. Moreover, the enormous quantity of pollens produced by parthenium cause allelohatic effects on cultivated plants like tomatoes, chillies and brinjals and inhibited their growth. Similarly, myracea (eucalyptus) and casurinacea (casurina) are the species introduced from Australia and Tropical America into our country. Their extremely fast growth made them a valuable source of rough timber. However, they tend to suppress the native inhabitants and hence are ecologically harmful. Another striking example is the introduction of Nile perch from north in Lake Victoria in Africa, because of which about 50% of the 400 original fish species of the lake were driven to near extinction.

(*d*) *Over-exploitation of selected species*: About ten thousand years ago, man used over 5000 species and varieties of edible plants for his sustenance. Now, several traditional varieties are neglected and only about 150 species of plants are cultivated on large scale and out of these only about 15 types provide nearly 90% of the world's food supply. Most of these important cultivars used today are synthetic varieties in which desirable traits from a number of varieties are pooled together to obtain better yields, tastes and flavours. This trend of neglecting traditional varieties and substituting them with a few selected strains for some immediate benefits has been causing considerable damage to the biological diversity.

Over-exploitation of certain types of species is leading to dwindling of some plants of scientific, medical, decorative and other values. Medicinal plants such as podophyllum sp., are fastly disappearing due to huge collection. Similarly timber-producing trees of economic value such as Dysoxylon malabaricum, santalum album and ptercarpus santalum, and orchids, productive decorative flowers are diminishing due to over-exploitation.

Further, only limited varieties of rice, wheat, corn, apples, potatoes etc. are widely cultivated today, neglecting totally the traditional varieties. Hybrid varieties have replaced old strains of wheat and rice to obtain higher yields. Such a reduction of genetic diversity among the cultivated species and disappearance of their wild relatives, may restrict the creation of new cultivars in future, which does not augur well for the posterity.

Elephants, tigers, rhinoceros, snakes, crocodile whales, minks and some select groups of animals and birds are mercilessly hunted for food, recreation, and monetary benefits from their horns, tusks, hides skins, etc. This has resulted in rendering certain types of selected group of organisms as endangered category.

(3) Increased concentration of greenhouse gases in our atmosphere which threatens our planet with rise in mean global temperature. This may cause extensive changes in environmental conditions and precipitation patterns, which may endanger the plant species as a whole. As the climatic changes occur, large-scale migrations of species take place to exist in favourable climate. Such species which are incapable of migration will be adversely affected. Further, global warming will inevitably lead to expansion of tropical belt, displacement of sub-tropical and temperate zones northward and soutward away from equator, while the temperate and polar regions may reduce in size. Analysis of fossil pollens indicate that under such a scenario, the forest will have to migrate 20 kms to 200 kms per century. Many species that cannot migrate at such a faster rate may perish. Even species that migrate will have to overcome barriers like oceans, mountain ranges and large human establishments and under unfavourable conditions, many of these species may become extinct. Further, since different species respond to changing climatic patterns in different ways, the entire global biotic spectrum will changes and in such a sweeping upheaval, the biological diversity is likely to be drastically reduced.

## CONSERVATION OF BIOLOGICAL DIVERSITY

Biotic resources represent the very essence of life on the earth and everything that is possible should be done to preserve the biological diversity. There is an urgent need to educate people to adopt environment friendly practices and re-orient our activities in such away a way that our development is harmonious with other life-forms and is sustainable.

The following steps will go a long way in achieving these objectives.

(1) *Biodiversity Inventories* : Current scientific understanding of ecological processes has to be strengthened by adequate research efforts aimed at improving methodologies, distributional and status information. Strategies based on sound information will ultimately provide the basis for pragmatic polices and management decisions. Better inventories and assessments are needed of current conditions, abudndances, distributions and management direction for genetic resources, species populations, biological communities and ecological systems. For this purpose, the following activities should be supported.

(*a*) Rapid assessment programs to provide a "Snap-Shot" or richness of species in an area.

(*b*) Surveys, to map out the distribution of earth's ecosystems.

(*c*) Broad-based inventory efforts, focused on poorly known habitats, degraded and multiple-use habitats as well as on vertebrates and vascular plants that can be used as " benchmarks" for habitat quality.

(*d*) Intensive inventories, to determine all the species present, from microbes to vertebrates, should be prepared at a few selected sites around the world. Although this requires intensive effort over a long time-frame, that will be highly useful to decipher processes such as how the biodiversity at species level affect diversity at the ecosystem and community levels.

Such an effort will lead to significant improvements in inventory processes, methodologies and technologies such as the following:

(*a*) Provide estimates of resources in specific geographical units and evaluate their reliability.

(*b*) Enable spatial display of the estimates and units.

(*c*) Enable the best use of existing information and new technologies, such as remote sensing and geographic information system (GIS).

(*d*) Provide a base-line for monitoring changes in the extent and condition of the resource.

(*e*) Eliminate redundant data collection, develop common terminology nd promote data sharing through corporate data bases.

(*f*) Provide up-to-date data bases using modelling techniques, accounting procedures, and re-inventories.

(*g*) Ensure optimum utilisation of information management systems to provide maximum flexibility for data integration, manipulation, and sharing.

(2) *Conserving Biodiversity in protected habitats* : Wildlife conservation efforts are mostly concentrated on protecting animal and plant life in zoos, sanctuaries, gardens, biosphere-reserves, etc. The two basic approaches to the wildlife conservation in protected habitats are as follows:

(*a*) *"Ex-situ" conservation* : It means the wildlife conservation in captivity under human care. In this, the endangered plants ad animals are collected and bred under controlled conditions in gardens, zoos, sanctuaries etc. wildlife management in captivity have the following advantages:

(*i*) The organisms are assured of food, water, shelter and security and hence can have longer life span and longer span of breeding activity, thereby increasing the possibility of having more number of offsprings.

(*ii*) The chances of survival of endangered species increase because of human care under secure conditions.

(*iii*) This offers, the possibility of using genetic techniques to improve the species concerned.

However, there are some disadvantages and limitations of wildlife management in capacity:

(*i*) Since maintenance and breeding of plants and animals under captivity is very expensive, it can be adopted only for a few selected species.

(*ii*) Wildlife captivity only under a set of favourable environment conditions deprives the organisms the opportunity to adopt to ever-changing natural environment. Therefore, new life forms cannot evolve and thus the gene-pool gets stagnant.

(*b*) *"In-situ" conservation* : This involves setting aside large portions of earth's surface of wild life. However, many protected habitats are used for logging, tourism and other profitable activities, thereby pushing wildlife conservation to a less important objective. Further, many protected habitats are not large enough, not maintained properly for want of adequate staff, and are not properly protected from environmental pollution. These factors, to some extent reduce and laudable advantages of "In-situ" conservation. Large pockets of protected zones are essential for not only conserving vast number of species of living organisms but also provide opportunities to evolve. Otherwise, man-made habitats may end up with static gene-pool.

Monitoring the effects of land management and conservation in captivity under human care. In this, the endangered plants and animals are collected and bred under controlled conditions in gardens, zoos, sanctuaries etc. wildlife management in captivity have the following advantages:

(3) *Conservation Biodiversity in Seed-Banks and Gene-Banks:* Most of the plant species form seeds with variable periods of dormacy after which they can be germinated to yield daughter plants. The seed can be stored in seed-banks or germ-plasm banks. The germ plasm of a plant is any of its parts from which new plants can be generated.

The International Board for Plant Genetic Resources was framed in 1974 with its Headquarter in Rome. By 1985, a chain of 43 gene banks were set up in different parts of the world.

Recent techniques of preservation have greatly improved the use of germ plasm banks. Seeds with low moisture content when cooled at –196°C under liquid nitrogen temperature can be reserved for almost indefinite periods, say for several centuries. By application of tissue culture and cryo-preservation techniques, most of the plants can be maintained in viable state for very long periods of time. Meristematic cells of plants of small parts of simple vegetative tissues to be preserved are incorporated into suitable synthetic medium under aseptic conditions and are allowed to grow into small clumps known as "*Callus*". These causes can be preserved under liquid nitrogen for almost unlimited periods of time. They can be multiplied as and when desired.

There is a dire need to establish more and more regional germplasm banks particularly in the megadiversity countries like India.

(4) *Restoration of Biodiversity* : Restoration of ecosystems and biological communities is an important means maintaining biodiversity of atleast to retarding its net loss. Biodiversity

is threatened not only by reduction of habitat area and of connections between habitats, but also by degradation of quality of the remaining habitats, restoration programs provide refuges for species and genetic resources that might be lost otherwise.

Several techniques are used to restore ecosystem, depending on the nature of the ecosystem and the impact type being addressed. These techniques include vegetation planting to control erosion, fertilization of existing vegetation to encourage growth, removal of contaminated soils, fencing to prevent cattle, reintroduction of extirpated species, restoration of hydrologic connections to wetlands, etc.

Rapid reforestation to reinstate green plant cover on barren lands, hill slopes, highways, roads, etc. taking care to ensure that the species planted are ecologically compatible to the region and acceptable and useful to the local people, will go a long way in restoring biodiversity.

However, we should remember that restoration is not a substitute for the preservation of good management and it is both expensive and time consuming.

(5) *Imparting Environmental Education* : Educating people from all walks of life regarding eco-friendly practices goes a long way in conservation of plant and genetic biodiversity. People at large should be motivated to conserve resources, to avoid extravagance, and educated properly regarding ecological issues. All efforts should be made to respect and conserve indigenous knowledge traditions and environment-friendly practices.

(6) *Enacting Strengthening and Enforcing Environmental Legislations* : Existing environmental laws against ecologically unsound practices should be strengthened and enforced ruthlessly. Simultaneously, voluntary oranizations should be motived to include protection of bio-diversity as a major and priority issue in their agenda.

(7) *Population Control* : Effective population control measures have to be taken as a top-priority issue in the national agenda by involving people of all political parties, religious faiths and social organisations. Suitable incentives and disincentives should be in-built into the strategies specially formulated for this purpose.

(8) *Reviewing the Agricultural Practices* : We should refrain from temptation of high yields and making a fast buck at th cost of substantiable development. We should try to infuse diversity in our agricultural practices by restoring to mixed cropping polyculture and tolerance to wild plants and other life forms around our agricultural fields. A healthy soil should be maintained by using more or organic manures and less of synthetic fertilizers and pesticides. All efforts should be made to maintain a balanced prey-predator relationship in our agro-ecosystems.

(9) *Controlling Urbanization* : Ever-increasing urbanisation and expansion of urban settlements should be controlled. Biological diversity should be infused into the urban localities.

## CONCLUSION

Maintaining global biological diversity should go much beyond crisis management of increasing number of endangered species. Long-term maintenance of species and their genetic management requires cooperative efforts across entire landscapes. Biodiversity should be dealt with at the scale of habitats or ecosystems rather than at species level.

## QUESTIONS

1. Discuss the importance of biodiversity and the strategies for its conservation.
2. What are the major factors that are responsible for the loss of plant and genetic biodiversity? What steps are needed to prevent them?
3. Write informative notes on any three of the following:
   (*a*) "Ex-situ" conservation.
   (*b*) Over-exploitation of selected species and its consequences.
   (*c*) Invasion by Exotic species and its consequences.
   (*d*) Bio-restoration.
   (*e*) Germ-plasm preservation.
   (*f*) Organisation of ecosystem
4. Write an essay on conservation of Biodiversity.
5. Define Biodiversity, give its importance.

# Bio-Technology and its Applications in Environmental Protection

## INTRODUCTION

Bio-technology encompasses a wide array of specialised disciplines, right from age old fermentation processes to the latest techniques of genetic engineering. Bio-technology has received tremendous attention in recent years due to its immense potentialities and applications in aquaculture, agriculture, immunology, forestry, chemical production, pollution control, etc.

Genetic engineering is the fundamental basis for the modern biotechnology. Genetic engineering involves the ability to transfer specific genes from one kind of organism to another thereby producing new biological properties.

Many types of bacteria possess small fragments of deoxyribonucleic acid (DNA) which are called plasmids. These plasmids exist in the cell cytoplasm and are not associated with the cell chromosome or nucleus. The plasmids generally contain genetic information for a relatively few cell properties. The plasmids are generally responsible for the ability of bacteria to become resistant to antibiotics. The plasmids are capable of being transmitted rapidly to bacterial populations.

Plasmids have been successfully isolated from one kind of bacterium and then inserted into a different kind of bacterium. The cell receiving the plasmids in this way often acquires properties from the donor cell due to the plasmid–borne genetic information. In order to engineer a cell genetically, the plasmid molecule is cut by using a specific enzyme called restriction endonuclease. Then, the genetic material of another cell containing information for formation of the desired product is cut into fragments by the endonuclease. The plasmid and genefragment molecules so produced are mixed together and treated with an enzyme that joins fragments of DNA together. This results in re-formation of the plasmids containing pieces of the genetic information of the cell. This recombinant plasmid is then introduced into a bacterium, usually the one from which it was originally isolated. The bacteria are now allowed to grow and multiply. Whether the desired property is being expressed in these bacteria or not, is analytically ascertained.

Genetic engineering offers unlimited possibilities for the creation of new kinds of bacteria tailored to perform some useful functions. For instance, the gene for producing insulin has been transferred from animal cells into the bacterium 'Escherichia coli' by the above technique which enabled commercial scale manufacture of insulin. The amino acids phenylalanine and aspartic acid that constitute the sugar substitute "aspartame" are produced by genetically engineered bacteria. Bio-technology has made possible the large-scale production of anticancer and antiviral agents. Similarly, some important hormones, which are present in the human body, in infinitesimally minute quantities can be now produced on a commercial scale by using genetically engineered bacteria.

## BIOINFORMATICS

Bio-technology, which was once considered as the science of the future has now become the technology of the present. During the last decade, voluminous amount of nucleic acid sequence data and protein sequence data has been collected, the management of which would have been impossible but for the computers. The various biomolecular sequence data-bases are available in India with the Bio-technology Information System (BTIS) which has been established by the Dept. of Bio-technology, Govt. of India. Data-bases on nucleic acid sequences, protein sequences, molecular structure, genetic information, restriction enzymes, vectors, microbial strains and cell lines, genetic maps and probes etc. are available to support further fundamental and applied research. Even databases describing other databases are available.

## BIO-TECHNOLOGY AND POLLUTION CONTROL

Bio-technology comprises of integrated application of theoretical and practical knowledge of biochemistry, microbiology, physiology, genetics and chemical engineering to exploit the properties of microbes and cell cultures for various beneficial technological uses. Environmental biotechnology is a very broad field which includes environmental monitoring and safety, waste treatment and recovery, restoration of environmental quality, substitution of non-renewable resource-base with renewable resources, research and development of various processes for the benefit of mankind with due regard to socio-economic, legal and environmental safety considerations. Environmental Bio-technology has a great potential in transformation and enhancement of resource-base, reduction of energy consumption, pollution abatement, enhancement of biomass production and conservation of non-renewable resources, generation of useful organic chemicals from bioconversion of biomass, etc. Some of these applications have indirect beneficial effects on wider problems such as global warming, acid rain, and climatic changes. The most important feature of bio-technology is that it is not capital intensive and is within the reach of the less privileged countries which have vast natural resources and abundant man power. However, uncontrolled or accidental release of genetically engineered micro-organisms (GEMs) or their hosts derived from recombinant deoxyribonucleic acids (DNA) technology into natural environment can be hazardous to natural ecosystems. Hence, due care should be taken regarding environmental safety and legal aspects with the help of legislative measures. At the same time, due care should also be taken to see that over-regulations do not discourage scientific and technological innovations on one hand and also to see on the other hand, that over-enthusiasm does not vitiate the legitimate concerns of environmental safety and social harmony.

Biological processes have been in vogue for over a century for the treatment of municipal and industrial wastes rich in organic matter. Anaerobic processes are widely used to convert concentrated sludges into biogas rich in methane. Biological processes for treatment of a variety of waste waters are extensively in use today. Genetically engineered micro-organisms (GEMs) are coming into vogue for the treatment of toxic and/or concentrated wastes and other refractory wastes such as those containing Xenobiotic compounds.

For removing specific types of pollutants selectively, new types of proteins, peptides and biomimetic adsorbers are being developed using biotechnology. Similarly, new types of micro-biosensors and control systems are being developed for the continuous monitoring and automation of water treatment processes.

Bio-processing of coal is mainly targeted on the following aspects :

(1) Beneficiation of coal by removing undesirable components *e.g.* Sulphur, Nitrogen and trace metals.

(2) Coal conversion processes such as microbial solubilization, gasification, liquefaction, etc.

Important applications of biotechnology in pollution control and waste management include :

(*a*) Improvement of existing processes

**Ex.:** Use of bio-sciences and biochemical engineering to obtain knowledge about the application of mixed cultures such as (*i*) improvements in sewage treatment, (*ii*) use of starter cultures for treatment processes, (*iii*) use of immobilised microbial cells in wastewater treatment.

(*b*) Treatment of toxic wastes using genetically improved organisms (GEMs)

**Ex.:** Genes which can bring about degradation of toxic environmental pollutants (*e.g.* toluene, chloro-organics, halogenated pesticides, etc.) have been identified. For degradation of toxic chemicals, enzymes are encoded by specific genes present on plasmids. However, one plasmid cannot degrade all the toxic compounds of different groups. The plasmids are grouped into four categories as follows :

(*i*) OCT plasmid which degrades oxane, hexane and decane.

(*ii*) XYL plasmid which can degrade xylenes and toluenes.

(*iii*) CAM plasmid which can decompose camphor.

(*iv*) NAH plasmid which can degrade naphthalene.

Dr. Anand Mohan Chakraborty (an Indian borne American Scientist) was successful in synthesizing an oil eating superbug by introducing plasmids from the four different strains as above by successive conjugation into a single cell of "P Putida". This superbug can degrade all the four types of substrates (groups of chemicals stated above) which otherwise would have required all the four separate plasmids. The superbug was first produced on a large scale in the laboratory mixed with straw and dried before storage. In 1990, these microbes were used for cleaning up of an oil spill in water in Texas (U.S.A.). When the straw containing the superbug microbes is spread over oil slicks, the straw gets soaked with the oil and the bacteria degrades the various chemical compounds constituting the oil into harmless products. Thus, the oil spill is cleaned up.

(*c*) Recovery of useful products from waste materials.

**Ex.:** Recovery of methane, metals etc. from waste materials.

(*d*) Development of new catalysts, new bioreactors, novel biosensors and automation of waste water treatment processes.

**Ex.:** (*i*) Use of immobilized organisms and heavy metals as new catalysts and bioreactors. (*ii*) use of traces of toxic organics as new sensors and (*iii*) design of cadmium-binding synthetic adsorbents which imitate cysteine-rich cadmium-binding peptides synthesized by plant and yeast cells which respond to cadmium stress. Development of such biosensors can help to automate the adsorber systems by detecting the saturation of the adsorber system and trigger a controller to divert the flow of the influent to a second adsorber system while regenerating the first adsorber.

## BIOREMEDIATION

Industrial operations generate complex and toxic pollutants which accumulate in natural systems. The toxic wastes thus generated are usually treated by physico-chemical processes. Bioremediation is the microbial clean-up approach. It employs biological agents to render hazardous wastes to non-hazardous or less hazardous wastes. Microbes can acclimatize themselves to toxic wastes and new resistant strains develop naturally. Such strains can be used for pollution control and environmental protection.

Bioremediation engineers from the Bio Trol Company in the USA used Flavobacterium to remove pentachlorophenol from contaminated soil. Aeration of the soil helped the process.

American microbiologists found microbes referred to as GS-15 which can metabolise uranium directly. The micro-organisms were used in a bio-reactor to treat the uranium-containing waste water from a nuclear weapon manufacturing plant. The organisms can be used for waste water treatment in uranium mining.

Scientists from Southampton University in England developed micro-organisms which when grown with added nutrients in the presence of metal-laden waste water were found to accumulate the metal ions in their coats, thereby reducing the metal ion concentration in the waste water.

Abandoned cars in Junk-yards (which run in millions) are posing disposal problem in Germany. The engines breakdown easily but their plastic shells made of duroplastics (not thermoplastics) cannot be destroyed easily. Burning them would emit toxic gases. Micro-organisms which will digest the resin constituent of duroplastics are being developed.

Various strains of bacteria and fungi are being experimented to treat pollutants such as Vinyl Chloride, PCBs, Oil Spills, etc.

Some fungi, even as dead biomass, were found to trap metallic cations in solution due to their peculiar cell wall composition. Many fermentation industries produce fungal biomass of an unwanted by-product which can be utilized for this purpose. For instance, the biomass of fungi, "Rhizopus arrhizus" can absorb about 130mg of cd per gram of the dry biomass. The Biorecovery Systems Company patented a product called Algasore for absorbing heavy metal ions from ground water or waste water. Algasorb is prepared by trapping dead algae in sillica gel. The trapping of the algal cells as above protects the cells from being destroyed by other microorganisms. Algasorb functions like a commercial ion-exchange resin and the sorbed heavy metal ions can be stripped out easily.

Some algae and bacteria can accumulate large quantities of metals. For example, pseudomonas aerugenosa can accumulate uranium and thiobacillus can accumulate silver.

Mixtures of microbes and enzymes are used to clean up chemical wastes such as detergents, pesticides, dioxin, paper mill wastes, oil, etc.

Microbial process was used for removing sulphur from coal. Sulphur present in coal used in thermal power Plants is known to contribute to acid rain.

White rot fungus phanerochaete chrysosporium, when placed in nitrogen-deficient medium, can degrade DDT, dioxane, halogenated aromatic rings, etc. producing $CO_2$.

## BIOLOGICAL DE-ODOURISATION

De-odourisation processes include physical, chemical and biological. In biological De-odourisation, foul smelling compounds are decomposed by exploiting the metabolic process of the micro-organisms. An exhaust gas treatment system for $H_2S$ and $SO_2$ based on Thiobacillus ferroxidan bacteria is already in practice in Japan.

Bioscrubbing involves scrubbing of waste effluents using microbiological methods to detoxify or de-odourise certain constituents in the waste effluents. For example, detoxification of cyanide can be affected by using enzyme Rhodanese of Bacillus Sterae othermophilus by converting the toxic cyanide to thiocyanate. Hypomicrobium SPP can be used to oxidise the malodourous dimethylsulfide aerobically or anaerobically.

Aerobic processes :

$$H_2S + 2O_2 \longrightarrow H_2SO_4$$

$$(CH_3)_2S + SO_2 \longrightarrow H_2SO_4 + 2\ CO_2 + 2\ H_2O$$

Anaerobic processes :

$$5\ H_2S + 8NaNO_2 \longrightarrow H_2SO_4 + 4\ Na_2SO_4 + 4\ N_2 + 4\ H_2O$$

$$(CH_3)_2S + 4\ NaNO_2 \longrightarrow Na_2SO_4 + 2\ CO_2 + 2\ N_2 + 2\ NaOH + 2\ H_2O$$

Table 10.1 Summarizes some micro-organisms which can degrade some odorous substances.

**Table 10.1. Microbial degradation of some odorous substances and the degradation products**

| Sr. No. | Substrate | Micro-organisms | Degradation Products |
|---|---|---|---|
| 1. | Phenol | Pseudomonas Putida | Aldehyde and Pyrovate |
| | | Trichosporon Cutaneum | Acetyl CO A and Succinate |
| 2. | Dimethyl Sulphide | Hypomicrobium SPP | — |
| 3. | Pyridine, 4-methyl Pyridine | Pseudomonas SPP | — |
| 4. | Dimethylamines | Pseudomonas aminovorans | Methylamine and formaldehyde |
| 5. | Aniline | Nocardia SPP | Pyrocatechol |
| 6. | Benzaldehyde | Acetobactor ascendens | Benzylalcohol and Benzoic acid |
| 7. | Indole | Chromobacterium Violaceum | Pyrocatechol |
| | | Neurospora crassa | Tryptophan |
| 8. | n-propyl amine | Mycobacterium convolutum | Propionate |

## BIOLOGICAL PURIFICATION OF CONTAMINATED AIR

Biological purification of waste air from garbage composting plants, carcass processing facilities, sewage treatment plants, fermentation plants, foundries etc. are available now. Purification of power plant waste gases containing sulfur compounds and binding of malodorous waste gases are also in the offing.

## BIOSENSORS

Biosensors using enzymes, micro-organisms, and antibodies are used in clinical, genetic, immunological and research fields.

**Ex.:** BOD sensor, Nitrite Sensor, Ammonia Sensor, Sulfite ion Sensor.

A bio-sensor is an analytical gadget comprising of an immobilised layer of a biological material (*e.g.*, enzyme, antibody, hormone, nucleic acids or whole cells) in conjunction with a transducer which analyses the biological signals and converts them into an electrical signal.

A bio-sensor works on the following principle :

A layer of a suitable biological material is immobilised on a permeable membrane which is kept in the close vicinity of a sensor. The substance to be measured passes through the membrane and interact with the immobilised material and yield a product (*i.e.* the monitored substrate which may be heat, gas *e.g.* oxygen, electrons, hydrogen ions or the ammonium ions). The product passes through another membrane to the transducer. The transducer converts the product into an electrical signal which is amplified and displayed or recorded.

## CONCLUSION

Any advanced technology is often accompanied by benefits to the mankind, as well as risks to the environment. Some of the environmental problems can assume trans-national dimensions and also may cause social and environmental imbalance. Hence, for the efficacious and safe utilisation of biotechnology which is a multi-disciplinary field of far reaching consequences, it is essential to carefully select safe and useful genetically engineered micro-organisms for applications to specific waste treatment targets of special environmental significance.

## QUESTIONS

1. Discuss the various applications of bio-technology in pollution control.
2. What do you mean by bioremediation ? Discuss the role of bioremediation in environmental protection.
3. Write informative notes on any two of the following :
   (*a*) Biosensors and their applications
   (*b*) Biological deodourisation
   (*c*) Bioremediation.
4. Write short notes on :
   (*a*) Biological purification of contaminated air
   (*b*) Bio-informatics
   (*c*) Bio-sensors.

# Environmental Management

## INTRODUCTION

The irrepressible human curiosity and the unquenchable thirst for knowledge are the fundamental basis for scientific development. A major part of innovations in scientific and technological development has been directed towards generation or elevation of human comforts, thereby increasing the standard of living in the society. This led to increase in industrialisation. Some of the important improvements to our standard of living that can be attributed to the application of science and technology include :

(*a*) Production of more and better quality food.

(*b*) Elimination of many infectious diseases

(*c*) Invention of new faster communication systems

(*d*) Creation of reliable and faster transportation

(*e*) Supply of safe water

(*f*) Invention of machines to replace human and animal power.

(*g*) Minimising water-borne diseases through improved water technology.

(*h*) Mitigation of bad effects due to natural disasters *e.g.*, droughts, floods, volcanic eruptions, etc.

Consequent to these improvements, disturbing side effects such a environmental pollution, deforestations, urbanisation, loss of arable land, evolution of new organisms resistant to control, etc., have emerged. These effects are considered as potential threats to environment and to humans.

In agrarian society, people lived essentially in harmony with nature raising food, gathering firewood and making clothing and tools from the land. The wastes from animals and humans were returned to the soil as fertilizer. Hence there were no appreciable problems of air, water or land pollution. For the small settlements which grew up, the supply of water, food and other essential goods and the disposal of wastes had to be kept in balance with the changing community, but no serious environmental problems were created.

The cities of ancient times, especially those of the Roman Empire, had built systems for water supply and waste disposal. However, the municipal technologies of ancient cities seem to have been forgotten or ignored or neglected for many centuries by those who built cities throughout Europe. This resulted in the outbreaks of water-borne diseases like cholera, typhoid and dysentery. Until the middle of the 19th century, it was not realised that improper waste disposal polluted water supplies with the disease-carrying organisms. The industrial revolution in 19th century particularly in Britain, Europe and North America aggravated the environmental problems due to increased industrialisation and urbanisation. These 2 factors caused such levels of water pollution and air pollution which the cities of that time could not handle effectively. However, rapid advances in technology for the treatment of water and the partial treatment of wastewater took place in the developed countries, over the next few decades. This led to a considerable decrease in the incidence of water-borne diseases.

After the Second World War, the industrialized countries had an economic boom due to burgeoning population, advanced technology, and a rapid increase in energy consumption. These activities increased considerably from 1950 onwards thereby increasing the variety and quality of wastes discharged into the environment. New chemicals, including pesticides and insecticides, used without adequate data regarding their environmental and health effects caused and continued to cause, enormous environmental problems. Alarmingly, these problems are aggravating further since the types and amounts of such pollutants discharged to the environment are increasing inexorably as against the limited capacity of our water, air and land systems to assimilate these waste materials.

The financial crunch since 1970 forced changes in the priorities of many countries. Issues like unemployment inflation, energy, effects of globalization, resource crunch, high technology, war threats, social and political compulsions became major concerns. Thus it appears that concern about the public health and safety aspects of hazardous wastes will continue to increase for a long time. It is in this backdrop that development and implementation of effective environmental management strategies assume paramount importance.

The United Nations Conference on Human Environment was held during June 5-11, 1972 at the Swedish capital, Stockholm. This Stockholm conference was the first major international event which created global awareness about the environment by placing environmental concerns on the agenda of major international topics. The Stockholm declaration containing 26 principles reflects the range of environmental concerns of that period. The set of 109 recommendations of the Stockholm conference, widely considered to be the "magna carta" of environment, laid down the principles and action plan to control and regulate human environment. To commemorate this great event, June 5 of every year is rightly observed as the "World Environment Day". The historic Stockholm declaration called for the world community to bear a Golemn responsibility to protect and improve the environment for present and "future generation" and gave the concept of "eco-development".

The United Nations Environment Programme (UNEP) convened a meeting in 1982 to commemorate the 10th anniversary of the Stockholm conference, to assess the State of environment and changes in the perception of environment since 1972 conference. In 1983, the U.N. General Assembly appointed the "World Commission on Environment and Development". This commission, having 21 countries as its members, was entrusted with the task of formulation of a "global agenda for change" and to prepare an environmental perspective upto the year 2000. After deep study, the committee submitted its report entitled "our common future" in 1987. The committee recognized that the "sustainable development" should become a central guiding principle of the United Nations. Sustainable development implies "meeting the needs of the present without compromising the ability of the future generations to meet their own needs".

The United Nations Conference on Environment and Development (UNCED), known as the "Earth Summit" or "Eco-92", held during June 3-12, 1992 at Rio de Janeiro, Brazil, is considered as another historic event involving virtually all member states of the U.N. About 170 countries, 500 non-governmental organizations and about 2000 Journalists all over the globe participated. This conference identified the practical environmental and developmental challenges and opportunities and their inter-linkages upto the end of this century and even beyond. This conference also paved the way for creation of new partnership among the various countries. The "*earth charter*" prepared at the conference enunciates the principles, setting out rights and obligations of all nations in relation to environment and guidelines to nations in their quest for ecologically sustainable development. "*Agenda-21*" evolved at this conference provides a resume of global environment and a blue-print of action plan for solving environmental problems and the tools and resources desirable for implementing the action plan. Besides the "earth charter" and the "Agenda-21", the earth summit also came up with "*Convention of climate change*" and the "Convention on biodiversity".

The blame for deterioration of environmental quality cannot be laid only on industrial activities. Wastes from residential and commercial sources also cause environmental stress. Further, both affluence and poverty also cause pollution in their own ways. Natural calamities, illiteracy, and lack of environmental ethics are also partially responsible. Therefore, skillful and equitable environmental management often requires balancing of a number of factors some of which, at times, may be conflicting with each other.

## OBJECTIVES OF ENVIRONMENTAL MANAGEMENT

(*i*) Regulating the exploitation of natural resources.

(*ii*) Protecting environmental degradation and maintaining environmental quality.

(*iii*) Balancing the ecosystem.

(*iv*) Preserving the biological diversity.

(*v*) Regulation of exploitation of natural resources.

(*vi*) Adopting engineered technology without creating adverse effects on environment.

(*vii*) Formulation of suitable environmental laws and regulations and effective implementation of the same.

## COMPONENTS OF ENVIRONMENTAL MANAGEMENT

The major components of effective environmental management are :

(*i*) Control of atmospheric pollution and environmental degradation.

(*ii*) Adopting technologies which ensure sustainable development.

(*iii*) Conducting environmental impact assessment to review the existing technologies and making it mandatory for clearing major projects of environmental concern.

(*iv*) Institute environmental perception among people by conducting awareness programmes.

(*v*) Environmental education and training at schools, colleges and universities. The importance of environmental education was highlighted at the Environmental Education Conference (EEC) held at Belgrade in 1975 (called Belgrade Charter) by UNESCO and at Tblisi, U.S.S.R. in 1977 by the United Nations Education Programme (UNEP).

(*vi*) Controlling over-population.

(*vii*) Controlling over-consumption and craze by inculcating sublime human values such as service to society, non-material enrichment and spiritual solace.

## (A) ENVIRONMENTAL IMPACT ASSESSMENT (EIA)

"Environmental Impact Assessment (EIA) is an activity designed to identify and predict the impact on the biogeophysical environment and on human health and well-being of Legislative proposals, policies, programmes, projects, and operational procedures, and to interpret and communicate information about the impacts."

EIA is widely accepted as a tool in environmental management. It has been adopted in many countries with different degrees of enthusiasm and evolved to varying levels of sophistication.

## ENVIRONMENTAL IMPACT STATEMENT (EIS)

"EIS is a public document written in a format specified by authorized national, state, and/or Local agencies".

"*Environmental Inventory* is a description of the environment as it exists in an area where a particular proposed action is being considered".

## HISTORICAL BACKGROUND

Whenever a new development project is planned which is likely to affect environmental quality, it is useful to carry out Environmental Impact Assessment (EIA). In many jurisdictions, EIA is mandatory before according permission to proceed with development projects such as power plants, dams, smelters, petrochemical industries, paper industries, iron and steel mills, mining, oil exploration, flood-control systems, etc.

The first comprehensive environmental legislation in the United States came into force on 1st Janurary, 1970, in the form of National Environmental Policy Act (NEPA). The Act contained the following 3 main sections:

(1) Declaration of National Policy defining the environmental goals to be pursued by the Federal Government.

(2) Specification for the preparation of Environmental Impact Statement (EIS) on actions which significantly affect the environmental quality.

(3) Institutionalization of the Environmental Impact Assessment (EIA) in the Executive office of the president through the establishment of the council on Environmental Quality.

In Canada, the environmental assessment and review process started in 1973 for planning, desicion making and implementation of new projects, programmes and activities with potentially significant environment effects.

Today, the EIA process has been accepted in many industrialized countries including United States, Canada, Japan, Australia, Netherlands, Columbia, India, Thailand, Philippines and even in some African countries like Rwanda, Botswana and Sudan. In India, the Central Ministry of Environment and Forests issued a Notification on 27th January, 1994 making EIA Statutory for 29 specified activities falling under various sectors such as Industries, Mining, Irrigation, Power, Transport, Tourism, etc. This Notification was amended on 4th May, 1994, and the amended version includes a self-explanatory note detailing the procedure for obtaining environmental clearance, technical information, documents required to be submitted at the time of applying for environmental clearance from the Ministry of Environment and Forests (MOEF), so as to enable the submission of complete application for environmental appraisal.

The EIA process, when it started in the early 1970s, laid emphasis on evaluating the adverse/beneficial effects of development projects/activities on Physical factors, particularly those for which codes and standards are available (*e.g.*, air quality, water quality, solid waste disposal), and to incorporate suitable remedial measures at the project formulation stage. After a few years, EIA began to include biological and ecological factors, although they were difficult to quantify. Later on, the scope of EIA was further broadened to include socio-economic factors (*e.g.*, creation of employment opportunities, recreational factors, cultural impacts, etc.) In some cases, EIA process was used to evaluate class actions such as banning a particular pesticide or regulating the lead content of gasoline.

The EIA system has been welcomed, in principle, by many people. However, in practice, some sections of people feel that it is often too technical and it does not deal with the environment in a holistic way.

## ELEMENTS OF THE EIA PROCESS

The first step in the Environment Impact Assessment (EIA) process is to determine whether the project under consideration falls within the Jurisdiction of the relevant Acts/Regulations, and if so, whether it is likely to create a significant environmental disruption. If so, an EIA is undertaken and the Environmental Impact Statement (EIS) is prepared. In some countries, the EIS is open to public scrutiny and reviewed at public hearings. Eventually, a political decision is taken so as to whether the development project is (*a*) accepted or (*b*) accepted with amendments or (*c*) an alternative proposal is accepted or (*d*) rejected.

**Participants in EIA process:** The following persons/groups/agencies usually are involved in EIA process.

(1) Proponent : Government or Private Agency which initiates the project.

(2) Decision Maker : Designated individual or Group or Body.

(3) Assessor : Individual or Agency responsible for the preparation of EIS.

(4) Reviewer : Individual/Agency/Board entrusted with the responsibility for reviewing the EIS and assuring compliance with the relevant guidelines/regulations.

(5) Other Government Agencies having special interest in the project.

(6) Expert advisers.

(7) Media and Public at large.

(8) Special interest groups : Environmental Organisations, Professional Societies, Labour Union, Local Associations.

*Contents of EIS* : The EIS should contain the following information/ data:

(1) Description of proposed action and alternatives including that of no action:

It should include details of the construction phase, operation phase and the shut-down phase wherever applicable. Selection of alternatives to the proposed action *e.g.*, different ways of building and operating the project, alternative sites, etc.

(2) Estimation of the nature and magnitude of the likely environmental effects of the various alternatives proposed: This is mainly done under the following 3 broad categories.

(*a*) Physical factors (*e.g.*, possibility of earthquakes, possible effects on surface and groundwater quality, soil and air quality, etc.).

(*b*) Biological factors (*e.g.*, effects on vegetation, wild life, sport and commercial fish species, endangered species, etc.).

(*c*) Socio-economic factors (*e.g.*, economic, demographic, social values and attitudes).

(3) Identification of the relevant human concerns.

(4) Criteria to be used in measuring the significance of environmental changes including the relative weightages to be assigned in comparing different types of changes.

(5) Estimating the significance of the predicted environmental changes and thereby the impacts of the proposed action.

(6) Recommendations for acceptance of the Project/Remedial Action/Acceptance of one or more alternatives/rejection of the project.

(7) Recommendations regarding monitoring procedures to be followed during and after implementation of the project.

Determination as to which environmental changes (including even those which are not covered by laws/regulations) are relevant is critical to the validity and credibility of the EIA process. It is also important to limit the scope of the assessment in time, space, and number of factors in order to enable the usefulness of the EIS document.

When once the factors to be included in EIS have been determined, their future magnitudes must be predicted, though qualitatively. A weighting system may then be devised to enable comparison of different impacts.

One of the single approaches to provide a visual assessment of the effects and magnitudes of the selected factors in EIA is the Leopold matrix (Fig. 11.1). The steps for using the Leopold matrix are given below:

(*i*) Identify all such actions that are part of the proposed project and locate them across the top of the matrix (*e.g.*, *a*, *b*, *c*, vide Fig. 20).

(*ii*) List the relevant environmental characteristics or conditions down the side of the matrix. (*m*, *n*, .....).

(*iii*) Under each of the proposed actions (namely, *a*, *b*, ...), place a slash at the intersection with each item on the side of the matrix, if an impact is possible.

(*iv*) After completing the matrix, write, in the upper left-hand corner of each box with a slash, a number in the scale of 1 to 10, which indicates the *magnitude* of the possible impact. In the scale of 1 to 10 chosen above, number 10 represents the *greatest magnitude* of impact whereas number 1 represents the least. Before each number so placed, put + if the impact would be beneficial.

(*v*) In the lower-right-hand corner of each box with a slash, write a number, in the scale 1 to 10, which indicates the *importance* of the possible impact (*e.g.*, Regional vs. Local). Here also, number 10 represents the *greatest importance* of the impact, while number 1, represents the least importance.

(*vi*) The text that accompanies the matrix should provide discussion of the significant impacts, of those columns and rows with large numbers of boxes marked, and of those individual boxes which have larger numbers.

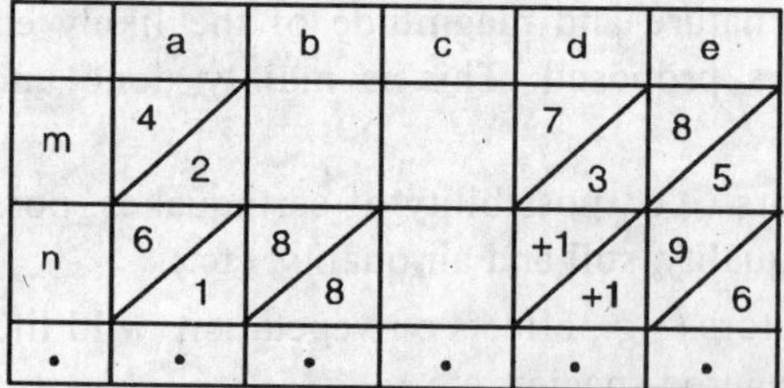


**Fig. 11.1.** Leopold matrix (Leopold *et al.* 1971)

## DESIGN OF EIA

There are several approaches for EIA documented in literature which should be consulted.

Further, it will be useful to study Environmental Impact Statements that have been prepared for similar developments/projects. Some of the important aspects that should be considered for the design of an EIA process are listed below :

**(A) Project Design and Construction**

(*i*) Type of project under consideration.

(*ii*) Physical dimensions of the area being considered.

(*iii*) Whether there is an irretrievable commitment of land ?

(*iv*) Whether there will be serious environmental disruptions during construction ?

(*v*) Whether the resources will be used optimally ?

(*vi*) Whether the project is a critical phase of a larger development ?

(*vii*) What are the long-term plans of the proponent ?

**(B) Project Operation**

(*i*) How will the hazardous waste products be handled ?

(*ii*) What provisions have been made for training the employees for environmental protection ?

(*iii*) What are the contingency plans developed to cope up with the possible accidents ?

(*iv*) What plans have been made for environmental monitoring ?

(*v*) What provisions have been made to check the safety equipment regularly ?

**(C) Site Characteristics**

(*i*) Whether the terrain is creating problems in predicting groundwater characteristics, air pollution, etc. ?

(*ii*) Whether the site is susceptible to floods, earthquakes and other natural disasters ?

(*iii*) How many people are likely to be displaced because of the project ?

(*iv*) Whether the local environment is conductive for the success of the project ?

(*v*) Whether the project will interfere with the movements of fish population and important migratory animals ?

(*vi*) What are the main attributes (*e.g.*, protein content, calorie content, weed or pest status, domesticity, carnivorousness, rarity of species, etc.) of the local fauna and flora ?

(*vii*) Whether any historic sites or traditional thoroughfares are likely to be endangered because of the project ?

**(D) Possible Environmental Impacts**

(*i*) What are the possible short-term and long-term environmental impacts for this type of projects during construction and after construction ?

(*ii*) Who would be effected because of these impacts ?

**(E) Socio-Economic Factors**

(*i*) Who are the expected Gainers and losers by the projects ?

(*ii*) Where are the expected trade-offs ?

(*iii*) Will the project interfere (blend, increase or reduce) the existing inequalities between occupational, ethnic, sex, and age groups ?

(*iv*) Will it effect the patterns of local/regional/national culture ?

**(F) Socio-Political Factors**

(*i*) What are the relevant political factors that have to be considered ?

(*ii*) What will be the difficulties in implementation of construction and operation phase of the project ?

(*iii*) What are the relevant governmental regulations and procedures ?

**(G) Availability of Information and Resources**

(*i*) Whether local and outside experts are available to consult regarding specific impacts of the project ?

(*ii*) Whether the relevant guidelines, technical information, and other publications are available to identify and deal with the possible impacts of similar projects.

(*iii*) Whether relevant environmental standards, by-laws etc. are considered ?

(*iv*) Whether the sources of relevant environmental data are identified and whether they are accessable ?

(*v*) Whether the views of the specialist groups and general public regarding the project have been ascertained and considered ?

(*vi*) Whether the competent technical manpower available to handle the project and the possible impacts ?

When once the magnitude and the significance of the impacts have been determined, the EIA may be considered to be essentially complete. Then follows the political decision as to whether the project be accepted, accepted with alteration or rejected.

The EIA is a potentially useful component of good environmental management. It represents a reconciliation of environmental and socio-economic factors with respect to the proposed development. However, some people still seem to have reservations about this system on the plea that the EIA is expensive, it delays the projects and the predictions are uncertain.

## ENVIRONMENTAL AUDIT FOR SUSTAINABLE DEVELOPMENT

Environmental Management is absolutely essential for sustainable development because it minimizes the environmental disturbances and ensures unhampered pace of industrial development and economic growth. Environmental audit is an important tool for environmental management because it enables the environmental Pollution Control Agencies to ensure the compliance with the environmental protection laws. It also motivates the mining and processing industries to demonstrate their concern and greater overall awareness towards their social obligation for environmental protection and to adapt eco-friendly technologies.

*Environment Audit* is an important management tool comprising of a systematic, periodic, objective and documented evaluation and assessment as to how well the environmental management systems are organised to facilitate control of environmental practices and how well the company policies are complying with regulatory requirements. *Environmental review, environmental surveillance* or *environmental assurance* are the other synonyms used for environmental audit.

Environmental auditing comprises of the following steps :

(*i*) Compiling of all relevant information on environmental management.

(*ii*) Evaluation of the information collected.

(*iii*) Formulation of conclusions and identification of areas that warrant improvement.

(*iv*) Active follow-up of the points raised and recommendations made to achieve the objectives of environmental management.

Environmental auditing may be practised in different organisations having different audit objectives. However the following interrelated characteristics are identified for an effective audit programme :

(*i*) Scope (*ii*) Organisation (*iii*) Objectives (*iv*) Resources (*v*) Approach and (*vi*) Coverage.

Environmental audit may have flexible methodology depending on the type of industry and its situation. Environmental audit should be preceded by well-defined pre-audit activities, on-site visits and post-visit activities, followed by preparation of report and suitable follow-up action plan.

Environmental audit is regarded as a voluntary responsibility of an industry or organisation. The following benefits will be accrued for an industry/organisation that practises environmental audit :

(1) It helps in assessing whether the existing environmental practices being followed are satisfactory and whether the environmental protection regulations are complied with.

(2) It provides an opportunity for comprehensive review of environmental policies, management systems, organisations and practices and to assess whether introduction of new innovative practices are necessary to comply with the stringent regulations from time to time.

(3) It protects against possible penalties, litigations or regulatory risks.

(4) It contributes its modest share towards sustainable development and gives due credit for environmental management to the Management.

(5) It provides an up-to-date environmental data base which may be useful in emergencies and also while making decision on plant modifications.

In India, environmental audit system has been introduced for the assessment of development projects on a regular basis. The system envisages submission of an Environmental Audit Report in a prescribed form for the financial year ending 31st March on or before 15th May starting from 1993 to the State Pollution Board by every person carrying on an industry, operation or process requiring approval under the water (Prevention and Control of Pollution) Act, 1974 or the Air (Prevention and Control of Pollution) Act, 1981 or both or an authorization under the Hazardous waste (Management and Handling) Rules, 1989. The report should clearly spell out the steps taken or proposed to be taken for adoption of clean technologies, waste minimization, recycling and reuse, pollution control investment in environmental protection and resource conservation measures taken.

Environmental audit is an effective tool for environmental management and sustainable development. The concept of sustainable development is based on

(*i*) Symbiotic relationship between consumer human race and producer natural systems, and

(*ii*) Compatibility between ecology and economy.

The principle of sustainable development is basically related to the carrying capacity of the ecosystem. The carrying capacity may be defined as the rate of maximum resource consumption and waste discharge into the environment that can be sustained indefinitely in a defined region with progressively impairing bioproductivity and ecological integrity. Thus, "carrying capacity" provides the physical limits to economic development.

United Nations World Commission on Environment and Development defined sustainable development as a new path of progress in which the needs and aspirations of the present generation can be met without compromising the ability of the future generations to meet their own needs and aspirations. Sustainable development is a process in which the exploitation of natural resources, the pattern of investments, and the institutional changes are all made keeping in view of the present as well as future needs of the human race on our planet.

In order to minimize the cost of environmental protection while maintaining the cost of natural resource exploitation within acceptable limits, it is essential to make structural economic changes to raise the levels of ecological and economic efficiency. This involves the following steps :

(1) Resorting to eco-friendly technologies which conserve natural resources, generate least pollution, promote sustainable industrial growth and provide direct economic benefits to society.

(2) Pursuing environmentally compatible areas of economy.

Some examples of such structural changes in different facets of human activity are as follows:

(1) *Energy Sector*

(*a*) Rational use of energy

(*b*) Conservation of energy and (*c*) harnessing renewable sources of energy.

(2) *Agriculture Sector*

(*a*) Development of land-use plans that are compatible with the ecosystem.

(*b*) Eco-cultivation.

(*c*) Harnessing modern development in biotechnology and genetic engineering in agricultural practices.

(*d*) Use of organic manures, and

(*e*) Use of biopesticides.

(3) *Transport Sector*

(*a*) Providing efficient public transport system.

(*b*) Replacing diesel and petrol with LPG.

(*c*) Reduction in the specific energy consumption of automobiles.

(*d*) Convincing people to use private vehicles selectively for long distances and for emergencies.

(*e*) Encouraging sharing of vehicles among people going to same or nearby destinations.

(*f*) Proper maintenance of vehicles for high efficiency and less pollution.

(4) *Manufacturing Industries*

(*a*) Utilizing highly efficient and eco-friendly technologies with built in waste treatment systems.

(*b*) Recycling and Reuse of raw materials.

(5) *Construction Sector*

(*a*) Use of environmentally compatible building materials.

(*b*) Efficient use of land, water and energy.

(*c*) Use of renewable energy sources.

(*d*) Replacing energy-intensive design with labour intensive designs.

Apart from the above, other measure such as imparting environmental education and environmental consciousness among masses and inculcating the values of non-material enrichment and spiritual solace among people, go a long way in our endeavours to achieve sustainable development.

## ECO-LABELLING OF ENVIRONMENT FRIENDLY-PRODUCTS

The scheme of eco-labelling of environment-friendly products provides accreditation and labelling for consumer products which meet certain environmental criteria along with quality requirements of the Indian standards for that product. This scheme was introduced by the

Ministry of Environment and Forests by a resolution dated 20th February, 1991, which was published in the gazette of Indian Extraordinary. According to this, an "environment-friendly product" is defined as any product which is made, used or disposed off in a way that significantly reduced the harm it would otherwise cause to the environment. Such products which meet the stipulated environmental and quality requirements are labelled with "Eco-mark" of the notified design. The major objectives of the scheme are as follows :

(*a*) To provide incentive for manufacturers and importers to minimise adverse environmental impact of products.

(*b*) To reward genuine initiatives by manufacturers to reduce adverse environmental impact of their products.

(*c*) To encourage people to purchase products having less harmful environmental impacts, thereby contributing to their environmental responsibility.

(*d*) To contribute towards improvement of quality of the environment and to encourage sustainable resource management.

The Resolution regarding Eco-labelling also provides for the following mechanism for implementation of the scheme :

(*i*) A Steering Committee of the Ministry of Environment and Forests to determine the category of products to be covered under this scheme and priority wise criteria to be adopted.

(*ii*) A Technical Committee constituted by the Central Pollution Control Board (CPCB) to identify the specific product to be selected and criteria to be adopted, including "inter se" priority between the criteria.

(*iii*) The Bureau of Indian Standards (BIS) would assess and certify the products and draw-up a contract with the manufacturers, allowing the use of the label, on payment of a fee.

The scheme also includes launching of countrywide mass awareness campaign to encourage consumers to purchase products having less harmful environmental impacts. The scheme also envisages assistance to consumer organisations for comparative testing of products and dissemination of information to the public. Obviously, the major purpose is to direct the market forces to bring about a change in the manufacturers to produce products that are environmentally more being.

Sixteen consumer product categories have been identified for this purpose in the first phase and the criteria for awarding Eco-mark to products like toilet soaps and detergents have already been notified.

## JUDICIAL RESPONSE FOR ENVIRONMENTAL PROTECTION

The judiciary in our country has also been exhibiting exemplary concern and appreciation towards environmental protection and ecological conservation. The Supreme Court of India, in its historical Judgement on 12th March, 1985, in a conflict between environmental conservation and industrial development, observed as follows :

"This is the first case of its kind in the country involving issues relating to environment and ecological balance and the questions arising for consideration are of grave movement and significance not only to the people residing in the Mussorie Hill range forming part of the Himalayas but also in their implications to the Welfare of the generally of people living in the country. It brings into sharp focus the conflict between development and conservation and serves to emphasize the need for reconciling the two in the larger interest of the country.

The consequence of this order made by us would be that most of time stone quarries would be thrown out of business. This would undoubtedly cause hardship to them, but it is a price that has to be paid for protecting and safeguarding ecological balance and for protecting their cattle, homes and agricultural land.

We are conscious that as a result of this order made by us, the workmen employed in the lime stone quarries will be thrown out of employment. We would, therefore, direct that immediate steps shall be taken for reclamation of the areas forming part of such lime stone quarries with the help of the already available Eco-task Force of the Department of Environment, Government of India and the workmen who are thrown out of employment in consequence of this order shall, as far as practicable and in the shortest possible time, be provided employment in the afforestation and soil conservation programmes to be taken up in this area".

In another case, the Rural Litigation and Entitlement Kendra, Dehradun brought to the notice of the Supreme Court that in Doon Valley, the mine owners and lessees were carrying on indiscriminate quarrying of limestone. The mining operations led to deforestation resulting in stripping the landscape base of its Verdant cover. The waste produced due to digging was carried down by rain-water to the villages and the agricultural lands located at lower levels. Some of the naturally formed streams were blocked. Blasting of the lime stone in the mines lossened the rocky structures by shaking the soil, thereby disturbing the entire ecology of the area. It was also apprehended that if the mining continued further, it may lead to dearth of water in the entire belt. The Supreme Court treated the letter rcccived from Rural Litigation and Entitlement Kendra. Dehradun as a Writ Petition. After hearing the case, the Supreme Court ordered that all fresh quarrying be stopped. It also directed that some of the quarries be closed down. The District Magistrate and the Superintendent of Police, Dehradun were directed to enforce the order.

In a landmark Judgement issued on July, 1987 (AIR 1987, AP 171), the Andhra Pradesh High Court observed as follows :

"The protection of Environment is not only the duty of the citizen but it is also the obligation of the State and all other State organs including courts. In this context, environment law has succeeded in unshackling man's right to life and personal liberty from the clutches of common law theory of individual ownership. The enjoyment of life and its attainment and fulfillment guaranteed by Article 21 of the Constitution embraces the protection and preservation of nature's gift without which life cannot be enjoyed. There can be no reason why practice of violent extinguishment of life alone should be regarded as violative of Article 21 of the Constitution. The slow poisoning by the polluted atmosphere caused by environmental pollution and spoilation should also be regarded as amounting to violation of Article 21 of the Constitution".

India's activist Judiciary has taken up the task of environmental protection. The Judiciary's concern for protection of the environment and its efforts to ensure prompt action by the Authorities are the remarkable features of Judicial activism in our country. In all such endeavours, the approach of the Judiciary is not that of agitating environmentalists, but that of an objective arbiter. This is how, the Judiciary is keeping a balance between environmental protection on one hand and development on the other.

**(B) Environmental Legislations in India**

Our country also demonstrated its concern for pollution control and environmental protection by enacting several legislations and constituting Statutory Bodies dedicated for this cause. A list of Environment related legislations is given in Table 11.1

## Table 11.1. Select list of Environment Related Legislations in India.

**A. Central Enactments**

1. *Air Pollution*
   (*a*) The Indian Boilers' Act, 1923
   (*b*) The Factories Act, 1948
   (*c*) The Mines and Minerals (Regulation and Development) Act, 1947
   (*d*) The Industries (Development and Regulation) Act, 1951
   (*e*) The Air (Prevention and Control of Pollution) Act, 1981, amended in 1987.
2. *Water Pollution*
   (*a*) The River Boards Act, 1956
   (*b*) The Merchant Shipping (Amendment) Act, 1970
   (*c*) The Water (Prevention and Control of Pollution) Act, 1974 amended in 1988
   (*d*) The Water (Prevention and Control of Pollution) cess Act, 1977 amended in 1991
3. *Pesticides*
   (*a*) The Factories Act, 1948
   (*b*) The Insecticides Act, 1968
   (*c*) The Poison Act, 1991
4. *Radiation*
   (*a*) The Atomic Energy Act, 1962
   (*b*) Radiation Protection Rules, Act, 1972
5. *Forests, Fisheries and Others*
   (*a*) The Indian Fisheries Act, 1897
   (*b*) The Indian Forest Act, 1927
   (*c*) The Prevention of Food Adulteration Act, 1954
   (*d*) The Ancient Monuments and Archaeological sites and Remains Act, 1958
   (*e*) The Wild Life (Protection) Act, 1972, amended in 1983, 1986 and 1991
   (*f*) The Urban Land (Ceiling & Regulation) Act, 1976
   (*g*) The Forest Conservation Act, 1980, amended in 1988
6. *Comprehensive Environment Protection*
   The Environment (Protection) Act, 1986

**B. State Enactments**

1. *Smoke Control*
   (*a*) The Bengal Smoke Nuisance Act, 1905
   (*b*) The Bombay Smoke Nuisance Act, 1912
   (*c*) The Gujarat Smoke Nuisance Act, 1963
2. *Water Pollution*
   (*a*) Orissa River Pollution Prevention Act, 1953
   (*b*) Maharashtra Prevention of Water Pollution Act, 1969
3. *Land Use*
   (*a*) The Bihar Waste Land (Reclamation, Cultivation and Improvement) Act, 1946
   (*b*) The Andhra Pradesh Improvement Scheme Act, 1949
   (*c*) The Acquisition of Land for Flood Control and Prevention of Erosion Act, 1955
   (*d*) The Delhi Restriction of uses of Land Act, 1964
4. *Pest Control*

(*a*) The Mysore Destructive Insects and Pests Act, 1917

(*b*) The Andhra Pradesh Agricultural Pests and Disease Act, 1919

(*c*) The U.P. Agricultural Disease and Pests Act, 1954

(*d*) The Assam Agricultural Pests and Disease Act, 1954

(*e*) The Kerala Agricultural Pests and Disease Act, 1958

## SALIENT FEATURES OF SOME IMPORTANT LAWS

### (1) The Wild Life Protection Act, 1972 (Amended in 1983, 1986 and 1991)

This was enacted for providing protection to wild animals and birds. The Act also provides for the constitution of a Wild Life Advisory Board, appointment of Chief Wild Life Warden, Wild Life Wardens and other employees by the State Governments for the protection of Wild Life. Regulation of hunting of wild animals and birds, laying down the procedures for declaring areas as sanctuaries, national parks and biosphere reserves, and regulation of trade in wild animals were also provided by the Act. List of endangered species, which is revised from time to time, is also included in the schedule of the Act.

As per the provisions of this Act, no one is permitted to hunt any wild animal, except Vermin, without a licence from the chief Wild Life Warden. A record of Wild Life animals hunted or captured has to be maintained. A special permit may be granted to hunt a wild life animal for education, scientific research, scientific management and collection of specimens for Zoological gardens, museums, etc. The Act provides for the establishment of sanctuaries, national parks, game reserves, and closed areas.

All wild life animals are the property of the Government. Trade or Commerce in wild animals and animal articles and trophies is strictly regulated. No person can cook or serve meat of wild animals in any eating house without a licence. Penalties for violating the provisions of the Act have also been laid down in the Act.

### (2) The Forest Conservation Act, 1980 (Amended in 1988)

As per this Act, no forest land or any portion thereof may be used for any non-forest purposes without the prior permission of the central Government. The Act has been amended in 1988 for incorporating more stringent penal provisions against violators of the Act. The scope of the definition of "non-forest purpose" was extended to include cultivation of tea, coffee, rubber, palms, oil-bearings plants, horticultural crops and medicinal plants. No State Government or other authority may issue order directing that any forest land or any portion thereof may be assigned by way of lease or otherwise to any private person or to any authority, corporation, agency or any other organisation not owned, managed or controlled by Government without prior approval of Central Government.

### (3) The Water (Prevention and Control of Pollution) Act, 1974 (Amended in 1988)

This Act provides for the prevention and control of water pollution and for maintaining or resorting of wholesomeness of water. The Act stipulates establishment of the central and State Boards for this purpose, and also stipulates how these Boards are to be constituted.

The Act defines terms like pollution, sewage effluent, trade effluent, stream and boards. The Act also assigns the functions to be carried out by the Central and State Boards.

The Water Boards have power to obtain information, to take samples of effluents from any industry/establishment and to make survey of any area and gauge and keep record of the flow or volume and other characteristics of any stream or well.

A person empowered by the Board has the right to enter, and inspect any place and examine any plant, record register, document or any other material object, or for conducting a search of any place where he has reason to believe that no offence of water pollution is committed. The Board has wide powers to prohibit the use of any stream or well for discharging any pollutant in it. The Board has powers to restructure the outlets for dumping pollutants.

The Act prohibits disposal of any poisonous, noxious or polluting mater or any matter causing obstruction to the proper flow of water in a stream. However, dumping of any material into a stream for the purpose of reclamation of land is not considered an offence.

The Act provides for severe and deterrent punishments for violation of the Act which includes fine and imprisonment.

### (4) The Water (Prevention and Control of Pollution) Cess Act, 1977 (Amended in 1991)

This Act empowers the Central Water Board to collect cess on water consumed by persons carrying on certain scheduled industries and by local Authorities responding for supplying water. The cess and the consent fees form the major sources of revenue to run the central and State Water Boards. The Act has been amended in 1991 with a view to augment the resources of the Boards by removing the lacunae in the Act and to provide rebate to the Industries for complying with the consumption and effluent quality standard.

### (5) Air (Prevention and Control of Pollution) Act, 1981 (Amended in 1987)

This Act was passed under Article 253 of the Constitution of India and in pursuance of decisions of Stockholm Conference. The objective of this Act is to provide for the prevention, control and abatement of air pollution in order to preserve the quality of air.

The Act defines relevant terms such as air pollution, air pollutant, automobile, industrial plant, etc.

Air pollution is defined as the "presence of any liquid or gaseous substances in the atmosphere in such concentration as may be or tend to be injurious to human beings or other living creatures or plants or property or environment".

All sources of pollution such as automobiles, diesel vehicles, industries, transport, railways and domestic fuels.

The Central and State Water Boards have been entrusted with the task of controlling and preventing air pollution and accordingly they have been redesignated as Central Pollution Control Board and State Pollution Control Board respectively. The functions of these Boards were clearly defined.

The Act provides the declaration of certain heavily polluted areas as "Air Pollution Control Area" and no industrial plant shall be operated in these areas without prior consent of the State Pollution Control Board.

The State Boards have to pay down and enforce standards for prevention and control of air pollution. The State Government in consultation with the respective Board may give instructions to the concerned Authority in charge of Registrations under the Motor Vehicles Act 1939, to ensure emission standards from automobiles. Failure to comply with the conditions prescribed for this purpose is punishable with fine and imprisonment.

The State Boards have powers to sue a polluter in a court of law to prevent him from polluting the air, and the expenses incurred by the Board for doing so will be recovered from the polluter. Further, the Boards have powers to authorize any person to enter and inspect the premises of the polluter and to collect samples of emissions from Chimneys, flues, ducts or any other outlets for analysis of the pollutants.

The Act has been amended comprehensively in 1987 to render it more effective and to include "noise" also under the definition of air pollutants.

### (6) The Environment (Protection) Act, 1986

The Environment (Protection) Act, 1986 was enacted as per the spirit of the Stockholm Conference held in June, 1972 to take appropriate steps for the protection and improvement to prevent hazards to human beings, living creatures and property. This is a landmark legislation to provide a single focus in the country for the protection of environment and to plug the loopholes in the earlier laws.

This Act ensures enforcement of several Acts/Regulations concerning pollution control and environmental protection/safety. This Act confers powers to the Central Government to take all such measures as it deems necessary or expedient for the purpose of protecting and improving the quality of the environment and preventing, controlling and abating environmental pollution. It empowers the Central Government to issue directions (*a*) for the closure, prohibition or regulation of any industry, operation and process and (*b*) for the stoppage or regulation of the supply of water, power or any other service, even without obtaining court orders.

The Act also empowers the Central Government to make rules for the first time for the (*i*) Standard of quality of air, water and soil for various areas and for various purposes (*ii*) Maximum permissible limits of concentration for various environmental pollutants (including noise) for different areas (*iii*) procedures and safeguards for handling of hazardous substances (*iv*) prohibition and restrictions on the location of industries and carrying out processes and operations in different areas (*v*) Procedures and safeguards for prevention of accidents which may cause environmental pollution and (*vi*) providing for remedial measures in case of accidents.

It is mandatory for persons carrying on any industry, operation etc. not to allow emission or discharge of environmental pollutants over and above the limits stipulated by the relevant standards. The Act provides for stringent penalties for defaulters. Any person can make a complaint of violation of provisions of the Act to the Central Government or Authority or Officer authorized for this purpose.

## (C) Environmental Management System (EMS) Standard – ISO 14000 Series

The preparatory work on the first international standards on environmental management systems was undertaken by the International Organisations for Standardisation (ISO). These norms come under the ISO 14000 series. The major objectives of these series is "to promote more effective and efficient environmental management in organisations and to provide useful and usable tools — ones that are cost-effective, system-based, flexible for gathering, interpreting and communicating environmentally relevant information". The intended end-result is the improvement of environmental performance.

ISO 14000 series of standards represents an opportunity for industrial organisations and enterprises in developing countries for technology transfer.

Further, it offers a source of guidance for introducing and adopting environmental management systems based on the best universal practices, in the same way that the ISO 9000 series on quality management systems, which is now widely applied, represents a tool for technology transfer of the best available quality management practices.

The subjects covered under the various ISO numbers are given below :

| ISO Number Range | Subject |
|---|---|
| ISO 1400 – 1409 | — Environmental Management Systems |
| ISO 14010 – 14019 | — Environmental Auditing |
| ISO 14020 – 14029 | — Environmental Labelling |
| ISO 14030 – 14039 | — Environmental Performance Evaluation |
| ISO 14040 – 14049 | — Life Cycle Assessment |
| ISO 14050 – 14059 | — Terms and Definitions |
| ISO 14060 | — Environmental Aspects in Product Standards |

The basic standard in this series is ISO 14001 Environmental Management Systems—specification with guidelines for use. It contains the requirements of a sound environmental management system, in the same way that ISO 9001 contains the requirements of a sound quality management system.

The ISO 14001 standard is applicable to any organisation (*e.g.*, a public or private enterprise, company, firm, institution or operational unit within an organisation) aiming at

(*a*) Implementing, maintaining and improving an environmental management system.

(*b*) Ensuring its conformance with its stated environmental policy.

(*c*) Demonstrating such conformance to others, either through an independent, third party certification or registrations of its environmental management system or a self-declaration or conformance with the standard.

ISO 9000 series on quality management system is focussed mainly on ensuring customer satisfaction and specifies 20 requirements in ISO 9001 to achieve this objective. Similarly, ISO 14000 series on environmental management systems addresses the needs of a wide cross-section of people in the society to protect the environment and for this purpose the following key requirements were specified under ISO 14000:

(1) Environment Policy

(2) Planning

  (2.1) Environmental aspects

  (2.2) Legal and other requirements

  (2.3) Objectives and targets

  (2.4) Environmental management programmes

(3) Implementation and Operation

  (3.1) Structure and Responsibility

  (3.2) Training, Awareness and Competence

  (3.3) Communication

  (3.4) Environmental Management System Documentation

  (3.5) Document Control

  (3.6) Operational Control

  (3.7) Emergency preparedness and response

(4) Checking and corrective action

  (4.1) Monitoring and Measurement

  (4.2) Non-conformance and corrective and preventive action

  (4.3) Records

  (4.4) Environment Management System Audit

(5) Management Reviwew

ISO 14001 exmplains each of the above requirements giving details of what should be done but not on how it should be done. for instance, under 2.3 (objectives and targets), it states as follows:

"The organisation shall establish and maintain documented environmental objectives and targets, at each relevant function and level within the organisation.

The objective and targets shall be consistent with the environmental policy, including the commitment to prevention of pollution.

When establishing and reviewing its objectives, an organisation shall consider the legal and other requirements, its significant environmental aspects, its technological options and its financial, operational and business requirements, and the views of interested parties:.

Adoption of ISO 14000 series of Environmental Management Standards by any organisation/ company is voluntary. However, exporters from developing countries are already subjected to pressure from buyers in industralized countries who are insisting on ISO 1400 certification for the products being exported. However, ISO 1400 should not be considered only as an export requirement but as a practical demonstration of environmental concern of an organisation. thus, the overall aim of the ISO 14000 environmental management standards is to support environmental protection in balance with socio-economic needs.

## OUESTIONS

1. Discuss the objectives of environmental management. What are the major components of environmental management?
2. discuss the advantages and limitations of Environmental Impact Assessment (EIA).
3. What are the important aspects in the design of an EIA Process?
4. discuss briefly the salient features of the Environment (Protection) Act, 1986.
5. What are the basic features of the water (Prevention and Control of Pollution) Act, 1974?
6. Discuss the basic features of the Air (Prevention and Control of Pollution) Act, 1981.
7. What ate the basic requirements for ISO 14001?

CHAPTER 12

# Social Issues and the Environment

Undoubtedly man is related with environment and he is solely dependent on nature. But since few years, over exploitation of natural resources disturbed the environment. Environment effects our life style, culture, social components etc. Technological development also have great impacts on social and natural resources. Development does not mean the increase in GNP(Gross National Product) but it should be visualized in a holistic manner, where it brings benefits to all. Though the natural environment undergoes continual change but man also produced some changes like domestification of animals, introduction of agriculture, development of industries etc. Increased population and higher consumption per head greatly effects the environment. The success of environmentally sound development depends on proper understanding of social needs, opportunities and of environmental characteristics

## FROM UNSUSTAINABLE TO SUSTAINABLE DEVELOPMENT

More and more natural resources were consumed in the process of satisfying the rapidly growing needs of the habitat. Every development activity has some impact on the environment. For meeting the needs, the human can not live without the developmental activities. Consequently, there is need to harmonize developmental activities in such a way that environment is polluted a least. Unsustainable development means the development of a few privileged nations both in science and technology. Such developments are at the cost of our life supporting systems like air, water, soil and over exploitation of our natural resources which may lead to the collapse of the interrelated systems of the earth. If growth continues in the same way, very soon we will be facing a "doom's day". (Meadow *et al.* 1972)

To be sustainable, development must process both economical and ecological sustainability. The concept of sustainable, development has received much recognition after the Stockholm Declaration 1972. The Brundtland (1987) has defined that sustainable development is development that meets the needs of the present without compromising the ability of the future generations to meet their own needs. The earth summit held at Rio-de-Janerio in 1992 put the world on the path of sustainable development which aim at meeting the needs of the present without compromising the ability of future generations to meet their own needs. The Rio declaration has taken cognizance of the fact that in order to achieve sustainable development, eradication of poverty is indispensable and thus development process and environment protection must go on simultaneously.

Freedom and sustainable development are mutually exclusive ideas. Freedom encourages people to do what they want to do, sustainable development dictates what people may and may not do. Freedom empowers people to control Govt., sustainable development empowers Govt. to control people.

There are two aspects of sustainable development :-

(*i*) *Inter-generational equity* – This emphasizes that we should stop over-exploitation of resources, reduce waste discharge and emissions and maintaining an ecological balance. It expects to hand over a safe healthy and resourceful environment to the future generations.

(*ii*) *Intra-generational equity* – This emphasizes that technological development should support economic growth of the poor countries so as to reduce the weather gas within and between the nations.

*Measures for sustainable development* – There are following major measures for sustainable development:-

1. *To promote environmental education and awareness* – From childhood, we should develop a feeling of belongingness to earth. This can be possible by introducing environment as a subject in education from primary stage. Media can also be helpful in developing such feelings. The transformations of policy making process from one that reflects and is controlled by the will of the people, to a process controlled by the selected elite who have found ways so impose policies fashioned by the international community to achieve what they believe to be sustainable development.
2. *Three 'R' approach* – Three 'R' means, Reduce, Reuse and Recycle. We should reduce the excessive use of natural resources, but use them again and again instead of passing it on to the waste stream. Recycle the materials to reduce pressure on our existing natural resources.
3. *Appropriate technology* – The technology should use less resources and produce minimum waste. It is over which locally adaptable, eco friendly, resource efficient and culturally suitable.
4. *To utilize resources as per carrying capacity of the environment* – Sustainability of a system depends largely upon the carrying capacity of the system. If carrying capacity of a system crossed, environmental degradation status and continues till it reaches a point of no return. Carrying capacity has two basic components-
   (i) *Supporting Capacity* – It is formed of productive and protective systems.
   (ii) *Assimilative Capacity* – It is formed of the systems which utilize the wastes produced by human activities.

The key to sustainable growth is not to less but to produce differently, offering solutions to a broad range of environmental problems. To achieve sustainability, economic growth can not be based on over exploitation of the resources but must be managed to enhance the resource base.

**Urban Problem Related to Energy**

Urban areas are developing very fast. In most of cities there is influx of populations from surrounding areas, mostly in search of employment and better living conditions. Therefore, it is difficult to accommodate all the industrial, commercial and residential facilities within limit. As a result, cities are spreading in to sub-urban or to rural areas. Uncontrolled population, irregular development are the main factor for receding facilities in urban areas. Energy is required in every walk of life like industry, transport, defence, agriculture, trade, education, domestic etc. Cities are the main centers of economic growth. Hence, energy is the most important input for development. The energy requirements of urban population are much higher than that of rural ones. Energy problems become more severe due to the limited amount of non-renewable resources of energy. Rapid utilization of fossil fuels produces increased production of wastes which causes environmental pollution. Energy problem day by day becoming serious. People are facing for 'power cut'. Energy demand is higher than production. There are following main causes of energy problems.

1. Increasing use of energy for domestic and commercial purposes (due to increased population and industrialization).
2. Industrial plants using big proportion of energy.
3. Non renewable resources of energy like coal, petroleum and natural gas are decreasing.
4. Increasing of transport means.
5. Decreasing production of Hydro electricity due to insufficient rains.
6. Transmission loss due to defected power distribution system.

There are following steps to solve the energy related problems.

1. To control urbanization.
2. To develop renewable resources of energy like solar radiation, wind power, hydel power, nuclear power, bio mass etc. These are pollution free also.
3. Non renewable energy resources should be used only when no non-conventional source of energy is available.
4. Welcoming the awareness programs to save energy.
5. Effective measures for transition loss and energy theft.

## WATER CONSERVATION

Water is needed in almost every sphere of human activity. Without water life is not possible. In many aspects the properties of water are unique. It is called universal solvent. No other liquid can replace it. The global distribution of fresh water on earth's crust including ground water and water present as it vapours in atmosphere.

### Table 12.1 Global distribution of fresh water

| | | *Water in Cubic KMS* |
|---|---|---|
| 1. | Water in snow caps, Ice sheets, Glaciers etc. | 24,000,000 |
| 2. | Surface ponds, Lakes and reservoirs | 280,000 |
| 3. | Water in Streams and rivers | 1,200 |
| 4. | Water present as soil moisture | 85,000 |
| 5. | Ground Water | 60,000,000 |
| | Total amount of fresh water on our planet | 84,366,200 |

Water is required for direct consumption or indirectly for washing, cleaning, cooling, transportation or even for waste disposal. Important sectors of human activity, which require water can be grouped as:-

1. Irrigation
2. Industries
3. Live stock management
4. Thermal power generation
5. Domestic requirements
6. Hydroelectric generation, fisheries navigation and recreational activities.

According to 1970 survey, about 3500 cubic kms of water are drawn for human use every year. Agriculture sector is the biggest consumer of fresh water. The amount of water drawn for human use is never used up completely. A large fraction is returned to the surface deposits or stream flows often in a polluted state. Water requirements have greatly increased due to rapid population growth, industrialization and agriculture. The demand for water is likely to exceed its supply by the first or second decade of the next country. The shortage of water shall make many localities barren, devoid of life. Fertile land become deserts. Conservation of water is, therefore an absolute necessity of today. Otherwise the tomorrow will be grion, drier and barren to live through. The following steps should be taken for conservation of water.

1. *Water economy, Re-use and Recycling.* If water meters are installed and charged properly, the consumption of water in domestic establishments, livestock management and industries shall drastically decline. The heated water from thermal power plants, where large amount of water is needed, may be utilized elsewhere after proper cooling. The same is true for many industries, water used once may be used again for another purposes.

2. *Agricultural runoffs from fields.* This can be used to irrigate cropland down the stream, while an efficient use of water with conditions of proper drainage can significantly reduce the agricultural runoffs.

3. *Efficient distribution system.* Water resources are not distributed evenly. Some localities have plenty of water and others have little. Many river basins have plenty of water, which flows down unused to the sea. Surplus of one basin can be used to make up the deficit at another.

4. *Enhancement of surface storage capacity.* About 27000 Cubic kms of fresh water which rush down to the oceans through stream and rivers are of no use to the mankind. We can store this water in tanks, reservoirs, dams for further use in drier seasons.

5. *Reduce evaporation losses.* Water losses through evaporation and seepage are enormous both from the reservoirs and distribution system. It should be reduced.

6. *Improvement of underground storage capacity.* The fresh water is stored in underground deposits. Every year about 10-15 % of the total precipitation enters the ground water table. These deposits regularly feed streams and rivers during the drier periods. These deposits are cheap and easily obtainable.

7. *Desalination of Sea Water.* A huge store of water exists in our oceans. If the salt content of the sea water is removed, we can use it. This can be done by desalination plants.

8. Afforestation and Reforestation of hill slopes to check loss of water in floods.

9. Artificial rain making and precaution of water pollution.

## RAINWATER HARVESTING

Water is an essential natural resource for sustaining life and environment. The available water resources are under tremendous pressure due to the increased demands and time is not far when water, which we have always thought to be available in abundance and free gift of nature, will become a scarce commodity. Conservation and preservation of water resources are urgently required to be done. Water management has always been practiced in our communities since ancient times, but today this has to be done on priority basis. The ministry of Water resources in India is endeavoring to make **rain water harvesting** a part of every day life in our villages and cities as a people's movement, and this will go a long way in the management of ground water as a sustainable resource.

While we have taken impressive strides in advance management of water resources, we also need to learn from the wisdom bestowed by our forefathers. One such wisdom is the concept of rain water harvesting. It is high time that the Government and the people join hands for creating awareness of the importance of rain water harvesting with the main objective of adopting these measures and techniques throughout the country. A judicious mix of ancient knowledge, a moderm technology, public and private investment, and above all, people's participation will go a long way in reviving and strengthening water-harvesting practices throughout the nation.

Rainwater harvesting is control or utilization of rainwater close to the point rain reaches earth. It is categorized into domestic rain water harvesting and rain water harvesting for agriculture, erosion control, flood control and aquifer replenishment.

Domestic rain water harvesting, also known as roof water harvesting or roof top rain water harvesting is the technique through which rain water is captured from roof catchments and stored in tanks or reservoirs.

*Rain water harvesting systems, both small and large, consists of six basic components.*

(*a*) Catchment area/roof, the surface upon which rain falls.

(*b*) Gutters and downspouts, the transport channels from catchment surface to storage.

(*c*) Leaf screens and roof washers, which are systems that remove contaminants and debris.

(*d*) Cisterns or storages tanks, where rain water is stored.

(*e*) Water treatment, the filters and equipment as well as additives to settle, filter and disinfect.

The system involves collecting water that falls on zinc, asbestos or tile roof of house during rein storms and conveying it by an aluminum, PVC, wood or plastic drain or collector to a nearby covered storage unit. The rain water yield varies with the size and texture of the catchment area. A smoother, cleaner and more impervious roofing material contributes to better water quality and greater quantity. The most common container now a days is a metal drum or barrel with a capacity of about 200 liters, set up at the foot of the fall pipe. This is enough to supply a family for 5 to 7 days with 6 liters per person per day, with possibility of using more than one storage barrel.

It is interesting to note that annually replenishable resources are assessed as 432 billion cubic meters (BCM) and by adopting water harvesting, an additional 160 BCM shall be available for use. The ground water levels in some areas are falling at the rate of one metre per year and rising in some other areas at the same rate. ***The main causes of fall in ground water levels are:***

(*a*) Over exploitation or excessive pumping either locally or over large areas to meet increasing water demands.

(*b*) Non-availability of other sources of water. Therefore, sole dependence is on ground water.

(*c*) Unreliability of municipal water supplies both in terms of quality and quantity, driving people to their own sources.

(*d*) Misuse of ancient means of water conservation like village ponds, baolis, percolation tanks and therefore, higher pressure of ground water development.

*The main effects of over exploitation of ground water resources are:*

(*a*) Drastic fall in ground water levels in some areas.

(*b*) Drying up of the wells/bore wells.

(*c*) Enhanced use of energy.

(*d*) Deterioration in ground water quality.

(*e*) Ingress of sea water in coastal areas.

You can capture and recharge 65,000 litres of rain water in Delhi from a 100 sq. m. size roof top and meet drinking and domestic water requirements of a family of four for 160 days. ***The method and technique include:***

(*a*) Roof top rain water harvesting and its recharge to underground through existing wells or bore wells or by constructing new wells, bore wells, shafts, spreading basins, storm water drains etc.

(*b*) Harnesting runoff in the catchments by constructing structures such as gabions, checkdams, bhandaras, percolation trenches, sub-surface dykes etc.

(*c*) Impounding surplus runoff in the village catchment and water sheds in village ponds and percolation tanks.

(*d*) Recharging treated urban and industrial effluents underground by using it for direct irrigation or through recharge ponds or wells etc.

*The main objectives of rain water harvesting are :*

(*a*) To restore supplies from the aquifers depleted due to over exploitation.

(*b*) To improve supplies from aquifers lacking adequate recharge.

(*c*) To store excess water for use at subsequent times.

(*d*) To improve physical and chemical quality of ground water.

(*e*) To reduce storm water run off and soil erosion.

(*f*) To prevent salinity ingress in coastal areas.

(*g*) To increase hydrostatic pressure to prevent or stop land subsidence.

(*h*) To recycle urban and industrial waste waters etc.

(*i*) To rehabilitate the existing traditional water harvesting structures like village ponds, percolation tanks, baolis, tanks etc.

(*j*) To convert the traditional water harvesting structures into ground water recharge facilities with minor scientific modifications and redesigning.

(*k*) To use the existing defunct wells and bore wells after cleaning and also the operational wells as recharge structures.

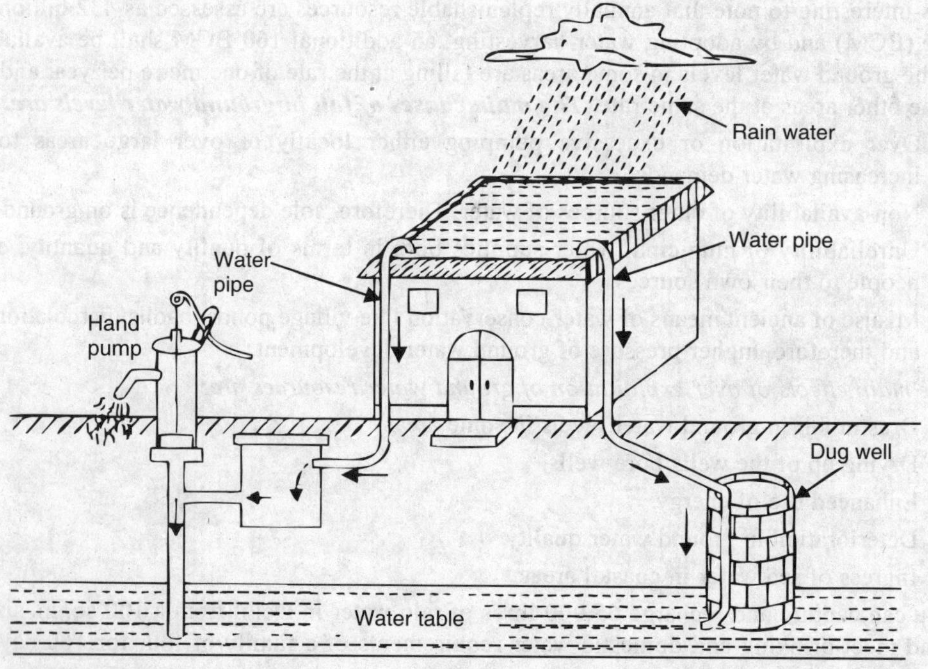


**Fig. 12.1.** Roof-top rainwater harvesting by recharging (*i*) through hand pump or (*ii*) through abandoned dugwell.

*The expected advantages of rain water harvesting are :*

(*a*) Rise in ground water levels in wells.

(*b*) Increased availability of water from wells.

(*c*) To prevent decline in water levels.

(*d*) Reduction in the use of energy for pumping water and consequently the costs. One metre of water level saves about 0.40 kwh assuming 10 hours of pumping per day for 365 days.

(*e*) Reduction in flood hazards and soil erosion.

(*f*) Improvement in water quality.

(*g*) Arresting sea water ingress.

(*h*) Assuring sustainability of the ground water abstraction sources and consequently the village and town water supply systems.

(*i*) Mitigating the effects of droughts and achieving drought proofing.

(*j*) Reviving the dying traditional water harvesting structures and their rehabilitation as recharge structures.

(*k*) Effective use of lakhs of defunct wells and tube wells as recharge structures.

(*l*) Upgrading the social and environmental status etc.

**Rain water harvesting** essentially means collecting rain water on the roof of buildings and storing it underground for later use. Not only rain water harvesting increases water availability, it also checks the declining water table. Every drop of water has to be saved and this will ensure that water is not wasted. Moreover, rain water harvesting is not only simple but economical too. The process of rain water harvesting is environmental friendly, helps improve ground water quality, helps meets increasing demand for water, particularly in urban areas and prevent flooding of roads.

In **Madhya Pradesh,** more than 1000 check dams, 1050 tanks and 1100 community lift irrigation schemes have been implemented in Jhabua district. The result-drought proofing achieved and food production increased by 38% in the past 5 years. Moreover, micro-watershed project with people's participation in Ghelhar Choti village, Jhabua district has been implemented. The result is that cultivated areas increased and yield per hectare doubled. Low cost small farm reservoirs along with improved crop and soil management systems tried in Chhattisgarh region of Madhya Pradesh and it augmented ground water storage, saved paddy from water stress during extended dry spells in 1990-91 and 1991-92. District administration in Dewas made roof top rain water harvesting mandatory for all houses having tube wells and banned tube well drilling. This resulted in improved soil moisture and recharged first-aquifer. This technique is also popular in other states like Maharashtra, Rajasthan, Andhra Pradesh etc.

## WATERSHED MANAGEMENT

Watershed is a drainage area on earth's surface from which runoff, resulting from precipitation flows past a single point in to a large stream, a river, a lake or the ocean. It is a geo-hydrological unit and drains at a common point, has been accepted world over as a scientific unit for area development. The watershed can range from a few square kilometer to few thousand square kilometer in size. Damodar Valley Corporation in 1949 adopted first Integrated Watershed Management. Watershed development is the rational utilization of natural resources of soil water and vegetation for increasing and stabilizing the productivity of land on a sustainable basis. The development of watershed will result in increase in sub soil water regime, recharge of wells. The watershed based development approach is undoubtedly an agreeable concept to set the goal. But this demands a massive people's movement to make the village community self reliant.

The watersheds are very often found to be degraded due to uncontrolled, unplanned and unscientific land use activities like over grazing, deforestation, mining, soil erosion, industrialization etc.

*Objectives of watershed management.* Watershed management is the rational use of land and water resources for optimum production causing minimum damage to the resources. The main objectives of watershed management are as :-

1. To increase agricultural production i.e. increasing the availability of fodder, fuel wood, timber and raw materials for industries.
2. The rational utilization of natural resources like water soil and vegetation.
3. To minimize the risks of floods, droughts and landslide.
4. To Manage the watershed for development activities like domestic water supply, irrigation, hydropower generation.
5. To develop the rural areas and their lifestyle.

Under the development of national policy, the watershed management was included in fifth Five Year Plan. Now a days, a number of national watershed development programmes are in progress. Various measures are necessary for watershed management. Some of them are :-

1. Scientific mining and quarrying must be done in the watershed areas because hills loose stability and get disturbed by improper mining.
2. Water harvesting in the watersheds to be used in dry season in low rainfall areas.
3. Afforestation and agro forestry (crop plantation) should be promoted to prevent runoffs loss and soil erosion and increase soil moisture. Woody trees like Eucalyptus and Lencaena should be grown in between crops to reduce the runoffs and loss of fertile soil in high rainfall areas.
4. Some mechanical measures like terracing, bunding, bench terracing, contour cropping etc. are used to minimize runoff and soil erosion in the sloppy regions of watersheds.
5. To promote soil binding plants like Vitex.
6. People's participation should be ensured including farmers and tribals in the water shed management programmes. This can be done by properly educating people about the campaign or paying some incentives.

The Himalayas are one of the most critical watersheds in the world. Most of the watersheds of our country lie in this region. Successful watershed management has done at Sukhomajri and Panchkula with the active participation of the local people.

**Resettlement and rehabilitation of people: Its problems and concerns**

Some times for the development of projects like construction of dams, mining, creation of parks etc. and during natural calamities like Earthquake, Landslides, Volcanos, Floods, Droughts, Cyclones, the problems of resettlement and rehabilitation arise. For example recently the Tsunami cyclone affected thousands of families and during construction of Indira Sagar dam in Khandawa district of Madhya Pradesh thousands families were displaced and rehabilitated near Chhanera and other places. This caused permanent loss of the benefits and facilities. This disturbed Socio-economic and ecological base of local community which are generally forest and tribal people. Families are disintegrated and also lost ancestral link between people and the environment. This can easily be seen in old HARSUD where people still go to remember old environment. More than 13000 families are displaced during Tehri dam construction. They rehabilitated in Dehradun and Haridwar regions. Various types of projects result in the displacement of native people are :-

1. *Displacement due to Dams.* Universe without energy is not imaginable. The most easily accessible and eco-friendly form of renewable energy is hydropower. Water is scarce natural resource and India is blessed with it. Hence it has to judiciously harnessed and managed for welfare of all living beings. India's exploitable hydropower potential is 84044 MW. River

Narmada has 3000 MW hydropower potential in the state of Madhya Pradesh. Sardar Sarovar Project (Gujarat), Hirakund (Orissa), Bhakra Nangal Dam (Punjab) Tehri Dam (Uttaranchal), Indira Sagar project (MP) etc. are some which displaced more than 25 million people.

**CASE STUDY**

**INDIRA SAGAR PROJECT (ISP)** - Indira Sagar Project (1000 MW) is in Khandwa district of MP. It is constructed, operate and maintain by Narmada Hydroelectric Development Corporation (NHDC). Narmada is bestowed with rich potential of 29 major, 135 medium and 3000 minor projects. The reservoir of ISP Dam is largest reservoir in India with storage capacity of 12.22 $BM^3$ of water for irrigation of 2.70 lakh Ha. A total of villages and people affected by this project. They were given plots, transportation grant, shifting facilities, agricultural land and other compensation. A separate township CHANERA was developed for displaced people. Under this project 26000 Ha area is proposed for irrigation, 564 villages to be benefited by irrigation. This will cause the production of 4.00 lakh tonnes of food grains and 10.55 lakh tonnes of other crops every year additionally. Other benefits of the projects are :-

(i) Pisciculture – 1500 tonnes of fish production every year.

(ii) Industrial development – Due to power and irrigation, development of new Industries in Nimar and Malwa region.

(iii) Supply of water to Thermal Power Plant – Water will be supplied to 2000 MW Thermal Power Plant proposed in village Bir.

(iv) Tourism – Resevoir of ISP spread over 913 Sq. km. area will be a boom for tourism development in Madhya Pradesh.

**TEHRI DAM** – Tehri Dam is on meeting point of River Bhagirathi and Bhilangana. It was controversial since 1991 i.e. from its construction. There was campaign against the project by Tehri people and noted Activist Sunder Lal Bahuguna, the propagator of CHIPKO MOVEMENT. The object of this dam is to produce power and provide irrigation facilities. Its length is 573 meter and 260.50 meter high. It is highest rock fill Dam of India. Its catchment capacity is 34.50 million $m^3$. Its hydroelectric capacity is 2400 MW. It provide 300 qu.sec drinking water to villages, towns of U.P and Delhi and irrigation facility to 2.70 m Ha. land. The project covered 467 sq. km. area of Tehri, which effected Tehri town and 112 villages. More than 1.25 m people were displaced. It is world's most controversial project. People displaced are called Eco-Refugee because project displaced them Tehri environment. They are rehabilitated in Dehradun and Hardwar. Will they adjust there? It is the question.

**SARDAR SAROVAR PROJECT** – It is situated in Bharuch district of Gujarat State on Narmada river. This project will submerge 245 villages of MP, Gujarat and Maharashtra and will displace more than one lakh people. It is also a controversial project. The project will also effect 56000 Ha agricultural and 60000 Ha forest land. It is multi-purpose project, and will fulfil the necessity of power drinking water and irrigation. This will result 900 crores of rupees additional in agriculture and 500 crores of rupees in water supply from this project.

(ii) *Displacement due to mining* – Mining is also one of important means in the field of prosperity. Mineral wealth has its own importance. Therefore, it covers large area for this purpose. Due to this developmental activity, thousands of people are displaced. People displaced due to this activity are poorest and mostly tribal. They lost much more than they get from the projects. Maximum persons lost their lives during settlement and transportation.

*Jharia Coal Field – a Case Study* – It is in Jharkhand. In this projects, underground fires are. Thousands of hectares land effected by this. People are asked to vacate the native place. Since 1976, about 115 crores of rupees have been spent to put out fires. The problem of resettlement and rehabilitation of the people is still there. Estimate for rehabilitation is much more than the availability.

(*iii*) *Displacement due to National Parks*. To conserve Flora and Fauna (natural resources), some times large forest area is covered under National parks and Sanctuary. It is declared as core area and entry of local dewellers, tribals and villagers (of nearby area) are prohibited. They are deprived of their right of access to the benefits of forests. Therefore, efforts should be made to provide proper rehabilitation and employment to the effected people.

**Case Study** – (*i*) Valmiki Tiger Reserve area in west Champaran districts displaced 142 villages of Tharu community tribals.

(*ii*) **Wayabanad Sanctuary** displaced about 53470 tribal families. Their rehabilitation is still incomplete. The Kerala people are fighting with officials of forest for their cause.

**Rehabilitation :** The United Nations Universal declaration on Human Rights (Article 25(1) has declared that "Right to housing is basic human right". This suggest better rehabilitation, adequate compensation, job opportunities, civic amenities and religious and cultural benefits. Therefore, National Rehabilitation Policy is needed to honour the human rights of the displaced people. Govt. under Land Acquisition Act 1894 has power to vacate the land from people by giving notice for Govt. use. Therefore, most of the displacements have resulted due to land acquisition by Govt. There is need of public awareness also in resettlement and rehabilitation plans. In general Govt. and other agencies provide a number of amenities for rehabilitated persons. Every landless person was provided 2 acres agriculture land, Rs 12-18000/- per acre cost of land to displaced person. Rs 20,000 or a plot for residence and 3-5000/- for transportation was also given with certain other compensations.

For displaced persons in case of Indira Sagar Dam, the following compensations for resettlement and rehabilitation were given :-

1. Developed plot or Rs 20,000 for purchase of plot for one family.
2. Rehabilitation grant of Rs 18700/- or 9350/- as per status of PAFs.
3. Transportation grant Rs 5000/- for shifting of one family.
4. Allotted 2-8 hectares of agriculture land per family or land compensation.
5. Attractive compensation for house, trees, wells and other structures.
6. At the plot sites, developed roads, water supply to lights, schools, health centres, worship places, panchayats, community centres, shops etc.
7. R and R work is being executed smoothly.
8. Additionally more benefits were given.
9. Professional training being given to project affected families at I + I Narmada Nagar.
10. Central school (Kendriya Vidyalaya started in June 01)
11. Different socio-economic upliftment programmes such as free medical check up, vaccination, training programmes are being organized.

## OZONE LAYER DEPLETION

**Stratosphere** – Troposphere is the part of atmosphere where humans live and other life processes also occur. The stratosphere is the region of space between approximately 15-50 kms above the earth's surface. The gas molecules in the stratosphere act as absorbing centres, moderating the transmission of the solar radiation to the earth. The qualitative as well as quantitative effect of this is an important determining factor with respect to life processes.

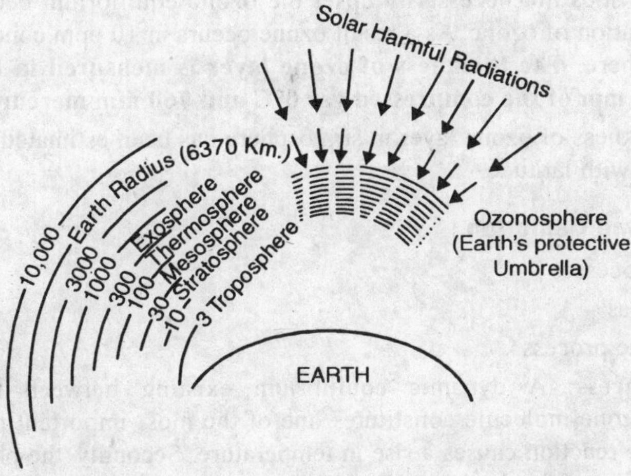


**Fig. 12.2.** Atmospheric ozone-earth's protective umbrella.

### Creation of Ozone Layer

Ozone is naturally occurring gas found through out atmosphere, with a maximum mixing ratio at the altitude ranging from 15-30 km above the earth. This region is known as ozone layer. Ozone can be toxic to plants and animals but increased concentration has a profound beneficial effect. Both, atmosphere and earth surface are subjected to radiation from sun. These certain radiations are absorbed by atmospheric gases leading to ionization or dissociation of gases. In the lower mesosphere, atmospheric oxygen gets dissociated and subsequently combines with molecular oxygen forming ozone in stratosphere.

The presence of ozone layer in stratosphere is of vital significance for all biota, because the harmful solar radiations such as ultraviolet rays, which are lethal to life on earth are not allowed to enter the earth's atmosphere by ozone layer (or ozone umbrella). In the absence of this ozone layer all the UV rays of sun will reach the earth surface to increase the temperature so that *biological furnace* will turn into *blast furnace*. **Thus the ozone layer strongly absorbs or blocks the short wave ionising ultraviolet rays and so protect the life on earth.** In stratosphere, ozone is an effective filter capable of absorbing UV radiations with wavelength between 200 nm and 315 nm.

### Formation of Ozone

In the lower mesosphere, the atmospheric oxygen absorbs UV radiation < 240 nm and photo-dissociates in to two oxygen atoms. These atoms subsequently combine with molecular oxygen of upper stratosphere producing ozone. Ozone is also capable of absorbing short wave length UV radiations releasing oxygen atoms.

$$O_2 + h\nu\ (\lambda < 240\ \text{nm}) \longrightarrow O + O$$

$$O + O_2 \longrightarrow O_3$$

Decomposition $$O_3 + h\nu\ (\lambda = 230 - 320\ \text{nm}) \longrightarrow O_2 + O$$

$$O + O_3 \longrightarrow O_2 + O_2$$

This mechanism does not necessarily upset the ozone equilibrium because ozone (loss) is compensated by creation of ozone. As a result ozone occurs in 10 ppm concentration in the form of layer in stratosphere. **The thickness of ozone layer is measured in Dobson Units (DU), where 1 DU = 0.01 mm of the compressed gas 0°C and 760 mm mercury pressure.**

The average thickness of ozone layer in stratosphere has been estimated to be about 230 DU. It varies marginally with latitude.

## Mechanism of Ozone Depletion

There are two processes –

1. Natural process
2. Anthropogenic process

**1. Natural process:** A dynamic equilibrium existing between the production and decomposition of ozone molecule constitutes one of the most important mechanism. The heat generated during the reaction causes a rise in temperature. Secondly, the photochemical process absorbs most of the harmful solar UV radiations.

Hence, the atmosphere is heated because of this absorption and the earth's biosphere is shielded from these lethal radiations. Without it, life on the earth would be completely destroyed. The ozone layer acts UV radiations from sun.

Ozone acts as powerful oxidising agent i.e. has ability to remove electrons from other molecules. Surface ozone is also produced by the action of UV radiations in sunlight involving both nitrogen oxides ($NO_x$) and volatile organic matter (VOM). Ozone also occur in air and water purification plants, oil wax, textiles and in synthetic industries.

**2. Anthropogenic process :** Some of the natural species moving in to stratosphere has been augmented in recent years by a number of human activities. Many of the processes, which are responsible for ozone layer depletion share a general mechanism of the type –

$$X + O_3 \longrightarrow XO + O_2$$

$$XO + O \longrightarrow X + O_2 \qquad \text{Where x = reacting species}$$

Net reaction $$O + O_3 \longrightarrow O_2 + O_2$$

The most of the common species (above x) have been identified to be free radicals like $HO_x$, $NO_x$ and $ClO_x$. Each of these species is capable of destroying the ozone layer. For example, near the stratosphere, the $HO_x$ radicals are responsible for about 70% of the total mechanism of ozone destruction, including the oxygen only process. Around 30 KM atmosphere, the $NO_x$ catalytic decomposition process dominates the ozone destruction.

## Effects of Ozone Depletion

(1) With the ozone layer depletion, there is danger of the increase in the flux of ultraviolet radiation over earth's biosphere. They are harmful for man's life.

(2) UV radiations effects biological systems in two ways – one is confined to patches of skin while the other develops in the immune system as a whole.

(3) These kinds of skin cancer, **Basal** cell carcinoma, squamous cell carcinoma and melanoma caused by UV rays.

(4) UV radiations cause sun burns, leukemia and breast cancer.

(5) UV radiations absorbed by cornea and lens in the eye leading to **Photokeratitia** and cataracts.

(6) Ozone at ground level (of low concentration) exerts its toxic effects directly on the lungs.

(7) Ozone exposure has been shown to be associated with lung cancer, DNA breakage.

(8) Photochemical smog is the measure cause of ozone exposure causing urban air pollution posing a threat to human health.

(9) Many micro – phytoplankton's would die because of their exposure to UV solar radiation. The marked reduction in the productivity of phytoplankton's would in turn adversely effect zoo planktons.

(10) The loss of fish population would directly effect the inhabitants of coastal areas.

(11) Ozone is reported to be highly toxic to fish.

(12) UV radiations increases the mortality rate of larvae of zoo plankton's.

(13) Ozone flecking is observed with the plants of grape, citrus and tobacco. At 00.02 ppm it damage tomato, pea, pine and other plants.

(14) Plant proteins are also susceptible to UV injury.

(15) In plants $O_3$ enters through stomata. It causes visible damage to leaves thereby reducing their photosynthetic rate.

(16) Due to ozone reduction, intense UV radiation causes greater evaporation of surface water through the stomata of the leaves and decreases the soil moisture content.

(17) Ozone reacts with many fibres, such as cotton, nylon and polyester, dyes etc.

## NUCLEAR ACCIDENTS AND HOLOCAUST

**Japanese towns of Hiroshima and Nagasaki.** The first atom bomb was exploded about 580 metres in the atmosphere over ill fated Hiroshima on August 6, 1945. The second atom bomb was detonated 507 metres high in air over Nagasaki. At least 100,000 people were reported killed, severely injured and missing in Hiroshima alone, where the bomb virtually demolished all structures and buildings in about 15 square km. area. In Nagasaki 49000 civilians were killed, injured and disappeared while an area of 6 to 7 km. was devastated.

The atom bomb exploded on Hiroshima used Uranium (U – 235) with a half life period of 8.5 $\times 10^8$ years, while the Nagasaki bomb had plutonium (Pu – 239) as an explosive man-made radio-nuclide with half life of 24,000 years.

Exhaustive studies conducted in Hiroshima show that heavily exposed hibakusha, bomb effected people. Have a 30% greater chance than normal of dying from cancer.

This lethal explosion caused millions to die, while thousands became unable to breathe, unable to see and eat. Public suffered from pulmonary oedema, anoxia, brain damage and increased risk of sterility. However, several cases reported high incidence of still births and congenital deformities among pregnant women. Excess cases of leukemia began appearing in 1948, but by the early 1970's the rate leveled off nearly to that of the general population.

**The first hydrogen bomb** was exploded in 1954 on Bikini Island in the Pacific. The radioactive fallout from this explosion severely affected the crew of a Japanese fishing boat, the Lucky Dragon about 150 km. away from the site of explosion. Several persons were hospitalized, killed and disappeared, while in Bikini Island the explosion caused the entire toll vanished.

In 1957 and 1958, the USA, Soviet Union and Great Britain detonated nuclear weapons whose total yield was about 85 megatons. These weapons were equal to 4250 Hiroshima sized atom bombs. They caused several dangerous effects on man.

In 1961, Russia detonated a bomb of 57 megatons that could obliterate a city more than three hundred times the size of Hiroshima.

## CASE STUDY

### Chernobyl Accident

**Chernobyl** was the first officially acknowledged nuclear accident in USSR and first reported to the world. April 26, 1986 was a sad day for nuclear power generation when a major accident occurred at 1.23 A.M. in the nuclear reactor at Chernobyl, in the Ukraine area of the Soviet Union. It resulted in clouds of radioactive smoke over a large area in Scandinavian countries which are 2000 km. away in the Russian region itself. There was a devastating fire in the reactor which caused few casualties and severe damage to the nuclear plant. On finding the fire uncontrollable the soviet authorities sought the help of West Germany and other nuclear nations to tackle the situation. Presumably, the core of the nuclear reactor had melted.

The explosion at the **Chernobyl power plant** in Soviet Ukraine, USSR confirmed the worst nuclear disaster. Poor design of the reactor magnified with operator negligence caused the havoc. The operators ignored warnings from various sensors and even disconnected the emergency core cooling systems.

*Neutrons went out of control and enormous steam built up in pipes. The explosion sent the graphite slabs of the reactor core through the roof, setting it a fire and spewing radioactive materials around the world.*

Twenty percent of the plant's radioactive iodine escaped along with 15 to 20% of radioactive caesium, hazardous plutonium and mixture of several radioisotopes. Radiation level reached 100 times than the normal. Samples of radioactive debris collected in Sweden indicated that some percentage of radiation in the reactor core actually escaped due to which air became frosty and the environment critically polluted. The USSR authorities claimed later that the fire had been controlled and put out.

The vast cloud of radiations caused considerable anxiety in Western Europe. The Soviet Union reported in a statement on April 29, 1986 that two persons were killed and a few injured. But the senior US official told that the death toll was probably much higher. On May 2nd, the Soviet authorities admitted that 18 persons affected by the leakage of Chronic radiation were in incident were credible. He made this remark after studying a data relayed by US spy satellite which over flew the Ukrainian area where the fire and radiation leak occurred.

A top Swedish nuclear official estimated that "thousands of people could have been killed" around the Chernobyl reactor, 100 km. north of the Ukrainian capital of Kiev in USSR, if all radiations had leaked into the atmosphere. He also reported that water supplies to the homes of 3.5 million persons of Kiev had probably contaminated with acute radionuclides.

Scientists have recently discovered that worms exposed to excess radiations have started copulating instead of reproducing asexually. The **Chernobyl nuclear disaster** has not only affected human being, but has even drastically changed the lives of worms. The finding is the clear evidence of how wildlife is affected by radioactive pollution. Although there is a wealth of knowledge on the impact of ionizing radiation on humans, the effects on wildlife have never been assessed. Many researchers are now focusing on how wildlife has been affected by the radioactive pollution that spewed from a reactor which exploded 17 years ago in Chernobyl. Researchers from Ukraine found that the worms from a lake near Chernobyl had received 20 times more radiation as compared to those from the lake situated far apart. They found some remarkable changes in the sexual habits of worms. According to the scientists, the worms have started sexually reproducing to protect themselves from radiation. Sexual reproduction allows natural selection of genes that offer better protection from radiation damage. A change in sexual behaviour may have increased the resistance of the worms.

The **Chernobyl power plant,** the world's worst nuclear accident best symbolizes the potential dangers of atomic energy, has been officially closed on December 15, 2000 – 14 years after the plant exploded and sent a cloud of radioactive dust over Europe. More than 4000 Ukrainians who took part in the hasty clean up effort since the 1986 disaster have died and 70,000 were disabled by radiation, according to government figures. About 3.4 million of Ukraine's 50 million people, including some 1.26 million children, have been affected.

## WASTELAND RECLAMATION

The area or land like salt affected, sandy, barren-hill-ridge snow covered or glacial which is economically unproductive suffer from environmental deterioration are called waste lands. These are for one reason or the other do not fulfill their life sustaining potential. Therefore, wasteland should be reclaimed and put to some productive use. About half of our country's geographical area is lying as wasteland. Maximum wasteland area in our country lie in Rajasthan, followed by Madhya Pradesh and Andhra Pradesh.

There are two modes of formation of wastelands :-

1. *Natural Process.* These include undulating uplands, snow covered lands, coastal saline areas, sandy areas etc.

2. *Anthropogenic* (*Man-made*) *activities*. These are deforestation, overgrazing, mining and erroneous agricultural practices.

Due to increasing population more land is required for agricultural and forestry. Opposite to it quality land is decreasing due to erosion, desertification, water logging, salinity, toxic effects of agrochemicals and industrial effluents. Now the only way to increase the land resources is by reclamation and developing degraded land. Govt. also planning and giving funding for conservation and regeneration of forest resources. Faulty irrigation specially canal, drainage practices, water logging are also main causes to form wasteland. Indira Gandhi Canal project in Rajasthan has changed water-starved wasteland to water-soaked wasteland.

Unscientific mining with no environmental protection have developed large tracts that lost productivity. In U.P, Bihar, M.P, Orissa, A.P, Punjab, a number of mining operations have started since long, which effects not only cultivated land but forest also. Ecological problems developed in coal mine areas in Ranchi, Bihar, Singrauli, M.P and other places due to high power and transmission lines, rails, road tracks and townships. Due to industrialization (Cement, Power, Auto etc.) and increased multinational companies in metros and other states, the land for cultivation is shrinking.

For reclamation of wasteland in 1985 National Wasteland Development Board (NWLDB) was established. The basic functions of this board, were to check the soil erosion, deforestation and plantation. But later on in 1989-90, the aims of board were re-examined and modified. Now the main objects of board are :-

1. To prevent soil erosion, landslides and flood.
2. To improve the quality, physical structure of marginal soils.
3. To increase the availability of Biomass.
4. To conserve the biological resources of the land for sustainable use.

Wasteland reclamation is development of degraded, mined and unused land for productive purpose. The following are some of the reclamation practices:-

1. *Soil Sodicity* : Sodicity in soil is due to excess of sodium. This can be reduced by use of Gypsom in soil. The calcium of gypsum will replace sodium by exchangeable process.

2. *Use of green – manures and Biofertilizers* – Dhaincha (Sesbania aculeate), Sun hemp (Corolania juncea) or guar etc. should be used as green manures to reduce salinity of soil. Farm yard manure and biofertilizers like blue green algae are also used to reduce salinity.
3. *Irrigation practices* – High frequency, thin and frequent irrigations have been found more useful. Surface irrigation with precise land leveling and smoothening also help to reduce water logging and salinity.
4. *Drainage* – Water logged soils are reclaimed by removing excess of water by artificial drainage. When water stands in fields after heavy rains, the ditches (30 – 40 cm deep at distances of 30 – 50 metres) are provided to runoff the excess of water. This is known as *surface–drainage.* But in *sub-surface drainage* method, perforated PVC pipes or open jointed pipes graved 2-3 metre below the land surface are used. This is better method.
5. *Leading and land development* – this is done in salt affected soil. After land levelling, apply excess of water, which pushes down excess of salts. This loss of extra salts by downward moving water is called leaching.
6. *Selection of tolerant crops* – Tolerance of crops to salt is found to range from sensitive, semi–tolerant, tolerant to highly tolerant. Date-palm, sugar barley are highly tolerant crops, which are not affected by high salinity. Soyabean, coconut, mustard, wheat etc. are salt tolerant. Rice, maize, pulses, sunflower, sugarcane, vegetables etc. are semi tolerant. The different combinations of crops should be grown in saline soil.

There are two successful cases of reclamation of mined area, one is Neyveli Lignite Corporation Ltd in Tamilnadu and other is Stone Quarries of Sayaji Iron Works in Gujarat. The NWLD Board and environment and forest ministry planned a programme for wasteland reclamation. In first phase they selected the five districts-Almorah, Purulia, Bellari, Dungarpur and Sundergarh. During 8$^{th}$ five year programme 13.75 crores of rupees sectioned for 234 projects to NGO's. In 1999-2000, 8 crores of rupees were sanctioned for 3000 hactare wasteland. With the help of retired military personals, Wasteland Development Task Force, 1994 in Morena district of M.P. was started to develop the 12000 hactare Chambal revines.

## CONSUMERISM AND WASTE PRODUCTS

Consumerism is the consumption of natural resources by the human beings. Exponential increase in consumption of natural resources are there now days as comparison to early period. This is due to increase in life style and population growth. Earlier much simpler life was but now in modern period needs and consumption of resources increased.

In different countries the consumerism of natural resources are not the same. On the basis of consumerism, population have two categories, one is people over population and consumption over population. This is nothing but population influence consumerism of natural resources and generation of wastes.

**(*a*) People over-population** – This occurs in less develop countries (LDC) where per capita consumption is less. Because as compared to resources population is high i.e. more people than available supply of food, water and other resources. India has 16% of the world population, hence per capita consumption is 3% waste generation has direct correlation with the rate of consumerism. In India wastes generated are about 3% very less than more developed countries (MDC).

**(*b*) Consumption over Population** – It occurs in more developed countries where population is less. Naturally the natural resources will be more as per population. Hence per capita consumption will be high. In other words more food, water and natural resources will be available, results very high wastes generation. The life style will be good as compared to less developed countries. USA, a more developed country has about 4.7% of world population and about 25% waste generation. About 400 million metric tonnes of industrial wastes (other than mining and mineral) and 180 million metric tonnes municipal wastes are in USA.

Thus tremendous amount of stuff thrown away. Much of the wastes can be reused or recycled in other useful products. But in practice easy to throw than environmentally responsible.

The relationship between population size, consumerism and environmental impact was explained by *Paul Ehrlich and John Hodlern (1972)* in following ways –

Environmental Impact = Population size × Per capita use of resources × Waste generated per unit of resource used

## ISSUES INVOLVED IN ENFORCEMENT OF ENVIRONMENTAL LEGISLATION

During the last hundred of years several laws have been enacted, which deals with environmental, eco-conservation and protection. Sixth plan documents emphasized the need to ensure conservation of environmental resources for sustainable development. Department of Environment was initiated in 1893. All this indicate that Government making conscious efforts to repair the damage and take a view to save the environment for future. But inspite of these, we are not able to achieve the target. It seems that there are some drawbacks in environmental legislations and problems in their effective implementation. In a number of countries the concern for disappearing wild life, deteriorating conditions of environment and over exploitation of resources led to the establishment of a body of laws. To these laws were added rules restricting practices which adversely affect human environment. It is fact that national action alone can not effectively serve the purpose. In third world countries, the national laws were environmentally blind. Environmental legislation in these countries was developed after the international legal framework had come into being.

Pity and compassion for animals can be seen in the common law of many countries which prohibit cruelty against animals. Animals are being used to pull over burdened carts or carriages without any regard to the capability or health of the animals. Such practices should be prohibited under the articles on Prevention of Cruelty against animals. There are provisions in law to restrict practices causing nuisance, and suffocating smoke in the atmosphere. But these provisions are rarely used for the purpose of protecting animals or restricting other environmentally damaging practices.

Till date the law of many countries did not possess provisions which could be effectively used to protect wild life, natural resources or the environment. This was simple because of the fact that the negative effects of the process of growth and development of a modern society had never been felt so deeply. The legal frame work, at national levels can be analysed in following categories –

1. **Legislation restricting hunting, killing or over exploitation of a species or a group of species.**

In addition to law made for wild elephants in 1873 AD and 1879 AD, the following important acts were framed –

(1) Madras Wild Elephant Preservation Act 1873.

(2) All India Elephant Preservation Act 1879.

(3) Wild Birds and Animals Preservation Act 1912.

(4) Bengal Rhinoceros Preservation Act 1932.

(5) Assam Rhinoceros Preservation Act 1954.

(6) Indian Wild Life (Protection) Act 1972.

These laws vested the state with the powers to act against those responsible for capture, injury or death of wild animals. Most of these laws were like a "Penel code" which specifies penalties for offences committed against wild life. *But the exploitation of wild animals and plants continued illegally or at times even under the state permit. Many Governments locked the will or adequate machinery to implement these laws. The lure of spectacular profits, vested interests and corruption often cause in the way of effective enforcement of these laws.*

**2. Legislation for the protection of wild life preservation of natural habitats and environment in general.**

In previous mentioned laws destruction of habitats, pollution of environment or factors which caused much more damage to wild life were ignored. Earlier some laws against ecologically damaging human activities were enacted. But it was the inconvenience caused to the general public which was responsible for their implementation. Some of the acts thus promulgated were –

(1) Bengal Smoke Nuisance Act 1905.

(2) Indian Ports Act 1907.

(3) The Motor Vehicle Act 1908.

(4) Maharashtra Prevention of Water Pollution Act 1953.

(5) Orissa River Pollution Act 1954.

(6) Gujarat smoke Nuisance Act 1963.

Most of these laws were ineffective due to Government inefficiency and apathy while human Society's obsession for economic growth and industrialization enfeabled public protests whatsoever. *In the light of population growth ecological issues are often ignored. Inspite of these Acts, ecologically human activities continued – after all there has to be a place for man also in the biosphere.*

**3. Legislation recognizing rights of people to a healthy environment and provision of compensation to people affected thereof.**

There is provision to fix responsibilities by Government for agencies responsible for ecologically damaging practices in laws but have no role for directly or indirectly affected people by damaged environment. Little help could be obtained from the judiciary as the law did not specifically recognize the right of the people to clean air for breathing, clean water for drinking or a healthy environment to live in. After amendment in Article 51-A, people could now go to the court to seek legal help to guard their rights concerning issues of the environment. The Indian judiciary on the other hand, also started assisting people in their effort to maintain a healthy environment and wild life by admitting litigations against state's continued neglect of ecological rights of its citizens. The principle that the "Polluter must pay" is on its way to become a reality now.

When matters concerning human rights are frontally posed, it becomes the duty of judiciary to face & resolve them. A bill called the National Environment Tribunal Bill 1992 was introduced in Parliament in August 1992 to decide the cases related with wild life and environment in addition to providing compensation to the people for death injury or damage to property or to the environment by constituting a high power body. Excessive centralization of powers and authority lindens efficient implementation. Various problems like the financial implications and lack of measurement data, appropriate instrumentation and technology creep in the path of enforcement of environmental legislation.

## Public Awareness

Environmental pollution, environmental degradation, environmental deterioration, environmental crisis etc. are few words which becoming day by day a subject of concern in every walk of life. This is all due to industrialization, rapid population growth, urbanization, changing life style etc. The formulation of various acts and legislations to control pollution and conserve or protect environment, underlines the will and concern of the Government. But incomplete knowledge, informations and ignorance about many aspects of environment has led to misconceptions. Therefore, it is necessary to make people aware about the laws and legislations and to save environment. There is no single subject by which we can have complete knowledge of environmental aspects. Simultaneously, it can not be done by single man, agency or institution. It is of the people, by the people and for the people. Thus public awareness means, making the people conscious about the physical, social and aesthetic aspects of environment.

**Methods**: To protect and conserve the environment is the basic duty of all sections of people because environment belongs to all and every individual matters. Instead of searching the solution, it is necessary to find the permanent solution of environmental and ecological problems. It can be done by following means –

**(1) Through mass-media :** There are various means of mass communication to educate, entertain & give informations, instructions etc. to people. Radio was the first. Next is T.V. in terms of its reach to the masses. In addition to these newspapers, magazines etc. are also there. Since media has an important role in masses hence awareness among people can be propagate by articles, important serial programmes, instructions, stories etc. using these media. Documentary and films on environmental awareness may be displayed in cinema houses. Posters can also be displayed in streets, on roads and at common places for environmental awareness.

**(2) Through Education :** Students are the back bone of a country. If environmental education is started from grass root level i.e. from childhood stage, it will give good results. It will be done through formal and informal environmental education. A welcome step to introduce environmental studies paper at collage level by Government by the directive of Supreme Court is good in the direction of awareness.

**(3) Through rallies, orientation and training programmes :** To promote environmental awareness, environmental rallies with posters, handbills, programmes may be organized on certain occasions like 5th June as world environmental day and 1st week of October as wild life week. Some training programmes, orientation, workshops, seminars, meetings based on environment awareness may be organized for decision makers, planners, leaders also, so that they can also spread the message to protect & conserve the environment.

**(4) Through voluntary organizations and NGOs :** Due to having link between people and Government some voluntary organizations and non Government organizations can play an important role in the direction of environmental awareness in people by organizing educational, religious, plantation, musical, Nukkad Natak, competitions (essay, drawing, oral) etc. programmes. These organizations can also advise the government to implement effective programmes for environmental awareness. They can organize certain public monuments for conservation of environment. Chipko Movement, Narmada Bachao Andolan, etc. are such public monuments were organized by Dasholi Gram Swaraj Mandal in Gopeshwar and Kalpavriksha respectively. Some voluntary organisations working in this field are –

(1) Bombay Natural History Society (BNHS).

(2) Wild Life Preservation Society of India. (WPSI)

(3) Worldwide Fund for Nature – India (WWF - India).

(4) Centre for Science and Environment (CSE)

In addition to these some scientific and technical societies like Indian Science Congress, National Academy of Sciences, Institution of Chemists, The Indian Chemical Society, Biological

Society of India, Pollution Research, CDRI, NEERI etc. are also doing work in the field of public awareness by organizing seminars, workshops, publishing articles, research papers etc.

NGOs usually consists of volunteers or group of individuals genuinely interested in the cause of conservation of wild life and protection of environment. They generate their own resources, contributory funds, grants etc. They play an important role in developing public awareness about environmental matters. These organizations can be grouped in to three categories –

(1) Specialized agencies with restricted membership consisting primarily of scientific and professional members. For example – International Council of Scientific Unions (ICSU), International Union for Conservation of Nature and Natural Resources (IUCN)

(2) Institutes and centres devoted to the task of collection of information, research and consultation. e.g. International Institute for Environment and Development etc.

(3) Activist organizations with open membership for the cause of raising support and consensus for wild life protection, are doing for conservation of resources and improvement in the quality of environment. For example Sierra Club National Audubon Society, Bombay National History Society etc.

## ROLE OF NGO'S IN PROTECTING ENVIRONMENT AND HEALTH

### INTRODUCTION

The past five decades have witnessed the difficult problems encountered in providing health care services to our poor people, the majority of who live in more than half-a-million villages and in the proliferating slums of our cities. Charitable and voluntary organizations since time immemorial have been contributing significantly towards the health care of the community. With the passage of time, Non-Governmental Organizations (NGOs) have equipped themselves adequately and come up enthusiastically in providing services like relief to the blind, the disabled and disadvantaged and helping the government in mother and child health care, including family planning programmes. As a result, all concerned have realized the potential of NGOs and their considerable merit compared to the public/private health sectors because of their staff's motivation, dedication and sympathy for the deprived sections of our society and their personalized approach towards the solution of problems.

The National Population Policy (NPP) 2000 and National Health Policy (NHP) 2002, states that there should be greater involvement of NGOs in the implementation of different health and family welfare programmes in the country. In recognition of the crucial role played by them, Government of India started granting financial aids to NGOs for various schemes. The important role played by the various national and regional level NGOs is briefly documented in the 'Encyclopaedia of Social Work in India – 1968', where special mention has been made of such organizations like All India Blind Relief Society, Family Planning Association of India (FPAI), Indian Medical Association, Indian Red Cross Society, National Society for the Prevention of Blindness, Sent Parmanand Blind Relief Mission, T.B. Association of India, Bombay Mothers and Children Welfare Society; to name a few. Greater roles for the NGOs was seen to ensure Health for All through the primary health care approach. Their role was also considered as most crucial to translate the concept of 'People's Health in People's Hands' into action.

### OBJECTIVES OF NGO'S

#### General Objectives

- To describe and discuss the common characteristics of health system functioning in the given socio-economic, socio-cultural, political and ecological settings.

- To highlight and delineate crucial factors responsible for the health sector reforms and to undertake, as the most challenging endeavour, effective and efficient health management and quality health and service provisions in the community.
- The fundamental objective is to act as a catalyst in bringing about local initiative and community participation in overall improvement in quality of life.

### Civic and Environmental Objective

This aims at developing civic and environmental consciousness among the public. Organisation of civic amenities and sanitary facilities on a self-help basis, enactment of suitable legislation for the betterment of civic standards, environmental protection is some means by which a cleaner and healthier environment can be achieved.

### Service Objective

This was formulated provide service to all segments of the society like the poor, women, children and youth through various schemes like drug and alcohol addiction programmers, organisation of rallies against child labour, medial camps, blood donation camps, etc. This also includes the protection and preservation of nature, wildlife, historical and heritage monuments.

## ACTIVITIES UNDERTAKEN BY NGO'S

### Solid Waste Management

This includes both municipal solid waste and bio-medical wastes. Civic Exnoras play a major role in assisting the Municipal Corporation in the collection of garbage from individual households and the deposition of the same in secondary collection points by appointing street beautifiers in the concerned areas. With its experience over the years, Exnoras has realised that this was merely a relocation of waste rather than management of solid wastes.

### Zero Waste Management

NGO's have now started focusing its efforts on the concept of *Zero Waste Management* by which practically all wastes can be converted into wealth through recycling. Exnora is also addressing the problem of handling and disposal of bio-medical wastes, and is trying to find a solution beneficial to all concerned.

### Citizen's Waterways Monitoring Programme (WAMP)

This programme was started with the sole purpose of developing clean and pollution free waterways in cities and for creating a healthy living environment for all city dwellers. WAMP was formed in 1991, as a joint programme with several NGOs and individuals dedicated to the cause of developing clean waterways in the city. The WAMP objectives are:

1. To stop pollution of waterways
2. To maintain the waterways of the city cleanly

A series of meetings were conducted with various governmental agencies, which has resulted in the drafting of a detailed action plan. If the Government and the public implement the plan properly this will result in achievement of clean waterways.

### Community Sanitation Improvement Projects

Inadequate sanitation facilities are a major problem to human health, especially so in the neglected low-income areas and slum settlements. NGO's concept of self-help is best displayed by the community sanitation improvement projects in these areas. Two of the most successful projects have been at the at Narikkurava (Gypsy) Colony in Indira Nagar, Chennai and at Giriappa Road in T. Nagar, Chennai.

**Student Environment Programme (STEP)**

This program has a dual role - of creating environmental awareness amongst the student community and to develop each child's mind resources through various personality development programs. A teacher's manual and an activity book that have been brought out as a part of this program are designed in the 'do and learn' format and provide an easy understanding of the problems faced by us and at the same time kindles the mind to find remedial measures.

**Tree Planting**

The Civic Exnoras in the city have been instrumental in planting trees for the purpose of beautification of roads, parks, playgrounds, burial grounds, etc. with the larger perspective of environmental protection.

**Vegetable Roof Gardening**

With agricultural land shrinking rapidly and deforestation rates rocketing, urban agriculture is the need of the hour. Exnora has been propagating and training youth in setting up and maintaining vegetate roof gardens in households of Tamil Nadu. The organisation is closely working with the Tamil Nadu Horticulture and Agriculture Departments on this project.

**Rain harvesting**

NGO's have propagated the system of rain harvesting in several residential areas in the city with the aim of exploiting one or another important water source, viz, and rainwater. Many cities suffer from perennial water problems every summer and therefore it is important that all avenues of water source be tapped. By using simple and inexpensive techniques the NGO (Exora) has arrived at a method to conserve a large part of the 110 cms of rain that we receive annually. A Water Conservation Committee constituted in Chennai by Metro Water Supply and Sewerage Board Extra is a core member.

**Aids Awareness**

NGOs are working diligently to educate the public about the prevention of AIDS and provide support and counselling for HIV-infected persons. The additional dangers of placing value judgements on how individual became HIV-positive. One key in the prevention of AIDS is to eliminate negative images of HIV-infected persons as projected by the media. NGOs should work toward educating the public to respect the dignity of all persons living with HIV, regardless of how they became infected.

## POLLUTION CONTROL

The task of pollution control in India is complex due to the large number of heavy, large and small-scale industries involved. Further, the rise in the number of vehicles coupled with poverty and the large population puts tremendous pollution pressure on air, water and land. A comprehensive approach to pollution control is being undertaken based on the following principles:

1. Prevent pollution at source.
2. Encourage, develop and apply the best available practicable technical solution.
3. Ensure that polluter pays for pollution and control arrangements.
4. Focus on protection of heavily polluted areas and river stretches.
5. Involve the NGO's in educating the public about pollution control measures to be taken.

### Noise Pollution

Noise pollution has become a major problem in the metropolitan cities and in other urban areas. With a view to regulate and control noise producing and generating sources, the Ministry of Environment and Forests has notified the *Noise Pollution* (*Regulation and Control*) Rules, 2000 under the Environment (Protection) Act 1986, for prevention and control of noise pollution in the country.

### Air Pollution

The sources of Air Pollution are industries like thermal power plants, sugar mills, distilleries, paper mills etc. Vehicular emissions are another source of air pollution. The Environment (Protection) Act, 1986 has classified the industries as red, orange and green depending on the degree of pollution caused by them. It further specifies the various pollution control measures to be adopted by these industries.

### Water Pollution

The rivers of India are part of its cultural heritage. Over the years, the quality of the water has deteriorated due to the uncontrolled release of effluents by industries into the rivers. A **National River Conservation Plan** (NRCP), which includes second phase of GAP also, has been formulated which aims to control the pollution of grossly polluted rivers o the country. A National River Conservation Authority has been setup to review the implementation of the programmes related to cleaning of rivers. The NRCP covers 141 towns located along 22 interstate rivers in 14 states. The total cost of the scheme is Rs 2013 crores. A **National Lake Conservation Plan** envisaging the conservation of lakes by prevention of pollution by catchments area treatment, desalting, weed control, based on the integrated watershed development approach, is under implementation.

## NGO's ROLE IN POLLUTION CONTROL

The success of India's environmental programmes depends greatly on the awareness and consciousness of the people. A National Environmental Awareness Campaign has been launched to sensitise people to the environmental problems through audio-visual programmes, seminars, symposia, training programmes etc. **Paryavaran Vahinis** have been constitued in 184 districts involving the local people to play an active role in preventing poaching, deforestation and environmental pollution. **4000 NGOs** have been given financial assistance for creating environmental awareness. An **Environmental Information System (ENVIS)** network has been setup to disseminate information on environmental issues. India has a large network of NGO's which are involved in spreading the message of sustainable development to the public.

## INSTITUTIONS INVOLVED IN ENVIRONMENTAL ACTIVITIES

### Environmental Training Institute

An integrated effort is required in environmental planning management and action. Interdisciplinary understanding, cooperations and coordination and increased public participation and support are important components in the process towards sustainable development. To move towards Sustainable Development, DANIDA has prepared a project document for establishing "Environmental Training Institute" in Tamil Nadu. A side letter was signed between Government of India, Ministry of Finance and the Kingdom of Denmark with a Danish grant of DKR 9.961 million. Government of Tamil Nadu has also given concurrence for setting up the Environmental Training Institute in Tamil Naidu with DANIDA assistance. Accordingly, the Environmental

Training Institute was established in the year 1994. ETI is a common platform that offers training to pollution control board staff, industries, the urban sector as well as NGOs.

Over the years the institute has conducted 124 Technical programmes, involving over 1965 participants and 36 special Environmental awareness programmes have been conducted for NGOs, Govt. official, Professionals, Universities and Educational Institutions. There are many similar Environmental Training Institute that have been established by the government in various states.

### Tata Research Institute

The Tata Energy Research Institute (TERI) has launched a major project, the first phase of which is nearing completion. This project called **"Growth with Resource Enhancement of Environment and Nature (GREEN India 2047)"** has rigorously estimated the reduction in India's key natural resources during the period 1947-97, and has computed economic values of consequent losses, which in some cases are alarmingly high. On the basis of past experience and a careful analysis of the cost behind the degradation that has taken place in the past, strategies for the future have been developed, whereby a fresh and creative approach can be taken in the next 50 years.

In the case of indoor air pollution, the major reason lies in the fact that the poorest sections o society are still dependent on low-grade biomass energy resources. As a result, women and children in particular are exposed to long hours of harmful emissions, which lead to a number of health problems including heart disease, pregnancy related problems and various upper respiratory diseases. The solution to such a problem will be larger production of biomass for increasing the availability of good quality fuel wood, and the development and use of superior technologies in its convention and end-use.

The use of modern biotechnology techniques, for instance, provides the only means to enhance production of biomass under soil and climatic conditions that have traditionally not been conducive to high growth. Concurrently, efforts to produce methane or other gases from organic matter, such that they can lead to clean combustion or cooking purposes poses not only a major scientific opportunity, but also a challenge in a managerial and organizational sense. The highest level of scientific expertise should be directed to the solution of this problem. Of course, the benefits of innovative solutions in this area would be confined not to India only, but would be of direct relevance to several other developing countries where inferior biomass fuels are used on a significant scale.

## CORPORATE SOCIAL RESPONSIBILITY

In addition to complying with regulatory standards and norms, industries have to adopt clean technologies and bring about implementation in management practices. This measure has been agreed upon on the understanding that a bank guarantee would be furnished by the concerned units indicating the commitment to the action plan. However, this is without any prejudice to the stipulations made in the existing standards and action already taken/initiated for non-compliance and area-specific requirements warranting stringent actions. The industrial units which are not complying with the **national standards notified under the Environment (Protection) Act, 1986,** will submit an action plan to meet the standards and bank guarantee to the respective State Pollution Control Board within 3 months (by Dec 2003).

## THE NATIONAL INSTITUTE OF HEALTH AND FAMILY

The National Institute of Health and Family Welfare is a premier institution in the country. It has kept pace with new thinking and contributed to the knowledge in the area of health and family welfare.

The main objective of Institute is to serve as an apex technical institution to promote national health and family welfare programmes in the country through education and training, research, evaluation and monitoring, consultancy and advocacy, and providing certain specialized services.

NIHFW is an autonomous Institute funded by the Ministry of Health and Family Welfare, Government of India. It is situated in south Delhi.

## CAF: AN ORGANISATION PROVIDING FINANCIAL HELP TO NGO's

CAF India's vision is to create a sustainable voluntary sector with resources contributed by relationships built on trust between NGOs and donors. CAF India has pioneered corporate community initiatives with several companies in India. CAF India has a registered public charitable trust, exempt under Indian tax laws, and it has offices in Delhi and Bangalore. CAF India is opening up new resources within companies by working with chambers of commerce, such as the *Confederation of Indian Industry*, on the development of *primary education* facilities and by providing fund raising support.

## LIMITATIONS IN THE PERFORMANCE OF NGO's

Governmental Organisations (NGOs) who came forward to shoulder several social responsibilities, are facing a credibility crisis with a number of cases of embezzlement and scandals involving some of them coming to the fore.

Recently, the Dutch government, funding a multi-crore water supply and sanitation project in Uttar Pradesh, terminated its contract with an NGO because of financial mismanagement. This particular NGO allegedly spent huge amounts of money on setting up a new office and on foreign tips for purposes other than those relating to the project. Bikaner, Rajasthan, another NGO involved in women's welfare reportedly pocketed around Rs two lakh by putting fictitious names on its rolls.

In another case, the CBI found that out of seven NGOs receiving funds to help women in self-employment, four had not done a thing and the rest existed only on paper.

About 400 such NGOs have been black listed by the Council for the Advancement of People's Action and Rural Technology (CAPART) while the Central Social and Welfare Board (CSWB) has black listed 3,000 NGOs for their poor performance in different fields in our country.

The NGOs which are really working at the rural areas don't even know how to write a project report necessary to get the money and those who don't work but know the skill of writing project report, manage to get the money sanctioned.

The CSWB, through which the government funds NGO's issues about Rs 40 crore every year in addition to thousands of dollars given by institutions like CAPART. After getting funds from the government, the NGOs default in fulfilling requirements like submission of audited accounts and details of programme implementation within a stipulated period of time. NGOs have become a money making machine and, being smart in public relations, they are more encouraged by the press who gives them coverage and don't cover the positive work done by the genuine NGOs, as good work is not an important news to the press.

## CONCLUSION

The total number of NGOs in the country is about 70,000 (Nagraj) while the actual number might be in lakh, the number of the registered NGOs may not be more than 10,000. Comercialisation of NGOs has no doubt led to their rapid growth but it does not mean that every one joining the field is coming because of money only.

A high official with CAPART says that there are more number of good NGO's than the bad NGOs but "unfortunately it is only the bad ones who get projected."

A code of conduct should be evolved to evaluate and rate of NGOs and the initiative should come from within the NGO sector.

The Insurance Regulatory and Development Authority (IRDA) is working on evolving a regulatory frame work to allow NGOs to sell micro insurance products particularly to tap insurance potential in the rural areas. Since NGOs are performing well the insurance sector also will be soon open to NGOs. These NGOs can capture the rural market easily.

## WHAT WE CAN DO FOR THE PROTECTION OF ENVIRONMENT

1. Educate the students about the pollution problem and the harmful effects of pollution.
2. We should minimize the use plastic cover for different purposes.
3. Buy only environment friendly products i.e. the products which are not reducing the natural resources.
4. Not waste water for various purposes.
5. To plant and grow trees in the house garden.
6. To support the NGO's financially which are rendering service to the Health and Environment sector.
7. To motivate research on different measures to be taken to solve environmental problems.
8. To support the initiate taken by the central and state government in protecting our environment.

## WOMEN EDUCATION

There cannot be an educated people without educated women. If general education had to be limited to men or to women, that opportunity should be given to women, for then it would most surely be passed on to the next generation.

General education for interesting and intelligent living and for citizenship in large part can be the same for men and women. We have heard frequent suggestions that women's education should run to pretty "accomplishments," such as drawing, painting or the like-skills which will enable well-to-do women to pass the time harmlessly while their husbands do the really important work. This point of view should be obsolete. Women should share with men the life and thought and interests of the times. They are fitted to carry the same academic work as men, with no less thoroughness and quality. The distribution of general ability among women is approximately the same as among men.

Yet, though men and women are equally competent ion academic work, and though many subjects are equally interesting and appropriate, it does not follow that in all things men's and women's education should be identical. Indian universities for the most part are places of preparation for a man's world. Little thought has been given to the education of women as women. Women must, share the same programme as men or go without. There are ways in which many women's interests or appropriate fields of work diverge from those of men, and educational programmes should take that fact into account.

In every country, no matter how far the "liberation" of women has gone, husbands and wives commonly play different parts. In general the man provides the income and the woman maintains the home. For many women who crave to achieve standards of excellence, the home provides an excellent setting. For a woman to give the home design, beauty, order and character, without being herself a slave to home-keeping and without imposing onerous prohibitions and restrictions on the freedom of movement of children, is a high art. It will not be acquired by chance, and for many women its acquisition will be impossible, except through education.

The home itself can be a Work of art, even though it must be maintained very simply and economically. In fact, simplicity may be the highest expression of beauty. In the Western world, homes tend to be crowded with belongings, as though the achievements of a, beautiful home, were a matter of acquiring furnishings or Other works of art. In Japan, great simplicity and restraint in furnishing, and ornaments' with skill and taste in selection and arrangement are the characteristics sought.

Women have demonstrated their ability to think and work alongside of men. Why not take that ability for granted and begin to recognize the ways which the education of women can well be differentiated from that of men? It is time to realize that the finest family relations result from the association of a mail and a woman who have had much of their education in common, but each of whom has developed according to his or her own nature, and not in imitation of the other.

**Preparation, for Home and family life:** Wise education for a woman will not leave her preparation for home and family to the bitter and wasteful school of experience. Here education as a woman should include practical "laboratory" experience in the care of a home and family. Equipment for a girl's education might well include:

1. A baby home.
2. A nursery school, which incidentally would relative nearby mothers during a part of the day.
3. A club for school children and adolescents.
4. A little home for convalescents.
5. A small home for old people.
6. A home setting where students may have experience home Maintenance and operation, and where they may act as hostesses.

A woman should learn something, of problems that are, certain to come up in all marriages, and in the relations of parents and children, and bow they may be met. Her education should make her familiar with problems of home management and skilled in meeting them, so that she may take her place in a home with the same interest and the same sense of competence that a well trained man has in working at his calling.

To make such a suggestion for women's education is to emphasize the fact that there has been inadequate study of that field in India, inadequate training of teachers for women's education, and inadequate provision for full education for women in our college and universities ; through some institutions such as the women's Christian College and Queen Mary's College for Women, both at Madras, and the Isabella Thoburn college at Lucknow, are doing pioneering work to that end.

The greatest profession of women is, and probably will continue to be, that of home maker. Yet her world should not be limited to that one relationship. There are varied conditions which may properly lead a woman to seek fulfilment of her life in other fields. Among the great contributors to human welfare have been some men who determined to forego home and family in order to commit themselves wholly to the chosen work of their lives. Women should have this same opportunity. The place of wife and mother offers opportunity for exercise of the highest qualities and skills, yet for a woman to decide that she can best fulfil her aims by living a single life should not put her under a social disability. Sometimes, also, there is a, period before marriage during which a young woman can do useful work, such as teaching or nursing. Sometimes the loss of a husband makes here the bread winner for the family, When children are grown, there often remain ten to twenty five years of vigorous life in which a woman may wish to have a useful career sometimes husband and wife wish to share a common occupation through the years. Sometimes with women, as with men, the needs homes and family leave time for useful and interesting occupations. For all such circumstances educational opportunities should be available.

One of the desirable development, of Indian life, and education for both men and women is a great increase in the kinds of work open to them. If only a few callings are recognized there is excessive competition for the available places, many kind of ability find no opportunity for expression and many kinds of needs remain up unsupplied.. A wholesome and interesting society will have many and varied occupations and professions. The educational system at all level should prepare men and women for such varied callings.

## IMPORTANCE OF WOMEN'S EDUCATION FOR NATIONAL LIFE

"If Government by the initial exclusion of the masses accentuated the segregation of the masses from the privileged few, by their initial restriction of their (educational) efforts to the male population, they brought a line of division where it had never existed before, within the household." While the movement for equal education for men and women began in Great Britain about a century ago with such serious thinkers as Frederick Maurice, who was a founder of Queen' College for woman in 1848, and Johan Stuart Mill, whose "Subjection of Women" was published in 1869, it did not reach India until several decades later. Even today the inequality is evident. According to the statistics issued by the Indian Ministry of Education for 1945-46, there were six and a half times as many boys and men in secondary schools and colleges as there were girls.

The underlying habits of men and women are largely fixed in the early years, and these years are spent chiefly with I the mother. If she is open minded, inquiring and alert, looking behind rumour and tradition to find the facts, concerned with the course of events, informed about the nature of the world around her and interest in it, and acquainted with history and literature and enjoying them, then her children will learn these interest and attitudes from her. The educated, conscientious mother who lives and works with her children in the home is the best teacher in the world of cloth character and intelligence. Much of what she learned at school her children yet unconsciously as second nature by living in her company. In a society made up of such homes children starting to school already have a background of information, understanding and culture which result in their getting more benefit from school than otherwise would be possible.

## THE FUTURE OF WOMEN EDUCATION

It is very clear that there are certain steps which urgently need to be, taken to make the conditions of women's education more tolerable.

There are few truly co-educational colleges in our country. Rather, there are men's colleges to which women have been admitted as students, which is a very different matter. Quite frequently in "Co-educational" colleges nearly all the amenities are for men, and women are little more than tolerated. Often sanitary facilities for women are totally inadequate, and sometimes wholly lacking. Recreation space and facilities for women similarly are inadequate or lacking. One of the most frequent suggestions made to us was that where women and men attend colleges that a, physical directress for women be provided.

In many co-educational colleges women have little or no share, in college life Women's hostels usually accommodate but few women, and sometimes none at all. According to the comments received, in one city where there are few or no hostels some women students must ; leave home at 5 A.M. in order to attend classes, and not provisions for refreshments are provided, although the students do not finish their work until mid-afternoon. The commission received comments to the effect that the attitudes of men students left, much to be desired. Examinations are a severe nervous strain to women in co-educational institutions.

**Suggestions for Improvement:** Among the many suggestions received are that women in seclusion should be allowed to appear for examinations; that special consideration be given to the status and needs of women; that in the employment of women staff members, there should be equal pay for equal work as compared with men.

We were informed that while a considerable number of women take medical courses, relatively few actually practise medicine. Since medical education is very expensive, and student enrolment in medical schools necessarily small, to train a person who will not practise is a social loss. To the extent that is the case, women's interests might well be redirected to nursing, medical laboratory technology, dietetics, and other callings where the period of training and the expenses are less. Social taboos are a large, element in tetarding such redirection of interests. Here as elsewhere, a democratic spirit does not necessary follow a democratic constitutions. Such redirection of interest should be a matter of counsel and advice, rather than of compulsion. The medical profession should remain open to women.

**Co-education:** The question of the wisdom of co-education was frequently raised,\and the most divergent opinions were expressed. There seems to be a definite preponderance of opinion that from the thirteenth or fourteenth year of age until about the eighteenth, separate schools for boys and girls and desirable. Whether this opinion is chiefly based on custom or upon experience is not wholly clear.

One of India's prominent educators states "The modern trend is for equality of opportunity for women in all spheres, and it can not be arrested. There should be no distinctions of any kind of women from men, after the matriculation stage".

On the other hand many, probably a majority of those who Commented, favour separate colleges for women when that is feasible, though not to the extent of denying women educational opportunity by excluding them from existing colleges organised Primarily for men.

Some of the arguments given are that a women cannot develop her personality in men's college: that here is no need for women to undergo the nerevous strain of examinations; that woman's education should be more in keeping with the temperament and needs of women as wives and mothers and that overcrowding is more serious for women than for men. "A pleasing feature of colleges for women has been the intimate relations of students and teachers." Some of these arguments have greater weight in the absence of truly co-educational colleges where the needs of men and women would be given equal weight in designing the programme.

As the age of entry to degree colleges would, on our recommendation, be approximately eighteen, college education may be co-educational, as it is present in many medical colleges. Separate institutions at this level would demand unjustified increase in expense. To maintain separate institutions for men and women side by side, duplicating equipment, even when it is very inadequate, would be an undue tax upon limited financial resources. Separate women's colleges commonly have poorer buildings, poorer equipment, and less able teachers. So far as possible co-educational institutions should be encouraged at the degree level.

**Appraisals of Women's Education:** The results of women's education under existing conditions has not been entirely satisfactory. The principal of a college wrote: "Women's present education is entirely irrelevant to the life they have to lead. It is not only a waste, but often a definite disability." Another wrote, "The present system of women's education, based as it is upon man's needs, does not in any way may them fit for coping with the practical problems of daily life. Their education should give them a practical bias, especially from the point of view of families, for making them good mothers, teachers, doctors and nurses".

One experienced woman educator wrote to the Commission "The modern educated Indian woman is neither happy nor contented nor socially useful. She is a misfit in life. She is highly suppressed, and needs opportunities for self-expression, the new education must provide this opportunity".

To some extent this maladjustment is the price of pioneering. The educated woman is ahead of her time. She is lonely and "out of place" while creating a new social atmosphere in which her daughter or grand daughter can be natural and at home. But to some extent her discontent is due to her effort, not only to be equal to men, but to be like them in all her interests and activities.

One educator wrote: "It is too late in the day to suggest that women should not have the same courses as men. The remaining question is, what additional opportunities shall be provided?"

It is the duty of those in charge of women's education to face these problems, and to design education for women which will enable them to have full and adequate lives. One educator wrote: "There has been a no planning of women's education. It has just happened".

## RECOMMENDATIONS

(1) that the ordinary amenities and decencies of life should be provided for women in colleges originally planned for men, but to which women are being admitted in increasing numbers;

(2) that there should be no curtailment in educational opportunities for women, but rather a great increase;

(3) that there should be intelligent educational guidance, by qualified men and women, to help women to get a clearer view of their real educational interests, to the end that they shall not try to imitate men, but shall desire as good education as women as men get as men. Women's and men's education should have many elements in common, but should not in general be identical in all respects, as is usually the case today;

(4) that women students in general should be helped to see their normal places in a normal society, both as citizens and as women to prepare for it, and college programmes should be so designed that it will be possible for them to do so;

(5) that through educational counsel and by example the prevailing prejudice against study of home economics and home management should be overcome;

(6) that standards of courtesy and social responsibility should be emphasized on the part of men in mixed colleges;

(7) that where new colleges are established to serve both men and women students, they should be truly co-educational institutions, with as much thought and consideration given to the life needs of women as to those of men. Except as such colleges come into existence there are no valid criteria for comparing segregated education with co-education;

(8) that women teachers should be paid the same salaries as men teachers for equal work.

Women constitute almost half of the population in the world. But the hegemonic masculine ideology made them suffer a lot as they were denied equal opportunities in different parts of the world. The rise of feminist ideas have, however, led to the tremendous improvement of women's condition through out the world in recent times. Access to education has been one of the most pressing demands of these women's rights movements. Women's education in India has also been a major preoccupation of both the government and civil society as educated women can play a very important role in the development of the country.

**History of Women's Education in India:** Although in the Vedic period women had access to education in India, they had gradually lost this right. However, in the British period there was revival of interest in women's education in India. During this period, various socio religious movements led by eminent persons like Raja Ram Mohan Roy, Iswar Chandra Vidyasagar emphasized on women's education in India. Mahatma Jyotiba Phule, Periyar and Baba Saheb Ambedkar were leaders of the lower castes in India who took various initiatives to make education available to the women of India. However, women's education got a fillip after the country got independence in 1947 and the government has taken various measures to provide education to all Indian women. As a result women's literacy rate has grown over the three decades and the growth of female literacy has in fact been higher than that of male literacy rate. While in 1971 only 22% of Indian women were literate, by the end of 2001 54.16% female were literate. The growth of female literacy rate is 14.87% as compared to 11.72% of that of male literacy rate.

**Importance of Women's Education in India:** Women's education in India plays a very important role in the overall development of the country. It not only helps in the development of half of the human resources, but in improving the quality of life at home and outside. Educated women not only tend to promote education of their girl children, but also can provide better guidance to all their children. Moreover educated women can also help in the reduction of infant mortality rate and growth of the population.

**Obstacles:** Gender discrimination still persist in India and lot more needs to be done in the field of women's education in India. The gap in the male-female literacy rate is just a simple indicator. While the male literacy rate is more than 75% according to the 2001 census, the female literacy rate is just 54.16%.

Prevailing prejudices, low enrollment of girl child in the schools, engagements of girl children in domestic works and high drop out rate are major obstacles in the path of making all Indian women educated.

Women hold a prominent position in the Indian society as well as all over the world. However since the prehistoric times women were denied opportunities and had to suffer for the hegemonic masculine ideology. Thus this unjustifiable oppression had resulted into a movement that fought to achieve the equal status of women all over the world. Women Education in India is the consequence of such progress and this led to the tremendous improvement of women's condition through out the world. Nevertheless eradication of female illiteracy became a major concern after the long time since the commencement of society.

Women education in India gradually became a major preoccupation of both the government and civil society and proved the educated women can play a very important role in the development of the country. In the recent era, the Indian society has established a number of institutions for the educational development of women and girls. These educational institutions aim for immense help and are concerned with the development of women.

Picture of women education

In ancient India, women and girls received less education than men. This was due to the set social norms. Interestingly, in the Vedic period women had access to education, but gradually they had lost this right. *Women Education in Ancient India* prevailed during the early Vedic period. In addition to that Indian scriptures Rig Veda and Upanishads mention about several women sages and seers. Women enjoyed equal status and rights during the early Vedic period. However, after the 500 B.C, the status of women began to decline. The Islamic invasion played a vital role in restricting freedom and rights of the women. A radical change attended and there was a terrific constraint for Women education in India.

**Women Education in Medieval India** further deteriorated with the introduction of Purdash system. Various customs and traditions of the Islam, Hindu and Chriwomen hold a prominent position in the Indian society as well as all over the world. However since the prehistoric times women were denied opportunities and had to suffer for the hegemonic masculine ideology. Thus this unjustifiable oppression had resulted into a movement that fought to achieve the equal status of women all over the world. Women Education in India is the consequence of such progress and this led to the tremendous improvement of women's condition through out the world. Nevertheless eradication of female illiteracy became a major concern after the long time since the commencement of society.

Women education in India gradually became a major preoccupation of both the government and civil society and proved that educated women can play a very important role in the development of the country. In the recent era, the Indian society has established a number of institutions for the educational development of women and girls. These educational institutions aim for immense help and are concerned with the development of women.

## ENVIRONMENTAL EDUCATION

### REALITY

I am a big boy now
I have left school;
But I am a fool still
A poor fool
With books and blackboards
Casting dark shadows
On me;
I look for a little opening
Of light
I hear people laughing
What are you going to do now
With your education and all?
There is the market place
Where the people
Sell their wares
Women chasing flies
From day-old pies.
I cannot do that
I have a certificate
I passed with honors
English, Maths, History and Geography

I learnt them all.
The laughter gets louder
I see my teacher
Sitting on a sterile rock
Near the beach
Selling green coconuts
What do I do now?
An old man close-by whispers,
"Come fishing with me today
For you have a lot to learn yet".
**-Konai Helu Thaman (1974:17)**

The 'reality' of the earth's environmental resources is that they are limited and can be depleted or damaged very easily. Without their protection, there can be no long-term sustainable development. This 'reality' is perhaps even more obvious in the resource-poor and ecologically fragile islands of the Pacific Ocean than it is in industrialised countries.

Similarly, the 'reality' of formal education in the Pacific Islands is that school children learn very little about their own environments and even less about the sophisticated cultural strategies that generations of Pacific Islanders have evolved in order to live successfully in those environments. Moreover, there is limited educational focus on the impact of modernization on both the island environment and traditional resource-use strategies, English, mathematics, history, and geograpny are 'learnt', but very little is understood about the natural and cultural environments outside the classroom... very few of our development-planner and policy-maker graduates ever having been fishing (or planting) with their elders!

## ENVIRONMENTAL EDUCATION AND ECOCULTURAL DEVELOPMENT

The survival of any society depends on the evolution of ecologically-sound cultural strategies for the provision of basic material and non-material needs. Such strategies, which have resulted from the complex and everchanging relationships between human populations and their environments, including the nature and quality of their environmental education, whether traditional or modern, have been referred to as 'ecocultures' (Schulte-Westrum, 1982). Those ecocultures which are not ecologically sustainable must, be definition, ultimately *fail*.

To evolve lasting ecologically sensitive strategies for development, it is imperative that environmental education which rekindles environmental awareness be made a priority at all levels, in both formal and non-formal education systems.

## FUTURE ECOCULTURES

In stressing the importance of capital preservation, Schumacher (1973) argue that any system of development which treates its resource base as income to be spent, rather than as irreplaceable capital for maintenance or growth must ultimately fail. He identifies three categories of "irreplaceable capital" : (1) finite or non-renewable resources (particularly fossil fuels); (2) the "tolerance margins of nature" (i.e. does a given system of development pollute or put so much stress on renewable resources, such as water, soils, forests, reefs, fish, people, etc., that they are pushed beyond their tolerance margins and their ability to recovery and be 'renewable'?); and (3) "human substance" (e.g. man's self-respect, health, spiritual well being, cultural traditions, resource-use strategies, etc.).

Similarly, Clarke (1977) argues that traditional Pacific Island ecocultural systems seem to have been guided by what he refers to as seven "principles of permanence". These principles or characteristics include: (1) little or no dependence on extra-system subsidies of energy, fertilizers, seeds, etc.; (2) little pollution i.e. they were not self-poisoning; (3) strongly positive energy yields (*i.e*, for every calorie of energy invested, 10 to 20 calories of food energy were received in return ; (4) dependence on renewable inputs, such as trees, crops, pigs, soils and humans, that could be reproduced within the normal human life span ; (5) fairly even distribution of the products and gains of the system, instead of their being concentrated in the hands (or bank accounts) of a few ; (6) resources being seen as productive capital to be protected for future generations, rather than being treated as expendable income; (7) subsistence production being based on polyculture and diversity rather than on monoculture of single crops.

In short, most traditional Pacific Island ecocultural strategies seem to have had much greater regard than modern systems do for such irreplaceable capital and for principles of permanence.

## THE ECOSYSTEM APPROACH

An ecosystem approach, which views human beings and their cultures as central to all ecosystems, must be the basis for environmental education, which focuses on human activities and the physical environment as interdependent parts of a single environmental system. In this way, students can focus on the relationships between selected human activities and biological and physical processes by including them in a single analytical system, the ecosystem (Geertz, 1970:3). Such an approach can help students to understand better the structure, functions and benefits that can be derived from human ecosystems and how they are affected by existing and potential ecocultural development strategies.

## THE HOLISTIC APPROACH AND ENVIRONMENTAL EDUCATION

Cohen (1976), in stressing the need for a more holistic approach to environmental resource-use analysis, suggests that many negative impacts or social costs arising out of modern urban-industrial development are because most decisions relating to the use or modification of the natural and cultural environment are based almost exclusively on economic and political grounds. He stresses the need to look beyond these grounds, and suggests that *technical, social, and environmental benefits or deterimental effects related to a given resource-use decision must be considered, even if they are difficult to quantify in monetary terms*. Some 'difficult to quantify' detrimental effects, such loss of traditional knowledge, soil erosion, and pesticide pollution, may have the greatest long-term social price.

Cohen argues that the total social price of a society's resource use strategies represents a fair estimate of the chances of survival of that society in all its aspects – physical, institutional and cultural – whereas the internal composition of that total price (be it economic, political or ecological) indicates the areas of stress and possible future change in that society. Areas of environmental stress or social cost resulting from the social price of development are apparent in all societies and, as such, constitute a major issue to be addressed by holistic environmental education.

## STRATEGIES IN ENVIRONMENTAL EDUCATION

Launching the World Conservation Strategy in India, Prime Minister Indira Gandhi reminded the audience that "the interest in conservation is the rediscovery of a truth well known to our sages. The Indian tradition teaches us that all forms of life : animal and plant are so closely liked that disturbance in one gives rise to imbalance in the other" (1). The Indian Constitution laid down the responsibility of Government to protect and improve the environment (2) and made it a "fundamental duty of every citizen to protect and improve the natural environment including forests, lakes, rivers and wildlife" (3).

It was in this background that a Department of Environment was established by the Government of India in 1980 and a Ministry formed in 1985. The role of the Ministry was seen as a nodal agency in the Central Government for the "planning, promotion, co-ordination and overseeing the implementation of various environmental and forestry programmes". In order to achieve these tasks, the Ministry recognized the "creation of environmental awareness among all sectors of the country's population" [4]. The Ministry in turn recognized environmental education as a key to the success of any overall environmental strategy and decided to help in the setting up of a 'Centre of Excellence' in EE to play the vital role of setting the pace and the agenda for EE in the country. The Centre for Environmental Education, CEE, was set up in 1984.

### (1) The Challenge of India

With over a billion people and at least 17 major languages, with poverty and low literacy levels, with over 650,000 primary schools, with a rapidly increasing population, the development and environmental challenge is enormous. But in spite of all the human pressures India continues to be rich in it's biodiversity. With over 127,000 species and a variety of habitats, India is one of the world's mega biodiversity centres. While several of the goals of development remain elusive, with better nutrition and health care, the life expectancy at birth which was a mere 24 years at the turn of the last century and 34 years at independence in 1947, today is close to 70 years. The agricultural revolution has ensured that the food grain production has not only kept pace with the growth in population but has grown faster. A factor of four while population has grown three times over the fifty years. Industrial production has shown an even more rapid growth, about 15 times since independence. Per capita income has grown 45 times during this period.

India has made considerable strides in slowing down its population growth rate. But with all the efforts the current estimate is that our population will only stabilize somewhere the middle of the century, by which time we will have become the most populous country of the world. Till then we add a population of approximately one Australia to India every year. With about 16% of the world's population and a little over 2% of its land, there is already enormous pressure on our resources. But while the population increase puts pressure on our resources, the pressure of "development" is perhaps even greater.

### (2) SWOT Analysis for EE (Environmental Education)

**(*a*) Strengths.** The Constitution and the Government's commitment to the environment along with the rich tradition of environmentally sound practices are an important backdrop under which the countries EE strategy has been evolved. The Central Government and every state within India, now has a Ministry of Environment. All education departments recognize EE as an essential part of education. The law courts have been sympathetic to environmental issues and the Supreme Court has passed a directive that all students must go through a compulsory course on the environment and the media must show free of cost a certain amount of programmes to create environmental awareness [5]. India has a vast network of NGO's that are activity participating in the creation of awareness on development and environmental issues. Working on their own and with Governments they are the backbone of the strategy to create greater environmental awareness, especially that leading to environmental action.

**(*b*) Weaknesses.** Illiteracy which is estimated at around 48% [6] and a high dropout rate in the school system impede most educational programmes. There is also insufficient recognition of EE or for that matter education itself as an essential tool for development. While commercial communication in the form of advertising is recognized by the business community as essential, development communication and environmental education and communication still remains at a sub-critical level as far as funding is concerned. The slower visible impact of EE improvements in the environment is an inherent "weakness".

(*c*) **Opportunities.** The rapid growth of mass media and coverage of radio and TV offer a major opportunity for EE. The increasing availability of the internet is also an emerging opportunity. With the media giving more coverage to environmental issues and in particular highlighting environmental linkages with various calamities and degradation, people are more interested in learning about the environment.

(*d*) **Threats**. With opportunity, the new and powerful media with their increasing dependence on advertising revenues, have sparked off a consumerism which will have its own impact on people and their behaviour. The advertising budgets are huge compared to the total resources being spent on EE. This "alternative" consumer messages can be seen as a threat to EE.

### (3) Evolving a Government Strategy

(*i*) **Roles of MoEF, MHRD and other Ministries.** While the principal responsibility for Environmental Education through the formal educational system is with the Ministry of Human Resources Development, the parallel responsibility of EE through non-formal efforts lies with the Ministry of Environment. But these efforts are by no means exclusive. There is a large amount of co-operation between the Ministries and with the institutions associated with them.

(*ii*) **EE in the Curriculum.** The national system of education, as defined in the National Policy on Education, 1986, visualizes a national curricular framework which contains a common core including several elements having a direct bearing on the natural and social environment of the pupils.

The National Policy on Education, 1986 (NPE) states that the "protection of the environment" is a value which along with certain other values, must from an integral part of the curriculum at all stages of education. The policy states: "There is a paramount need to create a consciousness of the environment. It must permeate all ages and all sections of society, beginning with the child. Environmental consciousness should inform teaching in schools and colleges. This aspect will be integrated in the entire educational process."

There has been special emphasis on the need to give importance to environmental education, and this has been kept in view while designing curricula, framing the syllabi and developing text books. The philosophy behind the curricula is that environmental education could be infused into the curricula especially at the primary level.

The National Council of Educational Training and Research (NCERT), the apex body in the area of developing curriculum for the formal education system. Currently there is a national debate going on about how to make EE not only widespread but also effective at all levels of education. The strategy under discussion envisages a four pronged approach. The four elements of the strategy are:

1. Strengthening Infusion of EE
2. Teacher Training for effective EE
3. Introduction of Environment as a separate subject
4. Use of non-formal methods of EE through the involvement of NGOs

The success of the above strategy will depend upon the close synergy and partnership between the MHRD and the MOEF, their key Institutions, State Governments, NGOs as well as educational Institutions throughout the country. In addition, the four components are closely linked and one cannot be viewed in isolation from the other. The operationalization would need to keep this in view.

CEE is closely involved in this process, part of which is being supported by the World Bank. The first step which is ongoing is the analysis of curricula across India to review the degree and quality of infusion of EE. The exercise is being undertaken by a partner organization Bharati Vidyapeeth Institute of Environment Education and Research (BVIEER).

## ENVIRONMENTAL ORIENTATION TO SCHOOL EDUCATION (EOSE)

EOSE, most EE programmes are designed and implemented with National and State level perspectives, hence the full intent to environmental education will not be achieved because environmental concerns and problems may be locale specific and, therefore do not include consideration of global solutions. Therefore, the Ministry of Human Resource Development (MHRD) supports a centrally sponsored scheme under its financial assistance is provided to government and voluntary organizations and other institutions working in the field of environmental education to develop locale specific programmes and materials. Under this scheme, CEE is one of the designated nodal agencies. To date approximately projects to support development of locale specific programmes and materials by NGOs have been facilitated.

**(*i*) Centres of Excellence**. One of the strategies of the Department was to establish "Centres of Excellence" for research, training and education which would play a pivotal role in strengthening the country's infrastructure and thereby its ability to deal with these issues. The Centre for Environment Education was set up in 1984 at Ahmedabad under this programmes. A second Centre, the C.P. Ramaswamy Environmental Education Centre of Chennai was set up in 1988 under the same scheme. The main activities of both the Centres are the development of environmental resource materials, organization of training programmes and creation of environmental awareness among teachers, students and the general community.

**(*ii*) National Environmental Awareness Programme (NEAC).** The Ministry of Environment and Forests (MOEF), Government of India started National Environment Awareness Campaign (NEAC) in 1986 with the aim of creating environmental awareness at all levels of society. Under NEAC, the Ministry provides financial assistance to selected non governmental organizations, education and training institutes, community organizations, etc. to create massive awareness among citizens of India. Diverse target groups ranging from students/youth/teachers to rural and tribal population, women, professionals and the general public are covered under this campaign. The campaign programmes are basically composed of a spectrum of short duration programmes.

**(*iii*) Environmental Information System (ENVIS).** The Ministry of Environment and Forests has set up ENVIS for different aspects of environment. The objective of this system are to provide national environmental information service to the users of originators, processors and disseminators; to build up storage, retrieval an dissemination capabilities; to promote support and assist education and personnel training programmes designed to enhance environmental information processing and utilizing capabilities; to promote exchange of information amongst developing countries, CEE has been designated as the National ENVIS Centre for Environmental Education.

**(*iv*) Museums of Natural History.** Setting up of Natural History of Museums as unique facilities to promote non-formal environmental education and to create environmental awareness among people is a strategy of the Ministry of Environment and Forests. The National Museum of Natural History is located at Delhi. Regional Museums are also being established at Mysore, Bhopal and Bhubaneshwar.

**(*v*) Eco-clubs in Schools.** A non formal pre-active system of imparting environmental education to school children by involving them in various environmental activities through the scheme of Eco-clubs has been evolved by the Ministry of Environment and Forests.

An Eco-club may be set up in a middle/high school and should consist of a minimum of 20 members and a maximum of 50 members, particularly interested in the conservation and protection of the environment, and willing to dedicate time and effort on a regular basis towards this end. The members may be drawn from students belonging to classes from VI to X. Each Eco-club will be in charge of an active teacher in the school concerned. The Ministry provides some finance support per annum per Eco-Club.

***(vi)* Promoting Mass Media.** The Ministry provide financial assistance for the production of films, video-spots and other audio visuals on important environmental issues for creation of mass awareness.

These are normally telecasted by our National Television (Door Darshan) Channels. There are slots in our National TV Channel especially for environment.

***(vii)* The CEE Strategy.** CEE is a national centre in the field of environmental education (EE). Its main objective is to promote, across a wide range of target audiences, environmental awareness – an understanding of natural processes ad the interdependence between human beings and environment – and concern leading to environmental action.

CEE's target audiences primarily include children youth and the general community. Equally importantly it addresses its efforts to servicing the EE needs of special groups such as teachers and educators, journalists and communicators, business and industry, government departments and NGOs, planners and policy makers, in-service professionals and community-based organizations.

From its inception CEE's activities have been rooted in and guided by certain basic strategies for maximization of their effectiveness and impact.

## ENABLING TOOLS

Any attempt at developing uniform educational programmes and materials for a counry as diverse as India is bound to run into difficulties. Therefore, adaptability to different geographic, cultural, social and economic contexts built into the basic design of CEE's programmes and materials. CEE's educational designing is somewhat like a saree, more a tool than a finished product, a tool that comes alive in the hands of each user, its application being the combined effort of the designer and the user. "The saree is a designed piece of clothing worn all over India. Over the years very beautiful designs, patterns and textures have been printed and woven into the Indian Saree and yet, several thousand years of Indian history has not tried to stitch the saree. It is worn in many ways ad fits all sizes. It is equally good for working, dressing up or sleeping in. The final effort of the person who designs the cloth and the person who wears it – of the designer and the user. This is a very different concept from that of designing, say, a well-stitched dress. The garment either fits or doesn't fit and, where it is, leaves little room for the wearer to be innovative in its (Kartikeya V. Sarabhai) 1985, Strategy for Environment Education: An Approach for India, 1985).

## MULTIPLIERS

Strategies are designed with a built in multiplier effect to achieve maximum reach and impact. Strategies include building up of partnerships with key partners in government and voluntary sectors; using the mass media, seminars, workshops, etc. For e.g. CEE's communication strategy for media envisages optimal use of existing networks. The 20,000 odd daily newspapers and periodicals in the country constitute a powerful medium to reach a significant section of the urban population and opinion leaders in rural areas. CEE runs a network for environment-related news items, features and articles to newspapers, non-governmental organizations and individual subscribers. The material is a supplied ion English and Hindi, and translated and used by publications in many other languages. The multiplier effect is considerable. Exhibitions form another effective medium for environmental communication. CEE develops and fabricates exhibits on diverse environment-related themes. Television and video facilities are becoming more widely available in the country CEE recognizes the effectiveness of this medium in environmental education.

## THE AGENTS OF CHANGE, SELECTING TARGET GROUPS

While CEE's mandate is to reach out across a wide range of target audiences CEE has strategically selected from within these groups in order to achieve maximum reach and impact. For e.g. in order to reach a large number of school children, CEE works to capacity build teachers as primary agent for introducing and sustaining EE in schools. This is because a single teacher can reach out to at least 100 to 200 children every year over a period of 15-20 years of his/her teaching career. Another initial strategy was to target the middle schools (Std. V to VIII), age group 10-14 years as the entry point to introduce activity based learning by doing approaches for environmental education. This was keeping in mind that at this stage, while children are open to, and capable of grasping new ideas, they are also not too pressured by competitive examinations. Thus the school managements also are more willing to allow innovative approaches at this stage.

## PARTNERSHIP WITH NGO's

The areas and target groups requiring environmental education are enormous and cannot be handled by one single institution. CEE has, therefore, developed a strategy of working with others. Partnerships utilizing complimentary strengths of other organisations to avoid duplication of effort and network effectively for synergistic convergence of ideas and goals.

One example of such a partnership is a school cluster programmes. Developing a sustainable model of partnerships between CEE, government and non-government institutions to sustain and invigorate this network was a key strategy adopted from the start. To this end CEE developed a model of close partnerships between the local schools, local NGOs, State Departments of Education across the country. Today, under the school-cluster programme supported by MoEF year round programmes soon in schools through 55 NGO-school clusters covering all States. Over 1000 schools are part of this network.

## NOT REINVENTING THE WHEEL

CEE strongly feels that it is a waste of precious talent and resources to reinvent the wheel. CEE also strongly feels that blindly using materials developed elsewhere would defeat CEE's role as innovators. It would also possibly reduce the chances of successful and widespread use of the materials.

Successful adaptation involves a variety of inputs including a fair amount of imagination and creativity. It is putting the issue under discussion in the relevant environmental and educational context ; finding ways in which the material can retain its excitement and validity, but still work within the constraints of the local educational system ; it is finding ways in which the creative aspects of the material are not lost, but find expression in local forms. Some examples of such adaptation is Nature Scope India which is the adaptation of the American Nature Scope under an understanding with the National Wildlife Federation, USA. Another such collaboration in with the World Resources Institute, USA under which a series of manuals is being developed for college teachers to aid them in Introducing environmental concepts, issues and perspectives to undergraduate students.

## FIELD DEMONSTRATIONS

CEE has through its field programmes designed especially for communities living in and around national parks and sanctuaries developed appropriate programmes aimed at upgradation around protected areas in order to reduce dependence on the protected areas and their conflict with the forest department. The programmes are also aimed at creating awareness among the communities about the ecological significance of the forest areas around which they live. For e.g. CEE has launched the first Eco-development programme around the Ranthambhore National Park in Rajasthan. A field office was set up in 1987 at Sawai Madhopur near Ranthambhore for implementation of various educational and developmental programmes in selected villages around the park. The focus of the programme has been on environmental improvement by people themselves through environmentally sound technologies. Among the areas for which communication/education programmes were developed to this end, were animal husbandry. Another project – Hingolgadh Ecodevelopment Project was initiated in six villages around the Hingolgadh nature Education Park in Rajkot district of Gujarat. Today the project continues activities in 15 such villages and has extended its activities to 25 other villages not directly dependent on the sanctuary.

## EXPERIENCING NATURE

Sundarvan, a nature education park, is part of CEE. This facility seeks to expose urban children and adults to the beauty and harmony of nature and to awaken their love for flora and fauna. This is carried out through a number of educational programmes using small animals, nature camps and training camps. Regular snake shows are held in Sundarvan to dispel myths and educate visitors about the nature and behaviour of reptiles. Outreach programmes include taking small animals to schools for demonstrations.

Two permanent camp sites have been set up for nature education in the context of forest and marine environments respectively at Bakore and Beyt Dwarka in Gujarat. These are managed by Sundarvan.

In total about 13000 participants have been exposed to nature programmes through these camps. Some special camps for special groups such as deaf and dumb children and under privileged children have also been offered. CEE Regional offices also conduct nature camps as part of their activities participants include students. NGOs and nature enthusiasts.

## SUSTAINABLE DEVELOPMENT

Some new Initiatives were launched after the Rio Conference and in the light of Agenda 21 objectives of working towards sustainable development. Developing a comprehensive information system in key decision making areas such as industrial technology, conservation of protected areas and urban development was reinforced. With a view to resolve the increasing conflict between development and environment goals in a wide variety of areas, CEE launched in 1996 a special initiative that targets policy makers in every field who make resource use decisions that have an impact upon the environment.

Centre for Environment Education seeks to raise the level of general awareness and provide policy support on key environment and development issues through its initiative Media for Sustainable Development (MSD). MSD projects and programmes identify and articulate issue of current policy relevance in the environment-development interface and develop them further through a process of wide public consultation and discussion.

A major book series funded by the Ford Foundation, the Swiss Development Cooperation and CEE on policy issues relating to sustainable development is being brought out by CEE. The books aim to educate and inform decision makers and policy planners about a concern or issue related to ecologically sensitive and equitable development and to present alternative and innovative models of action for a variety of ecological situations. Work has also been initiated on an audio-visual series dealing with sustainable development alternative for decision makers in particular, and the community in general.

MSD also runs a monthly press feeder service CEE-NFS (CEE's News and Feature Service) which supplies well researched, ready to use articles, news items and features on a non-exclusive basis to over 600 newspapers and magazines across the country. CEE-NFS seeks to make optimal use of existing media networks for the improvement of environmental awareness. CEE-NFS also carries articles based on issues identified by the books under the Environment and Development series. As part of MSD's capacity building efforts, training in communicating environment and development concerns is provided through a specialized internship programme for graduate and post graduates.

## IT/WEB SITES

Information Technology (IT) advances in the last few years has thrown up an entirely new medium, which is fast and economical. The internet has also provided everybody an equal opportunity to put information on any topic irrespective of which part of the world one is geographically located in. The Internet has given us the World Wide Web, the E-mail and the Internet Relay Chat amongst many others. CEE decided to look at this as a potential medium of the future for EE.

CEE has already launched seven sites targeted at various groups. The sites and the groups they are targeted at are

http://www.ceeindia.org/ (about CEE)

http://www.cee.envirodebate.org/ (for decision makers)

http://www.cee.greenteacher.org/ (for educators)

http://www.cee.kidsgreen.org/for (for children)

http://www. cee.saseanee.org/(the south and southeast asia network site)

http://www.cywen.com/ (for youth)

www.ceeindia.org/greenhousegases (about climate change)

Each of the sites uses a different style. Highlights of some of the sites are:

- the site for decision makers has an online debate facility. The topic of the debate changes regularly and allows the visitor put across their views on Environment and Development issues.
- the site for educators not only gives visitors an overview of EE, but also the tips and activity ideas on how to make EE more effective in the classroom.
- the site for children aims at not only providing information about our environment, but also provide opportunities for children to interact with one another and also with experts.
- the site for environmental educators in south and southeast asia provides an overview of EE in the region and who are the key people their areas. It also gives a list of professionals trained in EE through a Regional Capacity building programme of CEE, IUCN and WWF.

In addition to the internet, CEE is using IT in its capacity building programmes. The Environmental Education Bank for example is a computerized database of activities, concepts, case studies and resources. The electronic storage of the vast illustrations resource built up over the years on CDs is another area. This has helped in reducing the risk involved in sending out original illustrations and also saving of time in retrieving data.

The future areas which CEE would like to look at is ways in which IT can be used for providing Computer Based Training on the web, as part of its capacity building activities. This becomes particularly important as the cost of getting participants is going to become increasingly expensive.

Networking all of CEEs offices to enable faster and reliable systems of management, thus providing all offices to participate in discussions and meetings without having to get everybody to one location.

IT has many advantages, but care is to be taken so that it does not replace interpersonal communication avenues completely.

## TRAINING AND CAPACITY BUILDING

Over the years, training to capacity build individuals and institutions to environmental education and communication has become an important thrust area of CEE. A range of short and long term training programmes are offered. Some of the programmes are : Training programme in Environmental Education (TEE), Internship Programme in Environmental Journalism, Training in Librarianship and Documentation (TLD), Training for Indian Forest Officers, Environmental Education Bank, Certificate Course in Environmental Education (CCEE) to fulfill the requirement of trained professionals in EE in South and South East Asia.

From its inception, all these efforts and strategies have been part of CEE's philosophy of networking. Networking is not only a key activity but is characteristic of all projects and programmes at CEE. It is seen as a way to gain access to a greater range of expertise and other resources, and avoid the duplication of efforts thus leading to the efficient use of available resources.

## REGIONAL & INTERNATIONAL APPROACH

Various International activities are carried out by the Centre. The activities include on-going training programmes, workshops in collaboration with other NGOs, Institute etc. The activities/programmes include Workshops/Seminars/Specific Focus Meeting, Training Programmes, Subject Expertise, Collaborative Projects and Partnerships and Networking.

SASEANEE is the Asian Regional Network of IUCN's Commission on Education and Communication.

SASEANEE is a network of agencies and individuals in or interested in networking, initiating, or supporting environmental education programmes in the Region. The Secretariat is located at CEE. The South Asia Cooperative Environment Programme (SACEP), has through the Ministry of Environment and Forests, designated CEE as it's Subject Matter Focal Point of Environmental Education and Training and National Focal Point.

A number of workshops have been conducted in collaboration with international Agencies, Institutions, NGOs. Resource materials brought out by CEE have been adapted and developed for use in neighbouring countries by agencies of those countries. CEE has entered into various collaborative projects with international partners whereby there is sharing of resource materials and programmes. Such collaborative projects help in sharing and enriching the perspectives of all the partners concerned. These include partnerships with the following: National Wildlife Federation, USA, State University of New York, US National Park Service, World Resources Institute, USA, Field Studies Council, UK, International Television Trust for the Environment (TVE), UK UNESCO-UNEP.

## CONCLUSION

Speaking in 1928, Mahatma Gandhi cautioned that "God forbid that India should even take to Industrialism after the manner of the west.... It would strip the world bare like locusts." [7] And yet development is a must P.V. Narasimha Rao, India's Prime Minister at the time of Rao emphasized the need of country to "take upon itself the challenge to deal with population-related issues which are inextricably linked to the total development of our nation." But with all these warnings, India like most other developing countries is seeking to find sustainable strategies to an unsustainable goal. India's Report to the United Nations Conference on Environment and Development. 1992 highlighted this dilemma that "while the challenge of development might seem to be 'how to get there', the real challenge is 'how not to get there'. [8].

With increasing globalization, rapidly increasing consumption levels fueled by big budget advertising and a rate of population growth that is yet to come down to sustainable levels, the challenge of EE in India are enormous. Several strategies are in place and new ones are being evolved. But whether these will be adequate and are able to act in time is a question that only time can tell.

## REFERENCES

1. Indira Gandhi, World Conservation Strategy for India, March 1988.
2. Constitution of India, Article 48-A.
3. Constitution of India, Article 51-G.
4. Ministry of Environment and Forests, Government of India, Annual Report 1999-2000.
5. Supreme Court of India, Writ Petition (Civil) No 860 of 1991, M.C. Mehta vs. Union of India.
6. Census of India, 1991.
7. M.K. Gandhi in Harijan, Vol 20, no. 12, 1928, p.422.
8. Ministry of Environment and Forests, Traditions, Concerns and Efforts in India, June 1992, p.8.

## QUESTIONS

1. Explain sustainable development.
2. What are the measures to attain sustainable development ?
3. Why it is essential to conserve water ?
4. In what cases, hunting of wild animals is permitted ?
5. Discuss the role of NGO's to protect environment.
6. Write about urban problems related to energy.
7. What do you understand by public awareness? Explain.
8. Discuss need of women education.
9. Discuss rain water harvesting. What are advantages from this technique ?
10. What do you mean by resettlement and rehabilitation of people ?
11. Describe environmental ethics.
12. Differentiate between nuclear accidents and Holocaust.
13. Describe women education in India.

14. Discuss the main issues involved in enforcement of environmental legislation.
15. Write short notes on :
    (*i*) Water shed management
    (*ii*) Indira sagar project
    (*iii*) Green house effect
    (*iv*) Ozone layer depletion.
    (*v*) Chernobyl accident
16. Explain forest conservation Act.
17. Write an essay on water conservation.
18. Write an essay on Environment Protection Act.
19. Describe the role of public awareness in environmental protection.
20. Describe environmental education.

# Human Population and the Environment

## INTRODUCTION

A population may be defined as a group of organisms of the same species occupying a given area at the same time. It is subdivided in to demes or local populations, which are groups of interbreeding organisms or the smallest collective unit of a population. A population may consist of either *unitary* or *modular organisms*. Insects, fish amphibians, birds, mammals are examples of unitary organisms where each individual is produced from a zygote and the form and development of individuals is highly predictable. But in modular organisms, the zygote develops in to a unit of construction or module, which produces further modules to form a branching structure, therefore form and development of individuals are unpredictable. Sponges, Corals and plants are examples of modular organisms.

## POPULATION GROWTH

The most important features of population is the growth i.e. the capacity of increase in individual members. By measuring the size or density of a given population from time to time, we can get rate of increase and can also predict future changes in its size. It can be defined in following ways—

***(a)* Logistic growth :** When a population is allowed to grow in a limited space (environment) it shows logistic growth. If we plot a graph between number of bacteria or cells against time, we find a typical S shaped sigmoid curve called population growth curve.

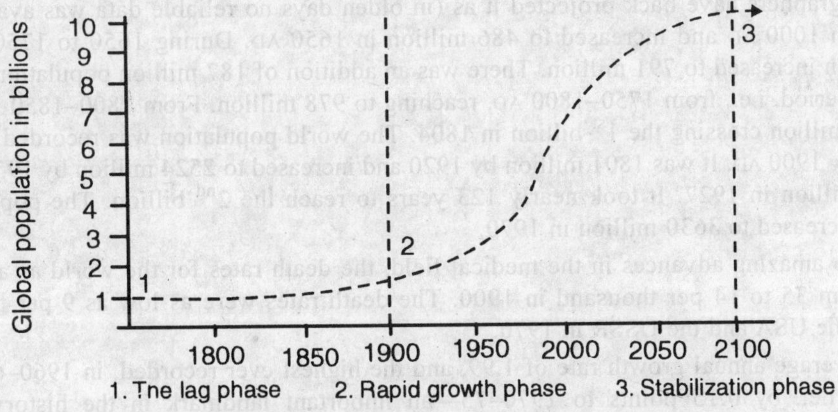


**Fig. 13.1.** The three phases of growth of human population

It has four phases i.e. Ist phase shows slow rate called lag period, second is accelerating stage followed by a phase of extremely rapid population. The last phase is accelerating multiplication followed by equilibrium phase where is essentially no net change in population called saturation level or carrying capacity. It is represented by letter K.

The logistic equation shows density dependent growth i.e. growth of a simple population in a limited space with limited resources. It may be written as—

$$dN/dt = \gamma N(1 - N/K)$$

where $dN/dt$ = rate of growth of population

$\gamma$ = intrinsic rate of increase (per individual of population)

$N$ = Population size (No. of organisms in population at time t)

$K$ = Carrying capacity of population

$(1 - N/K)$ = density – dependent factor.

**(*b*) Exponential Growth :** When a population growth curve quickly begins to rise very steeply, the population shows exponential growth. It is J shaped (Fig.13.1). A population growing exponentially increases accordingly to the equation

$$N_t = N_0 e^{rt}$$

where $N_t$ = The number of individuals in the population after t units of time

$N_0$ = initial population size (t = 0)

$r$ = exponential growth rate

$e$ = the base of the natural logarithm (about 2.72)

**(*c*) Geometric Growth :** Geometric growth may be defined as the population growth in which the rate of increase is proportional to the number of individuals in the population at the beginning of the breeding session. When young ones are added to the population only at specific times of the year during well defined reproductive periods, the population is said to have geometric growth. The equation for this is

$$N_t = N_0 \lambda$$

where $\lambda$ = the geometric growth rate.

## POPULATION - GROWTH

Demographers have back projected it as (in olden days no reliable data was available) 275 million in 1000 AD and increased to 486 million in 1650 AD. During 1650 to 1750 the world population increased to 791 million. There was an addition of 187 million population in another 50 year period, i.e., from 1750–1800 AD, reaching to 978 million. From 1800–1850 it increased to 1260 million crossing the $1^{st}$ billion in 1804. The world population was recorded to be 1650 million by 1900 AD. It was 1801 million by 1920 and increased to 2524 million by 1950, crossing the $2^{nd}$ billion in 1927. It took nearly 123 years to reach the $2^{nd}$ billion. The population has further increased to 3630 million in 1970.

Due to amazing advances in the medical field, the death rates for the world as a whole fell down from 35 to 14 per thousand in 1900. The death rates were as low as 9 per thousand in Europe, the USA and the USSR in 1970.

The average annual growth rate of 1.99, and the highest ever recorded, in 1960–65 declined to 1.84, i.e., by 0.15 points to 1970–75—an important landmark in the history of world population. The declining trend continued for 1980–1990, the rate of population growth was 1.7 per cent and further declined to 1.4 percent during 1995–2000.

The first billion (100 crores) was reached by 1804 AD (probably 29 lakhs year to reach the first billion on the earth).

The second billion was added only in next 123 years i.e., in 1927.

The third billion was reached in 1960, i.e., 33 years after 1927.

The fourth billion was added in 1974, in just 14 years.

The fifth billion was added in 1987, i.e., in 13 years.

The sixth billion was added in 1999 in just 12 years and it will be increased by 80 million people a year to reach 8th billion by 2025.

Doubling time : The concept of doubling time implies the time needed for the population to get doubled. It shows that more the growth rate, less will be the doubling time. (Table 13.1)

**Table 13.1**

| *Pop (billion)* from 0.8 (1750) | *Years* to 1.6 (1900) | *Doubling* 150 time (years) |
|---|---|---|
| 1.25 (1850) | 2.5 (1950) | 100 |
| 2.5 (1950) | 5.0 (1987) | 37 |
| 3.0 (1960) | 6.6 (1993) | 33 |

| **Growth rate %** | **Population doubling time** | |
|---|---|---|
| Slow growth | < 0.5 | 139 years |
| Moderate growth | 0.5–1 | 70–138 years |
| Rapid growth | 1–1.5 | 47–69 years |
| Very rapid growth | 1.5–2 | 35–46 years |
| Explosive growth | 2–2.5 | 28–35 years |
| | 3–3.5 | 20–23 years |

On July 11th 2000, the world population was 6.5 billion. According to United Nations estimates the world population will reach 7 billion mark in 2011, 8 billion in 2025 and 9 billion in 2042.

If the population explosion occurs in such a way the world will be more crowded, more polluted, warmer, and more unequal between north and south and more tension ridden. Now itself half billion persons of the world are starving or semi-starving. The world is facing grave problems like less (or unequal distribution of) food grains per person, less living space, greater unemployment and underemployment, higher rates of inflation, lower physical quality of life (PQLI) of the poorest of the poor and increasing lawlessness and corruption every where. Added to this are the problems of deforestation, green house gases, low level ozone, chlorofluorocarbons, nitrous oxide, and methane and carbon dioxide.

**Estimated world population**

| | |
|---|---|
| 3,000 BC — 2 crores | Date not available |
| 1 AD — 25 crores | Date not available |
| 1650 AD — 50 crores | Date not available |
| 1800 AD — 100 crores | Date not available |

## VARIATION AMONG NATIONS (DEVELOPED AND DEVELOPING COUNTRIES)

The population of Europe was 163 million in 1750 and increased to 276 million in 1850. The mortality conditions in Europe began to improve as a result of socioeconomic development, due to advances in medical technology and reforms in the field of environmental sanitation and public health. The population increased to 547 million in 1950 and to 728 million in 1999 and is projected to decline to 701 million in 2025.

The share of North America was 2 million in 1750 but by 1900 reached to 82 million and by 1999 to 303 million. This is mainly by immigration. The population of Latin America and Caribbean islands was 16 million in 1750, 38 million in 1850 and 166 million in 1950, further increasing to 512 million in 1999.

The population of Asia was 502 million in 1750 AD increased to 809 in 1850 AD and further increased to 1402 million in 1950. The population of Asia increased its share of world population between 1950–1999 from 56 percent to 61 percent. The two largest Asian countries (Asian elephants) China and India along added 28.6 million to the world population. The population of Africa was 106 million in 1750; fastest growth of population is seen in Africa.

The growth rates are not uniform in the world. In many countries (Europe) the growth rate is less than 0.5 percent per year. In developing countries the growth rates are excessive. For example, India, the second largest will soon overcome China, the present largest, as China is restricting population by rigorous population control and advocating one-child family norm to bring down the population. But impressive progress was seen in East and South East Asia and in Latin America.

The fertility rates of South Korea and Singapore are below the replacement level (NRR—1) and that of Thailand is 2.1, Sri Lanka 2.3, Indonesia 2.8. In India the fertility rate decline to 3.4, whereas in sub-Saharan Africa, The TFR was more than 6.0. In many developing countries, the key factors for fertility decline are changes in govt. attitude towards population growth, increased availability of contraception and the extension of services offered through family planning programme as well as marked changes in marriage patterns.

The population explosion is the major obstacle to the economic and social development of most of the under developed countries.

**Table 13.2. AD 2000 Crude birth rate and death rate (and Growth rate)**

| Country | BR | DR | GR% |
|---|---|---|---|
| India | 24 | 9 | 1.5 |
| China | 16 | 7 | 0.9 |
| Pakistan | 36 | 8 | 2.8 |
| Bangladesh | 28 | 10 | 1.8 |
| Sri Lanka | 18 | 6 | 1.2 |
| USA | 14 | 8 | 0.6 |
| Russia | 10 | 14 | -0.4 |
| Germany | 9 | 11 | -0.2 |
| UK | 12 | 11 | 0.1 |
| France | 12 | 9 | 0.3 |
| Afghanistan | 52 | 21 | 3.1 |
| Tanzania | 41 | 15 | 2.6 |
| Sierra Leone | 46 | 25 | 2.1 |

## POPULATION GROWTH IN INDIA

India's population was estimated as 120 million in 1800, 194 million in 1860 and 255 million in 1871. The modern population census was started in the year 1881; since then the population count has been taken regularly every 10 years. The population of India in 1891 was 267 million which declined to 238.4 million in 1901, due to severe famine, weak monsoons (failures) and epidemic of plague.

In 1911, the population of our country was 252 million, which fell to 251.3 million in 1921. After the world war, the entire country was swept by the distinct waves of the world wide pandemic of influenza. Due to epidemic there was a slight decline in the population. The year 1921 is called *the year of great divide* because it distinguishes the earlier period of declined population growth from the period of moderately increasing growth. From here the absolute number of population added in each decade increased. It is the turning point which marks the beginning of a regular growth in India's population and also the beginning of a rapid and massive population growth in India.

As a result of improved agricultural techniques, advancement in medical and health technology to control epidemics and diseases, improved sanitary and health services, improved transport, communication and infrastructural facilities, the population of India progressively increased from 251.3 million in 1921 to 361 million in 1951, a net addition of 110 million people. From 1951, population started growing at a phenomenal rate. The population of India was more than doubled during 1951 and 1991. The net addition to the population was 485 million during this 40 years period, reaching to 846 million in 1991. India's population touched 1 billion mark on 11$^{th}$ of May 2000 and according to 2001 census the population of India was 1027 million (i.e., 102.7 crore). Now, as per Census 2011, population is 121 crore.

The percentage decadal growth of the country as a whole declined from 23.86 during 1981–91 to 21.34 during 1991–2001, a decline of about 2.52 percent in the decadal growth rates.

Currently, India ranks second in size of the population, next to China. It is estimated that by 2025 AD India's population will touch 1414.3 million mark passing China's population and become the most populous country in the world.

India has only 2.4 percent of the world's land area, but it has to support 16.7 percent of the world's population with an annual growth rate of 1.93 percent (2001). Today for every 2 seconds, 1 baby is born in India and per year 17.6 million babies are born. The absolute addition to the population during the decade 1991–2001 was more than the estimated population of Brazil, the fifth most populous country in the world.

By state wise, Uttar Pradesh is the most populous state in the country with more than 166 million people (which is more than the population of Pakistan, the sixth most populous country in the world). Maharashtra (96.7 million), Bihar (82.8 million) are the second and third largest states in respect of population.

Almost half of the country's population lives in five states, namely, UP, Maharashtra, Bihar, West Bengal and Andhra Pradesh. Interestingly, in the last decade Andhra Pradesh achieved the sharpest decline in the annual growth rate of population (from 2.2 percent during the eighties to 1.3 percent during the nineties), mainly by increasing the level of contraception. However, some states like Bihar continue to register an increase in the annual rate of growth of population (from 2.1 percent in the eighties to 2.5 percent in the nineties).

## POPULATION EXPLOSION

As we have seen growth rate of a population is expressed as the number of individuals by which the population increases divided by the amount of time that elapses i.e.

$$\text{Growth rate } (r) = \frac{\text{No. of birth } (b) - \text{Number of deaths } (d)}{\text{average population in time interval}}$$

There are many cases where *b* is substantially large than *d* for a period of time, following which conditions change so that *d* becomes much larger than *b*. This sort of variations are exponential called "**Population explosion**" during favourable conditions, followed by a "**Crash**" when conditions change. For example, diatom populations in Lake Michigan USA undergo such exponential increases at different time of years.

In 20th century population growth increased too much. This is also called population explosion. Economist Malthus said, resources increases 1, 2, 3, 4, ... while population increases 2, 4, 6, 8, ... respectively. In India population growth is much more than twice. Fertility period is of 30 yrs (from 16 to 46 yrs of age). World population is also increasing day by day is 150 per minute 220,000 in one day. Growth rate is 2.2%, with this population will go 7 billion by 2010. World population increase by 9 crore 20 lac per year i.e. one Maxico every year. Do we have the resources and provisions for feeding, housing, educating and employing all those people being added every year. On 11th May 2000 we become one billion i.e. one person out of every 6 persons in this world.

Our resources like land, water, fuels, minerals, forests grasslands etc. are limited and due to population explosion these resources are getting exhausted. Social, economic, religious all type of reasons are responsible for the high rate birth in our country. The important reasons are lower marriage age, lack of education, joint family system, importance of male child, religious misbeliefs, decline in death rate, increased protection of life from natural risks, increase life span, better means of transport and other facilities.

Due to overpopulation some serious problems are like food supply, space (accommodation), unemployment, education, human health, energy crisis etc. There is a fierce debate on population explosion to reduce fertility rates through world wide birth control programmes. This can be achieved by proper education, mass media, educational institutions, raising the marriage age from 18 to 22, providing the facilities like contraceptives, intra uterine devices, birth control bills, sterilization etc. Family planning programme which is Govt. sponsored programme is also one of the effective means to reduce fertility. It was started in India in 1951.

## NATIONAL FAMILY WELFARE PROGRAMME

Previously this programme was known as National Family Planning Programme. In the year 1977 the name was changed to *National Family Welfare Programme.* Family planning programme was launched in India in 1952. India was the first country to do so.

Beginning of the programme was modest, i.e., establishment of few FP clinics, distribution of FP educational material, training of health functionaries and research. During the third 5-year-plan (1961-66) family planning was declared as *centre of planned development.* Then the emphasis was shifted from *clinic approach to extensive education approach* (i.e., motivating people about *small family norm*). A separate Department of Family Planning was created in 1966 in the Ministry of Health. In 1972, the MTP Act was passed. In April 1976, National Population Policy was framed.

During the emergency period (1976), forcible sterilisation campaign led to the defeat of Congress in 1977 elections. In June 1977, new Janata Government formulated *a new population policy* and made family planning as voluntary and renamed it as *Family Welfare Programme.*

The acceptance of primary health care approach as the key to the achievement of health for all by 2000 AD led to the formulation of National Family Welfare Programme in 1982.

### Importance of Family Welfare Programme

1. The family welfare programme occupies an important position in the nation's socio-economic development.

2. Indian population which was 34 crores in 1947 has crossed 100 crore mark by 2000 AD. India has only 2.4% of world's land area but it supports about 15.5% of world's population.
3. India's population is increasing by 1.8 crores every year. To check this galloping growth, the country has laid down long-term demographic goal of achieving an NRR of one by the year 2000 AD.
4. Acceptance of the family welfare services is made voluntary.
5. The programme was 100% centrally sponsored scheme. FP programme was integrated with the MCH services.

## ORGANISATIONAL SET UP

**1. Central level**

At central level Central Cabinet Subcommittee is present. It is headed by Prime Minister. Next level is Population Advisory Council. This is headed by Union Minister of Health and Family Welfare. Members are representatives of various professional bodies and some technical persons. Next level is Central Family Welfare Council, which is headed by union minister and ministers of health and family welfare of all states. It coordinates the work of the programme.

National Institute of Health and Family Welfare, situated in Delhi, is the apex institute. It undertakes research and training in family welfare. Directorate General of Health Services was the central programme officer for Family Planning. He advises Government of India on various aspects of family welfare.

**2. State level**

Ministry of Health and Family Welfare is the apex organization at the state level. This is headed by the minister of health and family welfare of the respective state. At the state level the family welfare work is organised by State Family Welfare Bureau. The State Family Welfare Bureau has three wings:

(*a*) Administrative wing (headed by state family welfare officer and associated by some officers)

(*b*) Education and information wing (headed by mass media information officer)

(*c*) Field operation and evaluation wing (headed by statistical officer).

**3. District level**

At district level the work of family welfare is organized by *District Family Welfare Bureau.* This has three wings like the state level. At some districts *Regional Family Welfare Training Centres* are present. These will undertake training of medical officers and para-medical staff.

**4. Peripheral level**

In rural areas the family welfare work is looked after by *rural family welfare centres* attached to PHC while in urban areas *urban family welfare centres* will took after this work.

**5. Village level**

At village level the MPHA(F) and MPHA(M) are mainly responsible for the programme. They will take the assistance of CHG, TBA and anganwadi workers..

## Goals of National Population Policy

1. NRR 1 (which implies two-child norm)
2. Birth rate 21/100 population
3. Death rate 9 per 1000 population

4. Raising couple protection rate to 60%
5. Reduction of family size to 2.3
6. Decrease the IMR to 60 per 1000 live births.

**Programme Strategies**

1. Integrated approach
2. Cafetaria approach
3. Welfare approach
4. At risk approach

**1. Integrated approach**

It is the integration of all the activities like *raising the age of marriage, increasing female literacy, empowerment of women, raising the overall economy of the country etc.* To achieve these objectives, various anti-poverty programmes have been started.

**2. Cafeteria approach**

The programme is made voluntary and all pressurizations are removed. It stresses more on motivation and education of the people about the benefits of family welfare programme.

**3. Welfare approach**

In 1977, the word *planning* was removed and changed to *family welfare programme.* This indicates the government's commitment to family welfare rather than family planning.

**4. At risk approach**

Like any other programme *NFWP stresses to ensure maximum support to those who are in need.*

Now family welfare has been given priority in the health development. The Government of India has evolved a comprehensive population policy.

## ENVIRONMENT AND HUMAN HEALTH

Environment is the main determinant of health status of a community. Poor housing is a contributor to low physical and mental efficiency. The relation of poor housing and prevalence of disease is easily recognisable. Certainly if we aimed at obtaining optimum conditions for physical and mental well being, in addition to preventing disease, we must include improvement of housing conditions in the programme. Since poor housing is related to poverty, public opinion has tended to consider it an unavoidable evil.

Environment is defined as, "all the external factors (living or non living, material or non-material) present around man." So it is the entire medium in which the population lives and interacts. The environment may be divided into four components:

1. Physical environment
2. Biological environment
3. Social environment
4. Cultural environment

**1. Physical environment:** Physical environment is defined as, "All those non living things and physical forces present around man". The important components of physical environment are water, air, housing, temperature, lighting, noise and vibration, radiation, refuse and waste such as human excreta etc.

**2. Biological environment:** Biological environment is defined as, "All those living things (plants, animals, rodents, insects, microbes and other human) present around us".

**3. Social environment:** Social environment is defined as, "Social interactions between the individual such as their socioeconomic status, religion and the way of living, standard of living and availability and utilisation of health care facilities."

**4. Cultural environment:** It is the culture in which the individual lives. It includes their knowledge, attitude, beliefs, practices, customs, behaviour etc.

Environmental sanitation is defined by WHO as, "the control of all those factors in man's physical environment which exercise or may exercise a deleterious effect on his physical development, health and survival". Dictionary meaning of sanitation is : "The science of safeguarding health". The world sanitation covers the whole field of controlling environment with a view to prevent disease. It is a known fact that in the countries where environmental sanitation is good (e.g., USA, European countries), there the communicable disease problem is less, whereas in countries where environmental sanitation is poor (like India, African countries etc.), there the communicable disease prevalence is very high.

## PHYSICAL ENVIRONMENT

The following factors

1. Housing and health
2. Water
3. Air pollution
4. Waste disposal
5. Excreta disposal
6. Ventilation and light
7. Noise
8. Radiation.

## HUMAN RIGHTS MOVEMENT

### (Early Political, Religious, and Philosophical Sources)

The concept of human rights has existed under several names in European thought for many centuries, at least since the time of King John of England. After the king violated a number of ancient laws and customs by which England had been governed, his subjects forced him to sign the *Magna Carta*, or Great Charter, which enumerates a number of what later came to be thought of as human rights. Among them were the right of the church to be free from governmental interference, the rights of all free citizens to own and inherit property and be free excessive taxes. It established the right of widows who owned property to choose not to remarry, and established principles of due process and equality before the law. It also contained provisions forbidding bribery and official misconduct.

The political and religious traditions in other parts of the world also proclaimed what have come to be called human rights, called on rulers to rule justly and compassionately, and delineating limits on their power over the lives, property, and activities of their citizens.

In the eighteenth and nineteenth centuries in Europe several philosophers proposed the concept of "natural rights," rights belonging to a person by nature and because he was a human being, not by virtue of his citizenship in a particular country or membership in a particular religious or ethnic group. This concept was vigorously debated and rejected by some philosophers as baseless. Others saw it as a formulation of the underlying principle on which all ideas of citizens' rights and political and religious liberty were based.

In the late 1700s two revolutions occurred which drew heavily on this concept. In 1776 most of the British colonies in North America proclaimed their independence from the British Empire in a document which still stirs feelings, and debate, the U.S. Declaration of Independence.

*We hold these truths to be self-evident; that all men are created equal, that they are endowed by their creator with certain unalienable rights, that among these are life, liberty and the pursuit of happiness.*

In 1789 the people of France overthrew their monarchy and established the first French Republic. Out of the revolution came the "Declaration of the Rights of Man."

The term natural rights eventually fell into disfavor, but the concept of universal rights took root. Philosophers such as Thomas Paine, John Stuart Mill, and Henry David Thoreau expanded the concept. Thoreau is the first philosopher I know of to use the term, "human rights", and does so in his treatise, *Civil Disobedience.* This work has been extremely influential on individuals as different as Leo Tolstoy, Mahatma Gandhi, and Martin Luther King. Gandhi and King, in particular, developed their ideas on non-violent resistance to unethical government actions from this work.

Other early proponents of human rights were English philosopher John Stuart Mill, in his *Essay on Liberty*, and American political theorist Thomas Paine in his essay, *The Rights of Man*.

The middle and late 190th century saw a number of issues take center stage, many of them issues we in the late 20th century would consider human rights issues. They included slavery, serfdom, brutal working conditions, starvation wages, child labor, and, in the Americas, the "Indian Problem", as it was known at the time. In the United States, a bloody war over slavery came close to destroying a country founded only eighty years earlier on the premise that, "all men are created equal." Russia freed its serfs the year that war began. Neither the emancipated American slaves nor the free Russian serfs saw any real degree of freedom or basic rights for many more decades, however.

For the last part of the nineteenth and first half of the twentieth century, though, human rights acitvism remained largely tied to political and religious groups and beliefs. Revolutionaries pointed at the atrocities of governments as proof that their ideology was necessary to bring about change and end the government's abuses. Many people, disgusted with the actions of governments in power, first got involved with revolutionary groups because of this. The governments then pointed at bombings, strike-related violence, and growth in violent crime and social disorder as reasons why a stern approach toward dissent was necessary.

Neither group had any credibility with the other and most had little or not credibility with uninvolved citizens, because their concerns were generally political, not humanitarian. Politically partisan protests often just encouraged more oppression, and uninvolved citizens who got caught in the crossfire usually cursed both sides and made no effort to listen to the reasons given by either.

Nonetheless many specific civil rights and human rights movements managed to affect profound social changes during this time. Labor unions brought about laws granting workers the right to strike, establishing minimum work conditions, forbidding or regulating child labour, establishing a forty-hour work week in the United States and many European countries, etc. The women's rights movement succeeded in gaining for many women the right to vote. National liberation movements in many countries succeeded in driving out colonial powers. One of the most influential was Mahatma Gandhi's movement to free his native India from British rule. Movements by long-oppressed racial and religious minorities succeeded in many parts of the world, among them the U.S. Civil Rights movement.

In 1961 a group of lawyers, journalists, writers, and others, offended and frustrated by the sentencing of two Portugese college students to twenty years in prison for having raised their glasses in a toast to "freedom" in a bar, formed Appeal for Amnesty, 1961. The appeal was announced on May 28 in the London Observer's Sunday Supplement. The appeal told the stories of six "prisoners of conscience" from different countries and of different political and religious backgrounds, all jailed for peacefully expressing their political or religious beliefs, and called on governments everywhere to free such prisoners. It set forth a simple plan of action, calling for strictly impartial, non-partisan appeals to be made on behalf of these prisoners and any who, like them, had been imprisoned for peacefully expressed beliefs.

The response to this appeal was larger than anyone had expected. The one-year appeal grew, was extended beyond the year, and Amnesty International and the modern human rights movement were both born.

The modern human rights movement didn't invent any new principles. It was different from what proceeded it primarily in its explicit rejection of political ideology and partisanship, and its demand that governments everywhere, regardless of ideology, adhere to certain basic principles of human rights in their treatment of their citizens.

This appealed to a large group of people, many of whom were politically inactive, not interested in joining a political movement, not ideologically motivated, and didn't care about creating "the perfect society" or perfect government. They were simply outraged that any government dared abuse, imprison, torture, and often kill human beings whose only crime was in believing differently from their government and saying so in public. They (naively, according to many detractors) took to writing letters to governments and publicizing the plights of these people in hopes of persuading or embarrassing abusive governments into better behaviour.

Like the early years of many movements, the early years of the modern human rights movement were rocky. "Appeal for Amnesty, 1961" had only the most rudimentary organization. The modern organization named Amnesty International gained the structure it has mostly by learning from mistakes. Early staff members operated with no oversight, and money was wasted. This led to establishing strict financial accountability. Early staff members and volunteers got involved in partisan politics while working on human rights violations in their own countries. This led to the principle that AI members were not, as a matter of practice, asked or permitted to work on cases in their country. Early campaigns failed because Amnesty was misinformed about certain prisoners. This led to the establishment of a formidable research section and the process of "adoption" of prisoners of conscience only after a thorough investigation phase.

The biggest lesson Amnesty learned, and for many the distinguishing feature of the organization, however, was to stick to what it knew and not go outside its mandate. A distinguished human rights researcher I know once said to me that, "Amnesty is an organization that does only one or two things, but does them extremely well." Amnesty International does not take positions on many issues which many people view as human rights concerns (such as abortion) and does not endorse or criticize any form of government. While it will work to ensure a fair trial for all political prisoners, it does not adopt as prisoners of conscience anyone who has used or advocated violence for any reason. It rarely provides statical data on human rights abuses, and never compares the human rights records of one country with another. It sticks to work on behalf of individual prisoners, and work to abolish specific practices, such as torture and the death penalty.

A lot of people found this too restrictive. Many pro-democracy advocates were extremely upset when the organization dropped Nelson Mandela (at the time a black South African anti-apartheid activist in jail on trumped-up murder charges) from its list of adopted prisoners, because of his endorsing a violent struggle against apartheid. Others were upset that Amnesty would not criticize any form of government, even one which (like Soviet-style Communism, or Franco-style fascism) appeared inherently abusive and incompatible with respect for basic human rights. Many activists simply felt that human rights could be better served by a broader field of action.

Over the years combinations of these concerns and others led to formation of other human rights groups. Among them were groups which later merged to form Human Rights Watch, the first of them being Helsinki Watch in 1978. Regional human rights watchdog groups often operated under extremely difficult conditions, especially those in the Soviet Block. Helsinki Watch, which later merged with other groups to form Human Rights Watch, started as a few Russian activists who formed to monitor the Soviet Union's compliance with the human rights provisions in the Helsinki accords. Many of its members were arrested shortly after it was formed and had little chance to be active.

Other regional groups formed after military takeovers in Chile in 1973, in East Timor in 1975, in Argentina in 1976, and after the Chinese Democracy Wall Movement in 1979.

Although there were differences in philosophy, focus, and tactics between the groups, for the most part they remained on speaking terms, and a number of human rights activists belonged to more than one.

Recognition for the human rights movement, and Amnesty International in particular, grew during the 1970s. Amnesty gained permanent observer status as an NGO at the United Nations. Its reports became mandatory reading in legislatures, state departments and foreign ministries around the world. Its press releases received respectful attention, even when its recommendations were ignored by the governments involved. In 1977 it was awarded the Nobel Peace Prize for its work.

Unfortunately, the Nobel Peace Prize didn't impress the governments Amnesty most wanted to get through to. That year the Argentine military dictatorship reportedly claimed that Amnesty was a front organization for the Soviet KGB. This supposedly occurred the same week that the Soviet government claimed Amnesty was run by the U.S. CIA, to the amusement of human rights activists and, presumably, embarrassment of certain people in Argentina and the Soviet Union.

## VALUE EDUCATION

**Introduction**. Man acts to satisfy his needs or wants. Any thing which satisfies a human need becomes thereby a thing of **Value.** It is the element of desirability and satisfaction that is common to all values, material or non material. In psychology the term value is generally employed to designate a dominant interest, motive or broad evaluative attitude. Value has been defined variously by different educationists, but on the whole it is interpreted to be either a set of feeling or an action. Human behaviour is governed by his values. These are socially approved desires or goals, conceptions or standards by which things are approved or disapproved. Value is a dynamic term used in different aspects. Indian philosophy has used it in sense of state free from pleasure and pain, psychologists in the sense of "psychic energy", sociologists in the sense of "use of time, energy and money" for certain ends. The last theory is named as "Integral theory".

The progress and development of a nation depends upon the quality of the values cherished by its citizens. One of the serious criticism against our educational system is that it lacks value orientation. Our 1986 National Policy on Education and its modifications have strongly advocated value education.

## IMPORTANT VALUES

Important values may be described as follows—

(*i*) **Religious Value :** It is defined in terms of faith in God. The outward acts of behaviour expressive of this value are going on pilgrimage, is linking in simple life, having faith in religious leaders, worshipping God and speaking the truth. Students (Higher studies) prefer least the religious value.

(*ii*) **Social Value :** It is defined in terms of charity, kindness, love and sympathy for the people, efforts to serve God through the service of mankind, sacrificing personnel comforts and gain to relieve the needy and affected of their misery.

(*iii*) **Democratic Value :** This value is characterized by respect for individuality, absence of discrimination among persons on the basis of sex, language, religion, caste, colour, race and family status, ensuring equal social, political and religious rights to all and respect for all democratic institutions.

(*iv*) **Aesthetic Value :** It is characterized by appreciation of beauty, from proportion and harmony, love for fine arts, drawing painting, music, dance, sculpture, poetry and architecture, love for literature, decoration and the surroundings. It is also the least preferred values in schools.

(*v*) **Economic Value :** This value stand for desire for money and material gains. A man with high economic value is guided by consideration of money and material gain in the choice of his job.

(*vi*) **Knowledge Value :** This value stand for love of knowledge or theoretical principles of an activity and love of discovery of truth. A man with this value considers a knowledge of theoretical principles underlying a work essential for success in it. He values hard work in studies.

(*vii*) **Hedonistic Value :** It is the conception of desirability of loving pleasure and avoiding pain. For a hedonist the present is more important than the future. He indulges in pleasure of senses and avoids pain.

(*viii*) **Power Value :** It is defined as the conception of desirability of ruling over others and also of leading others. A man with this value prefers a job where he gets opportunity to exercise authority over the others.

(*ix*) **Family Prestige Value :** It is defined as the conception of desirability of such items of behaviour roles, functions and relationship as would become one's family status. It implies respect for roles which traditionally characteristic of different castes of Indian society.

(*x*) **Health Value :** It is the consideration for keeping the body in a fit state for carrying out one's normal duties and functions. It also implies the consideration for self preservation.

## AIMS AND VALUES

People all over the world are going through "Value Crisis" due to science and technology. Future India is also most likely to suffer the 'Future shock' of value crisis. Because of this space age value crisis, the need for the preservation and inculcation of moral values of human life is keenly felt with a growing sense of urgency all over the world.

It is known, that aims of life are correlated with the aims of education. But, what aims are to be cherished? The aims of education is the value theory. Aims are value commitments. Aims which arise out of values are to be justified philosophically not based on dogmatic belief. It is suggested that modern civilization cannot survive unless it is inspired by ethical and spiritual values. Among these values the most important, which they stressed was love, greed, hatred, violence, exploitation of others are the order of present society.

Different educational commissions and committees appointed in India in the context of either education in general or value education in particular, have strongly recommended various direct and indirect measures, means and methods for the inculcation of human values through education.

## HIERARCHY OF VALUES

Modern India is passing through a transitional period. The old values are crumbling down without being substituted by the new ones. Tradition and modernity co-exist causing a good deal of duality in the minds of old and young alike. This confusion is most likely to be filtered down to the coming generations mainly through socialization practices and child-rearing techniques of the parents and instructors of the young learners. It is apprehended, that during this crisis period, the capacity for social interaction may be affected adversely. This is high time, to consider the priorities to be taken into account while passing on the values to the next generation.

Teaching is a value oriented activity. Teaching of value is unavoidable. All the activities in which teachers engage and governed by certain values. When we ask the students to read certain book, when the seating arrangement is done, when the topics are chosen, when we invite certain speakers to the school, when we hold the examinations, we are governed by the values we cherish. Every individual has got certain hierarchy of values in his mind. It is the duty of the teachers to enable the children develop proper priorities of values suited to the needs of Indian democracy. Freedom, equality, justice, brotherhood and secularism are to be regarded as top most values by the younger generation.

## THE CONTENT AND METHODOLOGY

After forming proper hierarchy of values, we look to the content. As far as subject matter is concerned, any subject is loaded with question and issues about values. To ignore Sanskrit, Indian culture, Indian Democracy, Indian Languages, for example, is to ignore much that is rich and vital.

Family factors and educational climate, the kind of experiences the child has had in early years and several other ingredients of the child's growth and development go a long way towards developing the values that he finally possesses. Parents and teachers need to attempt to all such aspects of the child's life.

Now comes the question of methodology. Even though many teachers have not been trained to discuss values systematically, several strategies and techniques are available for analysing and teaching values. Let us discuss the main components of approaches to the realization of values by teachers and students in the institutions.

The main components of the method of value-orientations are the following three :

1. By having formal or informal dialogue.
2. By giving some written exercises, *i.e.* values sheets.
3. By holding individual or group-discussions.

The first approach is the clarifying approach. According to Louis E. Raths and his associates (as mentioned in Value and Teaching, pp. 58-64) value should satisfy the following criteria :

1. It has been freely chosen.
2. It is chosen from one or more available alternatives.
3. It has been chosen after due reflection.
4. It is prized and cherished.
5. It has been affirmed to others.
6. It has been incorporated into actual behaviour.
7. It is repeated in one's life.

During formal and informal dialogue the values are known and the students come to know that their judgements are correct or not.

Then comes the technique of writing. At this stage value sheets are given. A value sheet consists of a provocative statement and a series of questions, e.g., why learning is said to be the best of all wealths?

Then comes the stage of holding individual or group discussions. Some problems may be given to the class, e.g., Freedom or worship may be the problem. The students may be asked to discuss this problem. Discussion on situations and solutions may also be held. Discussion on India's religions may be interesting to students.

The above three strategies are to be adopted by means of several devices, e.g., films, pictures, songs, poetry, documents, dramatization, parable, fiction, letters, diaries, interviews, socio-drama, cartoons, posters, field trips, quest speaking, or debate. The use of these devices will make the value orientation through all the three approaches more interesting.

## CO-CURRICULAR ACTIVITIES

The schools should plan their co-curricular activities also from the point of view of value-education. Through these activities many values can be inculcated profitably. Co-curricular activities form an integral part of the modern school curriculum. Various games and sports not only provide the best recreation but they have profound impact on the temperament and outlook of the players. They help in the sublimation of the personality of players. Training in physical education develops the high sense of obedience which leads a person to render his services sincerely. Besides games and sports, there are several outdoor activities like N.C.C., P.E.C., N.S.S., scouting and guiding, mountaineering, trekking which train individuals in team work, self discipline, courage, bravery, obedience, integrity and friendship. Students may learn how to serve others and how to respect labour through these activities.

Apart from the above activities literacy and cultural activities may be organised in the schools for inculcating certain values. Poem recitation, debate, symposium, essay competition, work parliament, science club, story competition, melo-drama, mono-acting, socio-dance, music, fancy dress show, painting competition, group dance etc. will offer opportunities to children to learn qualities of leadership, self-discipline, co-operation etc. Morning Assembly has been found to be very good tool for clarification and communication of certain values. If it is conducted properly, it may prove to be very purposive and fruitful for fostering social moral and spiritual values. The sanctity and dignity of the morning assembly should be maintained both by teachers and students and all must participate in it.

Students' sensitivity to the feelings of others are to be increased. They should be provided with the opportunities in the school campus to talk about feelings, to identify with the feelings of other people and to react emotionally themselves. In short, teachers should encourage and help students to participate in experiences that allow them to feel different kinds of emotions, to come into contact with many different people, to do different things and then to share their perception of these experiences.

## HIV/AIDS

**Keywords & Definitions :**

**AIDS** = Acquired Immuno Deficiency Syndrome

**Acquired** = which is not present since birth but acquired after birth.

**Immunodeficiency** = Deficiency of immune functional cells; deficiency to perform the immunological function.

**Syndrome** = A group of diseases and signs and symptoms of illness.

**HIV** = Human Immunodeficiency Virus

**HIV positive** = The presence of antibodies against HIV in human body is termed as HIV positivity and the person is called HIV positive (Seropositive). It takes 6-12 weeks after infection for antibodies to rise to detectable levels. There is thus a window period during which the infected person may transmit the infection despite being seropositive.

The latest killer disease that has created nightmares for the medical experts is the Acquired Immuno Deficiency Syndrome (AIDS). Its terrifying spread has earned it the title of the "Pandemic" or an epidemic, which is out of control. This disease has wreaked the social & economic devastation. In the absence of medical defence against AIDS, public education is the only weapon in the fight to limit the spread the infection. Only by influencing personal behaviour & life style can we hope to maintain the ravages of AIDS throughout.

**Virus :** HIV belongs to the family Retroviridae and subfamily Lentivirinae. Two different types of HIV have been recognized and are called as HIV-1 and HIV-2. Both differ in geographical distribution, biological & molecular characteristics and extent of transmissibility. These viruses store their genetic information as RNA. HIV-1 has three groups; HIV-1 major groups (HIV-1 M), outlier (HIV-1 O) and HIV-1 new (HIV-1 N) group. HIV-1 major groups can be further classified into different subtypes or clades designated A-K.

HIV is 120 nm icosahedral, enveloped, RNA virus. HIV comprises of an outer envelope consisting of lipid bilayer with uniformly arranged 72 spikes or knobs of gp 120 and gp 41. Inside the protein core surrounding two copies of RNA. Core also contains viral enzymes, Reverse transcriptase, integrase and protease, all essential for viral replication and maturation.

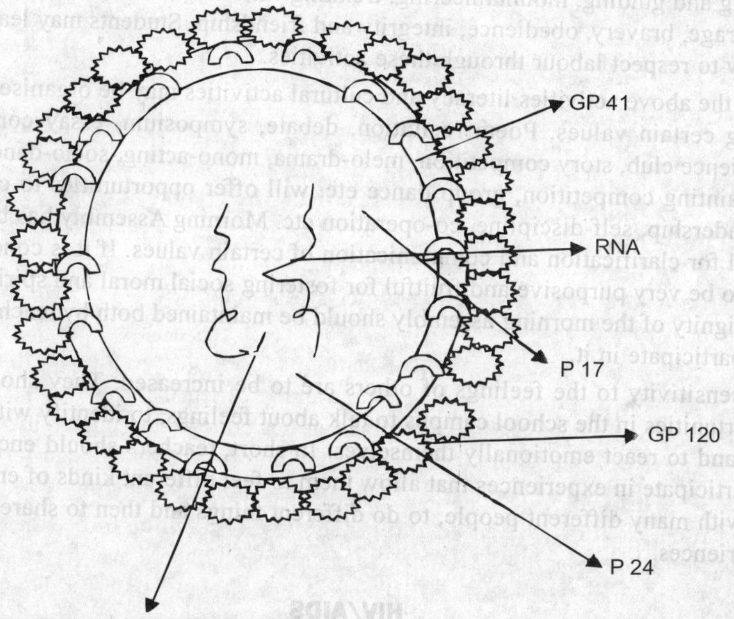


| **Major antigens of HIV-1 :** | | **Antigens :** |
|---|---|---|
| a. | Envelope antigen: | |
| | Spike antigen | gp 120 |
| | Transmembrane Pedicle antigen | gp 41 |
| b. | Matrix antigen | p 17 |
| c. | Core antigen | p 24 |

**HIV replication:**

Gp120 binds to host cell receptor

Reverse transcription

Proviral DNA synthesis

Integral with host cell DNA

Viral proteins synthesis

Virus assembly and budding

Maturation of core proteins

Transmission :

HIV infection can be transmitted by following modes and the efficiency of HIV transmission is:

| *Transmission Route* | *Percent Efficiency* | *Percentage of total cases* | |
|---|---|---|---|
| | | *World over* | *India* |
| Blood Transfusion | 90-95 | 5 | 7 |
| Perinatal | 20-40 | 10 | 10 |
| Sex | 0.1-1 | 75 | 80 |
| Injecting drug abusers | 0.5-1.0 | 10 | 7.3 |

**Transmission cycle of HIV:**

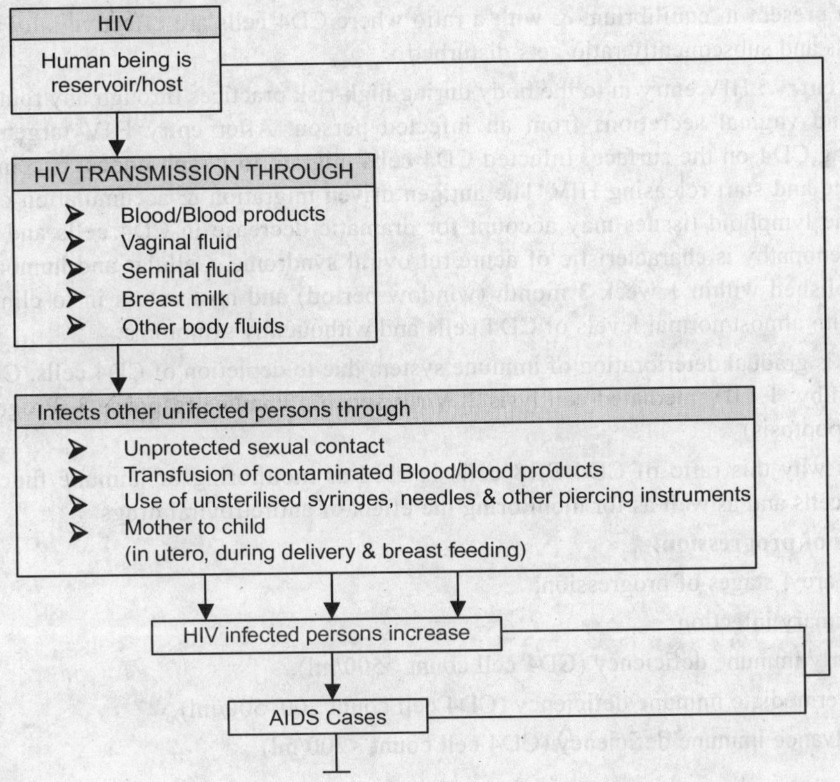

**Sex:** Homosexual and heterosexual contact with HIV infected partner.

**Parenteral:** Infected Blood, Blood products; infected needle & syringes, IVDU.

**Perinatal:** From infected mother to her child (Before 10-30%, during and after delivery 40-60%) and 30% through breast milk. Most common is the heterosexual contact.

**HIV is not transmitted by :**

Shaking hands, hugging, dry kissing, sneezing, coughing, mosquito bite, toilet sharing, sharing of telephones, offices, playing, traveling together, sharing cups, living in same room, donating blood aseptically.

## NATURAL HISTORY OF HIV INFECTION/AIDS

The natural history of any disease refers to the stages through which a disease passes in the absence of any intervention. Clear knowledge of natural history of a disease helps in identifying the stages of disease vis a vis appropriate intervention to prevent or control the disease.

Usually whenever human body comes in contact with any type of infection, our immune system fights against these agents. However, the horribleness of AIDS lies in fact that it destroys the immune system itself thereby making the body susceptible to the all kinds of diseases and infections.

Further, the major pillars of the immune system are antibody mediated (B Cell or Humoral immunity) and cellular (T cell or CMI) immunity. Of which important T cells are T helper cells (Cells with CD4 epitope/receptors), which cause immunity, and T suppressor cells (Cells with CD8 epitope/receptors), which suppress the immune response of TH cells. These cells are normally present in equilibrium & with a ratio where CD4 cells are effective but HIV destroys CD4 cells and subsequently ratio gets disturbed.

**Viral entry :** HIV entry in to the body during high-risk practices through any route via blood, semen and vaginal secretions from an infected person. After entry HIV targets onto cells expressing CD4 on the surface. Infected CD4 cells migrate to lymph nodes, become activated, proliferate and start releasing HIV. The antigen driven migration & accumulation of CD4 cells within the lymphoid tissues may account for dramatic decrease in CD4 cells and generalised lymphadenopathy is characteristic of acute retroviral syndrome. Cellular and humoral responds are established within 1 week-3 month (window period) and host enters in to clinical latency period with almost normal levels of CD4 cells and without any symptoms.

There is gradual deterioration of immune system due to depletion of CD4 cells. CD4 cells are destroyed by: 1. HIV mediated cell lysis 2. Virus specific immune response 3. Programmed cell death (Apoptosis).

That's why this ratio of CD4/CD8 cells is used as monitoring of immune functionality of infected cells and as well as for monitoring the effect of antiretroviral drugs.

**Stages of progression:**

There are 4 stages of progression:

1. Primary infection
2. Early immune deficiency (CD4 cell count >500/ml)
3. Intermediate immune deficiency (CD4 cell count 200–500/ml)
4. Advance immune deficiency (CD4 cell count <200/ml)

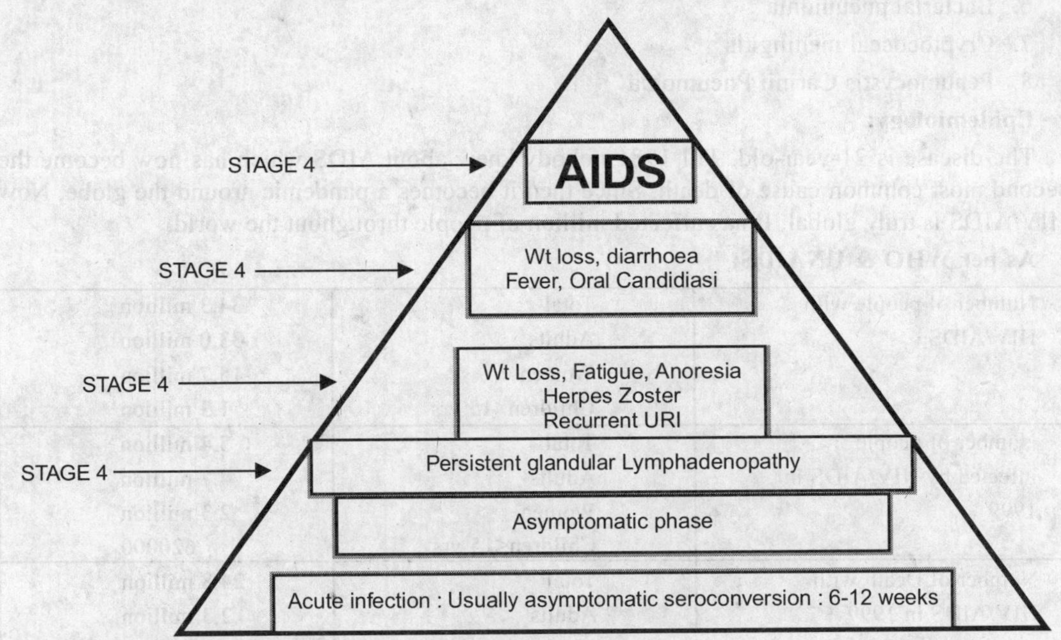


**Laboratory investigations for HIV infection:**

HIV infection can be detected in the laboratory either by detection of :

| | | |
|---|---|---|
| Antibodies to HIV | : | ELISA, rapid/simple, western blot (confirmatory test) |
| Antigen detection | : | p24 antigen |
| Direct detection of Virus | : | Microscopy |
| Detection of viral RNA | : | PCR |
| Isolation/culture of virus | | |

**The indirect predictors of HIV infection and disease**

CD4 cell count

Beta 2 microglobulin

These are used as monitors of immunity status of patients and for monitoring the progression of the disease.

HIV testing strategy :

Unlinked anonymous testing

Voluntary confidential testing

Mandatory testing

**Opportunistic Infections:** Are those infections, which are non, infectious to healthy immune competent human being but on immunosuppression or immunodeficiency status leads to them become infectious. Common Opportunistic infections are:

1. Tuberculosis
2. Candidiasis
3. Crytosporidiosis
4. Herpes Zoster
5. Toxoplasmosis

6. Bacterial pneumonia
7. Cryptococcal meningitis
8. Penumocystis Carinii Pneumonia

**Epidemiology:**

The disease is 21-year-old. Till 1981 nobody knew about AIDS which has now become the second most common cause of death. Since then it becomes a pandemic around the globe. Now HIV/AIDS is truly global. It has affected million of people throughout the world.

**As per WHO & UNAIDS:**

| | | |
|---|---|---|
| Number of people with HIV/AIDS | Total | 34.3 million |
| | Adults | 33.0 million |
| | Women | 15.7 million |
| | Children<15 yrs | 1.3 million |
| Number of people infected by HIV/AIDS in 1999 | Total | 5.4 million |
| | Adults | 4.7 million |
| | Women | 2.3 million |
| | Children<15 yrs | 620000 |
| Number of Death with HIV/AIDS in 1999 | Total | 24.8 million |
| | Adults | 2.3 million |
| | Women | 1.2 million |
| | Children<15 yrs | 500000 |

**In India:**

In India the first case was detected in 1986 in Tamil Nadu.

By the end of year 2000 it was estimated that 3.6 million Indians infected with HIV/AIDS.

More than 50% of all new infections take place among young adults below 25 years.

Every year 22837 newborn children are infected.

About 11,434 die due to HIV/AIDS.

Total 120,000 children are living with AIDS.

120,000 AIDS orphans living in country.

## CONCLUSION

The need of the hour is for dissemination of basic information in simple and intelligible terms about the causes of AIDS and the avenues of infection, the essential precautions to be taken, the facilities for treatment & all other related matters. The public must be alerted without generating hysteria about the real menace of AIDS epidemic & the measures required to combat it.

## WOMEN AND CHILD WELFARE

### History

The health of women and children began to receive separate attention early in this century, in recognition of their greater vulnerability, particularly to socioeconomic and environmental forces, and the interdependence of the child's health and that of the mother. In 1909 the first White House Conference on Child Health recommended the formation of the Children's Bureau, which proceeded to investigate the causes of infant mortality (more than 100 per 1,000 live births). The first direct support of health services for mothers and children came with the Shepard-Towner Act of 1921, which resulted in complete birth registration and the establishment of maternal and child health divisions in state and local health departments. Title V of the Social Security Act of 1935 extended services to handicapped children and further established the principle of public responsibility for the health of mothers and children.

In the 1960s and early 1970s a host of additional programs were initiated by Congress. These included Medicaid, Early Periodic Screening, Diagnosis and Treatment (EPSDT), Neighborhood Health Centers, Maternity and Infant Care, Family Planning, Children and Youth Projects, Head Start, Title I educational assistance, the Right to Education of the Handicapped (PL 94-142), and nutrition programs (WIC and School Lunch). While these laws expanded services at the state and local levels, the resultant programs were administered by a variety of different branches of government, diffusing responsibility and often leading to poor coordination and the undermining of maternal and child health (MCH) divisions.

MCH services were seriously weakened by the budget cuts of the Reagan years. Rather than cutting specific services, the programs were lumped into block grants that gave stage governments greater freedom to apportion the reduced funds as they saw fit. This has begun the trend of providing greater local autonomy in the establishment and administration of health programs for women and children. This trend may hold promise for the creation of innovative programs that more precisely address the community's needs, making greater use of local resources.

Managed care has emerged from the multiple efforts to control health care costs, and this new structure for the delivery of medical care has had a significant impact on maternal and child health services. Medicaid recipients are rapidly being shifted into managed care plans that restrict access to all but primary care services, causing health departments to lose their base of clients for immunization, family planning, and other health care service programs. Meanwhile the managed care plans that have enrolled these high-risk populations are not held accountable in the same manner as was the health department for service delivery. Cost control efforts led to postpartum hospital stays being reduced to 1 day until Congress intervened with a rule that allowed at least a 48-hour stay at the discretion of the mother and her physician.

Progress in MCH has been driven by the dual forces of research and advocacy. Multidisciplinary studies over the last 20 years have shed important light on the health problems of women and children and provided numerous examples of effective means to ameliorate those problems. Articulate and committed individuals and organizations have played a critical role in fostering public commitment to improve the lives of mothers and children. However, that support has been significantly reduced through the 1980s. As our knowledge of what to do has continued to grow, the political will to use that knowledge has shrunk.

## HEALTH INDICATORS

Various health indicators are used to assess the health status of mothers and children. The continuous monitoring of these indicators is in essential part of evaluating our progress in improving the health status of women and children.

Maternal mortality rates have reached such a low level that they are of little value now. Maternal health is better reflected in fertility rates and birth rates as well as in pregnancy-related morbidity rates. Pregnancy outcomes have become a more important measure of maternal health and quality of maternity services provided. Miscarriage, therapeutic abortion, stillbirth, and especially low-birth weight rates can be used to assess the success of the pregnancy. Prenatal care, place of delivery, attendant at delivery, type of delivery, complications, length of stay, and cost all measure the availability and quality of maternal health services.

The infant mortality rate remains an important though crude measure of MCH. Linking infant birth and death records has provided a far more precise way of assessing factors associated with pregnancy outcome, particularly when the causes of death are grouped by pregnancy-related conditions such as prematurity, rather than the organ system taxonomy of the ICD-9 codes. The new birth certificate form that was adopted in 1989 includes a wider array of information on both the mother and the child, offering opportunities for future exploration of the relationships between more extensive sociodemographic and medical information and various pregnancy outcomes. Childhood morbidity is less easily measured. Birth defects registries, neonatal intensive care use, discharge diagnoses, and national health surveys provide some estimates of morbidity. Immunization rates, school-based health data, and the data from such programs as EPSDT are also helpful indicators of child health, although they are not systematically collected at either the state or national level.

Larger social and demographic changes are also important indicators of the status of mothers and children. Over the last 20 years there have been dramatic increases in the percentage of mothers in the workforce, the percentage of marriages that end in divorce, the numbers of homeless mothers and children, and the percentage of children living in poverty. These social problems contribute directly or indirectly to most of the health problems of women and children.

## SERVICE DELIVERY

The goals of MCH services are (*a*) to encourage desired pregnancy, achieving the best possible outcome for the baby and the mother; (*b*) to promote healthy relationships within the family to nurture the growing child; (*c*) to optimize the normal developmental processes to allow the child to achieve his or her fullest potential; (*d*) to prevent child health problems and reduce the risks of adult health problems; and (*e*) to provide early intervention in the health problems of women and children so as to minimize morbidity and mortality in a cost-effective manner. To achieve these goals, attention must be paid to several basic principles of MCH. These principles are a product of the nature of women and children and the problems from which they suffer and therefore are a bit different from the general principles that underlie all health service delivery.

## TWO CLIENTS

Maternal services are unique in that they simultaneously provide care for two, equally important, clients, the mother and the foetus. Balancing the needs of both to achieve the best possible outcome requires a thorough understanding of the complex interdependence of the maternal/foetal unit and the implications of events and treatments for both mother and baby.

## FAMILY-CENTERED SERVICES

Because of the extreme importance of the family in the nurturing of the pregnant woman and the young child, it is essential that MCH services be delivered with attention to the family circumstances of the clients. The family influences growth, development, health-related behaviors, and lifestyle habits. Family resources influence the use and availability of health services and the ability to provide the care needed, particularly in chronic disease. The child is not merely the passive recipient of the influences of the family, but, rather, plays an increasingly interactive role in the family, shaping in part the environment in which he or she lives.

## DEVELOPMENTAL PERSPECTIVE

The fetus and child are being continuously shaped by the normal developmental processes that result in a reasonably predictable series of changes from conception through adolescence. Progress over this course is a sensitive measure of both health and disease. These developmental forces can be potent allies in the management of chronic health problems. However, continued disruption of normal development can have progressively magnifying adverse effects on the fetus or child. Because of the importance of development, the dimension of time and the continuity of care over time become critical elements in the provision of MCH services. Prompt identification of problems and early intervention, therefore, hold the greatest promise for achieving the best outcome.

## HEALTH PROMOTION AND DISEASE PREVENTION

There is great potential in childhood for health promotion and disease prevention to benefit both the current child and the future adult. However, careful attention must be paid to the immediate implications of interventions that are aimed at preventing problems in the distant future, making sure that the desired long-term benefits are not counter-balanced by short-term hazards.

## TIMELY, COST-EFFECTIVE TREATMENT

The early identification and proper treatment of common health problems is a critical dimension of reducing morbidity and mortality in women and children. Simple early treatments can often prevent very expensive and serious problems, such as adolescent pregnancy, a premature birth, or a handicapping condition.

## INTEGRATION OF PRINCIPLES

Perhaps the greatest challenge to delivery of MCH services is trying to integrate all the principles articulated above into the care of each client. It is difficult for the provider to attend simultaneously to the treatment and prevention needs while considering the family and development issues in the care of both the mother and the child. However, the greatest success is achieved when all these concepts are addressed together.

## TRENDS AND INNOVATIONS IN SERVICES

The United States lags way behind all other developed countries in the provision of most services to mothers and children. Western Europe and Japan offer extensive maternity benefits, prenatal care, day care, and well-child care to all women and children. In infant mortality rates, the United States ranks 20th in the world, a fact that many attribute at least in part to the inadequate provision of maternal and infant services. The last decade has seen little progress in this arena, but some interesting innovations in the delivery of MCH services have recently emerged and warrant brief description. Many of these have not yet come into general practice but hold promise for the future, once they have been more carefully evaluated. Social trends of the last decade have also had an effect on MCH services and must be considered in the process of recommending improvements for the future.

## FAMILY PLANNING AND ABORTION

Optimum health for both mother and child has long been known to be related to maternal age, spacing of children, and the balance between family resources and family size. The ready availability of birth control and the option for abortion have provided means of achieving family planning. Norplant was hoped to be an easy means of contraception that required no action for 5 years after its insertion, but it has turned out to be unpopular due to its side effects. RU-487 will soon be available as an early pregnancy abortion pill. Having such a non-surgical abortion option will make it very difficult to ever restrict abortion fully.

## PRECONCEPTIONAL HEALTH PROMOTION

Many of the critical phases of fetal development have already occurred before a woman is even aware that she is pregnant. Optimum fetal health, therefore, requires attention to maternal health and health-related behaviors even before conception. Efforts to counsel women before conception to avoid alcohol, drugs, tobacco and other fetal hazards are currently being tested to determine the impact on pregnancy outcome. The efficacy of periconceptional dietary folic acid in the prevention of neutral tube defects will likely soon result in routine folic acid fortification of bread.

## PRENATAL CARE

Improving access to and quality of prenatal services continues to be a challenge with no obvious solution. Expansion of Medicaid to include women up to 185 percent of the federal poverty level in comprehensive services has made prenatal care more accessible. Some states have developed innovative programs to improve quality and access for the poor, and many local community-based projects have also been created with these goals in mind.

## HUMAN IMMUNODEFICIENCY VIRUS

The vertical transmission of human immunodeficiency virus (HIV) from infected women to their babies will infect approximately one-quarter of these babies. However, treatment with azidothymidine (AZT) during pregnancy can reduce this transmission to as low as 8 percent. This creates the dilemma of whether pregnant women should be required to be HIV tested and, if positive, treated for the sake of their baby or whether they should merely be given the information and the option.

## IMMUNIZATION

The recently approved acellular pertussis vaccine offers hope that some of the complications of the whole cell vaccine might be avoided. The use of conjugated vaccine against *Haemophilus influenzae* in infants as young as 2 months of age has significantly reduced the incidence of meningitis in infant. The hepatitis B vaccine is now widely used in the prevention of both hepatitis and the consequent hepatic cancer—both major problems, particularly in developing countries. The eradication of polio from the Western Hemisphere has led to a recent switch from oral polio vaccine to a combination of oral and inactivated vaccines to reduce the incidence of paralytic reactions to the oral vaccine.

## DAY CARE

As maternal employment continues to rise, we are falling further behind in providing adequate, affordable day care for young children. Most states provide little regulation of day care facilities, particularly home-based centers with few children. This aggravates concerns about spread of infection, injury, and child abuse in day care facilities, as well as potentially being a poor substitute for parents.

## SEXUAL ABUSE

Over the last decade we have been forced to recognize that child sexual abuse occurs far more frequently than we would like to believe. As this problem has come out of the closet, innovative programs have been developed to teach children how to avoid sexual exploitation and to identify and treat more effectively those who are victims. Many school systems have adopted curricula in prevention of sex abuse for children as young as those in kindergarten, and the open discussion of this problem has made it easier to inquire about such incidents as a part of routine health care. Many states are also experimenting with creative ways of humanely dealing with the child witness in court without unduly infringing on the constitutional rights of the accused. However, we are probably still studying only the tip of this iceberg as we begin to explore the consequences of the more subtle forms of abuse and try to gain a better understanding of the abusers and the factors that lead to abuse in some families.

## COMMUNITY-BASED SOCIAL SUPPORT

Pregnant women and young children thrive best when they are surrounded by friends and relatives who provide companionship and assistance. As unwed motherhood becomes more common and the extended family further disintegrates, more young families face isolation and inadequate social supports. To remedy this, several programs have utilized home visitors to befriend and work closely with pregnant women and young families, offering the assistance and social support that are so often inadequate. Some of these programs have been able to demonstrate benefits in health and well-being associated with participation in the program.

## FUTURE DIRECTIONS

The 1960s and 1970s produced a number of centrally funded programs to help mothers and children. Through the 1980s the support for these programs eroded, and control over spending priorities for the diminished funding was shifted to the state capitals through the block grants. Over the same period new initiatives for the elderly received increasing support. Despite renewed outcries about the plight of America's children, Congress was able to pass sweeping welfare reform, which is predicted to have significant negative impact on women and children.

As managed care appears to be cutting costs, there is a rising pressure for shifting MCH services from public sector provider systems into managed care plans where demand can be controlled. In those states where many of these services are provided by health department, the work and the funding of these departments will be dramatically reshaped by this trend. Both the consumers and the managed care providers will need to make adjustments if this new system is to simultaneously save money and still meet the health needs of this population. The tightening of other welfare benefits for the poor will additionally compromise the basic well-being of poor women and their children. Over the next decade it will be increasingly important to monitor the well-being of women and children to ensure that the systems established to improve their health are indeed succeeding. A recently released report of the Institute of Medicine of the National Academy of Sciences addresses ways in which communities might improve health through performance monitoring.

## ROLE OF INFORMATION TECHNOLOGY IN ENVIRONMENT AND HUMAN HEALTH

Just as chemical or metallurgical or electrical technologies enable the processing of raw materials in to usable goods, to satisfy man's and societies needs, so does information technology (IT) help the storage, processing, transmission and exploitation of information to satisfy a person's, company's, society's or Govt's needs for information. Information covers voice as in telephony, text as in fax, images as in video and data as between computers.

Information is knowledge and knowledge is power. Knowledge plus experience is wisdom and it is the wise use of information that gives advantage to those who have information.

Information technology as commonly picturised by computers in extending man's mind or brain or intellectual power. Information technology devices like microprocessors and becoming mass appliances from pace makers for the heart, hearing aids and efficiency enhances in automobile engines and devices to steer space vehicles on the moon. Like banking, trading, learning and teaching, librarying and other sectors of human activity, Information technology has tremendous use in the field of environment and human health. The print media has also highlighted and participated in some of the important environmental issues and helped in making the people aware. Apart from the development of softwares for environment & healthy studies facilities like internet, world wide web geographical information system (GIS), information through satellites are also developed. These all are also helpful in environment and health studies.

Media has the capability to influence people's opinion. It has highlighted and participated in some of the important environmental issues and has helped in making the people aware. The Ministry of Environment and Forests Govt. of India has taken up the task of compiling a DATABASE on various biotic communities. Database includes wildlife database, conservation database, forest cover database etc. Since database is the collection of inter-related data on various subjects, also available for chronic diseases like HIV/AIDS etc. They are in computerized form. The Ministry of Environment & Forests, Govt. of India has created Environmental Information System (ENVIS). ENVIS has many centres all over the country for generating a network of database in different areas like pollution, environmental management, wildlife etc.

www.mhhe.com/environmental science and multimedia Digital Content Manager (DCM) in the form of CD-ROM are the on line learning centre, which provides the most current and relevant information on environmental science.

With the help of computers and internet, not only we can have knowledge of the patients at a glance within no time, but also we can get information about the diseases their medicines and alternative medicines. (Alopathic, Ayurvedic and Homoeopathic). CT-scanning (Computed Tomographic Scanning, CAT (Computed Axial Tomography) are such examples of information technology in human health. It also help in computer-aided monitoring, health testing, implantation of artificial body parts and in pharmaceutical designing of new drugs. IT is expanding rapidly with increasing applications in the field of environment and human health. For awareness, songs, drama, bhajans, advertisements in TV, radio, cultural programmes should be exhibited in urbans & rural sectors both.

# HUMAN ACTIVITIES AND ENVIRONMENT

## HOW DO HUMANS AFFECT THE ENVIRONMENT

Climate change, extinction of species and pollution of life supporting air and water has become a growing concern for nations all over the world. Knowing how do humans affect the environment will help us understand and address the problems better.

The Intergovernmental Panel on Climate Change (IPCC) was formed in 1988 to study the risk of climate change due to human activity. In the Kyoto Protocol of 1997, industrialized nations agreed to cut down their emissions of greenhouse gases considerably by 2012. Even the G8 committee that began as a response to the 1937 oil crisis and was primarily concerned with the economic issues of the members nations, now unfailingly have something related to the environment in their summits.

Years of thoughtless exploitation of nature by man has resulted in the effects staring right in our face now. The truth, that in the bid to improve our lives, we have put our own survival to stake, has finally hit us hard. Now as more and more studies and researches are being carried out to understand how do humans affect the environment, an increasing number of people are awakening to the fact that the well-being of the environment and survival are intricately woven into each other.

## EFFECTS OF HUMAN ACTIVITIES ON ENVIRONMENT

### CLIMATE CHANGE

Just about a year back, I came across an article in which environmentalists expressed their concern that if the Gangotri glacier (that feeds river Ganges) kept melting at its current rate, the river Ganges would soon dry up. River Ganges is the holy river of India, which has served as a lifeline for centuries for millions of people its banks. Rivers have played an important role in the survival of mankind, and many of them are fed by glaciers. Global warming refers to the increase in the temperatures of the earth due to release of greenhouse gases like carbon dioxide and methane from industries and vehicles. This phenomenon is causing the glaciers to melt at an alarming rate. Not just the Gangotri, but even the polar ice caps are melting at a faster rate they can form. The result is increase in the sea level, and it poses a danger of drowning the low lying areas. Some of the areas that may go under the sea if the sea levels kept increasing include Bangladesh, parts of Africa and even major cities like London and New York!

### DEPLETION OF OZONE LAYER

The stratosphere has a layer of ozone that protects us from the harmful ultraviolet (UV) rays of the sun. Exposure to these layers cause skin cancer and cataracts. However, the ozone layer filters out the dangerous UV rays from sunlight as it enters the earth's atmosphere. The cholofluorocarbons (CFCs) that are man-made chemicals are released in the atmosphere through CFC containing aerosols, refrigeration equipment, foam and as by products of certain industrial processes. As these chemicals are released, they rise up into the atmosphere and break down the ozone molecules that form the ozone layer. There is an ozone hole in the Antarctic stratosphere that is causing great concern to environmentalists all over the world. Not only is the depleting ozone layer harmful to human beings, but the UV rays can destroy a certain type of bacteria known are Cyanobacteria that are important for a number economically important crops. Researches are even predicting that excess level of UV rays could lead to the death of the phytoplanktons, that are an important component of the food web of the oceans.

## POLLUTION

Industrialization has been the hallmark of human progress. However, with industries have come a host of toxic gases that are being released into the atmosphere even as I write this article and you read it. The industries release gallons of liquid waste into the seas and rivers. Some of the effluents percolate down to the reach the ground water and pollute it to the extent that it can't be used by human beings for drinking or cooking. Besides adding to air pollution, the innumerable vehicle running on the roads add to noise pollution that has lead to increase in stress, anxiety and problems related to hearing. Pollution of water has led to the decrease in a number of large number aquatic life forms. Migratory birds are known to change their course due to pollution or change in weather. Respiratory diseases in human beings is another price that we are paying for polluting the environment.

## DEFORESTATION

Expanding population, industrialization and need of land for development of expanding cities have led man to cut down forests selfishly. Not only are the forests home to a large number of animals, trees are also an important component of the water cycle. The roots of plants hold the soil together and prevent soil erosion. The global forest cover has shrunk to half its area in the last 11,000 years. During the period between 1990 and 2000 itself, the yearly loss of natural forests 16 million hectares. Deforestation at such alarming rate has been a cause of constant worry for environmentalists the world over.

## EXTINCTION OF SPECIES

Man has been killing animals right since the time he acquired the skill of hunting. Although in those times hunting was the means for survival, human beings continued to kill animals even after he had learned to cultivate crops. The relentless hunting by human beings, sometimes for the hide of a cheetah or the tusks of the elephants, or simply to cook the tasty shark fin soup, has wiped out the existence a large number of animals in just a century. Besides hunting, human activities like environmental pollution and deforestation has led to the extinction of a large number of animals and plants due to loss of habitat. Recent studies have shown that in North America, 37 animal species have become extinct in the last 50 years due to human activities. Loss of habitat led to the extinction of the Bali subspecies of tiger in 1937. According to the 2008 annual IUCN report, there were 16,928 animal and plant species that are threatened of extinction and the list keeps increasing every year.

Despite the indiscriminate exploitation of nature by some people, there are a handful of those that are relentlessly working to reverse the deterimental effects of human activities on the environment. The World Wildlife Fund, European Environment Agency and the National Geographic are some organizations that run programs for the preservation of nature and educate and inspire people to conserve all life forms be it flora or fauna.

Now that we know how do humans affect the environment, let us not sit back leaving all the responsibility of conserving our planet on a handful of people. Each one of us can make a valuable contribution towards preserving our environment. Let us do our own bit, for, in the well-being of the environment lies the key to our survival.

**Antarctica is certainly the most pristine place on earth although it is not as unspoiled as may be imagined**. For more than a hundred years people have travelled to Antarctica and most parts have now been visited. More than just footprints have been left and more than just photographs have been taken.

**Some Antarctic species have been taken to the verge of extinction for economic benefit.** Others have been killed or disturbed, soils have been contaminated, sewage has been discharged into the sea and rubbish that will not decompose or break down has been left behind in even the remotest parts.

Recently attitudes have changed as we begin to realise that there are **few unvisited places left on earth** and that they are tremendously important of humanity. Antarctica's clean air, water and ice of are of importance to science for understanding how the Earth's environment is changing both naturally and as a result of human activity.

Tour operators are in beginning to trap a huge and ever increasing demand to visit the Earth's last great wilderness. Both science and tourism have the potential to damage the very qualities that draw them to Antarctica.

## GLOBAL IMPACTS

Antarctica is an important laboratory for research into the global impacts of the industrialized world.

Lakes on Signy Island in the Maritime Antarctic for instance have shown possibly the **fastest local response to regional climate found anywhere on Earth**. Average lake temperatures having risen by 0.9°C in 15 years while temperatures in the surrounding seas have stayed constant.

Global change may have effects that impact directly on the Antarctic environment and its fauna and flora. Global warming for example may contribute to break-up ice-shelves causing loss of habitat for animals dependent on the ice-shelf as well as the effect of increasing sea level on low-lying regions in the rest of other world.

**Increasing Ultra Violet (UV) radiation due to the ozone hole** may cause changes to phytoplankton communities and could have effects up the food chain.

**Antarctica is a sensitive indicator of global change.** The polar ice cap holds within it a record of past atmospheres that go back tens or even hundreds of thousands of years, allowing study of the earth's natural climate cycles against which the significance of recent changes can be judged.

## ENVIRONMENTAL MANAGEMENT

The concerns for the environmental management of Antarctica is how to make good past damage and how to reduce the current and future impacts.

### PAST SINS

In the early days of Antarctic programmes (well into the 1980's) waste was dealt with in one of two ways:

Metal scrap litter on ice of Antrarctica

**If it was flammable, it was burnt**

**If it wasn't flammable, it was thrown into the sea**

There was also a variation of the second approach whereby larger items such as broken and useless vehicles were taken out onto the sea ice and left. When the ice broke up in the spring, the rubbish on them would break out too and they would be dumped further out at sea as the floe they were on melted or tipped up.

Further inland, rubbish was often dropped down a crevasse or just left where it was. In the harsh Antarctic environment, even large items such as empty oil drums can be blown by the wind for miles from where they were originally dumped so spreading the problem. Broken down or crashed vehicles and aircraft were usually just left where they were.

A part of the Antarctic Treaty called **Madrid Protocol** (because it was agreed on at a conference in Madrid) requires that national Antarctic programmes clean-up abandoned work sites and waste tips so long as the process of clean-up does not cause greater adverse impacts or cause the removal of historic sites or monuments.

## A CLEANER FUTURE

**Environmental audits** are now carried out around bases, on land and in the sea to assess the impact that the base and its activities is having on the area around it. Any activity is bound to have some degree of disturbance to the environment, vehicle exhausts, wildlife disturbance, waste of various types being produced etc.

Some bases are now experimenting with **alternative energy** sources, the Australian Mawson base for example now has wind powered electricity generator - fairly obvious choice for one of the windiest places on earth! But potentially problematic as the wind is so strong so often that the windmills are at risk of damage.

**Protected areas** are being set up with various levels degrees of protection, no vehicles allowed, limitations to annual visitors, or even no people allowed at all in the some cases.

Expeditioners are now educated far better about protecting the environment and are frequently guided by a code of conduct while in Antarctica.

## HUMAN IMPACT ON THE NATURAL ENVIRONMENT

Natural environment is of crucial importance for social and economic life. We use the living world as

- a resource for food supply
- an energy source
- a source for recreation
- a major source of medicines
- natural resources for industrial products

In this respect the diversity of nature not only offers man a vast power of choice for his current needs and desires. It also enhances the role of nature as a source of solutions for the future needs and challenges of mankind.

## STATE OF ECOSYSTEMS, HABITATS AND SPECIES

In the past, human interaction with nature, although often having a disruptive effect on nature, often also enriched the quality and variety of the living world and its habitats, e.g., through the creation of artificial landscapes and soil cultivation by local farmers.

Today, however, human pressure on natural environments is greater than before in terms of magnitude and efficiency in disrupting nature and natural landscapes, most notably;

- intensive agriculture replacing traditional farming; this combined with the subsidies of industrial farming has had an enormous effect on western rural landscapes and continues to be a threat.
- mass tourism affecting mountains and coasts.

- the policies pursued in the industry, transport and energy sectors having a direct and damaging impact on the coasts, major rivers (dam construction and associated canal building) and mountain landscapes (main road networks).
- the strong focus of forestry management on economic targets primarily causes the decline in biodiversity, soil erosion and other related effects.

The clearest manifestations of the degradation of the natural environment are:

- *Reduction and fragmentation of habitats and landscapes*

  The expansion of humans activities into the natural environment, manifested by urbanisation, recreation, industrialisation, and agriculture, results in increasing uniformity in landscapes and consequential reduction, disappearance, fragmentation or isolation of habitats and landscapes.

  It is evident that the increasing exploitation of land for human use greatly reduces the area of each wildlife habitat as well as the total area surface throughout Europe. The consequences are :

  - A decreased species diversity, due to reduced habitable surface area which corresponds to a reduced "species carrying capacity".
  - The reduction of the size of habitats also reduces the genetic diversity of the species living there. Smaller habitats can only accommodate smaller populations, this results in an impoverished gene pool.
  - The reduction of genetic resources of a species diminishes its flexibility and evolutionary adaptability to changing situations. This has significant negative impacts on its survival.

The conditions under which the reduction of habitats often occur prevent living organisms making use of their normal ways to flee their threatened habitat. Those escape routes include migration to other habitats, adaptation to the changing environment, or genetic interchange with populations in nearby habitats. Of particular concern is:

- The abrupt nature of human intervention; human projects are planned and implemented on a much shorter time scale than natural processes;
- Furthermore human intervention, such as the construction of buildings, motorways or railways results in the fragmentation of habitats, which strongly limits the possibility for contact or migration among them;
- In extreme cases even the smallest, narrowest connections between habitats are broken off. Such isolation is catastrophic for life in the habitat fragments.

- *Loss of Species of Fauna and Flora*

Although relatively few species of Europe's fauna and flora have actually become totally extinct during this century, the continent's biodiversity is affected by decreasing species numbers and the loss of habitats in many regions. Approximately 30% of the vertebrates and 20% of the higher plants are classified as "threatened". Threats are directly linked to the loss of habitats due to destruction, modification and fragmentation of ecosystems as well as from overuse of pesticides and herbicides, intensive farming methods, hunting and general human disturbance. The overall deterioration of Europe's air and water quality add to the deterimental influence.

## AGRICULTURE

Europe's natural environment is inextricably linked with agriculture and forestry. Since agriculture traditionally depends on sound environmental conditions, farmers have a special interest in the maintenance of natural resources and for centuries maintained a mosaic of landscapes which protected and enriched the natural environment.

As a result of needs for food production since the 1940s, policies have encouraged increased production through a variety of mechanisms, including price support, other subsidies and support for research and development. The success achieved in agricultural production has however entailed increased impact on the environment.

Modern agriculture is responsible for the loss of much wildlife and their habitats in Europe, through reduction and fragmentation of habitats and wildlife populations. The drainage of wetlands, the destruction of hedgerows and the intensive use of fertilizers and pesticides can all pose a threat to wildlife. Highly specialised monoculture are causing significant loss in species abundance and diversity. On the other hand increased production per hectare in intensive areas, raising of livestock volume, and lower prices for agricultural products also caused marginalization of agricultural land, changing the diversity of European landscapes into the direction of two main types: Intensive Agriculture and Abandoned land.

## ENERGY

Abandonment can be positive for nature, but this is not necessarily so. Land abandonment increases the risk of fire in the Mediterranean Region, causes a decline of small-scale landscape diversity and can also cause decrease in species diversity.

All energy types have potential impacts on the natural environment to varying degrees at all stages of use, from extraction through processing to end use. Generating energy from any source involves making the choices between impacts and how far those impacts can be tolerated at the local and global scale. This is especially of importance for nuclear power, where there are significant risks of radioactive pollution such as at Chernobyl.

Shell Oil Company and IUCN have jointly drafted environmental regulations for oil-exploitation in Arctic areas of Sibera. Other oil companies are aware of this and use these environmental regulations voluntarily for developing oil fields.

Into the future the sustainability of the natural environment will be improved as trends away from damaging energy uses and extractive methods reduce and whilst real cost market forces and the polluter pays principle take effect.

*Now we can summarise the human activities to some areas and their impact as follows –*

## WATER ACTIVITIES

- irrigation:
- mannure management (e.g., collection, storage, disposal/spreading):
- fuel storage (tanks), distribution refuelling activities;
- waste disposal (hazardous and non-hazardous); and
- urban development (removal of vegetation, increase in hard surfaces).

## POTENTIAL IMPACT OF THESE ACTIVITIES

- reduced groundwater quality (e.g., pollutants/toxins, hydrocarbons, pathogens, bacteria);
- important of drinking water quality;
- need for increased water treatment;
- reduced groundwater quantity;
- surface water effects (quality and quantity);
- explosions (from petroleum products, fuels).

## COASTAL AREAS/MARINE

**Example of activities**

- discharges of sewage/wastewater;
- commercial fisheries, dredging;
- ocean dumping;
- energy production (exploration, drilling, distribution);
- boating/shipping (e.g., releases, refuelling, ballast water exchange);
- aquaculture;
- urban development (removal of coastal vegetation, including wetlands);
- spills and accidental releases.

## POTENTIAL IMPACT OF THESE ACTIVITIES

- alteration/degradation of quality of fish and other marine habitat;
- depletion of fish populations;
- increased disease and pathogens affecting fish;
- impairment of water quality – e.g., pollutants, (including petroleum products), pathogens, bacteria, nutrients;
- exotic, invasive species;
- socio-economic effects;
- reduction of tourism activity.

## LAND

**Examples of activities:**

- creation of transportation infrastructure (roads, highways, bridges);
- urban development (free and vegetation removal);
- forestry/logging (including access roads, herbicide spraying);
- mining activities;
- agriculture (e.g., Soil tilling, livestock grazing, application of fertilizers and pesticides, intensive farming practices);
- spreading of manure and sewage sludges;
- tourism and recreational development (e.g, ski resorts, golf courses);
- oil and gas production and distribution, and oil sands development;
- storage of fuels and other hazardous materials (e.g., storage tanks);
- landfilling – disposal of waste (hazardous and non-hazardous);
- spills and accidental releases (during transportation or from stationary sources); and
- military training (use of training areas).

## POTENTIAL IMPACT OF THESE ACTIVITIES

- depletion of renewable and non-renewable resources;
- soil and groundwater contamination;
- erosion/desertification;
- reduction/removal of wildlife habitat;
- removal/reduction of wetlands;

- reduction in biodiversity (soil organisms, plants, wildlife);
- increases surface water runoff/stormwater runoff;
- mining waste (tailings); and
- opening of remote areas.

## OPPORTUNITIES TO AVOID OR MINIMIZE NEGATIVE ENVIRONMENTAL EFFECTS

- consideration of environmental factors/concerns in the early stages of decision making (e.g., for projects, product development);
- reducing energy consumption and increasing use of renewable energy sources through
  - increased efficiency (e.g., enhanced fuel efficiency for vehicles, reduced electricity consumption by household appliances); and
  - building design (new buildings) or retrofitting;
- advancing, developing, and employing green technologies;
- reducing consumption of resources;
- increased reuse and recycling, thus decreasing resource consumption, and waste production and disposal;
- eco-efficiency;
- green procurement–purchasing more environmentally friendly goods and services;
- pollution prevention by
  - avoiding the use of hazardous/toxic materials;
  - Using cleaner fuels;
  - Using clean emissions technologies for engines; and
  - Using cleaner energy sources (e.g, solar, wind power); and
- improved emergency response and preparation.

## ACTIVITIES THAT HAVE AN IMPACT ON THE ENVIRONMENT

- **Energy:** development, distribution, processing, management, and/or consumption/use (oil, gas, nuclear, other)
- **Natural resources:** development, management, and/or harvesting, use (e.g., fisheries, aquaculture, forestry, hunting/trapping, mining)
- **Agriculture/food production:** land cultivation, animal husbandry, food processing (waste handling, treatment, and disposal)
- **Physical infrastructure:** creation or use of infrastructure, such as roads, housing, bridges, ports, storage facilities, railways, sewage, or waterworks
- **Transportation:** road, marine, rail or air transportation, and all related activities and infrastructure
- **Toxic/hazardous substances and materials:** generation/manufacture, use, management, regulation, transportation, or disposal (e.g., toxics and pesticides)
- **New substances and organisms:** development, deployment, and regulation (e.g., new chemicals, generally modified organisms)
- **New products and technologies:** development and deployment
- **Industrial activity:** e.g., resource processing and manufacturing
- **Urban development**
- **Military activities:** training, equipment, materials, natural disasters and other emergencies (e.g., preparation and response)

- **Waste generation or management** (including hazardous waste)
- **Movement of goods and services** (local, regional, national, international)
- **Occupational/workplace hazards**
- **Cleanup/rehabilitation of contaminated sites**
- **Procurement and consumption of goods**

## AUTOMOBILE POLLUTION

### INTRODUCTION

It is possible that no invention could effect the society as the passenger automobile. It did not take long after its introduction in the early part of this century for the auto to quickly become the primary means of transportation in the United States, where there are now 752 motor vehicles for every 1,000 people (World Almanac 211). While no other country can match the excessive automobile use of the U.S., it's not for lack of trying. Even in China, where the use of bicycles by its citizens is legendary, the number of cars has been doubling every five years for the past 30 years (World Resources Institute, hereafter "WRI" 172).

But reliance on cars is not without its problems and the most obvious being air pollution and energy consumption.

This is one way of looking back what damages we have done to the environment through our inappropriate actions and how to arrest/control further damages that we may inflict on the environment in future. In the last around 200 years the industrialization and the associated urbanization and population growth inflicted severe damage to the environment in several different ways. We consider environment as source that is available freely and after using it we dump the waste generated in the process to destroy the environment. This is a creation of man's greed at the cost of his own health. Among these the most hazardous waste is the "evil" pollution. The presence of the evil pollution in the atmosphere generated several hazards such as health hazard, climate change hazard this including the global warming, ozone hole, etc.

There are several agents that cause pollution, namely industry, vehicles, disposal of biomedical waste, disposal of bio-degradable waste, burning of non-degradable waste such as plastics and PVC pipes, modern agricultural inputs (fertilizer, insecticides/pesticides/herbicides, etc.), building activity, mining activity etc. According to Down to Earth report of January 31, 2001 of the twelve dirtiest states in the country that pollute the air and water, both with toxic pollutants and as well as metal pollution, Andhra Pradesh takes the 5th position with Tamil Nadu in the 9th position and Karnataka in the 11th position and in terms of top 10 polluting states Andhra Pradesh takes the 4th place.

### POLLUTION: GENERAL

Pollution by cars causes lung cancer, respiratory problems, urban smog, and acid rain (Brown 25). By 1970, after decades without government regulation, air quality had become a serious problem. The first federal Clean Air Act was passed during the Nixon Administration to curtail the ever-increasing amount of pollution caused by automobiles and industry, and Congress passed in updated version in 1990 (WRI 1982). However, the Clean Air Act didn't prohibit pollution; it simply defined an "acceptable" amount. Further, the legislation addressed only certain airborne contaminants, while ignoring others. Perhaps most significantly, although bad air was outlawed, it still exists. More than half of the people in the U.S. live in areas that failed to meet federal air quality standards at least several days a year (30 Simple Energy Things You Can

Do to Save the Earth, hereafter "30 Simple Things," 11), and around 80 million Americans live in areas that continually fail to meet these standards (WRI 63). Despite the Clean Air Acts, the reality is that air pollution continues to be a major public health problem.

As bad as the air is in the U.S., in other countries which have waited too long to address the pollution caused by cars, it's worse. Mexico City, Sao Paulo, New Delhi, and Bangkok are grappling with serious air problems. And much of that pollution is caused by private automobiles (Brown 25).

## POLLUTION: GROUND-LEVEL OZONE

One way cars create pollution is by contributing to the amount of ground-level ozone (not to be confused with the atmospheric ozone layer). In the atmosphere, the ozone layer shields the planet from harmful ultraviolet radiation rays. But on the ground, ozone is another matter, causing hazy smong and respiratory problems. Most ozone pollution is caused by motor vehicles, which account for 72% of nitrogen oxides and 52% of reactive hydrocarbons (principal components of smog) (30 Simple Things 11). The seriousness of ground-level ozone should not be underestimated. According to the World Resources Institute:

Ozone pollution has become widespread in cities in Europe, North America, and Japan as auto and industrial emissions have increased. ...Breathing ozone concentrations of 0.012 ppm & emdash ; levels typical in many cities & emdash ; can irritate the respiratory tract and impair lung function, causing coughing, shortness of breath and chest pain.... .

## HEALTH HAZARDS

The studies indicate that the constituents of vehicular pollution cause several health hazards in addition to modification of weather in and around the capital city. For example nitrogen oxide causes sore throat, running nose and in severe cases pulmonary dysfunction. Leukemia is caused due to continuous exposure to benzene. It is said that the synergy between ozone and nitrous oxides causes asthma. Particulate matter released due to adulterated fuel, some of which gets absorbed into the blood, is the cause for cardiovascular diseases, breathing problems and lung cancer in severe cases lead, released due to un-burnt fuel causes heart ailment and hypertension. Here age is not a barrier.

The small particulates, coasted with toxic gases in Auto Emissions are called Respirable Suspended Particulate Matter (RSPM). The particulate matter less than 10 microns in diameter (PM 10), is considered unsafe, since it is respirable. The fine particles PM 2.5 are considered to be more dangerous than PM 10, as they contain more sulphates and nitrates, having serious health impacts. As per the study published in the journal of American Medical Association, a mere increase of 10 micrograms per cubic meter of fine particles (smaller than 2.5 microns of PM 2.5), can increase the risk of lung cancer by 8% cardio pulmonary deaths by 6% and all deaths by 4%. In our country, CPCB started publishing PM 10 data only from 199, for the whole of the country for the first time. These data shows that PM 10 has recorded critical levels in 13 cities - Hyderabad is one, and can reach as high as four times the national annual standard of 60 micrograms per cubic meter. As against this the PM 10 levels are below 40 micrograms in almost 90% of the cities in the US. The PM 2.5, Polycyclic Aromatic Hydrocarbons (PAH) and Benzene, exposure to which may lead to cancer, is not even mentioned in India. APPCB data of Hyderabad present PM 10 peaking to dangerous levels up to 750 micrograms per cubic meter. Even the lowest peak value of RSPM recorded is around 150 marks. The high incidence of respiratory, cancer and cardio pulmonary causes in and around Hyderabad should be an eye opener to the state government! The high-tech PUC certificate does not address the deadly RSPM, P AH, Benzene, etc.

## POLLUTION UNDER CONTROL (PUC)

Check is serving the purpose of harassing the public to collect mamools. This does not really check pollutants that cause health hazards and also it is not a suitable testing procedure for diesel vehicles. This is another wasteful expenditure on the vehicle owners. In fact the particle size (the particulate matter size of 2.5 microns, Polycyclic Aromatic Hydrocarbons (PAH) and Benzene, are not even monitored in India) plays the major role. Nowadays, even now vehicles show pollution is not under control, basically because when you give your vehicle for service, there is no guarantee that it will come back without replacing new parts by old parts. It is easy to tamper the miles the vehicles had run! To date still more than 40% of the vehicles are not having PUC certificate and to make others to follow, the government planned to introduce from Ist December 2003 "Fuel with PUC" programme but not implemented.

## POLLUTION: LEAD

Cars also pollute by emitting lead from leaded gasoline. Although the use of lead in gasoline is banned in the United States, leaded gasoline is common in other countries. In fact, of the countries for which data is available, 43% use nothing but leaded gasoline. Many of the rest use at least some leaded gasoline in their energy mix. This is a definite cause for concern:

One of the oldest metals used by humans, lead is a cumulative neurotoxin that impairs brain development among children and has been connected to elevated blood pressure and resulting hypertension, heart attacks, and premature death in adults. Emissions from vehicles is the largest source of lead exposure in many urban areas (WRI 266-267).

The effect of all this pollution on human health are unsetting. A study of U.S. cities found that mortality rates were 17-26% higher in cities with the dirtiest air compared to those with the cleanest air. Not surprisingly, the study also found correlations between bad air and lung cancer and cardiopulmonary disease. The risks translate roughly to a two-year shorter life span for residents of dirty-air cities. On a global basis, estimates of mortality due to outdoor air pollution range from about 0.4-1.1% of total annual deaths (WRI 63-64). In the U.S., 30,000 people die every year from automobile emission ("Bicycling and Our Environment" 1). [Also see our separate page on lead.]

## POLLUTION: GLOBAL WARMING

Perhaps even scarier than the direct damage to our bodies from auto pollution is the fact that car emissions are contributing to an overall warming of the entire planet, which could destroy the world's food chain.

Cars emit carbon dioxide ($CO_2$), a heart-trapping gas. In fact, they emit a lot of it: 20 pounds per gallon of burned (NRDC 12, Zuckermann 29). Atmospheric concentrations of $CO_2$ having increased by 30% preindustrial times, and much of that increase is directly related to the burning of fossil fuels. According to the Worldwatch Institute: "$CO_2$ levels are now at their highest point in 160,000 years, and global temperatures at their highest since the Middle Ages" (Brown 26).

The effects of this global warning are frightening: rising sea levels, dying coral reefs, spreading of infectious diseases, and extreme weather conditions, including droughts, rare forest fires, historic floods, and severe storms. Even more frightening, these events are not just predictions & mdash ; they're happening right now (Brown 26).

## VEHICLE MAKE VS POLLUTION LOAD

The pollution emission load of vehicles manufactured after 1990 per vehicle was less than 10% of the vehicles manufactured prior to 1990. The pollution load (CO-carbon monoxide) from

12 cars produced after 1st April 2000, would be roughly equivalent to the pollution from one car produced before 1990, correspondingly 17, 2-wheelers are equivalent to one 2-wheeler and 10, 3-wheelers are equivalent to one 3-wheeler.

## FUEL

Fuel type and fuel adulteration play important role in determining the pollution load under different categories of pollutants.

## FUEL TYPE VS POLLUTION TYPE

From the available data from various sources, efforts are made to indicate the source and magnitude of the pollution loads ; 61.4% of CO, 34.0% of HC and 4.6% of $NO_x$ of particulates. As per pollution profiles of vehicles prepared by Indian Institute of Petroleum, Dehradun, the petrol two-stroke 2 and 3 wheelers are estimated to have to their credit, about 70% of total vehicular unbrunt Hydro Carbons (He) and 46% of the total r=O emissions with negligible $NO_x$, while the vehicles powered with diesel engines emit 22% of HC, 29% of CO and 94% of $NO_x$. The petrol four-wheelers, on the other hand, emit 7.5% of HC, 25% of CO and 5% of $NO_x$. The buses, trucks etc., all most all of them powered with diesel engines, operating in HUDA limits, are estimated to be about 50,000 (less than 5%), out of which 50% are know to be older than 15 years and contribution to about 39% of the total pollution load.

Diesel contains sulphur level of about 0.5 to 1.0% to be reduced to 0.25% by the year 2000. Three refineries are already supplying 0.05% diesel to the four metros. SPM levels are directly proportional to the sulphur content. Also sulphur is a poison for catalytic converters. Diesel engines are 20-30% more efficient and inherently cleaner than petrol engines and if low sulphur diesel is extensively available then new technology engines can be used to meet the EURO III norms. Indian gasoline's have high levels of Benzene which is a class I carcinogen Studies in Mumbai indicate that Benzene levels to be 40-70 times above WHO standards. Gasoline engines produce 19.4 g/ mile of benzene and diesel engines produce only 1.40g/mile. Sadly Benzene levels are not covered under new laws or by the Supreme Court orders. Benzene levels in India are currently at 3%. Reliance's new refinery is expected to produce at 2% level. International standards range from 1.1% to 2.5%. Sulphur content in Indian gasoline is 0.2% whereas European gasoline has 0.02% with planned reductions to 0.003%. Lead pollution is not a major problem any more since only unleaded fuels are being supplied across the country from the year 2000. The analysis of air pollution carried out in USA reveals those emissions from diesel engines account 78% of the total cancer risk.

## FUEL VS ADULTERATION

The principal problem with older lorries/ transport vehicles, 3-wheelers and a to a limited extent the 2-wheelers is one of adulterated fuel-mixing of kerosene with petrol, diesel adulterated 2-T oil. Based on the level of adulteration the load of different pollutants combination changes along with the technical load.

**Traffic jams are a familiar sight in city.**
**Courtesy : Eenadu**

## MINIMIZATION OF POLLUTION LOAD

Reduction of pollution has two parts namely reduction of pollution load and reduction in respirable particulate matter content in the load. This includes dispensing with vehicles older than 15 years from city roads, eradication of adulterated fuel, changing the fuel mode, traffic management & town planning etc.

## VEHICLES OLDER THAN 15 YEARS

In 1999 at a meeting convened by Ministry of Transport of Secretariat, Government accepted to withdraw vehicles older than 15 years in a phased manner. A year later the government announced dispensing all those vehicles older than 15 years from the city roads. The public opposed this and then the government withdraws the Order. Now, with the Hon'ble Supreme Court of India's directive in WP (c) No. 13029 of 1985, dated 14-8-2003 "that among other 13 cities, 'Respirable Suspended Particulate Matter (RSPM)' levels in Hyderabad city has become a serious problem and directed to draw a plan to bring down their levels" the government convened a meeting on 16.10.2003 and presented the proposal of 1999 suggested by us. However, at the meeting the other departments like Traffic Police & MCH were not represented.

## ADULTERATED FUEL

In one of the official meetings during 2000, where the Forum also participated, it was agreed to create a special task force to tackle the menace of fuel adulteration and accordingly it was constituted. However, no progress was evident in this direction. But, now the government says (at the 16.10.2003 meeting) that there is no law to punish the seller or user for adulterated fuel. This shows the indifferent attitude of the state government in controlling auto pollution.

## CHANGE OF FUEL MODE

Table 13.3 given below presents the pollution loads under different fuel modes. In the gas mode, the SPM's are low. In the case of RTC buses, the "fuel mode" can be converted into "gas mode". On 3rd November 2003 the first 'Auto LPG Dispensing Station' in Secunderabad, run by Margadarsi Service Centre, a dealer for HPCl was opened. HPCl runs two more, one at Attapur and the other at Bahadurpura. The other such facility in the city is the BPCl's lakdikapul outlet. 10Cl runs three: Thimmapur, Kompally and Moulali. The OMVI conversion kits are made in Italy imported by Ace LPG car Kits and distributed in AP, by Jeeyar Enterprises. The major deterrent in this was the cost of conversion kits - Rs. 14000 for autorickshaws, Rs. 19000 for cars with carburetors and Rs. 23000 for cars with multi-point fuel injection system. To minimize the burden, it was suggested to give some sops to those vehicle owners who convert from fuel to gas mode. However, it was cautioned not to encourage very old vehicles, which may pose more danger. At present the gas filling takes 4 minutes for an auto-rickshaw and slightly more for cars.

**Table 13.3. Alternate Fuels vs pollution: Gasoline, diesel, LPG and CNG**

| Mode | gm/km | | | |
|---|---|---|---|---|
| | Gasoline | LPG | CNG | Diesel |
| CO | 1.50 | 1.00 | 0.50 | 0.60 |
| HC | 0.20 | 0.10 | 0.50 | 0.10 |
| $NO_x$ | 0.25 | 0.20 | 0.25 | 0.60 |
| SPMs | 0.01 | 0.005 | 0.01 | 0.03 |

## ALTERNATE MODE OF TRANSPORT

The use of 2-wheelers can be brought-down if we can improve the public transport system. This includes the RTC bus network, MRTS (Mass Rapid Transport System) LRTS (Light Rail Transport System). The other important area to reduce the two-wheeler traffic on city roads is through decongestion of Hyderabad and development of counter magnets. Under this, the outer ring road of the HUDA draft master plan of 2003 can have both road and rail tracks. All these actions will also reduce, to a certain extent, the pollution load of Lorries in the city area and contribute to the free flow of vehicles on the road. However, it was reported during the 16.10.2003 meeting that the population using public transport system is coming down day by day.

## SUMMARY

Automobiles are responsible for a tremendous amount of air pollution and wasted energy. These problems impact people all over the world, both motorists and non-motorists alike, but affecting their health, their economies, and their communities. Legislation to address air pollution has been only partially successful, and air quality continues to be a major concern in countries all over the world. As for energy use, one can only hope that world leaders find a better way to address this problem than fighting wars over an increasingly shrinking supply of all.

## QUESTIONS

1. Differentiate exponential growth with geometric growth.
2. Discuss population growth in India.
3. How HIV/AIDS is spread ?
4. What do you mean by Indira Mahila Yojana and National Commission for women ?
5. Discuss the child development programmes in rural areas.
6. Write the causes of population explosion.
7. Explain the importance of family welfare programmes.
8. HIV and AIDS stands for.
9. Name any two organization concern with the women and child welfare.
10. Write brief note on human activities in various fields.
11. Define population growth in various ways. How it varies among nations ?
12. What is meant by "population explosion" ? Discuss it in Indian scenario.
13. What are family welfare programmes ? To what extent family planning is useful for population control.
14. Discuss the salient features of Human right movement.
15. What is value education ? How it can be described ? Give aims and methodology.
16. Discuss the role of information technology in environment and human health.
17. Write informative notes on :–
    (*i*) HIV/AIDS
    (*ii*) Zero population growth.
    (*iii*) Women and Child welfare.
18. What is population explosion ? What steps are taken to check the population growth?
19. Write an essay on "Human Rights".
20. Describe the role of information technology in environmental studies.
21. How human activities effects the environment, discuss in detail?
22. Write note on Automobile pollution.

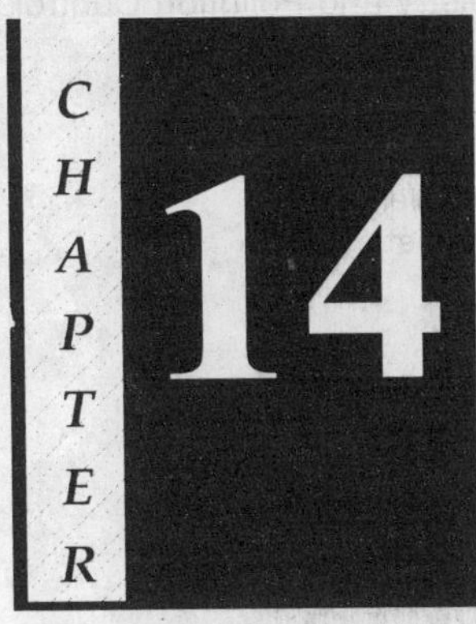


# Effects of Electric and Magnetic Fields in the Environment

## INTRODUCTION

The electromagnetic spectrum covers the entire range of radiations extending from cosmic rays (having wavelengths as short as $10^{-9}$ nm) all the way up to radio-waves having wavelengths (as long as 1000 km or more). Encompassed within these extremes and moving from shorter to longer wavelength lie the gamma rays x-rays, UV, visible, IR and microwaves. The nature of these radiations is the same and all move with the speed of light. They differ only in frequency, wavelength and the way in which they interact with matter.

The chemical and physical effects of various types of radiation are quite different and they can be understood in terms of the differing energies of the photons. In the radio-frequency range, the energy of the photon in very low, and the energy transitions are concerned with reorientation of nuclear spin states of substances in a magnetic field. In the slightly higher-energy microwave region, there are changes in electron spin states for substances with unpaired electron when present in a magnetic field. In the infrared region, absorption caused changes in vibrational and rotational energy. Changes in the visible and ultraviolet regions involve the electron energy of atoms or molecules; for molecules this accompanied by changes in vibrational and rotational energy. These changes in electronic energy involve the most loosely held or the so-called outer electrons, finally, at the high-energy end of the spectrum, X-rays will raise inner electrons to excited states. Gamma radiation originated atomic nuclei,

The various regions in the electromagnetic spectrum and the nature of changes brought about by the respective radiations are given below:

**Table.14.1. Schematic representation of the electromagnetic spectrum (The wavelength scale is non-linear)**

| Spectral Region | Wavelength, λ | Energy changes involved while the radiations interact with matter |
|---|---|---|
| Gamma rays | < 1 A° | Nuclear |
| X-rays | 1 A° A – 100 A° | Inner cell electron |
| Vacuum UV | 100 A° – 200 nm | Ionisation of atoms and molecule |
| Near UV | 200 nm – 400 nm | |
| Visible | 400 nm – 800 nm | Valence electron |
| Near IR (overtone) | 0.8 mm – 2.5 mm | Molecular vibrations stretching |
| IR (fundamental) | 2.5 mm – 25 mm | Bending |
| Far IR | 25 mm – 400 mm | |
| Microwaves | 0.04 cm – 25 cm | Spin orientation (in magnetic field) Electrons (ESR) |
| Radiowaves | > 25 cm | Spin orientation (in magentic field). Nuclei (NMR) |

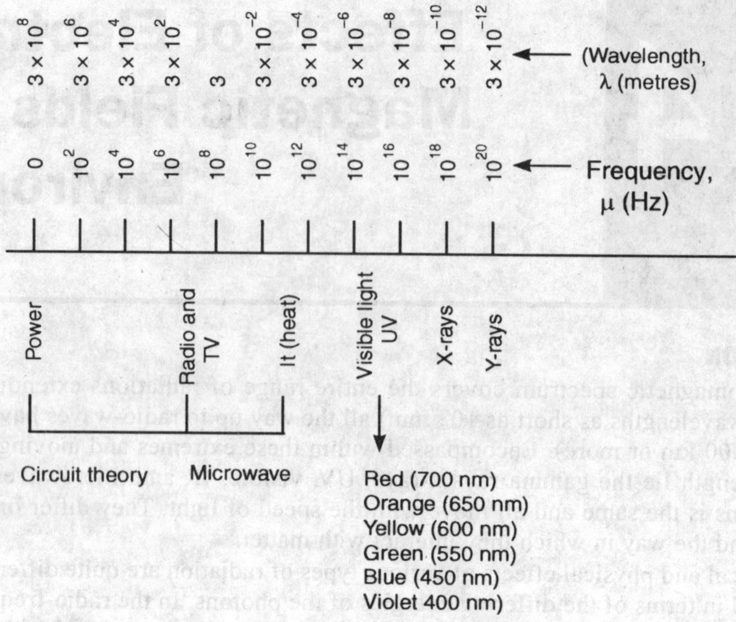


## HEALTH EFFECTS DUE TO ELECTROMAGNETIC FIELDS

Life has evolved on earth in an electromagnetic environment comprising natural terrestria sources such as quasi-steady state field due to thunder storms, lightening, low frequency Schumann resonance etc. The Sun and other extra-terrestrial sources also contribute to our electromagnetic environment. Example are the Sunspots and giant radio bursts.

The term 'electromagnetic" indicates the coexistence in time and space of electric and magnetic field components, the inter-relation of which depends upon the nature of the source and the frequency (v) of variation in the field amplitude. The electric and magnetic field components are coupled since a time-varying electric field will induce a magnetic field at the same frequency and a time-varying magnetic field will in turn induce an electric field in conducting objects such as the human body. The extent of electricmagnetic coupling is directly proportional to the rate of-change, or frequency, of the field components. At low-frequencies, such as in the extremely low frequency (ELF) range (i.e. 0 to $10^3$ H), the coupling is minimal and thus, either the electric or magnetic component will pre-dominate depending upon whether the source is a current source or a voltage source. Under these minimal coupling conditions, the components are referred to as the electric or magnetic fields. At higher frequencies, there is significantly greater coupling or energy exchange between the field components which result in propagation of the field through space as an electromagnetic wave or electromagnetic radiation. The basic relationship between the frequency (v) and the wavelength ($\lambda$) of electromagnetic radiation is : $\lambda = c/v$, where $c$ is the velocity of light ($3 \times 10^8$ m/s). Fields having frequencies of $10^3$ Hz or less are generally called electromagnetic fields (EMF) whereas as fields having frequencies higher than $10^3$ Hz are called electromagnetic radiation (EMR). Important man-made environmental EMF sources are associated with the generation, transmission, and use of 60 or 50 Hz power frequency fields. Important environmental sources of man-made EMR include radio-frequency (RF) and mincrowave (MW) radiations in the frequency range $10^5$ to $10^{11}$ Hz used for communications and heating. As the frequency increases, localized EMR energy coupling increases such that at frequencies of $10^5$ Hz or greater (corresponding to UV radiations), molecules or atoms are ionized. EMR in this range of frequencies which include UV, X- and gamma-rays, is referred to as ionizing radiation. Adverse health effects or ionizing EMR include somatic and genetic mutations, have been characterized as a result of several studies over the past 60 years. However, the health concerns resulting from exposure to non-ionizing electromagnetic radiations (viz. ELF, EF, RF and MW) are less well-defined. The present position regarding the possible health effects due to exposure to non-ionizing EMR are reviewed in the following sections.

Until relatively recently, the adverse health effects recognized due to ELF and EMFs were only shocks or burns resulting from inadvertent human contact with electrical conductors. Adverse health effects due to exposure to high frequency EMR (such as microwave), were generally attributed to radiation energy absorption in tissue resulting in excessive heating and consequent thermal damage. In either of the above cases, the acute nature of exposure and the immediacy of the injury resulted in obvious cause-effect relationships.

Recent reports indicating elevated incidence of cancer in human populations exposed to chronic long-term low-intensity EMF and EMR fields in the workplace or in the home have focussed greater attention on health effects due to such fields. Unlike in case of acute, high-intensity-exposure effects, it is difficult to establish cause-effect relationship resulting from chronic low-intensity exposure. This problem is because of several reasons including the non-availability of adequate data and the lack of theoretical basis to explain how such low-intensity fields after living systems. In contrast to the effect of higher intensity EMFs that involve significant field/tissue energy exchange, low-intensity field intersections in several substances occur at levels well below those associated with energies involved in classical physico-chemical processes in living systems. Further difficulty arises due to the dependence of such effects on field frequency or modulation, particularly in the ELF range. We should also remember that in addition to potentially adverse exposure effect, beneficial aspects of the biological activity of electromagnetic fields have also been demonstrated. For instance, the capability of RF and MW electromagnetic radiations to include localized heating has been exploited in tumour therapy and low-intensity pulsed magnetic fields have been used to enhance the rate of tissues healing. In view of the ubiquitous and ever-increasing levels of human exposure to EMR and EMFs in our environments, it is absolutely essential to define the potential health effects more precisely and at the earliest.

**Effect of exposure to extremely low frequencies (ELF) on humans and animals**

Principal sources of ELF residential magnetic fields include high-voltage transmission lines, distribution lines, electrical service wiring and household electric wiring.

Results of epidemiological studies published in 1979 indicate a positive association between residential 60 Hz magnetic field exposure and childhood cancer.

Under conditions of low average home power usage, there was a moderate association of magnetic field exposure with childhood leukemia risk, whereas there was a more obvious association when exposure was estimated by the use of wire code configurations. However, parameters such as (*a*) instantaneous of peak magnetic of electric field intensity (*b*) frequency or rate of occurrence of field transients (*c*) rate of change of electric of magnetic fields (*d*) long-term average exposure intensity, also play significant roles. Lack of information regarding relevant magnetic field exposure parameters lead to apparent inconsistencies in the results of epidemiological studies. Further, there is only a limited theoretical understanding of basic intersection mechanisms responsible for the effects of low intensity ELF magnetic fields on living systems.

## OCCUPATIONAL EXPOSURE

Epidemiological studies have suggested an association of occupational exposure to ELF magnetic fields and cancer incidence. A dose response relationship for the association of cancer incidençe and occupational exposure to ELF has indicated that there was three-fold increase in chronic lymphocytic leukemia (CLL) and a 60% overall increases in leukemia in workers exposed to 4.1 mG. However no association was found for acute myeloid leukemia. The study also provided some evidence of elevated risk of brain tumours, but the dose-response relationship was less evident.

## ELF MAGNETIC AND ELECTRIC FIELD ON EXPERIMENTAL ANIMALS

Harmonal alterations have been reported to occur in laboratory animals exposed to ELF fields. The effect of ELF-EMF exposure on melatonin levels in experimental animals suggested neuro-endocrine involvement in the observed increase in breast cancer in occupationally exposed males. The mechanism could be involved in exposure of females as well as males and could relate to other tumour types.

Epidemiological studies have also indicated and association of ELF-EMF exposure and other physiological alternations such as altered biorythms, behavioural and neurological disorders and reproductive outcomes.

Exposure to residential magnetic fields from electric blankets, heated water-beds and ceiling cables that produce fields in the range of 4-15 mG have been associated with seasonal variations in fetal growth and fetal loss.

The results of a recent epidemiological study indicated an association of residential 50 Hz magnetic fields and early pregnancy loss (EPL). Magnetic field measurements in residences indicated that exposure to fields of 6.3 mG or more resulted in an EPL odds ratio of 5.1. Animal studies have also indicated that pulsed low-intensity magnetic fields may increase the incidence of malformations in some cases.

Studies on experimental animals revealed that low-intensity ELF electric or magnetic fields are likely to act as tumor promoters rather than tumor initiators.

Effect of ELF electric and/or magnetic fields on cats, dogs, swines and non-human primates with respect to behavioural changes, reproductive outcomes, harnone levels, brain-neurochemistry, cardiovascular response etc., have been studied. Although these data do not reveal well-defined dose-responses, they do indicate general sensitivity of mammalian systems.

"In-vitro" studies of the effects of ELF-EMFs on normal as well as transformed (cancerous) mammalian cells have indicated biological sensitivity on end-points such as (*a*) cell proliferation, (*b*) membrane signal transduction (*c*) bio-molecular synthesis (*d*) ion-fluxes and binding (*e*) immunoresponses and (*f*) energy metabolism.

"In-vitro" data led to the generation of some interesting hypothetical mechanisms which may help further our knowledge on the biological effects of ELF-EMFs.

Trends in electric power consumption and cancer incidence are indicated to be consistent with the possibility of age-dependent cause-effect relationship, especially for some specific cancer types such as childhood leukemia, breast cancer for multiple myeloma. However, there are multiple cancer risk factors in industrial societies. Therefore, the extent of contribution of ELF magnetic field to cancer risk cannot be accurately determined.

## RADIO FREQUENCY (RF) AND MICROWAVE RADIATIONS (MR) AND THEIR BIOLOGICAL EFFECTS

The frequency of RF radiation extends from approximately $10^4$ Hz to $3 \times 10^8$ Hz, whereas the EMR in the frequency range $3 \times 10^8$ Hz to $3 \times 10^{11}$ Hz is referred to as microwave radiation (MR).

Mircowaves intersect with matter by coupling energy from the electromagnetic field by electric and magnetic dipole interactions. They penetrate materials with a low dielectric constant such as plastics, glass ceramics and composite materials, even when several centimeters thick but are reflected by metal surfaces.

Natural sources of RF and microwave energy include movement of cold fronts and solar radiations. Peak fields intensities of over 100 V/m are produced at ground level by the movement of cold fronts. Solar radiation intensities range from $10^{-18}$ to $10^{-17}$ watts per square meter per Hz (W $m^{-2}$ $Hz^{-1}$). However, the integrated intensity at the earth's surface for the frequency range of 0.2 – 10 g Hz is approximately $10^{-8}$ mW/$cm^2$. This value is to be compared with an average of $10^2$ mW/$cm^2$ on the earth's surface attributable to the entire (UV, Visible, IR and microwave) solar spectrum.

Apart from communications and navigational technology, other applications of micro-waves include microwave ovens, diathermy, materials drying equipment or in the pulsed mode in radar systems. Typical source of microwave energy are Klystrons, Magnetrons, backward wave oscillators and semiconductors transmit time devices (impatt diodes). Such sources may operate (*a*) continuously as in the case of communication systems or (*b*) intermittently as the case of microwave ovens.

Traditional microwave spectroscopy for providing information on molecular structure in only applicable to molecules in the gas phase at low pressures. Hence it is not very useful for non-invasive analysis. However, since gases at high pressure resemble the condensed phase in their interaction with microwaves, microwave absorption has been used as a process monitoring tools. Due to the high dielectric constant of water, the microwave absorption of water is used for measurement of moisture content. Reported applications include profiling of water content of sheet materials, monitoring water content of oil shale, measurement of water content in cotton, determining water content in crude oil, etc. Microwaves are widely used for liquid and solid level detection in tanks and happens. The majority of such applications are invasive with microwave antennae installed within the vessel. For metal vessels, microwave window materials such as plastics (e.g., Teflon) and ceramics must be used. The microwave sources used ar process applications are almost always the Gunn or tunnel diodes.

Because of increasing number of commercial application of microwave, there is a widespread interest in the possible new application as well as an increased concern regarding their potential hazards.

Typical sources of RF radiation include AM, television, and FM broadcast signals. The most common environmental source of microwave radiation are microwave ovens in the home that operate at a frequency of $2.45 \times 10^9$ Hz.

Radio frequency (RF) and microwave radiation have physical characteristics and hence their interaction mechanisms are quite different from those of ELF electromagnetism fields (such as 60 Hz fields associated with electric power transmission). The contrast between ELF and RF/mirowave EMR is most clearly illustrated by considering differences in wavelength. The wavelength of 60 Hz electric field is measured in thousands of miles compared to RF/microwave radiations having wavelengths of yards to fractions of an inch. Despite the differences in physical characteristics there are some similarities in biological response resulting from exposure to ELFs and EMR.

**EMR and Cancer :** A possible association of RF and MW-EMR exposure and human cancer incidence emerged recently. The use of hand-held cellular telephones operating in the frequency range of $8.25 \times 10^8$ to $8.9 \times 10^8$ has been reportedly linked with brain cancer. Exposure or police officers to higher-frequency microwave radiation ($1 \times 10^{10}$ to $2.4 \times 10^{10}$ Hz), emanating from traffic speed detecting radar units, has been associated with increase in testicular, brain and eye cancer. Exposure to hand-held cellular telephone radiation and police radar both involve lacalized radiation absorption. Tumors reportedly occurred at or near the apparent site of maximum microwave energy absorption in either case. For instance, parietal lobe brain tumors were reported to occur in the brain region closest to the location of the hand-held cellular phone antenna.

However, neither in the case of hand-held cellular telephone nor police radar have epidemiological studies been conducted to establish the validity of the association with cancer incidence. Only a few studies have been conducted on experimental animals or mammalian cells "in vitro" regarding the effects of these microwave frequencies. Hence, in the absence of a relevant and adequate scientific database the possible association of microwave sources with cancer incidence must be only viewed in terms of effects reported from exposure to other RF or microwave frequencies.

The results or, epidemiological studies, studies involving exposure of laboratory animals, and studies on the effect of RF or MW radiation on mammalian cells, provide evidence for a possible association of such exposure with cancer incidence. However the data available at present is incomplete and hence precludes from conclusions. It can only be said that under certain conditions of exposure (presently not well defined), RF and MW radiation exposure may act as a cancer promoter, rather than cancer initiator.

**Epidemiological Studies:** The photon energy in RF and microwave radiation is considered to be too low to cause photochemical reactions in biological matter. However, microwave radiation is absorbed in biological systems and ultimately dissipated in tissue in the form of heat. Irradiation of the human body with a power density of 10 mW/cm$^2$ will result in the absorption of about 5gW with a resultant elevation of body temperature by 1°C. This value is considered acceptable from the personal hazard point of view.

Microwave wavelength of less than 3 cm are absorbed in the outer skin surface, 3–10 cm wavelengths, penetrate more deeply into the skin, and at wavelength of 25-200 cm the penetration is great with potential of causing damage to internal body organs. The human body is considered to be essentially transparent to wavelength greater than 200 cm.

In order to described absorption of radio frequency electromagnetic energy, the National Council on Radiation Protection and Measurement (NRCP) introduced the term "Specific absorption rate" (SAR). This is defined as the rate at which electromagnetic energy is absorbed at a point in a medium per unit mass of the medium, and is expressed in W/kg.

Milham observed a statistically significant excess number of leukemia cases among amateur radio operators. However, the elevated leukemia risk in the study group could have been due to other risk factors but the exposure to RF radiation should be considered as an etiological factor.

An unexpectedly large number of cancer cases, including lymphomas, chronic myelocytic leukemia and pancreatic cancer was observed among military personnel exposed to mircowave radiation. A retrospective epidemiological study of cancer morbidity in polish military personnel aged 20-59 during the period 1971-1980 was observed. Cancer morbidity in personnel occupationally exposed to RF/microwave EMR was three times greater than the control group.

Evidence of an association of occupational EMR exposure and brain cancer was reported in an epidemiological study of 951 cases of brain tumors among white male residents of Maryland during the period 1969-1982. Fifty cases of glioma and astrocytoma were observed among electricians, electrical engineers and high voltage transmission line-men. Compared to the expected 18 cases, there was a statistically significant increases in cases of brain tumors in EMR workers.

Investigators at Johan Hopkins University have suggested a possible relation between mangolism (Down's syndrome) in offspring and previous exposure of the male parent to radar. However, such association must be taken with caution because of the probability of a variety of exposures to environmental agents (including ionizing radiation) while in military service.

Soviet investigators claim that microwave radiation produces a variety of effects on central nervous systems and without a temperature raise. Other reported effects include disturbances of the menstrual cycle, biochemical changes and changes in RNA at power density levels of about $10Mw/cm^2$.

Epidemiological studies, though limited in number, provide evidence of an association of long term exposure to RF/MW and low frequency EMFs and cancer incidence. Leukemia and brain cancer seem to be the most prevalent as compared to other cancer types.

**Studies on animals:** Carpenter and Van Ummersen (1968) investigated the effects of microwave radiations on the production of cataracts in rabbit eyes and conclude that microwaves may exert cummulative effect on the lens of the eye if the exposure are often repeated. The interval between exposures in an important factor in that a repair mechanism seems to act to limit the damage to the lens if adequate time has elapsed between exposures.

Power and frequency "windows" have also been reported indicating enhanced biological response within narrow bands of incident power and radiations frequency.

Effects of RF and mirowave radiation on various psycho-physiological responses e.g.. Changes in behaviour in experimental animals such as rodents, dogs and non-human primates were reported. Results of more recent long-term or chronic low intensity exposure studies involving the relationship or microwave exposure to cancer promotion have indicated the necessity to reconsider the health effect issue.

The potential tumor-promoting effect of microwave exposure was investigated by several workers. They suggested that long-term exposure to low intensity ($5mW/cm^2$) microwave radiation had an effect on tumor growth in mice similar to the effect of chronic stress.

Animal studies have provided evidence that low intensity EMR alter brain neurochemistry under conditions not involving tissues heating. The results show that the low-intensity EMR can alter brain function, resulting in behavioural neurological and neuroendocrine alterations.

Certain other biological effects of microwave radiation such as "pearl chain effect" where particles align in chain when subjected to an electric field.

While describing the biological effects of microwave radiation, one should remember that such radiations have also produced beneficial effects. Controlled or Judicious exposure of humans to diathermy or microthermy is widely in practice. The localized exposure level in diathermy may be as high as $100mW/cm^2$.

**"In Vitro" Studies:** Studies on effect of EMR on a variety of cellular end-points have been conducted "in vitro". These include (*a*) membrane structure (*b*) membrane cation transport and binding (*c*) single ion channel kinetics (*d*) energy metabolism (*e*) neuronal activity (*f*) proliferation and activation, and (*g*) transformation. "In vitro" studies on the effects of low-intensity RF and microwave radiation indicate dose-rate dependent increases in neoplastic transformation frequency and proliferation. However, due to the limitations on the extrapolation of "in-vitro" results to "in vivo" response, these results cannot be related directly to cancer incidence in human populations exposed to such radiation.

**Control measures:** Excessive exposures to microwave radiation can be controlled by engineering measures such as restriction of azimuth, elevation setting on radar antennas, and completely enclosing the magnetrons in microwave ovens. The use of personnel protection devices have their place but are of lower priority importance as compared to engineering controls. Personnel protective devices such as protective suits, goggles and mesh have been used for special problems. Susceptibility of cardiac pacemakers has to effectively reduced by providing improved shielding so that they can withstand adequately, power densities of at least 10mW/cm$^2$ without interference with their functioning.

**Conclusion :** There is increasing evidence of possible health effects of environmental exposure to EMFs and EMR in the home and at the workspace. Epidemiological evidence indicates possible associations of long-term exposure and cancer incidence, adverse reproductive outcomes and behavioural and neurological changes. However, inherent limitations on exposure assessment, common to epidemiological studies, provide imprecise information regarding time or exposure intensity thresholds for these effects. Although the results of animal experimentation and cellular studies of ELF-EMFs and EMR effects are generally consistent with the results of the epidemiological studies, the data provided by them is inadequate for precise risk assessment. Obviously, there is greater need to conduct more intensive investigation. Until then, it is prudent to avoid undue exposure to these radiations.

## BIOLOGICAL EFFECT OF ULTRAVIOLET (UV) RADIATION

The UV region may be divided on the basis of wavelength, λ as follows:

| Wavelength, λm | UV region | Photon energy range, Electron volts (ev) |
|---|---|---|
| < 0.16 | Vacuum UV | > 7.7 |
| 0.16 – 0.28 | Far UV | 7.7 – 4.4 |
| 0.28 – 0.32 | Middle UV | 4.4 – 3.9 |
| 0.32 – 0.4 | Near UV | 3.9 – 3.1 |

The major natural source of UV radiation is the Sun. Due to the absorption by the stratospheric ozone layer, generally, wavelengths >0.29 μm reach the earth's surface. The man-made sources of UV radiation include low pressure mercury vapour discharge lamps (over 8.5% of the radiation emitted is at 0.2537 μm, i.e. the germicidal wavelength), quartz lamps (radiations at about 0.38 μm), xenon discharge lamps, lasers and relatively new types of fluorescent tubes (which emit radiation at wavelength about 0.315 μm).

The recent data show that a maximum erythemal effect (i.e. reddening of the skin) is produced at 0.266 μm with secondary peak at about 0.29 μm. Depending on the total UV dose, the latent period for erythema may range from 2 to several hours, the severity may vary from simple erythema to blistering and desquamation. Although the long term effects of the process of melanin migration is not completely understood, the available data seem to indicate that some regions of the UV may produce or initiate carcinogenesis in the human skin.

High incidence of Photosensitisation of the skin and skin cancer was reported in outdoor workers who are simultaneously exposed to chemicals such as coal tar derivatives, methyl cholanthrene, and other anthracene compounds raises the question as to the role played by UV radiation in these cases.

UV radiation at 0.36 mm (with photon energy of about 3.5 eV) may excite the lens of the eye or cause the aqueous or vitreous humor to fluoresce thereby producing a diffuse hazyness inside the eye followed by visual activity or produce eye fatigue.

Depending on the severity of exposure, a latency period from half an hour to 24 hrs exists for the development of photokeratitis. A sensation of "Sand in the eyes" accompanied by varying degrees of photophobia, lacrimation and blepharospasm. Blepharospasm is a reflex-protective mechanism characterized by an involuntary tight closing of the lids, usually over a damaged cornea.

## BIOLOGICAL EFFECTS OF LASER RADIATIONS

Lasers (light amplification by stimulated emission of radiation) are at present used for cutting, welding, micro-machining, sealing, spectroscopy, holography, optical alignment, interferometry, surgery and as communication media.

Lasing action has been obtained in gases, crystalline materials, semiconductors and liquids. Stimulated emission in gaseous systems was first reported in 1961 in a helium-neon mixture. Since then, lasing action has been reported at hundreds of wavelengths stretching from the UV to the far IR. Helium-neon systems can be operated in a pulsed mode or continuous wave (CW) at wavelengths of 0.6328, 1.15 or 3.9 μm depending upon the design of the resonator. Typical power for He-Ne systems is of the order of 1–500 μW. The carbon dioxide gas laser system operates at a wavelength of 10.6 mm in either continuous wave, pulsed or Q-switched modes. The power output of $CO_2$-$N_2$ systems may range from several watts to > 10 kW. The $CO_2$- laser is preferred for terrestrial and extra-terrestrial communications due to its low absorption window in the atmosphere between 8 and 14 qm. From the personal hazard point of view, the significant fact is that enormous power may be radiated at wavelengths invisible to human eye. The argon ion gas system predominantly operate at 0.488 and 0.515 μm in continuous or pulsed mode. Power generation is greatest at 0.488 mm at < 10W. Laser systems involving solid state crystalline materials such as neodymium ($Nd^{3+}$) in garnet or glass and chromium ($Cr^{3+}$ in aluminium oxide are noteworthy. Gallium arsenide laser operating at 0.84 mm is a well known example of semi-conductor type laser. Semiconductors operating over the range 0.4-5.1 mm have been developed.

$CO_2$-lasers and helium-lasers are used in surgery. The latter is particularly useful for percutaneous myocardial revasculation (PMR).

The body organ most susceptible to laser radiation seems to be the eye. Skin is also susceptible. The degree of risk to the eye depends upon the types of laser beam used, its wavelength, beam divergence, pulse repetition frequency and output power.

In case of UV lasers, the expected response is similar to that produced by non-coherent sources, e.g., photophobia, erythema, exfoliation of surface tissues and possible stromal haze. The damage to epithelium results from the photochemical denaturization of proteins.

In case of IR laser radiation damage results exclusively from surface heating of the corrnea followed by absorption of the incident energy by tissue water in cornea.

In case of the visible lasers (wavelength of 0.4 – 0.75 mm), the organ at risk is the retina and more particularly the pigment epithelium of the retina.

The chronic and long term effects of laser radiation have not been fully investigated.

The most common effects of the skin range from erythma to blistering and charring dependent upon the wavelength, power, and exposure time of the radiation. At extremely high powered laser radiation (e.g.. Q-Switched pulses), de-pigmentation of the skin and damage to underlying organs may occur.

The threshold for producing retinal lesions at all visible wavelengths were considered to be 5 to 10 $W/cm^2$. A much greater sensitivity to blue wavelengths was discovered recently, which may be due to photochemical factors rather than thermal effects.

Important control principles to protect from exposure to laser systems include imparting awareness to the people concerned regarding potential hazards, applying primary engineering and other controls e.g., enclosure, beam stops, beam enlarging systems, shutter, interlocks and isolation of laser systems as well as shielding and safety goggles.

All living organisms have internal electric field due to electrolytic processes for cell and nerve functions, as well as magnetic fields from the magnetic moments of molecules. Further, we live in natural electromagnetic fields originating from atmosphere electricity and earth's magnetism. These natural fields are superimposed by man-made fields such as those from electric power lines, wiring and electric appliances. The interaction of electromagnetic fields with living systems depends of the basic modes of intersection of electric and magnetic fields with matter.

All types of matter is composed of atoms which in turn comprises of negatively charged electrons and positively charged particles called protons. The unit of charge in SI units, taken from rationalized MKS A systems, is a coulomb.

$$1 \text{ coulomb (C)} = \text{IA.S. (ampere seconds)}$$

The charge magnitude on an electron and proton is $1.6 \times 10^{-19}$ coulombs and 1 coulomb of electricity contains about $6.24 \times 10^{18}$ elementary charges. Each of these charges is associated with an electric field, E which is expressed (in SI units) in volts per meter (V/m). The electric field emanates radially from a point charge and is proportional to the force that the charge exerts on other charges as per coulomb's law, where opposite charge attract and like charges repel.

When charges move, they constitute a current, and they give rise to the magnetic flux density B, which has SI units of Teslas (T). Since 1-Tesla of magnetic field is very large and uncommon, a more common unit, namely gauss (G) is used to express the magnetic flux density (B).

$$1 \text{ Tesla (T)} = 10{,}000 \text{ Gauss (G)}$$

A related quantity, the magnetic field intensity (H) is also often used to describe magnetic fields, and it has units of amperes per meter (A/m). The magnetic flux density (B) is related to the magnetic field intensity (H) as per the following relationship:

$$B = \mu_0 H$$

Where $\mu_0 = 4\pi + 10^{-7}$ H/m and is known as magnetic permeability of free space. The magnetic field from a small element of current as given by the Biot-Savart law, is in the direction perpendicular to the current element, and is proportional in magnitude to the force that the current element exerts on other current elements as given by the Lorentz force law where opposite flowing currents repel and like flowing currents attract. The magnetic Lorentz force on a current element is in the direction perpendicular to both the direction of current flow and the direction of the magnetic field acting on the current element.

The earth's core comprises of iron. An iron atoms behaves as if the protons in its nucleus were spinning. This moving charge creates a dc magnetic field distribution around the earth, which at ground level has values of about 0.25-0.6 G. Since motion through a magnetic field creates an electric field, walking through the earth's magnetic field can induce body currents that add to biologically generated body currents.

Human body is composed of electrolytic materials. The body functions depend upon the ion concentrations and membrane potentials that are of order of the thermal voltage (Kr/q), where K is the Boltzman's constant, T is the absolute temperature in degrees Kelvin and q, is the charge of an electron which is nearly equal to $1.6 \times 10^{19}$ Coulombs. At room temperature (T@300 Kelvin), the thermal voltage is about 25 mV. The brain communicates with the rest of the body by electrical impulses. A typical nerve cell fires at about 50-100 mV. It is possible to monitor the health of the body by measuring these electrical signals using electroçardiography (ECG) or the heart and electroencephalography (EEG) of the brain. The electrodes of heart pacemaker in contact with heart muscle provide current pulses of the range of 0.1-10mA a few milliseconds long to synchronize the firing of heart cells. Magnetic resonance imaging (mri) can image internal organs from the magnetic moments of molecules in the body using d-c magnetic fields up to 2 Teslas (2-T)

Electricity generated by human adds additional electric and magnetic fields to the environmen. which can differ from natural electromagnetic fields in amplitude, direction and frequency. In the united states, electric power uses 60 Hz a-c, whereas Europe and some other parts of the world employ 50 Hz. Table (14.2) given below lists representative values of generated electromagnetic fields.

**Table 14.2: Some Typical Steady-State Electric and Magnetic Power Frequency Fields.**

| S.No. | Field | Electric Field, E(V/m), rms | Magnetic Field, B.(mG), rms |
|---|---|---|---|
| 1. | Under-high-voltage transmission line | 1000-7000 | 25-100 |
| 2. | Near electrical appliances | 30-30 | 5-3000 |
| 3. | Under domestic distribution lines | 10-60 | 1-10 |
| 4. | Domestic wiring | 1-10 | 1-5 |
| 5. | Inside railroad cars on electrified lines | – | 10-200 |

**Note:** r.m.s is root mean square, which is equal to the square root of the average of the square of periodic voltage or current over a period.

Typical household background at 60 Hz is 1–10 V/m for the electric field and 0.1 – 10 mg for the magnetic field. In the United States, maximum exposure standards for 60 Hz electromagnetic fields at transmission lines are typically in the range of 1-11 KV/m rms for the electric field, and about 250 mg for the magnetic fields. At other places, exposure standards are not prescribed for 50/60 Hz magnetic fields, presumably because they are nearly similar to those of natural electromagnetic fields and hence may not cause any health hazards,

## ELECTROMAGNETIC FIELDS

The electromagnetic field intersection are of the following three types:

(1) **Electroquasistatics,** a low frequency range for which charges and voltages are the sources of the electric field coupled to dielectric and conducting media with negligible magnetic field.

**Ex:** Electroquasistatic applications include capacitors, vacuum tube, and semiconductors in circuit application, microphones, electrocardiography and electroencephalography, and electrostatic precipitators, printers and copiers.

(2) **Magnetoquasistatic,** a low frequency range for which currents are the source of the magnetic field coupled to magnetizable and conducting media with electromagnetic induction generating an electric field.

**Ex:** Magnetoquasistatic applications incudes inductors and transformers in circuit and power system applications; audio speakers; magnetic resonance imaging (mri) devices;

and motors and generators. The proposed new train systems of magnetically levitated trains (Maglev) and perhaps electric cars provides further urgency to resolving adverse health issues in low level magnetic fields.

(3) **Electrodynamics,** a high frequency range for which the electric and magnetic fields are of equal importance, resulting in radiating waves that travel at the speed of light ($c \cong 3 \times 10^8$ m/s) in free space.

**Ex:** Electrodynamic fields are present in radio, television, communications, optics and microwave ovens.

Quasistatic fields occur when the wavelength of electromagnetic waves at frequency $f$ ($f = c/\lambda$) is much larger than the size of the system. For power frequency of $f = 60$ Hz, $\lambda = 5 \times 10^6$ m, so that the usual small electrical appliances may be considered quasistatic. This power frequency regime is known as **extra low frequency (ELF)** range. Quasistatic field are confined to the immediate vicinity of the electrical device. However, radio frequency and higher frequencies have propagating electromagnetic waves that travel at the speed of light away from a source.

Photon is the fundamental equivalent particle of an electromagnetic field. The energy (w) in each photon is given by the relation:

$$W = hf$$

Where $f$ is the frequency of the electromagnetic waves and he is the Planck's constant which is equal to $6.6256 \times 10^{-34}$ J.S. ($f < 10^{15}$ Hz) are usually called as non-ionizing radiation, because the energy W is too low to break chemical or molecular bonds. The energy of a 60-Hz photon ($W = 4 \times 10^{-32}$ J) is 11 orders of magnitude smaller than the thermal energy at room temperature (T = 300 K), $KT = 4 \times 10^{-21}$ J. However, ultraviolet, X-ray and Y-ray radiations have sufficiently high energy to break, chemical and molecular bonds, and can cause cancer and other health problems, depending on the extent of total exposure. On the other hand, radio- and micro-waves do not have enough energy to break bonds, but they are absorbed by water in the body tissues to cause heating. This is the principal involved in cooking with a microwave oven and a medical treatment by heating body tissue with diathermy machines.

Imposed charges and voltage generate electric field and imposed currents generate magnetic fields. However, these fields induce further charges and current in dielectric, conducting and magnetizable materials which in turn create additional electric and magnetic fields. Further, time-varying magnetic and time varying electric fields create magnetic fields (Ampere's Law). Electric fields in conducting materials cause current flow, voltage differences, and heating.

Quasistatic field from small electric appliances increases linearly with the source strength (charge or voltage for the electric field and current for the magnetic field) and source size (spacing and area), and decrease with distances as $1/r^3$ (r = distance). Small appliances and motors have approximate point dipole fields at large distance compared with their size. Thus the near fields depend directly on the source size, but a doubling of the distance from the device centre would approximately decrease the fields by a factor of eight. This indicates that by moving electric clocks, lights, motors, and other small electrical devices as far as possible from the body, the electric and magnetic field exposure would greatly decrease.

The electrical and magnetic field values of some appliances and power lines in a typical residential environmental are shown in Table 14.3 given below:

**Table 14.3. Typical residential electrical and magnetic field value of power lines and small electrical appliances.**

| S.No. | Field source | Electric Field, E. (V/m), rms | Magnetic Field, B. (mG), rms |
|---|---|---|---|
| 1. | **Power lines** | | 5 |
| 2. | **Appliances at a distance of 25 cm.** | | |
| | Refrigerator | 60 | 5 |
| | Clock radio (analogue) | 15 | 24 |
| | Fluorescent light | — | 21 |
| | Colour T.V. | 30 | 12 |
| | Electric range | 4 | 20 |

**Note :** rms is the root mean square.

At immediate vicinity of these small electrical appliances, the maximum electric field, E is about 100 V/m and the maximum magnetic field is typically about 2 G. For instances, a 325-W soldering gun and hair dryers have localized magnetic fields from 10 to 25 G, whereas a cloths drier, toaster, vacuum cleaner, dishwasher, electric iron, and refrigerator have magnetic fields less than IG. Electric blankets are thought to cause high exposure to power frequency electric and magnetic fields because they are close to human body for a long period of time. Older type of designs of electric blankets have magnetic flux densities of about 22 mG and electric fields of 10-2000 V/m. However, newer designs using partial magnetic field cancellation of parallel conductors with current flow in opposite direction have the magnetic field strength reduced by about 30 times while the electric field remains essentially unchanged. Twisting the two wires would further decreases the magnetic fields.

## FIELDS FROM POWER LINES

An electric power system consists of generating stations, transmission lines and distribution systems. Transmission lines connect generators to the distribution systems, which in turn connect to individual loads of the end users. A generator can typically produce 100–1000 MW power at 11–35 KV rms at typically 3,000-50,000 A rms. Since the transmission lines losses are primarily due to series resistance in the cables, these losses can be minimized by minimizing the line current. Since the transmitted power is proportional to the product of line voltage and line current, the same power can be transmitted at reduced current by increasing the voltage. A voltage step-up transformer can increase the generator voltage at the primary winding up to the typical range of 69-765 KV rms at its secondary winding, thereby reducing the secondary current from the primary current by the inverse ratio of the voltage step-up.

The step down of the voltage from the transmission levels at a bulk power station typically reduces the voltage to the range of 35-138 KV rms. The next step down in voltage is done at the distribution sub-station, reducing the voltage to 5-35 KV rms. The first epidemiological study indicated a higher incidence of childhood leukemia for those homes near pole distribution transformers. When the voltage is stepped down from 5-35 KV rms to 120/240 V rms (*i.e.* about a factor of 40-300), the secondary distribution current from the transformer similarly increases by the same factor thereby increasing the magnetic field strength by the same factor. It was hypothesized that children living and sleeping near such a distribution transformer with high magnetic fields may suffer from adverse health effects.

## ELECTRIC RAILWAYS

Conventional electric railways have typical steady-state magnetic fields within the compartments of the order or 10-25 G. The modern German Maglev System (i.e. magnetically levitated system) has a magnetic field below 0.5 G.

## RADIATION FIELDS

Radio, television, communication, and radar systems operate at much higher frequencies ($f$ = 500 KHz to 1000 GHz).

(1 KHz = $10^3$ Hz; 1 GHz = $10^{12}$ Hz)

Measurements of FM radio and UHF and VHF broadcast signal field intensities indicate that typical population exposure is a follows:

Electric field, E = 0.14 V/m

Magnetic field, B = 4.6 μG

If one uses transmitted radio or telephone commendations, such as with CB radio or cellular phones, the electric and magnetic fields decreases by $1/r^3$ (r = distance), in the vicinity of the user. However, the electric and magnetic fields from large commercial transmitting and radar antennae die off as 1/r, if one is more than few wavelengths away. Low power radar is used in aircraft for monitoring weather, as navigational aids on small boats, and to determine vehicle speed by the police. In such cases, the electric fields of electric field, E lie in the range of 19 to 580 V/m and the magnetic field, B lie in the range 0.6 to 20 mG.

A microwave oven is a microwave resonator at about 2.45 GHz with typical power levels of about 500 W. Most of the microwave ovens have leakage below 50 W/m$^2$, which is the emission limit from the ovens at a distance of 5cm, as stipulated bay the U.S. Bureau of Radiological Health. This corresponds to microwave rms fields, E = 137 V/m and magnetic field. B = 4.6 mG. The 60 Hz magnetic field from a microwave oven at 25.4 cm is 1-120 mG

## FIELD CONCENTRATION

Dielectric and conducting objects placed into an electric field change the electric field distribution in the vicinity of the object, often tending to enhance the electric field at selected positions along the surface of the medium. Similarly, conducting and magnetizable media tend to enhance magnetic fields at selected locations.

## POWER FREQUENCY FIELDS AND THE BODY

At 60 Hz the human body is a good conductor for electric fields but a poor conductor for magnetic fields. Therefore, the introduction of a body into an electric and magnetic field will locally change the electric field distribution but it will have negligible effect of the magnetic field distribution.

## TYPICAL STANDARDS

The international Non-Ionizing Radiation Committee (INIRC) constituted by the International Radiation Protection Association (IRPA) formulated limits of exposure of 50/60 Hz electric and magnetic fields, which are given in Table 14.4.

### Table 14.4. Limits of exposure to 50/60 Hz Electric and Magnetic fields as prescribed by IRPA/INIRC

| S.No. | Class | Electric Field, E in KV/m rms | Magnetic Field, B in G (rms) |
|---|---|---|---|
| A. | Occupational Workers | | |
| | Whole working day | 10 | 5 |
| | Short-term | 30 | 50 |
| B. | General Public | | |
| | Whole day | 5 | 1 |
| | Few hours per day | 10 | 10 |

**Note :** rms = root mean square

Table 14.5 summarizes the IRPA/INIRC guidelines for exposure limits at Radio and Microwave frequencies. They assume no adverse health effects for energy deposition rates below 4 W/kg. Further, a safety factor of 10 for long-term exposure and a further safety factor of 5 for general public limits are used.

### Table 14.5. Guidelines for exposure limits at Radio and Microwave Frequencies.

| F(MHz) | Occupational Exposure | | Exposure for general public | |
|---|---|---|---|---|
| | E | B | E | B |
| | (V/m, rms) | (mG, rms) | (V/m, rms) | (mG, rms) |
| 0.1-1 | 614 | 20/f | 87 | $2.9f^{1/2}$ |
| 1 -10 | 614/f | 20/f | $87/f^{1/2}$ | $2.9f^{1/2}$ |
| 10 - 400 | 61 | 2 | 27.5 | 0.9 |
| 400 -2000 | $3f^{1/2}$ | 0.1 | $1.37f^{1/2}$ | $0.465\ f^{1/2}$ |
| 2000-300,000 | 137 | 4.5 | 61 | 2 |

rms = root mean square

## CONCLUSION

Although Epidemiological, laboratory, and human and animal studies show possible biological effects from non-ionizing electromagnetic fields, there are no well-established physical mechanisms or conclusive proof of adverse health effects. Nevertheless, pending further conclusive scientific studies, it is essential that people should minimize their exposure to electromagnetic fields. In the meanwhile more intensive research has to be directed to enable us to understand to potential health effects due to human exposure to electromagnetic radiation fields. This is more urgent because of the ubiquitous and ever-increasing levels of human exposure to electromagnetic fields (EMFs) and electromagnetic radiation (EMF). For example, the increasing use of appliances e.g., microwave ovens, cellular phones, etc. on one hand the proposed new applications such as magnetically levitated trains (Maglev) and electric cars provide greater urgency to resolve adverse health issues due to exposure to electromagnetic fields.

## QUESTIONS

1. Discuss the health effects due to electromagnetic fields.
2. Discuss the effects of exposure to extremely low frequencies (ELF) on humans and animals.
3. Discuss the biological effects because of exposure to radio frequency and microwave radiations.
4. What are the biological effects of the following:

   (*a*) UV radiation (*b*) Laser radiation (*c*) Microwave radiation

# Sustainable Development—New Approaches

The World Commission on Environment and Development in its report to the United Nations in 1987 defined the "sustainable development" as "meeting the needs of the present without compromising the ability of future generation to meet their own needs".

Owing to the unprecedented population explosion and worldwide industrial expansion, earlier practices of controlling pollutants at the "end-of-the pipe" (i.e. at the point where pollutant enter the environment) are no longer effective in controlling the present industrial effluents, both in quantity and complexity. Although the prescriptive command-control approach for controlling pollution has been effective to the extent of controlling religious pollution, it suffers from the major drawback of shifting pollution from air to land to water and back again. Henceforth, "end-of-pin" treatment as a means of pollution control can no longer keep pace with increasing complexity and quantity of pollution from "non-point success" and the expanding scale of worldwide economic activity. For "susceptibility" to occur, further advances in technology in process control, product design and monitoring are essential for efficient use of resources in production while simultaneously abating pollution. The ultimate objective of all these efforts is the quest for zero-emissions from the industrial processes. Two important programs arising from this new strategy are:

(1) Pollution prevention and (2) Design for Environment.

(1) **Pollution prevention:-** It is a proactive multimedia management approach to pollution to achieve "front-end" reduction of pollutants in the waste streams by strictly controlling the industrial processes with respect to process upsets, undesirable emission and by improving the product quality, reducing loss of raw materials to the waste effluents and recycling the by-products. Pollution prevention also eliminates the transfer of pollutants from one media to another because pollutants are not generated in the first place. The raw materials are used more efficiently and also the costs and liabilities of waste disposal and clean-up are avoided. The success of pollution prevention strategies depends upon a multi-disciplinary approach where efficient process control and optimization are achieved by the combined efforts of chemical engineers and process analytical chemists.

**(2) Design for Environment:-** Apart from strict process control during the manufacture or products, it is also essential to review the overall synthetic sequence in the production of chemical substances and to consider the substitution or elimination of hazardous chemicals wherever possible. In view of the new emphasis of pollution prevention by both the industry and regulatory agencies and also due to the huge costs of treatment, disposal and compliance, selection of any synthetic scheme solely on the basis of yield is no longer, valid. All the scientific, environmental and economic impacts should be incorporated into the synthetic scheme being selected to ensure environmentally benign synthesis of a chemical substance. Thus, a synthetic chemist too will have to play a major role in achieving pollution prevention.

In the industrial world today, "end-of-the-pipe' control of pollutants is being superceded by a policy shift to "front-end" control emphasized by "pollution prevention", and also a shift to less hazardous chemical as advocated by the "design for environment" program as well as promotion of good operating practices and the concept of industrial ecology. Thus, the paradigm shift for the future emphasizes pollution prevention by ensuring that industrial processes run under tight control, substitute less hazardous chemical wherever possible and recycle by-product of design new production processes which are environmentally benign. The future challenge, therefore, it is switch over from the present environmental policy based on pollution controls and Government mandated regulations to a new and more realistic environmental policy based on pollution prevention, source reduction, recycling and waste minimization.

## INDUSTRIAL ECOLOGY

Industrial ecology is based upon an analogy of industrial systems to natural ecological systems. It offers a unique systems approach within which environmental issues can be comprehensively addressed.

In nature, an ecological system operates through a web of connections in which organisms live and consume each other and each others wastes products. The system seems to have evolved in such a way that no useful material or available energy is lost. Some or the other organism manages to sustain its living with the help of waste product that provides available energy or usable material. Ecology seems to involve a food web that provides an interconnection of uses of both organisms and their wastes. In the industrial context, we may visualize such an interconnection of the use of product and waste products.

From the stand point of industrial ecology, unit processes and industries should be considered as interacting systems rather than isolated components. Therefore, different waste-producing processes, plants, or industries should be connected into an operating web that minimizes the total amount of industrial material that goes to the waste disposal sinks or lost in intermediate processes. Thus, the focus shifts from merely minimizing waste from a particular process or facility (i.e. pollution prevention), to minimizing waste produced by the larger system as a whole It gives further emphasis on the already known idea that a waste should be considered a potentially useful resource.

An industrial complex established in the city of Kalundborg. (Denmark) provides an excellent example of industrial ecology concept. In this, a refinery supplies gas to a power plant and plastic board company for their energy needs. The steam from the power plant is sent to a biotechnology plant and into a direct heating system of the city. Lower temperature energy goes to an experimental fish farm. The gypsum produced in the desulfurization unit of the power plant is used by the plaster board manufacturing unit. By refusing energy and material, this "industrial ecosystem" at Kalunborg saves 19,000 tons oil, 30,000 tons of coal and 60,000 $m^3$ of water every year, thereby saving about 14 million US$ per year. Thus, they have gone beyond traditional compliance monitoring to process monitoring to avoid pollution. They have literally shifted from "end-of-the-pipe" pollution control to "front-end" pollution prevention by creative "design for the environment".

Industrial ecology promotes a system view of industrial products and process and processes that considers the total materials and energy cycle to minimize adverse environmental effects. For effective implementation of industrial ecology concept, several changes in industrial practices are essential, which include improving the efficiency and productivity of industrial systems, minimizing wastage of raw materials, substituting environmentally benign and abundant raw materials wherever possible, developing useful applications for waste products and reusing/ recycling of the manufactured products after use. For achieving this, we should investigate and understand the web of waste recycling and reuse found in natural systems and try to explore the possibility of applying them in industrial context. Moving towards such a system involves radical changes in the various and complex interacting systems by improving the current information systems or creating new ones and devise attractive economic incentives.

## GREEN CHEMISTRY

The green chemistry program initiated by the United States Environmental Protection Agency is another initiative under the "Design for Environmental Program". The pollution prevention Act of 1990 in the United States of America established a national policy to prevent or reduce pollution at its source wherever feasible. In the year 1991, the Office of the Pollution Prevention and Toxic (OPPT) launched a model research grants programme known as "Alternative Synthetic Pathways for Pollution Prevention". This Green Chemistry program provided unprecedented grants for research, development and implementation of innovative chemical technologies that prevent pollution in a scientifically sound and cost-effective manner. The green chemistry programme of the United States Environmental Protection Agency (USEPA) in non-regulatory and works with many partners in industry, academic, other government agencies, scientific societies, trade organisations, national laboratories and research centres to promote pollution prevention through green chemistry. International initiatives through partners such as international Union of Pure and Applied Chemistry (IUPAC) and the Organisation for Economic Cooperation and Development (OECD) are under way.

Green chemistry is the use of chemistry for pollution prevention by environmentally-conscious design of chemical products and processes that reduce or eliminate the use or generation of hazardous substances.

The basic principles of green chemistry are as follows:

**1. Prevention :** It is better to prevent waste than to treat or clean up waste after it has been created.

**2. Atom Economy:** Synthetic methods should be designed to maximize the incorporation of all materials used in the process into the final product.

**3. Less Hazardous Chemical Syntheses:** Wherever practicable, synthetic methods should be designed to use and generate substances that posses little or no toxicity to human health and the environment.

**4. Designing Safer Chemicals:** Chemical products should be designed to effects their desired function while minimizing their toxicıty.

**5. Safer Solvents and Auxiliaries :** The use of auxiliary substances (*e.g.*, solvents, separation agents, etc.) should be made unnecessary wherever possible and innocuous when used.

**6. Design for Energy Efficiency:** Energy requirements of chemical processes should be recognized for their environmental and economic impacts and should be minimized. If possible, synthetic method should be conducted at ambient temperature and pressure.

**7. Use of Renewable Feedstocks:** A raw material or feedstock should be renewable rather than depleting whenever technically and economically practicable.

**8. Reduce Derivatives:** Unnecessary derivatization (use of blocking groups, protection/deprotection, temporary modification of physical/chemical processes) should be minimized or avoided if possible, because such steps require additional reagents and can generate waste.

**9. Catalysis :** Catalytic reagents (as selective as possible) are superior to stoichiometric reagents.

**10. Design for Degradation:** Chemical products should be designed so that at the end @ of their function they break down into innocuous degradation products and do not perish in the environment.

**11. Real-time analysis for pollution Prevention:** Analytical methodologies need to be further developed to allow for real-time, in-process monitoring and control prior to the formation of hazardous substances.

**12. Inherently Safer Chemistry for Accident Prevention :** Substances and the form of a substances used in a chemical process should be chosen to minimize the potential for chemical accidents, including releases, explosions and fires.

**Emerging concepts**

Several new concepts and industrial practices are emerging to facilitate the integration of industrial production with environmental factors.

**(*a*) Concurrent Engineering:** Environmental factors can be considered as an element of the concurrent engineering "Design for X" (DFX) process, where X represents a product characteristic such as reliability or manufacturability that the company wants to maximize in its product design.

**(*b*) Total Quality Management (TQM):** The adoption of TQM programmes, has rendered the quality as the responsibility of everyone in the company, from top to bottom.

**(*c*) Life Cycle Approach:** This approach mandates that environmental impacts, broadly including relevant safety, health and social factors, be considered across the life time of products, process, material, technology, or services. This approach is meant to minimize to overall environmental impact of a product or process, and not simply address only or a few aspects of that impact. Life cycle assessment (LCA) provides a comprehensive profile of energy and material inputs to, and environment impact of, manufacturing, using and retiring of products.

The design for environmental (DFE) methodologies can be developed as a module of concurrent engineering practices to facilitate the integration of environmental objectives and constraints in the design process directly. Simultaneously, it is essential to improve tangential systems such as total environmental cost accounting and selling function of the products or supplier management systems, which indirectly influence the design decisions.

**(D) Preventive Environmental Management (PEM)**

Preventive Environmental Management (PEM) involves minimization or elimination or wastes and pollutants at their sources, rather than at the end-of-pipe stage. This can be achieved by a package of activities such as

(*i*) Redesigning the products that cause less wastes or less pollution during their manufacture, use or disposal.

(*ii*) Altering the production process to minimize the utilization of toxic chemicals.

(*iii*) Implementing better housekeeping practices and devising better handling procedures to minimize leaks and fugitive releases from the manufacturing process.

(*iv*) Taking proper steps to ensure efficient use of energy and materials, with a spirit and commitment to conserve natural resources.

The hierarchy for preventive Environmental Management can be summarized as follows:

1. Pollution should be prevented or minimized at the source wherever feasible.
2. Pollution that cannot be prevented should be recycled in an environmentally safe manner wherever feasible.
3. Pollution that cannot be prevented or recycled should be treated in an environmentally safe manner wherever feasible.
4. Disposal or other release into the environment should be used only as a last resort and it should be done in an environmentally safe manner.

Preventive environmental management (PEM) offers a dependable way for protecting the environment and ensures more efficient use of resources. However, it should be appreciated that PEM is not something that can be just added to the existing systems and practices. It involves identification of the underlying causes of waste and pin-pointing the ways and means to minimize its generation, often by using materials and energy more prudently and efficiently. Since pollution occurs from all human activities, PEM poses a challenge to all sectors of society, at all levels of activity and decision making. Thus, environmental awareness becomes vital. PEM requires acquirement of skills, creativity and a mind-set to holistically identify the various options for improvement and innovation.

PEM involves sincere implementation of a constellation of strategies, tools and concepts. The tools contemplated can be broadly categorized as follows:

**(*i*) Tool related to product**

(*a*) Design for environment (DEE)

(*b*) Industrial ecology (IE)

(*c*) Life cycle assessment (LCA)

(*d*) Extended producer responsibility (EPR), etc.

**(*ii*) Tools related to process**

(*a*) Pollution prevention

(*b*) Cleaner production

(*c*) Cleaner technology

(*d*) Waste minimization

(*e*) Toxic use reduction

(*f*) Best operating practices

(*g*) Eco-efficiency, etc.

**(*iii*) Tools related to Management**

(*a*) Environmental performance evaluation

(*b*) Environmental audit

(*c*) Environmental Management System

(*d*) Environmental Reporting and Communication

(*e*) Eco-industrial development

(*f*) Trade and environment

(*g*) Environment economics.

(*h*) environmental economics, etc.

Industry, being a major contributor to environmental problems, becomes the major focus for PEM. A number of big companies are developing structured programmes to facilitate the implementation of PEM into their operations and decision-making and ensuring active participation of their employees. Obviously, the success depends upon top-level commitment, educating and motivating the employees, and adopting long term planning strategies. Implementation of PEM in industry will ensure increased cost saving, improved workers's safety, reduced legal liabilities and enhanced corporate image.

**(E) Regulations on Occupational Safety and Health (OSH)**

Many countries including India make efforts to strength occupational Health and Safety (OSH) by enacting laws and acts to regulate the measures taken by the companies towards the health and safety of their employees. Despite comprehensive legislations, a large number of accidents still take place. This may be because of the lack of manpower needed to enforce the laws, inadequate penalties, or unrealistic provisions of the acts vis-a-vis local work situations. Further, only 8% of the Indian workforce is employed in the organised sector and hence most of the workers do not fall under the ambit of legislative jurisdiction.

Hazardous and unhealthy workplaces result in huge expenditure for the employer. The cost of treatment of the injured or sick worker has to be met by the employer. If the worker is not in a position to resume work after his recovery, a substitute has to be trained and it will take a lot of time for the new worker to attain the productivity levels of the previous workers, Indemnification of the injured workers or those who have died and their families may cause considerable expenses which can impose heavy burden especially for small and medium scale industries. Therefore, apart from moral and legal responsibility of the employers towards their employees, it is in the economic interest of the industry to take adequate steps to prevent accidents and occupational diseases in their factories. The expenditure towards occupational health and safety should therefore be considered as an investment for a profit-maximizing industry.

Many countries try to improve OSH by regulatory measures through labour laws. However, economic incentives together with the legal provisions will be more effective in this realm. In countries where enforcement is lax, the companies tend to ignore regulations on OSH. Moreover, regulations prescribe a minimum level of OSH measures. Once this level is reached, there is no appreciable incentive for further improvement. Also, the adoption the laws to the new risks takes time. Further, economic incentives measures the outcome of OSH but not the means. Regulations prescribe certain means which are intended to be effective.

Standards on OSH are formulated on the basic of consensus arrived at by the representatives of all the economic stakeholders concerned. Standards represent a means to set a certain level of OSH without fixing minimum requirement. Most countries have their own "standards" bodies. For instance, India has a "Bureau of Indian Standards (BIS)", U.S. has "American National Standards Institute (ANSI)", and so on. These bodies publish standards in order to respond to customer demands, who want a certification from their suppliers for certain aspects such as product quality or environmental protection. The best known standards all over the world are ISO 9000 series for quality management, and ISO 14000 series for environmental protection, both formulated by the International Standardisation Organisation (ISO), Geneva. These can be used by companies for voluntary certification of implemented management systems to enable them to distinguish themselves from their competitors.

Occupational Safety and Health (OSH) is another area where standards could be applied effectively to render to workplace environment in industry more attractive for workers. ISO 9000 and ISO 14000 do comprise some aspects of OSH but only marginally. ISO 9001 obligues the employers to communicate to the organisation the importance of meeting statutory and regulatory provisions, while environmental issues go along with the safety of the plants and machinery. A specific ISO standard on occupational safety and health does not exist, not it is being planned for the present, because, differences in values, culture, and local requirements preclude formulation of a common standard suitable for all countries. It is for this reason that some countries have developed their own standards on Occupational Health and Safety Management Systems (OHSMS) as per their own requirements. A group of 13 European certification companies and the British Standardisation Institute (BSI) have developed the occupational Health and Safety Assessment Series (OHSAS) 18000. It corresponds to the structure of ISO 14000 and hence can be implemented concurrently without any incompatability. On the basis of OHSAS 1800 and adopted to our requirements, India has formulated IS 1500: 2000 Indian Standard on "Occupational Health and Safety Management System-Specification and Guidance for Use". It comprises four phases of the improvement process, viz., planning, implementation and operation, measurement and evaluation (corresponding to checking and corrective action in

OHSAS 18001), and management review. IS 15000 comprehensively describes risks assessment process under its Annexure-C. It comprises six activities viz. classifying work activities, identifying hazards, determining risks, deciding if risk is tolerable, preparing risk control action plan, and reviewing adequacy of action plan. OSH can be integrated into the overall business management process.

**(F) Energy Utilization**

## INTRODUCTION

Production of energy from the various sources and utilization of the energy produced are the two sides of the human energy equation. Energy is needed to support multifarious activities of the civilized society. However, the actual amount of energy required depends on the efficiency with which it is used, which in turn depends on technology, the integration of energy systems, energy conservation practices, and on our living patterns.

When energy is converted from one form to another, there is always some loss of useful energy as waste heat. Minimising the waste heat represents a major challenge in energy conversion.

## BASIC PRINCIPLES

The first law of thermodynamics is simply a statement of energy conservation, viz., energy can neither be created nor be destroyed, provided we remember that heat is a form of energy. In fact, it is the sum of matter and energy that is conserved. Energy and matter are related according to Einstein's formulated, $E = mc^2$, where E is energy, $m$ is mass and $c$ is the speed of light. However, this conversion of mass into energy become significant only in nuclear reactions.

In chemical systems, the conversation of energy law is most easily handled in terms of three quantities viz., the work that is performed, the heat that flows, and the energy that is stored in the system. To be more specific, let us consider any process that takes a system from an initial state '$a$' to a final state '$b$'. The system may be a mechanical device, or a container holding a chemical compound or anything like that. The process may be a change in pressure or temperature, performing of a mechanical work, or occurrence of a chemical reaction and so on. The first law of thermodynamics can be stated by the following two equations:

$$\Delta E = q - w, \qquad \text{... (1)}$$

$$\Delta E = E_b - E_a \qquad \text{... (2)}$$

where $q$ = heat absorbed by the system,

$w$ = work done by the system,

$\Delta E$ = change in internal energy of the system.

If heat is absorbed by the system, $q$ has positive value. If heat is given by the system, $q$ has a negative value.

Similarly, if the system does on the surrounding, $w$ has a positive value. If the surroundings do work on the system, $w$ has a negative value.

Eqn. (1) shows that any net energy in the form of heat and work, $(q - w)$, is not created or destroyed but must correspond to the change $\Delta E$ in the stored energy.

Eqn. (2) states that, for a process taking the system from state '$a$' to state '$b$', the internal energy change depends only on state '$a$' and '$b$' and not on the way in which the system was transformed from one state the other.

The Enthalpy, $H$ is defined as:

$$H = E + PV$$

where $E$ is the internal energy, $V$ is the volume and $P$ is the pressure of the system. Since $E$, $P$ and $V$ are state functions, enthalpy is also a state function. Therefore, the change in enthalpy, $\Delta H$ between any pair of initial and final states is independent of the path between them.

The maximum efficiency that can be attained in energy conversion processes is set by the second law of thermodynamics. The second law of thermodynamics recognizes two classes of processes - the spontaneous and non-spontaneous. The second law of thermodynamics is expressed in several equivalent ways. The one formulated by Kelvin is that

"No process is possible in which the sole result is the absorption of heat from a reservoir and its complete conversion into work".

The signpost for spontaneous change is to look for the direction of such change that leads to more disorderly dispersal of the total energy of an isolated system. The second law of thermodynamic, which is used to identify the sign-post of spontaneous change, may also be expressed in terms of another state function, viz. the entropy, $S$ as follows:

"The entropy of an isolated system increases in the course of a spontaneous change:

$$\Delta S_{tot} > O$$

where $S_{tot}$ is the total entropy of the system and its surroundings. Thermodynamically irreversible processes (such as cooling to the temperature of the surroundings, and the free expansion of gases) are spontaneous processes, and hence they must be accompanied by an increase in total entropy.

The thermodynamic definition of entropy focuses on the change in entropy, $dS$, that occurs as a result of a physical or chemical change (in general, as a result of any "process"). The definition is motivated by the idea that a change in the extent to which energy is dispersed in a disorderly way depends on how much energy is dispersed as heat. Heat stimulates disorderly motion in the surroundings. Work, which stimulates uniform of atoms in the surroundings, does not change the degree of disorder and so doesnot change the entropy.

The thermodynamic definition of entropy is based on the expression:

$$dS = \frac{dq_{\text{rev}}}{T} \quad \text{... (3)}$$

where $dS$ is the change in entropy, $dq$ is the energy supplied as heat and $T$ is the temperature. For a measurable change between two sites $i$ and $f$, this expansion integrates to

$$\Delta S = \int_i^f \frac{dq_{\text{rev}}}{T} \quad \text{... (4)}$$

That means, in order to calculate the difference in entropy between any two states of a system, we find a reversible path between them, and integrate the heat supplied at each stage of the path divided by the temperature at which the heat is supplied. Entropy is a measure of disorder and as per the second law of thermodynamics, it increases naturally.

One measure of entropy (S) is the ratio of the heat energy (Q) to the absolute temperature (T):

$$\text{S} = \frac{\text{Q}}{\text{T}} \quad \text{... (5)}$$

At absolute temperature, the Kinetic energy of the molecules is zero; in other words, there will not be any motion of the molecules. According to the above equation, a given quantity of heat energy produces greater disorder for a cold sample of matter (small value of T) than for a hot sample (high value of T). As per the second law of thermodynamics, Q/T increases in a spontaneous process.

Where a quantity of heat is transferred from the hot liquid to the cool surroundings, then

$$\frac{Q_{transf}}{T_{hot}} < \frac{Q_{transf}}{T_{cool}} \quad \text{... (6)}$$

where $Q_{transf}$ is the quantity of heat transferred.

Thus, entropy increases during the cooling process

## HEAT ENGINES

Consider a steam-engine power plant shown in Fig. 15.1. The maximum efficiency of such a power plant or any other heat engine can be obtained with the help of eqn. (5). Steam generated by burning a fuel in a boiler turns a turbine to generate electricity. The spent steam is then condensed to water which is returned to the boiler. In this process, the energy derived from the combustion of the fuel is converted to work in the turbine. The work energy in this case is the difference between the quantity of heat used to boil the water ($Q_h$) and the quantity of heat transferred to the cooling water in the condenser ($Q_c$).

$$W = (Q_h - Q_c) \quad \text{... (7)}$$

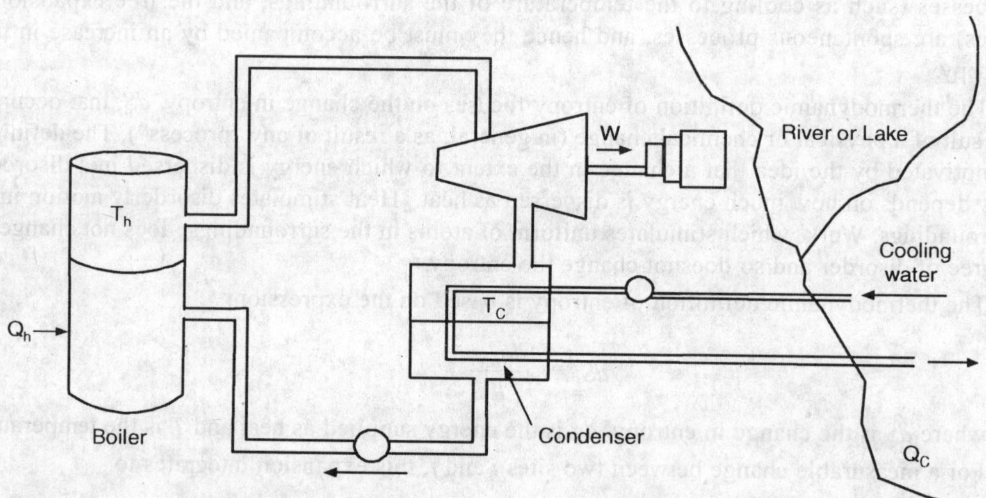


**Fig. 15.1.** Maximum work and waste heat from a steam engine power plant.

$Q_h$ = quantity of heat delivered to boiler

$Q_c$ = waste heat discharged to water body

$T_h$ = heat source at temperature, $T_h$

$T_c$ = heat sink at temperature, $T_c$

This equation is nothing but an expression of conservation of energy stated in the first law of thermodynamics.

By dividing both sides of eqn. (7) by $Q_h$, we have

$$\frac{W}{Q_h} = 1 - \frac{Q_c}{Q_h} \quad \text{... (8)}$$

$\frac{W}{Q_h}$ is the efficiency with which heat is converted to work.

However, as per the second law of thermodynamics, entropy must increase in such as process, (vide eqn. 6). Therefore, we have

$$\frac{Q_c}{T_c} > \frac{Q_h}{T_h} \qquad \text{... (9)}$$

where $T_h$ is the temperature of the heat source; and $T_c$ is the temperature of the condenser. Multiplying both sides of eqn. (9) with $\frac{T_c}{Q_h}$, we get

$$\frac{Q_c}{Q_h} > \frac{T_c}{T_h} \qquad \text{... (10)}$$

Now, substituting eqn. (10) in eqn. (8), we get

$$\frac{W}{Q_h} < \left(1 - \frac{T_c}{T_h}\right) \qquad \text{... (11)}$$

This important equation shows that irrespective of how the power plant is designed, its efficiency is constrained by the value of $\frac{T_c}{T_h}$, the ratio of the condenser temperature to the temperature of heat source, which in this case is a boiler. The lower the ratio, the closer the efficiency approaches unity. Since the condenser temperature is constrained by the available cooling water i.e. the temperature of the environment, the ratio mainly depends upon how hot the boiler can be heated.

Eqn. (11) holds good for any heat engine, whether it is a steam plant or an internal combustion engine. It emphasizes the critical importance of the temperature at which the heat energy is delivered. Although energy is always conserved, the efficiency with which heat can be converted to work is fundamentally constrained by its temperature. The maximum conversion efficiency increases with the temperature of the heat source. Therefore, high temperature heat is high-quality energy, because its entropy (per unit of energy) is low.

## HEAT ENGINE EFFICIENCY

The modern society depends on converting heat to work for various facets of human activity such as mining, manufacturing, transport, construction and even domestic functions. Energy of fossil fuel is used in internal combustion engines. Electricity is produced in power plants where heat derived from combustion of fossil fuel or from nuclear reactors produces steam to turn turbines. Various domestic and industrial appliances offer tremendous scope for enhancing energy efficiency.

In spite of the fact that there has been a substantial improvement in the design of power plants, the heat-to-electricity conversion efficiency is still about 40% only. In a modern coal-fired thermoelectric power plant, the boiler temperature reaches about 550°C (823 K), whereas the condenser temperature depends on the availability of cooling water, mostly at about 27° C (300 K). Therefore the maximum theoretical efficiency (vide eqns. 10 and 11) will be $\left(1 - \frac{300}{823}\right) = 0.64$ or 64%.

Moreover, additional heat losses from the boiler, turbine, electric generator, etc. reduce the theoretical efficiency still further. With the result, about 60% of the heating value of coal is lost as waste heat. The theoretical efficiency for a nuclear power plant is still lower because the temperature of the boiler is restricted for safety reasons. With the result, the overall efficiency of a nuclear power plant is about 30%.

## COMBINED CYCLE POWER PLANTS

Higher efficiencies can be achieved by increasing the temperature of the heat source. A boiler cannot be heated much above 550°C. However, a gas turbine can be run at temperature up to 1260°C. Gas turbines were originally developed as Jet engines for aircraft but recently they are being adopted to electricity generating turbines. However, in spite of the high operating temperatures, the temperature of the exhaust gas is as high as about 500°C. The theoretical efficiency calculated on the basis of $T_c$ (773 K) and $T_h$ (1533 K) works out to be about 50% i.e. $\left(1 - \frac{300}{823} \approx 0.5\right)$. Indeed, the actual efficiency is only about 33% (considering other heat losses). However the hot gases can themselves be used to run a steam turbine (fig. 15.2).

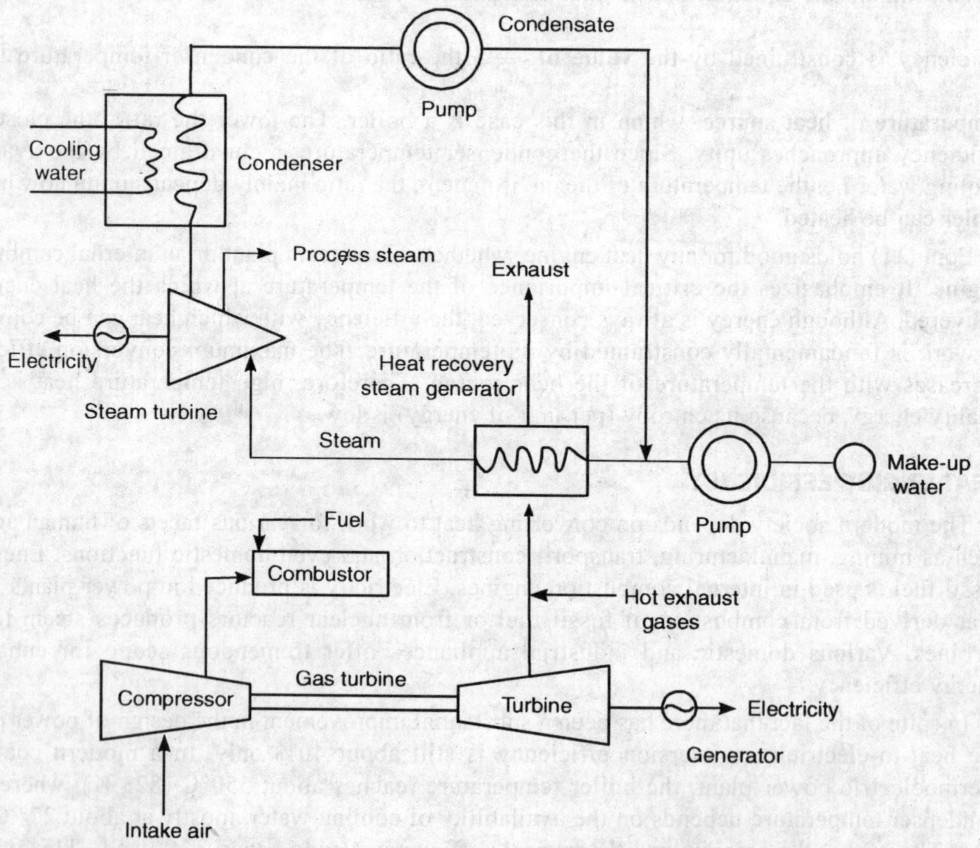


**Fig. 15.2.** Schematic representation of gas turbine and steam turbine combined-cycle power plant.

This second stage operates at about the same conversion efficiency similar to that of a regular steam plant. By combining these two stages, one can extract about 80% of the energy in the fuel. Many coal fired plants are now being fitted with a natural gas turbine "topping cycle" to boost their efficiency.

## HEAT PUMPS

Similar thermodynamic consideration as above apply when energy is consumed. A highly efficient gas furnace can heat a house more efficiently than electrical heating because, in the latter case, about two-thirds of the energy was lost at the power plant in producing electricity which is then converted back for heating the house. In contrast, most of the energy from the fossil fuel can be transferred in case of a good gas furnace. However, electricity can be converted to heat more efficiently if it is made to run a "heat pump". A heat pump is a device which is a small version of a power plant run in reverse (Fig. 15.3). Mechanical work is used to condense a working fluid at the temperature of the heat sink. The fluid is then allowed to evaporate, thereby absorbing heat from the heat source. Thus, heat can be pumped from lower to higher temperatures by expending the work. This is the way by which air-conditioners and refrigerators work. A house can be heated in cold whether by reversing the direction of the air-conditioner.

The conversion of work to heat is governed by the same entropy considerations as the reverse of the process. The maximum theoretical efficiency (or the maximum theoretical degree of conversion) is given by the inverse of eqn. (11):

$$\frac{Q}{W} < \frac{T}{T_h - T_c} \qquad \text{... (12)}$$

but $T_c$ and $T_h$ are reversed, since the heat source ($T_c$) is now colder than the heat sink ($T_h$).

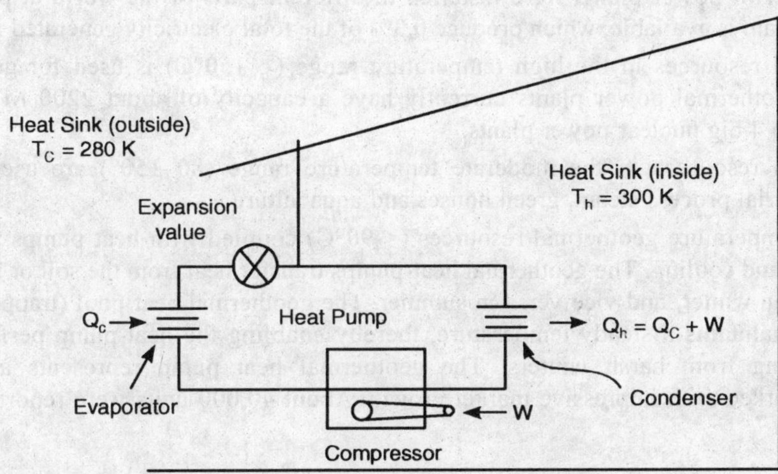


**Fig. 15.3.** The heat pump: An efficient device for residential heating and cooling.

If the inside temperature is about 300 K, and the outside temperature is 280 K, then the amount of pumped heat is theoretical 15 times greater than the amount of work:

$$\text{Theoretical efficiency} = \frac{Q_h \text{ (heat delivered)}}{W \text{ (work)}}$$

$$= \frac{T_h}{(T_h - T_c)} = \frac{300}{(300 - 280)} = 15$$

The means that 1 KJ of electricity can theoretically the transfer as much as 15 kJ of heat. This should not be misconstructed as a violation of First law of thermodynamics, because we must consider both the quality as well as the quantity of the energy. The higher the temperature, the higher is the quality. A high grade heat can be converted to a larger quantity of low-grade heat, thereby increasing the entropy in the process. Although electricity is a very high quality energy, the quality is largely wasted by converting it directly to heat at low temperature. It is precisely here that the heat-pump is efficient because it allows the quality energy to be more fully utilized. However, in practice, the theoretical efficiency cannot be achieved due to the resistances to heat transfer. All the same, a heat-to-work ratio of 2 can be easily achieved. Thus, a large fraction of the calories obtained from the fossil fuel that are lost at the power plant are regained. Heat pumps were found to perform best in moderate weathers.

## GEOTHERMAL HEAT PUMPS

In cold winter days, heat pumps may required back-up heating. Geothermal reservoirs can improve the heat pump performance appreciably.

Each itself generates heat from its molten core and from the decay of naturally occurring radioactive isotopes. In some parts of earth's crust, geological anomalies cause concentrated upwelling of heat from below. Volcanoes and hot springs are such examples. Romans used the hot springs for heating their bath houses. A modern geothermal industry for extraction of boric acid from hot springs started in 19th century in Italy. Geothermal steam replaced firewood as a fuel for concentrating boric acid. The first geothermal power plant was started in 1913 in Italy. Since then, a number of power plants were installed in different parts of the world at places where geothermal steam is available, which produce 0.3% of the total electricity generated in the world.

Geothermal resources in the high temperature range (> 150°C) is used for generation of electricity. Geothermal power plants currently have a capacity of about 2200 MW, which is nearly equal to 4 big nuclear power plants.

Geothermal resources in the moderate temperature range (80–150°) are used for space heating, industrial process steam, green houses and aquaculture.

The low-temperature geothermal resources (< 90°C) coupled with heat pumps are used for home-heating and cooling. The geothermal heat-pumps transfer heat from the soil or groundwater to the houses in winter, and vice versa in summer. The geothermal heat pool (trapped at depths of 30–50m) maintains a steady temperature, thereby enabling the heat-pump perform well in places suffering from harsh winters. The geothermal heat pump represents an emerging technology marked with impressive market growth. About 40,000 units were reportedly sold in 2000.

## COGENERATION

Energy saving can also be achieved through *cogeneration,* which involves a combination of power and heat generation. The waste heat from a power plant can be used by piping it into the heating systems of the nearby building. Cogeneration plants may be considered for big factories, large institutions and high density housing areas.

Further saving of energy in domestic space heating can be achieved through more efficient insulation and through sensors and control systems to deliver heat and light precisely as per requirements.

## ENTROPY AND CHEMICAL ENERGY

Chemical reactions are associated with changes of entropy and energy. Entropy change limits are amount of work that can be extracted in case of heat engines. Similarly, entropy change also limits the amount of energy that can, in principle, be extracted from a chemical reaction, consistent with the second law of thermodynamics. The amount of energy available for work is called the free energy change of the reaction ΔG. if ΔH is the enthalpy change of the reaction and ΔS is the entropy change, then

$$\Delta G = \Delta H - T \cdot \Delta S \qquad \text{... (13)}$$

The entropy change (ΔS) is multiplied by the absolute temperature (T) in order to express the energy deficit that it produces.

(Vide eqn. 5 according to which $S = \frac{Q}{T}$, so that $T \times S = Q$, the heat value of the entropy).

Since ΔG is the maximum energy that can be converted to work, the maximum efficiency for energy conversion is

$$\frac{\Delta G}{\Delta H} = 1 - \frac{T.\Delta S}{\Delta H} \qquad \text{... (14)}$$

Since T.ΔS increases with increasing temperature, the efficiency decreases with increasing temperature; in contrast to the efficiency of a heat engine. However, ΔS also can change with temperature, particularly when there is a phase change (*e.g.*, when water vaporizes). Further, ΔH too is not entirely independent of temperature. Therefore, it is obvious that the relation of efficiency with the temperature depends on the particulars of the reaction.

It is also interesting to note that for those reactions for which entropy increases, the theoretical efficiency exceeds 100%.

For exothermic reactions, the reactants have higher energy than the products, and hence ΔH is negative. Similarly, if the reaction proceeds spontaneously as written, ΔG is negative. Further, ΔS is negative if the entropy of the products is lower than of the reactants. For an exothermic reaction, a negative ΔS makes ΔG smaller than ΔH.

For the reaction:

$$2H_2 + O_2 = 2\,H_2O,$$

$$\Delta H = -476 \text{ KJ/mol at 1000 K, and } T.\,\Delta S = -84 \text{ KJ.}$$

$$\therefore \quad \Delta G = \Delta H - T.\Delta S$$

$$= -476 - (-84) \text{ KJ}$$

$$= -392 \text{ KJ}$$

∴ Theoretical energy conversion efficiency,

$$\frac{\Delta G}{\Delta H} = 1 - \frac{T.\Delta S}{\Delta H}$$

$$= 1 - \left(\frac{84}{-476}\right)$$

$$= 1 - 0.18 \approx 0.82$$

i.e. about 82%

At 300 K, the entropy change is, greater, because is vapour condenses to liquid water.

∴ T Δ S = 116 KJ. However, ΔH too increases to – 590 KJ/mol. Therefore,

$$\Delta G = \Delta H - T\,\Delta\,S = -590 - (-116) = -474 \text{ KJ}$$

$\therefore$ Theoretical energy conversion efficiency $= 1 - \frac{T\,\Delta S}{\Delta H} = 1 - \frac{(-116)}{(-590)} = 0.8$

i.e. about $\approx$ 80 %

## FUEL CELLS

By combustion of chemical fuels, useful energy can be obtained. However, with the help of fuel cells electricity can be directly generated. We know that burning of hydrogen produces heat. The same reaction can also be carried out at two electrodes, with an electric current flowing between them (Fig. 15.4). The reactions taking place in this case are as follows:

**At anode**

$2H_2 = 4\,H^+ + 4e^-$

**At cathode**

$O_2 + 4H^+ + 4e^- = 2H_2O$

**Overall reaction**

$2H_2 + O_2 = 2H_2O$

Thus, hydrogen is oxidized at the anode, and the electrons generated pass through the external circuit to the cathode where oxygen is reduced. However, in this case, the energy released is mostly electricity instead of heat.

Production of electricity in a fuel cell is just the reverse of electrolysis where $H_2$ and $O_2$ are produced at the corresponding electrodes immersed in water when current flows between them. Since the reaction energy in the fuel cell is transformed directly into electricity, the efficiency is not limited by the heat engine formula. The theoretical efficiency in this case is given by the ratio of $\frac{\Delta G}{\Delta H}$, which works out to about 80% for the reaction: $2H_2 + O_2\ 2H_2O$.

($\Delta G$ and $\Delta H$ are free energy change and enthalpy change respectively).

Theoretical efficiency can be obtained only if no current is actually drawn from the fuel cell. When a current flows through the external circuit, several resistance build up that reduce the efficiency and increase the fraction of the energy that is converted to heat. These resistances include the electrolyte resistance, mass transfer resistance and resistance to the chemical reactions themselves due to slow reaction kinetics (called activation barriers). These resistances cannot be eliminated but can be minimised by proper optimization of the fuel cell design. The electrical resistance can be minimized by connecting the electrode compartments with electrolytes having a high conductance. Mass transfer rates can be improved by increasing the surface area of the electrodes (such as by using porous electrodes). Reaction kinetics can be accelerated using catalysts (e.g., platinum). Further, overall reduction in all the three types of resistance can be achieved by using elevated operating temperatures. Important characteristics of some successful fuel cell technologies are given in Table 15.1.

Amongst the various types of fuel cell designs, the proton exchange membrane (PEM) fuel cell, which operates at a relatively low temperature range (80-100°C), emerged as the most satisfactory for transport purposes. The essential features of a PEM fuel cell system are illustrated in Fig. 15.4.

A thin electrolyte membrane made of a fluorocarbon polyerm allows the proton conduction from anode to cathode through the membrane. It is because of this that the cell can operate on hydrogen fuel at a relatively low temperature range. The polymer membrane contains covalently attached sulfonate ($-\,SO^-{}_3$) groups which form a network of water-filled channels, through which protons can easily pass. Anions are excluded by the negatively charged sulfonate groups. The PEM cells start up quickly because of the high proton conductivity. The PEM cell can have a high power density. Its output can quickly respond to variations in power demand. These excellent characteristics are responsible to render to PEM cells most suitable for automotive transport applications. The efficiency of PEM fuel engines approach 60% in contrast to 25% for an internal combustion engine. Since there are no emissions from a hydrogen fuel cell except $H_2O$, PEM fuel cell technology holds a promise and potential for zero-emission transport vehicles.

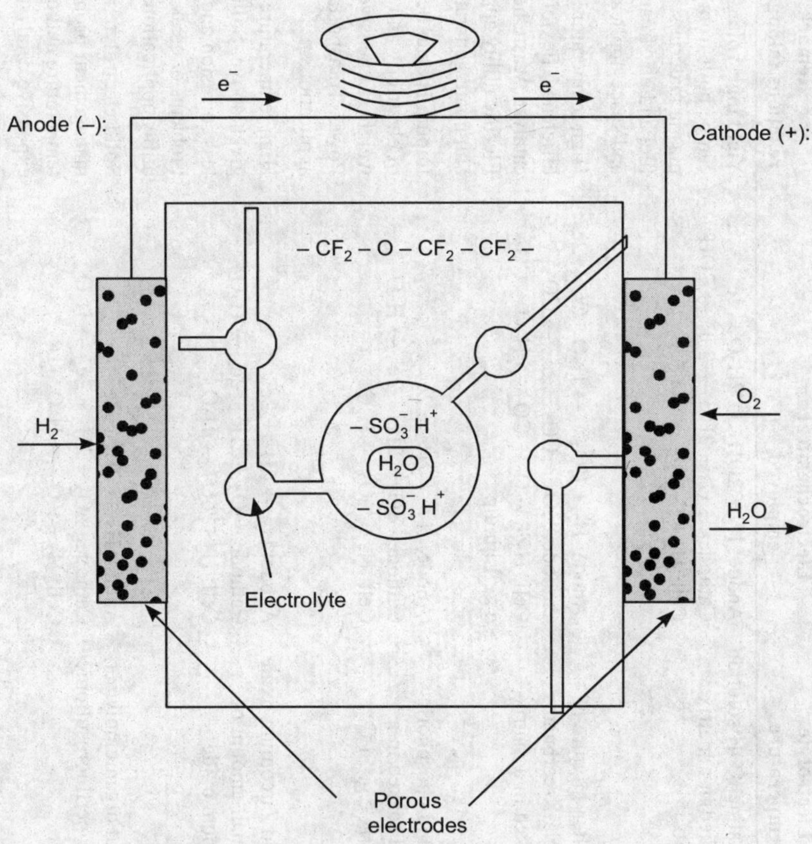


**Fig. 15.4.** Essential features of a proton exchange membrane (PEM) fuel cell.

Anode reaction : $H_2 \longrightarrow 2H^+ + 2e^-$ . $E° = 0V.$

Cathode reaction $\frac{1}{2} O_2 + 2 H^+ + 2e^- \longrightarrow H_2O$ $E° = 1.229$ V.

Cell reaction: $H_2 + \frac{1}{2} O_2 \longrightarrow H_2O$ $E° = 1.229$ V

The limitations with PEM fuel distribution include

(*i*) Lack of hydrogen distribution system.

(*ii*) Problem of on-board hydrogen storage.

**Table 15.1. Important characteristics of some fuel-cell technologies**

| Fuel cell type | Operating | Electrolyte temperature | Electrochemical employed | Remark reactions taking place |
|---|---|---|---|---|
| Alkaline fuel cells (AFC) | 90-100°C | Alkaline KOH solution soaked in a matrix | Anode: $H_2 + 2OH^- \rightarrow 2H_2O + 2e^-$<br>Cathode: $\frac{1}{2} O_2 + 2H_2O + 2e^- \rightarrow 2OH^-$<br>Cell: $H_2 + \frac{1}{2} O_2 \rightarrow H_2O$ | Yield high voltage, long operating life: can be built from low cost carbon and plastic materials. Removal of $CO_2$ from fuel and air streams is essential. Military and space applications |
| Molten carbon fuel cells (MCFC) | 600–1000°C | Molten mixture of Li-Na-K-carbnates soaked in a matrix | Anode: $H_2 + CO_3^{2-} \rightarrow H_2O + CO_2 + 2e^-$<br>Cathode: $\frac{1}{2}O_2 + CO_2 + 2e^- \rightarrow CO_3^{2-}$<br>Cell: $H_2 + \frac{1}{2} O_2 + CO_2 \rightarrow H_2O + CO_2$<br>e.g., Lithium | High temperature advantages eg., high efficiency, flexibility of fules, inexpensive catalysts. Waste heat is useful for concentration. Electric utility applications and disposed power aluminate generation. Problems of corrosion. |
| Phosphoric acid fuel cells (PAFC) | 175-200°C | Liquid phosphoric acid soaked in a matrix e.g. Si C. | Anode: $H_2 \rightarrow 2H^+ + 2e^-$<br>Cathode: $\frac{1}{2}O_2 + 2H^+ + 2e^- \rightarrow H_2O$<br>Cell: $H_2 + \frac{1}{2}O_2 \rightarrow H_2O$ | Impure $H_2$ can be used as fuel. Cogeneration of electricity and heat possible. Large size and weight. Expensive Pt catalyst. Low current and Power. Transportation and electric utility applications. |
| Solid oxide fuel cells (SOFC) | 600-100°C | Solid Zirconium oxide + small amount of Yitrium oxide | Anode: $H_2 \rightarrow 2H^+ + 2e^-$<br>Cathode: $\frac{1}{2}O_2 + 2H^+ + 2e^- \rightarrow H_2O$<br>Cell: $H_2 + \frac{1}{2}O_2 \rightarrow H_2O$ | High temperature advantages as in MOFC given above. Solid electrolyte advantages eg., less corrosion and lesser management problems. Problems of break down due to high temp. Electric utility applications. |
| Polymer electrolyte membrane/ proton exchange membrane (PEM) fuel cells | 60-100°C | Solid organic polymer, poly perfluorosulfonic acid. | Anode: $H_2 \rightarrow 2H^+ + 2e^-$<br>Cathode: $\frac{1}{2}O_2 + 2H^+ + 2e^- \rightarrow H_2O$<br>Cell: $H_2 + \frac{1}{2}O_2 \rightarrow H_2O$ | Solid electrolyte reduces corrosion and management problems. Advantages of low temperature and quick start up. Requires expensive catalysts sensitive fuel impurties. Transportation, por table power and electric utility applications. |

(*iii*) Need for highly active platinum catalyst because of low operating temperature. The Pt-catalyst is sensitive to poisoning by chemicals e.g., carbon monoxide.

(*iv*) Generation of hydrogen by electrolysis of water is expensive due to the cost of electricity. Atleast for the foreseeable further, hydrogen, has to be produced from fossil by steam reforming. In such cases carbon monoxide contamination is inevitable. Decontamination of hydrogen from CO (to trace level, lest it will poison the catalyst used in PEM fuel cell) is expensive.

(*v*) Generation of $H_2$ as fuel for PEM fuel cell vehicles from methanol via steam reforming is another option. Methanol is more easily reformed as compared to hydrocarbons and the reaction can be conducted at sufficiently low temperature. Therefore it is possible to fit methanol reformers on the vehicle itself. This would eliminate the need for storage or distribution units for hydrogen. However, this increases the complexity and cost of the vehicle. Moreover, methanol is corrosive. Bigger fuel tanks are needed. Each methanol fuel tank needs and underground tank.

(*vi*) Hydrogen can also be absorbed on metals e.g., Pd, thereby reducing its volume but not the weight of the storage container on-board. Carbon fibers show promise for this purpose.

## STORAGE OF ELECTRICITY

Efficient electricity storage is essential to tackle the problem of mismatch between supply and demand. One approach followed by a few companies is by using the excess electricity to pump water uphill to reservoirs. When peak demands are to be met, the water is allowed to run back downhill to turn the turbines. Other energy storage systems that the under consideration include storage of compressed air in caverns, mechanical energy storage in flywheels, and direct electrical storage in large super conducting magnets.

At the small end of electricity utilization, electricity storage is also vital for the development of electric vehicles, where electricity is required not only for auxiliary needs but also for locomotion itself. Today, electric cars run on current from lead-acid storage battery. In this battery, electricity is stored in the chemical conversion of $Pb^{2+}$ ions to Pb at one electrode and Pb $O_2$ at the other electrode.

The overall cell reaction is:

$$2Pb^{2+} + 2H_2O \;=\; Pb + Pb\,O_2 + 4H^+$$

which is energetically uphill. When the current is drawn, both the electrode reactions are reversed and the reaction is allowed to run downhill. The battery can be charged and discharged several times. It is used in all automotive vehicles as a portable store of electricity for auxiliary requirements. However, when it has to be used in electric vehicles, electricity is needed as a main source for locomotion itself. For this purpose, the lead acid battery has two major limitations: (1) It adds considerably to the weight of the vehicle thereby lowering its efficiency and driving range. (2) It takes several hours for recharging.

Efforts are in progress to develop alternative batteries that give more power with lesser weight and easier recharging. Promising alternatives that have been developed or under development include nickel-cadmium batteries, nickel-metal hydride batteries, sodium-sulfur batteries, lithium-iron disulfide batteries and lithium-polymer batteries.

The hydrogen fuel cell offers an attractive to any of the storage batteries. In fact, it itself may be considered as a battery in which the storage medium is hydrogen and oxygen, instead of Pb and $PbO_2$.

Fuel cell can be used for electricity storage once appropriate systems are commercialized.

## HYDROGEN ECONOMY

According to this concept, hydrogen gas can be used as the main energy currency. It can be directly used for electricity generation and heating, either with fuel cells or by combustion. It can also be used directly for transportation via fuel cell electric vehicles or it can be used to synthesize liquid fuels in the same way as coal conversion to methanol and liquid hydrocarbons. Power companies can use hydrogen for electricity storage when once the fuel cell technology for this purpose is commercialised. Electricity is convertible to hydrogen through electrolysis of water (the reverse of the fuel-cell reaction), with efficiency reaching 85%. Thus, a relatively efficient load-leveling capacity can be achieved by the combination of electrolysis and fuel cell, with tank storage of $H_2$. Moreover, hydrogen can be transported more cheaply and efficiently through pipelines. The concept of hydrogen economy is being tested in Iceland.

## DEMATERIALIZATION

Development of new materials better suited for their real-life application would ensure lesser consumption of materials. Stronger light-weight materials and smaller equipment and appliances are the present trends. Laptops and note book computers have greater computing power than big machines used earlier. Countering this, is the new trend to procure more appliances. Sustainability also means that less materials have to be extracted from earth and less of them go into manufacturing, thereby conserving materials and energy.

Despite the problems involved, recycling has considerable impact on energy saving. For instance, the production of steel, aluminium and paper consumes over 20% of the total industrial energy. Recycled aluminium requires only 5% of the energy to process aluminium from its ores, which is energy intensive. Substantial energy conservation can also be achieved by recycling steel scrap and paper.

Transportation represents an important energy sector not only because of its high rate of energy consumption (it accounts for over 25% of the total energy consumption) but also because of the international economics and politics. Automobiles dominate all other modes of transport. In the conventional automobile only a small percentage of the energy present in the gasoline serves to move the car while the rest of the energy goes as waste heat from the thermodynamic cycle of engine, to overcome frictional losses (e.g., aerodynamic drag, rolling resistance) or to power the auxiliary equipment.

In order to improve automobile fuel efficiency, several auto manufactures including Ford, Honda and Toyota have developed the hybrid electric vehicle (HEV). It combines the internal combustion engine with the electric motor and battery storage of an electric vehicle. This combination offers extended range and rapid refueling, with a significant fraction of the energy and environmental benefits of an electric vehicle. The HEVs use regenerative braking, strong light-weight materials and aerodynamic body shape to minimize the losses. The HEV is nearly twice as efficient as a conventional vehicle. The internal combustion engines tested for HEV include compression-ignition direct-injection (CIDI), the spark-ignition direct-injection (SIDI) and the turbo-changed direct-injection Diesel engines. However, the HEV is not a "Zero-emission vehicle". Diesel-based designs emit $NO_x$ and PM (particulate matter). Advances have been made to reduce the emissions and the CIDI engine is far better than older diesel designs. However, increasingly stringent standards may warrant alternative fuels (e.g., methanol and dimethyl ether) for CIDI engines.

The ultimate fuel-efficient, zero-emission vehicle seems to be the one powered by a fuel-cell. PEM fuel-cell engine can give a thermal efficiency of about 60% which corresponds to a fuel efficiency of 38.3 KM/L.

The green chemistry perspective and industrial ecology approach for materials and energy conservation and environmental protection have already been discussed in earlier sections of the book.

The scenarios for a clean energy future seems to be

(1) Efforts to reduce the energy intensity of the economy (such as more efficient cars, industrial processes, lighting, heating etc.).

(2) Measures to reduce the carbon intensity of the energy used (such as by more efficient use of fossil fuels and greater use of renewable energy sources, etc.)

## QUESTIONS

1. Discuss the recent approaches to achieve sustainable development.
2. What do you mean by "Green Chemistry". What are its basic principles?
3. Write informative notes on
   (*a*) Industrial ecology
   (*b*) Life cycle assessment
   (*b*) Preventive Environmental Management
   (*d*) Occupational Safety and Health Regulations
   (*e*) Energy utilization
   (*f*) Fuel cells
   (*g*) Dematerialization
4. Write short notes on
   (*a*) Proton exchange membrane (PEM) fuel cell
   (*b*) Cogeneration
   (c) Heat pump
   (*d*) Combined cycle power plant
5. Discuss the various new approaches being made to achieve better energy efficiency.

# Green Chemistry for Clean Technology

## INTRODUCTION

An ideal chemical reaction should have a number of attributes such as

(*a*) Safety
(*b*) Simplicity
(*c*) Selectivity
(*d*) High yield
(*e*) Energy efficiency
(*f*) Use of renewable or recyclable raw materials and reagents.
(*g*) Absence of hazardous byproducts or at least minimizing or containing them.

In practice, it is impossible to achieve all these attributes simultaneously. Indeed it is a challenge for chemists and engineers to identify environmentally preferable reaction pathways that optimize the balance of all the desirable attributes.

The concept of green chemistry was coined by Paul Anestas of America. He enunciated 12 principles of Green Chemistry in 1994 towards ideal synthetic methods to save natural resources. Green chemistry is the use of chemistry for pollution prevention by environmentally-conscious design of chemical products and processes that reduce or eliminate the use or generation of hazardous substances.

## GOALS OF GREEN CHEMISTRY

The goals of "green chemistry" perspective include the following :

(1) To reduce adverse environmental impacts by appropriate and innovative choice of materials and their chemical transformations.
(2) To develop processes based on renewable (plant-based) rather than non-renewable (fossil carbon-derived) raw materials.
(3) To develop processes that are less prone to obnoxious chemical releases, fires and explosions.
(4) To minimize byproducts in chemical transformations through redesign of reactions and reaction sequences. In other words, to achieve better "Atom economy".

$$\% \text{ Atom economy} = \frac{\text{Formula weight of the product}}{\text{(Sum of formula weights of all the reactants)}} \times 100$$

(Good atom economy means most of the atoms of the reactants are incorporated in the desired products and only small amounts of unwanted byproducts are formed and hence lesser problems of waste disposal or waste treatment).

(5) To develop products that are less toxic or which require less toxic raw materials/ feedstocks.

(6) To develop products that degrade more readily/rapidly in the environment than the current products.

(7) To reduce the requirements for hazardous or environmentally persistent solvents and extractants in chemical processes.

(8) To improve energy efficiency by developing low temperature and low pressure processes by using new/improved catalysts.

(9) To develop efficient and reliable methods to monitor processes (*e.g.* monitoring reactions and releases) for improved control.

## SIGNIFICANCE OF GREEN CHEMISTRY

The chemical industry releases about 5 billion tons of chemical wastes annually to the environment. It also spends over 300 billion dollars annually for treatment, control and disposal of these chemical wastes. This poses a formidable challenge or chemists and chemical engineers to review the design, manufacture and use and then to further accomplish improvements or changes in the design of new products and processes. The innovations in technology must be planned and implemented in such a way that they are sustainable both economically and environmentally. From the stand point of environmental protection, many industrial sectors have made significant studies in reducing emissions. Yet, even with these improvements, the impact of industrial processing, use and disposal of chemicals on environment is staggering. The environmental protection in many countries was based on promulgation of statutes and regulations and implementing them using command and control approach. In order to implement the environmental mandates, industries had to install a variety of waste handling, treatment, control and disposal systems, which require new equipment with high capital costs. On a societal level, this means increase in the price of the consumer goods. Further, the true costs of the environmental impacts due to the manufacture, processing, use and disposal have not been fully incorporated into the price of the goods produced. These costs include site remediation, ecosystem destruction and health care. It is precisely in this context that green chemistry provides the best opportunity for manufacturers, processors and users of chemicals to carry out their work in the most economical and environmentally beneficial way.

Many countries have been contemplating to levy environmental tax on industries towards payment of environmental impacts ostensibly caused because of their operations. By using the principles of green chemistry in the design and manufacture of chemical products and processes, many industries in the world have been successful in lowering the overall costs associated with environmental safety and health. Environmental expenditures are now being thought of as the cost of doing business. Green chemistry is already demonstrating the potential and promise to develop new techniques and methodologies to enable industries to pursue their traditional innovations and at the same time minimizing environmental impacts. The funds which they might have to pay towards environmental tax (as and when levied) may now be diverted for R & D efforts.

## BASIC COMPONENTS OF GREEN CHEMISTRY RESEARCH

The major research efforts in green chemistry may be broadly classified into the following four areas :

(1) Alternative feedstocks or starting materials,
(2) Alternative reagents or transformations,
(3) Alternative reaction conditions,
(4) Alternative final products or target molecules.

These four areas are interrelated and at times overlapping. Although incremental innovations may be made in any of these areas to achieve a more environmentally benign process, the final judgement should be made on the basis that whether this specific change will result in the net improvement in the overall synthetic process from the standpoint of environmental impact. Examples of some important innovations made in each of the above areas are given below :

### (1) Alternative feedstocks or starting materials

Utilizing environmentally benign feedstocks reduces the risk to human health and environment. This can be done by different methods as follows :

(*a*) reducing the amount of the feedstock used
(*b*) reducing the intrinsic toxicity of the feedstock through structural modification
(*c*) replacing the feedstock

Eventhough the risk can be reduced through protective gear and control technologies, they are often very expensive. Green chemistry offers economically viable solutions, as illustrated by the following examples.

**Ex. :**

**(*i*) Synthesis of aromatic amines :** Chlorinated aromatic are used in the synthesis of a variety of aromatic amines. However, chlorinated aromatics are known to be environmentally hazardous and are known to be persistent bioaccumulators. Stern and coworkers at the Monsanto Corporation successfully proved that by using nucleophilic substitution for hydrogen, the use of chlorinated hydrocarbons can be avoided. This research on feedstock replacement helped in removing the environmental concerns in the commercial preparation of aromatic amines.

**(*ii*) Synthesis of Adipic acid :** Significant advances have been made by Frost at Michigan State University in using alternative feedstocks coupled with biosynthetic methodologies. New synthetic pathways have been reported for the manufacture of adipic acid, catechol and hydroquinone using glucose as the starting material in place of traditionally used benzene which is carcinogenic. Traditional and alternative synthetic pathways for the manufacture of adipic acid are given in Fig. 16.1.

**(*A*) Traditional pathway**

Benzene $\xrightarrow[\text{370-800 psi}]{Ni.Al_2O_3}$ Cyclohexane $\xrightarrow[\text{120-140 psi}]{CO,\ O_2}$ Cyclohexanone + Cyclohexanol $\xrightarrow[HNO_3]{Cu,\ NH_4VO_3}$ $HO_2C$–$(CH_2)_4$–$CO_2H$ (Adipic Acid)

**(B) Alternative greener pathway**

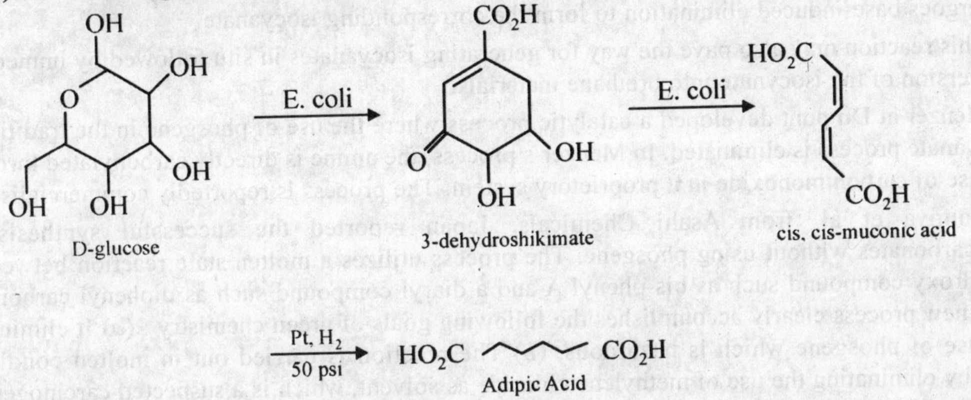

D-glucose —E. coli→ 3-dehydroshikimate —E. coli→ cis, cis-muconic acid —Pt, $H_2$, 50 psi→ $HO_2C$–$(CH_2)_4$–$CO_2H$ Adipic Acid

**Fig. 16.1.** Traditional and alternative synthetic pathways for the manufacture of adipic acid.

(*A*) The traditional pathway which uses benzene which is a fossil-fuel based and carcinogenic feedstock. (*B*) The alternative pathway which uses glucose, a renewable feedstock, which is absolutely safe.

***(iii)* Synthesis of isocyanates, urethanes and polycarbonate polymers without using hazardous phosgene :** Riley, Mc Ghee and coworkers at Monsanto Corporation have been successful in eliminating phosgene in the manufacture of isocyanates and urethanes by designing a process for direct reaction of $CO_2$ with amines.

The largest industrial use of phosgene is in the preparation of isocyanates. Isocyanates are mostly used for the manufacture of polyurethanes and pesticides. Polyurethanes are used in the manufacture of cushions, mattresses, insulation, car bumpers, paints, adhesives, etc. Hence, isocyanates are produced on the billion pounds scale annually and are mostly produced so far via phosgene technology. The reaction of an amine with phosgene has several problems such as high toxicity of phosgene and generation of HCl. The use of $CO_2$ as a phosgene replacement has been explored. Experimental conditions for the highly selective synthesis of isocyanates from carbamate anions (generated by the reaction between amine and $CO_2$) have been found using various "dehydrating agents" (e.g., acetic anhydride, benzoic anhydride, trifluoroacetic anhydride, benzene sulphonic acid anhydride, etc). The initial work was based on the use of electrophilic dehydrating agents for the generation of isocyanates from amines and $CO_2$.

$$RNH_2 + CO_2 + Base \longrightarrow RNH\,CO^{\bullet}_2\ {}^{+}H\ Base$$

$$R\,NHCO^{\bullet}_2\ {}^{+}H\ Base + Dehydrating\ agent \longrightarrow RNCO + Salts$$

These reactions were shown to proceed with high selectivity and high yield using phosphorous containing electrophiles (*e.g.*, $POCl_3$, $PCl_3$ and $P_4O_{10}$) and also offer a relatively low cost and mild reaction route to obtain new materials. However, a large amount of salt waste is generated. To solve this problem, the possible use of non-halide and potentially low to no waste reagents capable of achieving the same high selectivities and yields has been explored. It is during these efforts that the use of O-sulphobenzoic acid anhydride as the dehydrating agent has been found to be successful. This reagent has not only provided selective conversion of amines and $CO_2$ into their corresponding isocyanates (thereby eliminating the need to use toxic phosgene) but also provided a process by which the dehydrating agent can be recycled (thereby eliminating the huge amount of salt waste). This reaction is supposed to proceed via nucleophilic attack of the carbamate anions at the anhydride giving rise to the formation of a mixed anhydride, which then undergoes base-induced elimination to form the corresponding isocyanate.

This reaction may also pave the way for generating isocyanates in situ followed by immediate conversion of the isocynate into urethane materials.

Menzer at Du pont developed a catalytic process where the use of phosgene in the traditional isocyanate process is eliminated. In Menzer's process, the amine is directly carbonylated through the use of carbon monoxide in a proprietory system. The process is reportedly commercialised.

Komiya et al. from Asahi Chemicals, Japan reported the successful synthesis of polycarbonates without using phosgene. The process utilizes a molten state reaction between a dihydroxy compound such as bis-phenyl A and a diaryl compound such as diphenyl carbonate. This new process clearly accomplishes the following goals of green chemistry : (*a*) It eliminates the use of phosgene which is hazardous, (*b*) The reaction is carried out in molten condition thereby eliminating the use of methylene chloride as solvent, which is a suspected carcinogen.

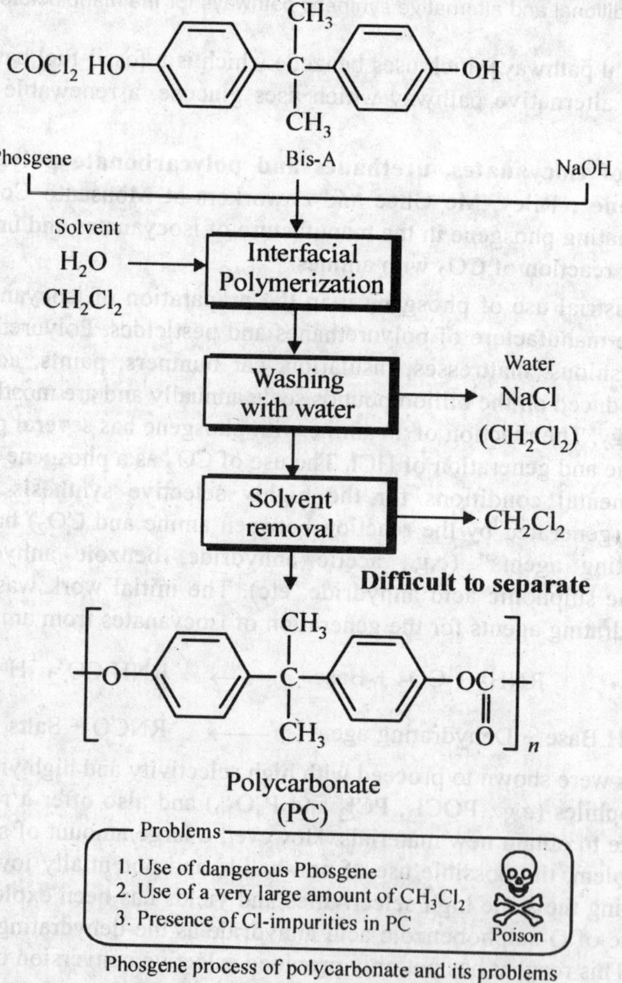


**Fig. 16.2.** Synthesis of polycarbonate
(A) Phosgene process (Hazardous traditional route)

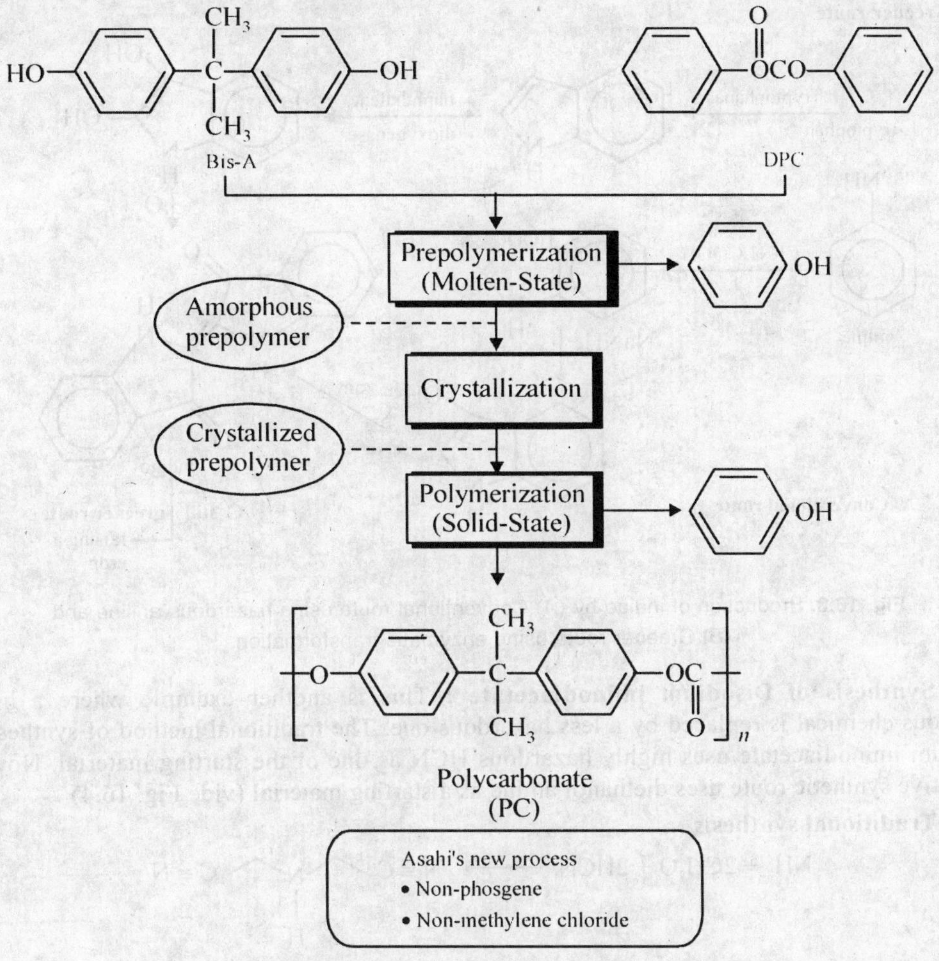


Solid-state polymerization process of polycarbonate

**Fig. 16.2.** (B) Asahi's solid state polymerization process without using phosgene and methylene chloride as solvent (Benign green chemistry route).

***(iv)* Synthesis of indigo :** Indigo, the dye used to colour blue Jeans, can be made enzymatically by removal of the side chain of tryptophan to give indole, which can be dehydroxylated enzymatically, and then oxidized with oxygen to indigo. The presently used commercial process starts with aniline which is highly toxic. Moreover, it produces considerable amount of waste salts, thereby causing disposal problem.

An even greener way to produce indigo would be to raise it is a crop as practised in colonial America. A new commercial strain yields 5 times more indigo than the traditional plant.

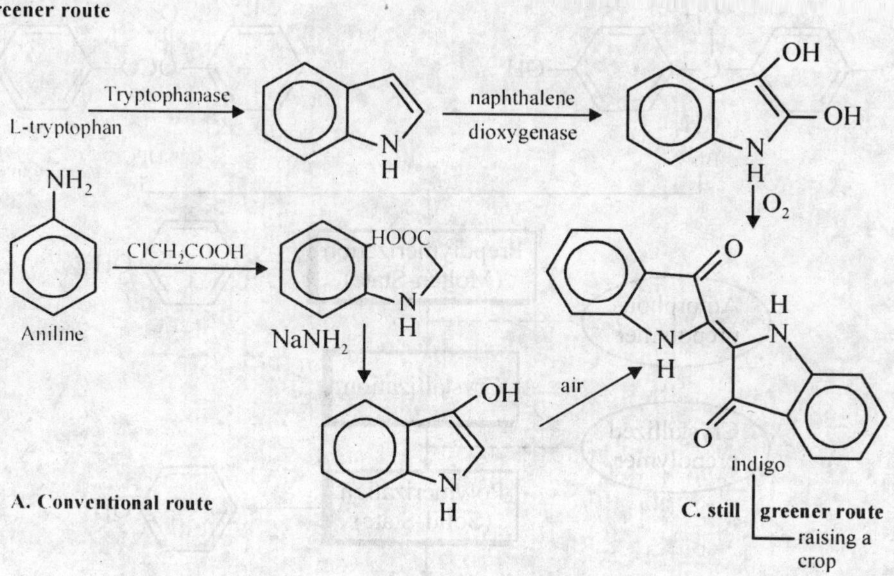

Fig. 16.3. Production of indigo by (*A*) Conventional route using hazardous aniline and (*B*) Greener route using enzymatic transformation.

**(*v*) Synthesis of Disodium iminodiacetate :** This is another example where a highly hazardous chemical is replaced by a less hazardous one. The traditional method of synthesis of disodium iminodiacetate uses highly hazardous HCN as one of the starting material. Now, an alternative synthetic route uses diethanol amine as a starting material (vide Fig. 16.4)

**(A) Traditional synthesis**

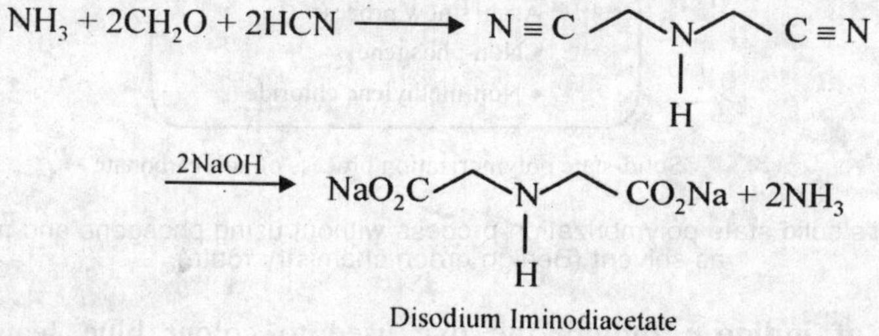

$$NH_3 + 2CH_2O + 2HCN \longrightarrow N\equiv C\text{–}CH_2\text{–}NH\text{–}CH_2\text{–}C\equiv N$$

$$\xrightarrow{2NaOH} NaO_2C\text{–}CH_2\text{–}NH\text{–}CH_2\text{–}CO_2Na + 2NH_3$$

Disodium Iminodiacetate

**(B) Alternative synthesis**

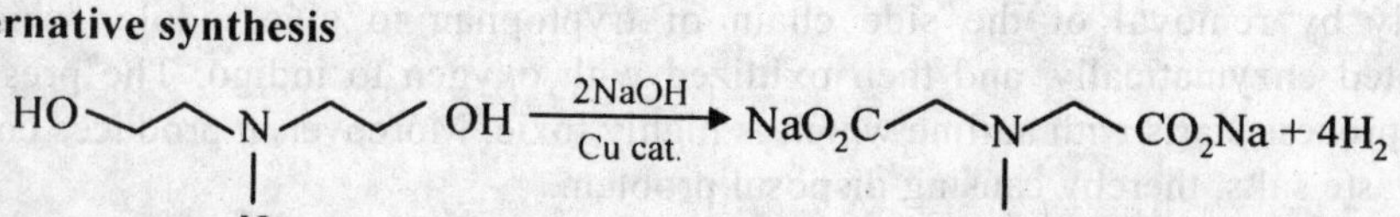

$$HO\text{–}CH_2CH_2\text{–}NH\text{–}CH_2CH_2\text{–}OH \xrightarrow[\text{Cu cat.}]{2NaOH} NaO_2C\text{–}CH_2\text{–}NH\text{–}CH_2\text{–}CO_2Na + 4H_2$$

Diethanolamine    Disodium Iminodiacetate

Fig. 16.4. Synthetic pathways for disodium imminodiacetate. (A) Traditional route using hazardous hydrogen cyanide. (B) Alternative route using diethanol amine with copper catalyst.

**(*vi*) Synthesis of polymeric substances from biological/agricultural wastes :** Gross at the university of Masachusettes prepared new polymeric substances from biological/agricultural wastes such as polysaccharides using biocatalytic transformation. This process simultaneously addresses several green chemistry objectives and environmental concerns such as

(*a*) use of environmentally benign feedstocks

(*b*) use of biocatalytic transformation rather than the conventional monomers and harsh experimental conditions

(*c*) addresses the environmental concerns of persistent polymers and plastics by paving the way for the manufacture of biodegradable gross polymers.

**(2) Alternative reagents or transformations**

There can be several opportunities to substitute more benign chemicals for the toxic reagents used at present to carry out some transformations. The following examples illustrate how such efforts lead to reduce environmental and human health risks.

(*a*) Paquetta of Ohio State university reported the use of Indium, instead of metals of environmental concern being used so far, to bring out several transformations. Such transformations also demonstrated increased selectivities. Further, these transformations could be carried out in aqueous media instead of volatile organic solvents.

(*b*) In the synthesis of several important chemicals, methylation is a vital step. However, many strong methylating agents such as dimethyl sulphate are acutely toxic and often corcinogenic. Fundo at the university of Venice reported the use of dimethyl carbonate for methylation of acrylonitriles and methylacrylacelates, which are important reagents in the synthesis of a class of antidepressant drugs. This advance in green chemistry serves dual purposes — (*i*) It eliminates the use of toxic dimethyl sulphate. (*ii*) It also eliminates the problem of high production during the transformation because this procéss requires only catalytic amount of the base to proceed.

However, dimethyl carbonate is traditionally synthesized by the reaction of highly toxic phosgene with methanol. Therefore, the overall benefit accruing from replacement of dimethyl sulphate with dimethyl carbonate becomes questionable unless the latter is synthesized by using a more benign reaction pathway. Indeed, such an effort resulted in the successful work of Rivetti and coworkers of Enichem Chemical Company. These workers synthesized dimethylcarbonate by the oxidative carbonylation of methanol using carbon monoxide.

(*c*) Epling and coworkers at the university of connecticut reported alternative methodology, using visible light as a "reagent", for the cleavage of a variety of dithiane and oxathiane ringe systems commonly used as protecting groups. Traditionally, rings of this type are cleaved by toxic heavy metal-catalyzed reduction. Epling's work has the potential to accomplish important transformations useful in dye and pharmaceutical industries. This methodology addresses the health and environmental concerns associated with the use of toxic heavy metals for such reactions.

(*d*) Kraus from Jowa State University studied the widely used Friedel Craft's acylation reaction which is catalyzed by the Lewis acids. He found that by using quinolic moieties and aldehydes, formal synthesis of a number of target molecules (such as diezapam and analogues), which are of vital importance in the pharmaceutical industry, has been achieved.

**(3) Alternative reaction conditions**

During the manufacture, processing, formulation and use of chemical products, a variety of associated substances contribute to environmental pollution. Examples are the solvents used in reaction media, separations and formulations. Many solvents, particularly the volatile organic solvents (VOC's) came under severe scrutiny and regulatory restriction due to their toxic contributions to air and water pollution. Therefore, much of the R & D in green chemistry reaction conditions is focussed on alternative solvents. Some of the main alternatives to traditional solvents include super-critical fluids, aqueous solvents, polymerized or immobilized solvents, ionic liquids, solventless systems and reduced hazard organic solvents.

**(a) Super-critical fluid (SCF) systems :**

Solvent systems such as supercritical (SC) carbon dioxide and supercritical $CO_2/H_2O$ mixtures are under investigation for use in various types of reactions. In general, super-critical fluids (SCFs) appear to offer promise for low cost, innocuous, solvents that can supply "tuneable" properties.

Super-critical $CO_2$ (SC $CO_2$) is non-toxic, non-inflammable and inexpensive. Moreover, since the solubility of most of the solutes in super-critical fluids changes considerably near the critical point, it is often possible to recover the solute from super-critical $CO_2$ merely by reducing the pressure to below the critical point. In this manner, several applications of SC $CO_2$ such as de-caffeinating coffee (where supercritical $CO_2$ has now replaced the earlier solvent viz. methylene chloride), strictly depend upon the physical, solvating properties of this solvent. However, all materials are not soluble in super-critical $CO_2$. For materials such as high molecular weight hydrocarbons, which are not highly soluble in supercritical $CO_2$, the advantages of super-critical fluids can be obtained by adding a surfactant to SC $CO_2$. Addition of the appropriate surfactant generates a micelle phase in which materials that are not normally soluble in SC $CO_2$ can also be suspended. The following reaction illustrates this concept.

Catalytic copolymerization of $CO_2$ with epoxides :

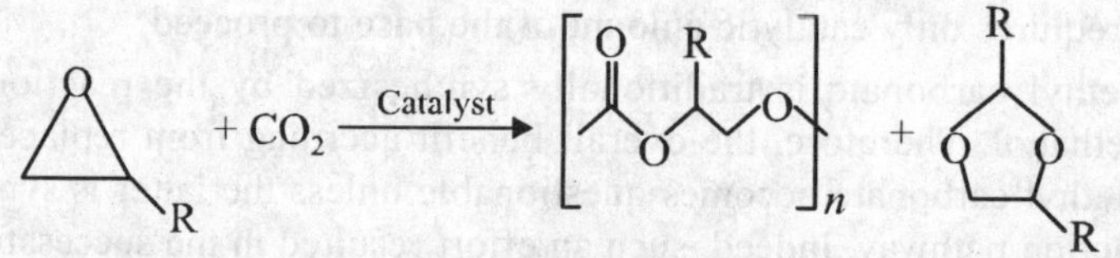

**Fig. 16.5.** Example of the use of supercritical $CO_2$ to replace conventional solvents. A co-solvent is used along with SC $CO_2$ to allow a polymerization reaction to take place.

In this case, a polymerization reaction, which produces high molecular weight materials, is conducted in a surfactant-super-critical $CO_2$ system. Thus, the need for conventional solvents is eliminated.

Tanko, at Virginia polytechnic Institute demonstrated successfully some free-radical halogenations in SCF. He proved that when compared with standard bromination reactions of alkylated aromatics as the model system, the yields and selectivities of the reactions of SCFs are equal or even superior to those conducted in conventional solvent systems.

De Simone at the university of North Carolina demonstrated that a variety of polymer types can be synthesized with different monomeric systems in super-critical fluids. His studies in methyl methacrylate polymers showed many definite advantages in using SCFs as a solvent system over conventional halogenated organic solvents.

Tumas *et al.* at the National Laboratory at Los Almos showed reactions such as polymerization of epoxides, oxidation of olefins and asymmetric hydrogenations in SCF systems were comparable and even superior to the conventional halogenated organic solvent systems.

In many cases, solvents are used strictly for their physical, solvating properties. However, in some situations, such as chemical reactions occurring in solvents, in solvents play a specific role in the chemical synthesis. In such cases where a reaction takes place in a solvent, supercritical fluids (SCFs) may enhance or inhibit the desired reaction. The effect of SCFs on reaction chemistry is an active area of research. A particular type of reactions, known as asymmetric catalytic reductions have been studied in super-critical $CO_2$. It was found that the selectivities achieved in these reactions are comparable or even superior to those obtained in conventional solvents.

**(*b*) Aqueous solvent systems :**

Several projects are underway on the use of aqueous solvent systems in place of organic solvent systems in chemical manufacturing. Many of these show great promise and efficacy.

Although water is non-toxic, non-inflammable, renewable and inexpensive solvent, the limited solubility of many hydrocarbon reactants in water has been a limitation in its utility. However, Anastas and Williamson (1998) and Li and Chan (1997) reported some case studies showing the innovative use of water as a reaction medium. Water with an alcohol co-solvent is used in Diels Alder reactions.

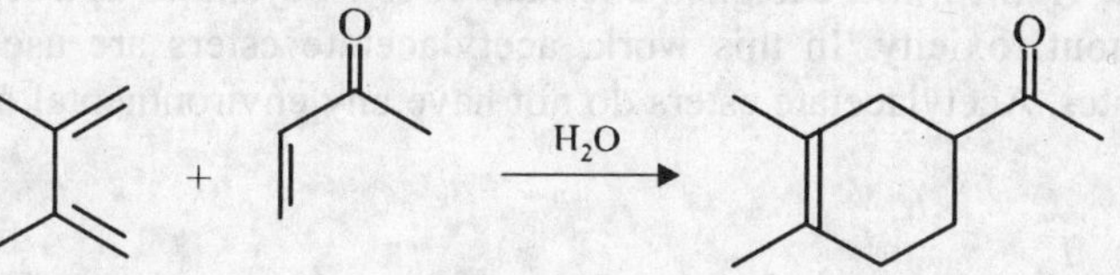

**Fig. 16.6.** Use of water as a solvent for certain Diels Alder reactions to increase reaction rate.

Some Diels Alder reactions, such as the dimerization of 1, 3 cyclopentadiene, were found to be accelerated in water. This may be due to favourable packing of hydrophobic surfaces in the transition state of the reaction.

Many other organic reactions that were traditionally carried out in organic solvents have now been carried out in aqueous media. Examples of such reactions include Barbier-Grignard reaction, pericyclic reactions, and transition metal-catalyzed reactions (Li 1998).

**(*c*) Derivatized / Immobilized solvent materials**

Hurter and Hatton, 1992 and U.S. EPA 1996 used a novel concept of using derivatized / immobilized solvent materials with the twin-objectives of reducing emissions and promoting the recovery of hazardous solvents. In this case, the hazardous substance under consideration, viz. tetrahydrofuran (THF) is attached to a polymeric backbone using chlorinated styrene derivative. The THF then remains relatively mobile, but because it is attached to a polymeric backbone, it is less likely to volatilize and is easily recoverable using methods such as ultrafiltration.

Derivatized/Polymeric solvent replacement for THF

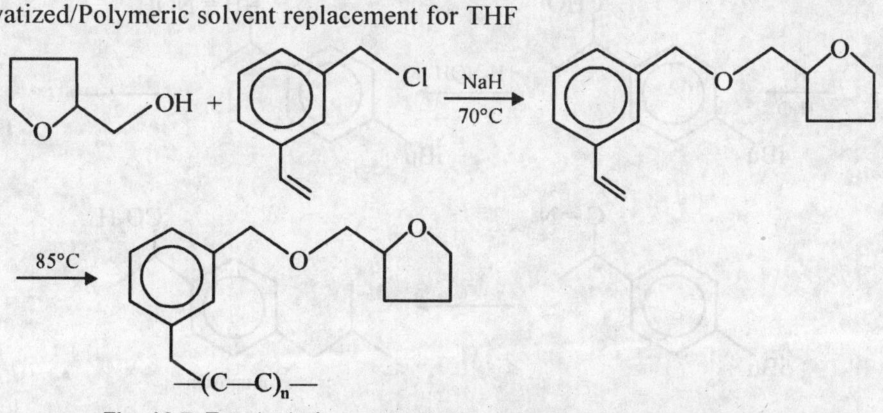

**Fig. 16.7.** Tetrahydrofuran, attached to a large polymeric backbone, to make it less volatile.

**(4) Alternative products and Target Molecules**

In many cases, the part of a molecule which provides its intended activity is separate from the part responsible for its toxicity. Thus it is possible to greatly reduce the hazards associated with certain chemical products while maintaining their performance and innovation. This is particularly of interest in pharmaceutical and pesticide industries.

(*a*) De Vito at the United States Environmental Protection Agency (US EPA) found that by blocking the alpha position, the ability of nitriles to form an alpha radical is prevented. Thus the toxicity of nitrile can be decreased by several orders of magnitude without adversely affecting the ability of the nitrile to function as a cross-linking agent. The nature of the hazard posed by the release of cyanide depends upon the ability of nitrile to form an alpha radical.

(*b*) De Pompei at Tremco, Inc. designed alternatives to isocyanates as a sealant by maintaining their function without toxicity. In this work, acetylacetate esters are used as alternatives to hazardous isocyanates. Acetylacetate esters do not have any environmental or health hazards.

## ATOM ECONOMY

Any synthetic design has two important goals namely reaction efficiency and product selectivity. However, even when a synthetic transformation achieves hundred percent selectivity to the desired product, still a substantial quantity of waste can be generated, if the transformation is not "atom economical". Atom economy is defined as the ratio of the formula weight of the target molecule to the formula weights of all the starting materials and the reagents. It indicates the interinsic efficiency of the desired transformation. The example given below illustrates the atom economy achieved in the synthesis of ibuprofen by the traditional alternative routes.

**Traditional Synthesis of ibuprofen**

$(CH_3CO)_2O$ / $AlCl_3$ → (iBu) → $ClCH_2CO_2C_2H_5$ / $NaOC_2H_5$ → $O—CHCO_2C_2H_5$ (iBu)

$H^+$ / $H_2O$ → CHO (iBu) → $H_2NOH$ → CH = NOH (iBu) →

C = N (iBu) → $CO_2H$

ibuprofen

**Alternative Synthesis of ibuprofen**

$(CH_3CO)_2O$ / HF → ; $H_2$ / Catalyst → ; CO, Pd → ibuprofen

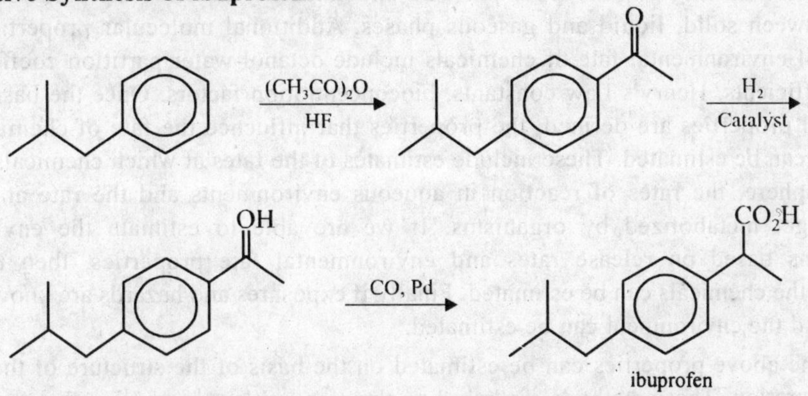

**Fig. 16.8.** Synthesis of ibuprofen. (A) Traditional method involving larger number of steps with atom economy of 40%. (B) Alternative method which is simpler, and employs recoverable strong-acid as catalyst. The atom economy is 77%.

## FUNCTIONAL GROUP APPROACHES TO GREEN CHEMISTRY

There are several tools in the design of more environmentally benign chemistry, which include the following :

## STRUCTURE-ACTIVITY RELATIONSHIP

Today, about 15,000 chemicals are commercially produced and every year, a thousand or more new chemicals are developed. Every chemical in use may possess a number of potential risks to human health and the environment. It is difficult to evaluate all possible environmental impacts rigorously and precisely. However, preliminary screening of the potential environmental impacts of chemicals is possible. It is also essential to identify potential risk reduction opportunities with the limited amount of available information about the chemical structure. Table 16.1 identifies the chemical and physical properties that may influence each of the processes that determine environmental exposure.

**Table 16.1. Physical and chemical properties required to perform Environmental Risk Screenings.**

| *S.No.* | *Environmental process* | *Relevant chemical and physical properties* |
|---|---|---|
| 1. | Toxicity and health effects | Dose-response relationship |
| 2. | Persistence in the environment | Atmospheric oxidation rate, aqueous hydrolysis rate, rate of microbial degradation, rate of photolysis, adsorption, etc. |
| 3. | Uptake by humans | Transport across dermal layers, rate of transport across lung membranes, degradation rate within human body. |
| 4. | Uptake by organisms | Volatility, lipophilicity, degradation rate in organisms, molecular size, etc. |
| 5. | Dispersion and fate | Density, volatility, melting and boiling point, water solubility, effectiveness of waste water treatment, etc. |

Melting point, boiling point, vapour pressure and solubility describe the partitioning of the chemical between solid, liquid and gaseous phases. Additional molecular properties used in assessment of environmental fate of chemicals include octanol-water partition coefficient, soil sorption coefficients, Henry's Law constants, bioconcentration factors. Once the basic physical and chemical properties are defined, the properties that influence the fate of chemicals in the environment can be estimated. These include estimates of the rates at which chemicals will react in the atmosphere, the rates of reaction in aqueous environments and the rate at which the compounds get metabolized by organisms. If we are able to estimate the environmental concentrations based on release rates and environmental fate properties, then the human exposures to the chemicals can be estimated. Finally if exposures and hazards are known, the risk to humans and the environment can be estimated.

Each of the above properties can be estimated on the basis of the structure of the chemical under consideration. These methods are based on the reasonable presumption that a molecule is composed of a set of functional groups or molecular fragments and that each of these fragments contributes in a well-defined manner to the properties of the molecule. These methods are therefore known as group contribution methods, structure activity relationships (SARs), or Quantitative structure Activity Relationship (QSAR).

Although the precise mechanism of action of a chemical may not be known, many a times, the structure activity relationships can be used to identify structural modifications that may improve the safety of the chemical under consideration. For example, if we know that the methyl-substituted analog of a substance has a very high toxicity and that the toxicity decreases as the substitution moves from ethyl to propyl, then we can try to reduce the toxicity of a chemical by increasing the alkyl chain length. Thus, structure-activity relationship can be used as a powerful tool to design safer chemicals.

## ELIMINATION OF TOXIC FUNCTIONAL GROUP

Chemicals of any particular class are defined by their structural features, such as alcohols, aldehydes, ketones, nitriles or isocyanate functional groups. In case that the toxicity of any chemical or the mechanism by which it generates toxicity are not known, one can reasonably assume that certain reactive functional groups will react similarly in the body or in the environment. Thus if the toxicity regarding other compounds in a chemical class is known, one may suspect toxic effect of any chemical having the same functional group. Thus, we can design a safer chemical by removing the toxic functionality (*i.e.*, the functional group that is suspected to produce toxicity). In cases where the functionality of toxicity is the same as the functionality of the desired activity (*i.e.* if the functional group responsible to cause toxicity and the functional group that provides the necessary properties required for a specific application are the same), then, we can mask the functional group to a non-toxic derivative form. The parent functionality can be released or regenerated as and when necessary.

This technique can be best illustrated by the following example. The vinyl sulfone functionality is highly electrophilic and reacts with cellulosic fibers, making it an effective component of dyes. However, several toxic effects are attributed to this functionality. The sulfones can be made safer by masking the functional group by synthesizing a relatively less hazardous hydroxyethylsulfone. This will make the manufacturing, storing and transporting the chemical safer. The hydroxyethyl sulfone can be readily converted to vinyl sulfone as and when required.

$$R-\overset{\underset{\|}{O}}{\underset{\overset{\|}{O}}{S}}-CH_2CH_2OH \xrightarrow{H_2SO_4} R-\overset{\underset{\|}{O}}{\underset{\overset{\|}{O}}{S}}-CH_2CH_2OSO_3 \xrightarrow{\text{Strong base}} R-\overset{\underset{\|}{O}}{\underset{\overset{\|}{O}}{S}}-CH=CH_2$$

Hydroxy ethyl sulfone — Not Reactive — Vinyl Sulfone Highly Reactive

**Fig. 16.9.** Masking of highly reactive and hazardous vinyl sulfone as safer hydoxyethyl sulfone. The latter can be converted to vinyl sulfone as and when required.

## REDUCING THE BIOAVAILABILITY

If a potentially toxic substance cannot reach the target of toxicity due to some features of structural design, then for all practical purposes, it may be considered as innocuous. If we do not know what structural features of the molecule have to be modified to render it less hazardous, still we can exercise the option to minimize its bioavailability. This can be achieved by manipulation of the water solubility / lipophilicity relationships that govern the ability of a substance to pass through biological membranes such as skin, lungs or the gastro-intestinal tract.

The same principle also applies to designing safer chemicals for the environment such as ozone depleting substances (ODS). For any chemical to exhibit a considerable ozone depleting potential, it must satisfy two basic conditions : (1) It should be able to reach the necessary altitudes, and (2) It should have adequate life time in those altitudes so as to cause the damage. Keeping this in view, several substances are now being designed which possess the same useful properties, for their applications as those of ozone depleting substances but at the same time are not available to the target of hazard, viz. stratospheric ozone layer.

## DESIGNING FOR INNOCUOUS FATE

It is generally desirable to design substances that are robust and long lasting but at the same time they should not be toxic and bioaccumulating. They should be designed to degrade after their useful life. They should not persist in the environment or in a landfill for ever. Therefore, safer chemicals have to be designed not only from the point of manufacture and use of the chemicals but also keeping in view of their disposal and ultimate fate at the end of their life cycle.

## OPTIMIZATION OF FRAMEWORKS FOR THE DESIGN OF GREENER SYNTHETIC PATHWAYS

In order to design green chemical synthesis pathways, identifying the alternatives is not only vital but also challenging. Combined approaches are generally used for this purpose. The following steps are usually followed :

(*a*) Select a set of molecular or functional group building blocks from which the target molecule can be constructed.

The following guidelines are suggested by Buxton, *et al.* (1997) for selecting a group of starting materials :

(*i*) The building blocks should include the groups that are present in the product.

(*ii*) Groups present in any available industrial raw materials or byproducts or coproducts may be included.

(*iii*) Such groups which provide the basic building blocks for the functionalities of the product or of similar functionalities may be included.

(*iv*) Sets of groups associated with the general chemical pathway employed (*e.g.*, cyclic, acyclic or aromatic) may be selected.

(*v*) Groups that violate the property restrictions should be rejected.

The functional group building blocks thus selected can be used to identify a set of potential molecular reactants. This exercise will generate a very large number of potential reactants. To restrict this number of potential reactants, constraints based on chemical principles and intution are generally imposed. For example, if the product contains aromatic monosubstituted group, it is better to start with a monosubstituted aromatic molecule. Similarly, such reactants for which, the carbon skeleton will have to be altered (by ring condensation, etc.) to get the product should be avoided as far as possible. By using such criteria and assumptions, one can arrive at a limited set of reactants for further consideration.

(*b*) Having selected a limited set of potential reactants, a set of stoichiometric constraints have to be applied to describe how the reactants can interact to form the desired molecule. For instance, if the desired product molecule contains 7 aromatic carbons bound to hydrogen and 2 aromatic carbons bound to other aromatic carbons, the reactants that we select must provide sufficient number of aromatic carbons of various types to synthesize the desired product molecule. Simultaneously, thermodynamic constraints too have to be identified for selectivity. These constraints help to reduce the number of possibilities that might be considered further.

(*c*) Finally, a set of criteria have to be used to identify the reaction pathways followed by economic and environmental ranking.

Such approaches have helped in designing alternative synthetic procedures which are relatively safer as compared to the traditional methods. Synthesis of carbaryl provides an example.

The Bhopal tragedy occurred on Dec. 3, 1984 was attributed to the catastrophic release of methyl isocyanate. The Union Carbide was manufacturing a pesticide called carbaryl (1-napthyl-methyl carbamate) by the traditional method using 1-naphthol and methylisocyanate as reactants. Now, an alternative reaction pathway is suggested to minimize the use of hazardous materials. In this method, 1-naphthol and phosgene are used as reactants to produce 1-naphtholenyl chloroformate, which when treated with methyl amine yields carbaryl. Although the reactants used are same, this alternative process eliminates the formation of methyl isocyanate, which is highly hazardous. Further insights and innovations in synthetic organic chemistry are needed to avoid the use of phosgene which is also toxic. The traditional and alternative synthesis pathways are illustrated in Fig. 16.10.

**Traditional Synthesis of Carbaryl**

$$\underset{\text{Methyl Amine}}{CH_3NH_2} + \underset{\text{Phosgene}}{COCl_2} \longrightarrow \underset{\text{Methyl ISO cyanate}}{CH_3-N=C=O} + 2HCl$$

$$\text{1-Naphthol (O—H)} + CH_3\text{—}N=C=O \longrightarrow \text{Carbaryl (O—C(=O)—N(H)—}CH_3\text{)}$$

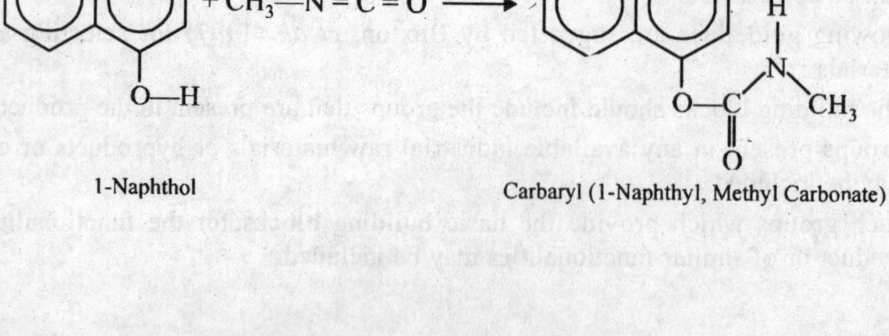

1-Naphthol

Carbaryl (1-Naphthyl, Methyl Carbonate)

**Alternatively Synthesis of Carbaryl**

1-Naphthol + $COCl_2$ ⟶ 1-Naphthyl Chloroformate + HCl

1-Naphthyl Chloroformate + $CH_3NH_2$ ⟶ Carbaryl + HCl

**Fog.16.10.** Synthesis of carbaryl by traditional and alternative routes.

The basic purpose of this exercise is to incorporate systematic decision rules into the search for alternative synthesis pathways.

Further details on Green Chemistry Expert System (GCES) can be downloaded from US EPA website (http://www.epa.gov/greenchemistry/tools.htm).

## INDUSTRIAL APPLICATIONS OF GREEN CHEMISTRY

Nature produces globally around 170 billion tonnes of plant biomass annually in the form of wood, cereals and other crops. However, not more than 4% of this is exploited by the economy. As per the predictions of the U.S. National Research Council, the market share of renewable resources will reach 25% by the year 2020.

## PRODUCTS FROM NATURAL MATERIALS

The oleochemicals branch of Cognis Company (U.S.A.) offers complete range of chemical products produced from natural materials to its customers all over the world. Using palms, coconuts, palm seeds, soya oils etc., as starting materials, the company produces a variety of products such as fatty acids, their methyl esters, glycerol and long-chain alcohols. These products, in turn, are used as starting materials for various items used in bodycare, pharmaceutical, and food industries.

Palm oils can now compete with mineral oils even in the production of petrol and natural gas. The new drilling fluid based on plant-derived esters of palm oil is an example. The resulting rock powder can be returned to the seas without any concerns.

The biotechnological process for the production of adipic acid, which is the precursor of nylon, from glucose is already described earlier in this chapter.

The Bioactive polymer systems (Biopos) at Teltow is developing one of the first biorefineries in Germany. Their initial target is to couple this refinery to an existing drying facility for green materials, where the biomass is separated into a liquid extract and solid residue. This solid residue is processed to produce animal feed, composite fibre materials and a starting material for producing lerulinic acid. The liquid extract is used for producing proteins and fermentation media.

The Cargill-Dow Company (U.S.) is producing 1,40,000 tonnes of eco-plastics annually from corn waste in its biorefinery in Nebraska. These eco-plastics can be used for routine packaging like foil or plastic pots and even T-shirts. For this purpose, the starch is degraded to a glucose syrup which is eventually converted to lactic acid. This is polymerised from which the thread is spun.

**Green Solvents :** Supercritical media are fast emerging as promising alternatives for organic solvents in various processes. The initial industrial application was the removal of caffeine from coffee and tea. Encouraged by this technology other applications such as extraction of other natural products, decaffeination via hop extraction, fat reduction in cocoa, extraction of peanut oil, etc. are also being actively pursued. Supercritical fluids are also being used in the production of well-defined polymer particles. For example, polyamides used in paints and varnishes for automobile industry must be possessing precise characteristics such as particle diameter. If the particles are too fine, the spray painting requires special efforts to exude dust. On the other hand, if the particles are large, they will appear as defects in the finished lacquer. The traditional procedure followed for achieving well-defined distributions of grain sizes is the repeated sieving and sorting, which is expensive and cumbersome. However, such a narrow distribution of grain size can be achieved with supercritical fluid media by rapidly changing the solubility by suddenly relaxing the system (*i.e.*, by reducing the pressure to below critical point) to end supercriticality, which leads to precipitation of the solvated particles. This is because the solubility of most solutes in supercritical fluids dramatically change around the critical point and hence by just reducing the pressure below the critical point, desired materials can be recovered.

Dupont has developed a process for producing Teflon using supercritical $CO_2$ instead of trichlorotrifluoroethane.

Supercritical fluids are also finding application in nano-electronics too. In nano-electronics, supercritical $CO_2$ can replace ultra-high purity water because it does not affect nanometric structures. Moreover, the controllable evaporation rates observed when supercritical $CO_2$ is gradually moved out of supercriticality render it possible to obtain extremely thin homogeneous films of photo varnishes. These coating methods are found to be superior to the traditional methods.

**Ionic liquids :** Ionic liquids are coming to be used as "green solvents". Ionic solvents have the advantage of low vapour pressures. Researchers at the university of Carolina are studying the use of chloroaluminate compounds for electrophilic alkylation, acylation and cationic polymerisation.

The first commercial process based on ionic liquids has been developed by Ludwig Maase and coworks at BASF, Ludwigshaten, Germany for the production of alkoxyphenylphosphine, which is a precursor for photoinitiators. This process has the advantage that the ionic liquid is generated in the course of the reaction itself, and it absorbs HCl, which is an unwanted byproduct. The product is claimed to give higher yields and it has the advantage of recycling and hence is cost-effective.

**Green fuels and E-Green propellants :** The use of biodiesel as an environmentally friendly fuel is already established. Now, the space industry is also looking for alternative environment friendly fuels known as "Green propellants". The challenges are to replace the present engines with those that can work with environmentally benign fuels, and to understand the relevant parameters of the selected green propellants.

Among the alternative fuels under consideration, hydrogen peroxide ($H_2O_2$) is the current favourite. It decomposes without producing any harmful substances and it also qualifies as the "green propellant" from the point of view of energy balance. Some companies have already developed the expertise to produce ultrapure $H_2O$, which is suitable for this application. Green propellants and green rocket engines offer great promise and potential to replace the existing ones in near future.

**Zeolites :** Zeolites are crystalline inorganic polymers made of aluminosilicates and have tetrahedral structures of the type $XO_4$ in which each X atom is linked by shared oxygen ions. X may be Al, B, Ga, Ge, Si or P.

Zeolites have been extensively used in heterogeneous acid-catalysis involving hydrocarbon transformation. Their high acidity is derived from protons that are required to maintain electrical neutrality as shown below :

$$\text{O}\diagdown\text{Si}\diagup\text{O}\diagdown\overset{\text{H}^+}{}\text{Al}\diagup\text{O}\diagdown\text{Si}\diagup\text{O}\diagdown\overset{\text{H}^+}{}\text{Si}\diagup\text{O}$$

**Fig. 16.11.** Super acid sites in zeolites used as solid acid catalysts.

Zeolites have dramatically improved the performance of fluidized catalytic cracking units. For example, in the alkylation of butene, large amounts of corrosive acids like HF and $H_2SO_4$ are used conventionally. In the alternative greener process, the corrosive acids are replaced by *Solid Acid Catalysts* such as Zeolites.

Base-catalyzed and bifunctional (acid / base) applications of zeolites have been reported for manufacturing bulk chemicals. Base catalysis may play major role in the synthesis of fine and speciality chemicals. For example, in the synthesis of 4-methyl thiazole (a systemic fungicide), the basic sites of the caesium zeolite catalyst were found to work very efficiently without using $Cl_2$, $CS_2$ or NaOH.

Chemical modifications of zeolite allow selectivity based on shape. Thus, these catalysts offer promising and versatile technology for the future.

**Biocatalysis :** Biocatalysis mainly includes enzyme catalysis, antibody catalysis and biomimetic catalysis. Among them, enzyme catalysis has been more widely studied. Enzyme catalysts are no longer confined to biochemical reactions only. They are having impact on complex organic synthesis of bioactive molecules and in biotechnology in general. Enzyme catalysts exhibit chemoselectivity, regioselectivity and sterioselectivity. Enzyme catalysis is applicable not only in aqueous media but also in non-aqueous solvents, including supercritical fluids. However, enzymes require water to function as catalysts. Usually, a small amount of water, corresponding to a monolayer of the enzyme molecule is sufficient.

Enzyme having broad substrate selectivity, while still retaining other features of selectivity can prove to be powerful and versatile catalysts. The remarkable properties of enzyme catalysts and their selectivity make them desirable in many synthetic organic pathways. Synthesis of acrylamide is an example :

**Conventional method**

$$CH_2 = CH - CN + H_2O \xrightarrow[2.NH_3]{1.H_2SO_4} \underset{\text{acryl amide}}{CH_2 = CH.C(=O)\ NH_2} + (NH_4)_2\ SO_4$$

**Alternative methods**

$$CH_2 = CH - CN \xrightarrow[\text{hydritase}]{\text{nitril}} \underset{\text{acryl amide}}{CH_2 = CH.C(=O).NH_2}$$

In the year 1981, Degussa developed the concept of an enzyme membrane reactor. It is a milestone in the development of modern biocatalysis because it combined the advantages of both enzyme reactor and homogeneous catalysis.

The "projektaus katalyse" centre established by Degussa has been working on the entire spectrum of new catalyst development. In an effort to discover enzymatic methods to the existing synthetic pathways, scientists of that centre developed several methods. One such example is the synthesis of l-tert-lencine (l-Tle). This amino acid does not occur in nature but serves as a chiral auxiliary compound in symmetric synthesis and as a building block for synthesis of drugs against cancer, inflammation and viral infection. The redox enzyme leucin dehydrogenase reduces a keto-carbonic acid precursor to form the l-Tle. The problem is that the natural cofactor NADH is required as a hydrogen donor in stoichiometric amounts, and the reactor does not include the NADH regeneration system that is in place in the cell. The high cost of such cofactors came in the way of its industrial application so far. However, Maria-Regina Kula at the Institute for Enzyme Technology of the university of Dusseldorf was able to find an enzyme which can generate the precious redox cofactors during the reaction. She isolated the enzyme formate dehydrogenase from the yeast "candida boidinii" and discovered that it is suitable for a large number of redox systems requiring NADH as a cofactor.

Another important enzymatic method that can compete with a chemical process is the production of L-Dopa; this drug is used for the treatment of parkinson's disease. In this process, catechol, pyruvate and ammonia are allowed to react in presence of the enzyme L-Tyrosine phenol lyase. Dehydroserine is formed as an intermediate during the course of a reversible $\alpha$, $\beta$-elimination, which is subsequently converted to L-Dopa by stereoselective addition of catechol. The enzyme accepts catechol as a mimic of its natural substrate phenol. The process is now used for industrial production by Ajinomoto Company, Japan.

Biocatalysis, by virtue of its merits such as selectivity, sustainability and environmental safety, holds a strong promise to discover new and novel production pathways for very useful chemical compounds.

## CONCLUSION

There have been several occasions in the history of chemistry when the chemistry community focussed on goals (such a synthesizing natural products), developing different reaction types (such a stereospecific reactions) to specific applications (such as anti-cancer agents). Now, it is the turn of green chemistry towards sustainable development via pollution prevention and resource conservation. Goals merely provide an objective on which energy, knowledge and resources are focussed. More often than not, pursuit of a goal provides the avenues for several other accomplishments. The same is expected to be true for green chemistry as well. Every project on green chemistry may not achieve its goal of innocuous feedstocks or reagent or benign conditions or products, but striving for this sublime goal will certainly result in excellent chemistry. What else can be more satisfying for the chemistry community as a whole ?

## QUESTIONS

1. What is green chemistry and what are its principles ?
2. What are the goals of green chemistry perspective ? What are the various approaches that lead towards those goals ?
3. What are the basic components of green chemistry research ? Discuss with suitable examples.
4. Discuss the significance of green chemistry towards cleaner production and sustainable development. Give suitable examples.

5. Write informative notes on any two of the following from the point of view of green chemistry approach :
   (*a*) Supercritical carbon dioxide (*b*) Ionic liquids
   (*c*) Atom economy (*d*) Structure-activity relationship.
6. Discuss with the help of suitable examples the various Functional Group Approaches to Green Chemistry.
7. Write short notes on any three of the following :
   (*a*) Optimization of frameworks for the design of greener synthetic pathways.
   (*b*) Green solvents
   (*c*) Solid acid catalysts
   (*d*) Biocatalysis.
8. Discuss the various applications of green chemistry for achieving sustainable development.

# Ethics and Moral Values

## INTRODUCTION

"Ethics" is a english word derived from the Greek word **"Ethos"** literally means the character of a person as described by his/her actions. In Latin it is called "Ethics". Overall its meaning is character or manners of a person in which he/she behaves. But profesionally it is conduct of behaviour and practice when carrying out the work. The work includes consulting, researching and limiting. The study of ethics was developed further by Epicurus and the epicurean movement and by Zeno and the stoics.

The subject of ethics was of great concern to the writers of Bible and major world religious. There was great debate between ethics based on natural law and divine law, which gained a new importance. Therewere several inters like John Rawls, Elliot M.Dorff who wrote seriousphilosophical writing or ethics.

The word **"morality"** derived from the Latin word **"mores"** which means **"habits and customs".** Morality therefore is the character of being in accord with the principles or standards of **right** conduct. The **"right"** word is derived from the Latin word **"rectus"** which means the straight.

Hence, moral or ethical philosophy includes right conduct ethical duty, virtue & behaviour.

**Ethics**– It is said that the ethics is the science of proper conduct. It is also deals with happiness, and obligations to other creatures and comming generation.

Ethics may be defined as –

"Ethics is generally viewed as the system or code by which attitudes and action ·e determined to be either right or wrong" (Sackson)

OR

"Ethics is a science of conduct and the fundamental problem determining what constitutes proper conduct" (Carnell)

OR

"Ethics is a code of values which guide our choices, actions and determine the purpose and course of our lives" (Ayn Rand)

OR

"Ethics is concerns with principles of human conduct." (Baumhert)

As we have discussed earlier, ethics deals with contain standard of conduct and moral. It is an governing force believed every human conduct, irrespective individual or organisational. It differentiate between right/wrong, proper/improper, or fair/unfair actions. Thus we can say that ethics is a branch of philosophy that deals with morals. It is an action concept. It is not simply an idea to think and argue. Ethics really has to do with – acting ethically as individuals, creating ethical institutions and government and malling our society ethical as a whole. Ethics includes the issues like academic honesty, profesional accountability, impartiality in profession & adherence to condidentiality, arrangements. So that it protect both client and profession, improve the profile of the profession, and set ideals and responsibilities of the profession. It also provides guidance on accetable conduct and helps to raise awareness and consciousness of issues.

There are two aspects of ethics – (*i*) the ability to discern right from wrong, good from evil and propriety from impropriety (*ii*) the commitment to do what is right, good and proper. It is an action concept and operates at the level of individuals, institutions, societies. For practical decision making, it is necessary to understand what ethics is not ? Ethics is not the feeling, religion, following the laws and culturally accepted norms. Ethics is not science, though social and natural sciences provide important information to help individuals and communities. Ethics is defined as "the discipline dealing with what is good & what is bad and with moral duty and obligations".

The personal ethics is "the rules by which an individual lives his/her personal life, while accounting ethics pertains to "the code that guides the professional conduct of accountants. Business ethics is concerned" with truth & justice and has different aspects. Ethics is the study of the characteristics of morals.

**Engineering ethics** constitutes the rules and standards governing the conduct of engineers in their role as professionals. Engineering ethics encompasses the more general definition of ethics, but applies it more specifically to situations involving engineers in their professional lives. Thus, engineering ethics is a body of philosophy indicating the ways that engineers should conduct themselves in their professional capacity.

The **science of ethics** addresses issues having both moral and practical significance. Decisions are not simply reactive, such as choosing which bus to catch, but involve concepts of study and responsibility, and of conflicts between rival goods or ills. Reflections upon the ethical significance and correctness of any situation require the use of judgemental reason which is not always familiar to engineers, fine-tuned as they are to the use of deductive logic and the pursuit of the 'correct' answer to any technical problem.

The study of ethics can be of two ways:

1. Ethics involves the discipline of systematic enquiry into moral norms or standards of behaviour and their underlying values and justification. It can help us to understand and justify these values.
2. **Applied ethics looks into the ways in which moral value applied to particular areas of concern such as business. Applied ethics reminds us that while the discipline of ethics** may be treated as separate from other disciplines (such as mathematics or economics) when it comes to practice, the ethical dimensions is a fundamental part of all judgements and decisions. What becomes important then is the development of **'ethical thinking,** which is a part of the decision making process.

## MORAL VALUES

Morality is often described as natural in humans. Moral values also acquired through teaching and inprinting from parents and society. These (values) can be observed in cultural evolution. For moral conduct, the moral values are enforced by parenting, guidance, law, slumming & disapproval. Moral percepts are concerned not with what is, but with what it ought to be. Moral behaviour is often held to be essential to the practice of that religion. Which is directed so: Morality is also closely related to law of the land. Some aspects of the morality are not covered by law. However law differs from humanility without physical and financial sanctions enforce the law, but only the sanction of conscience and reputation enforces morality. Morality should be the basis of the law. The law is not as deep as morality.

Moral behaviour can only be judged according to its cultural standards. It (moral values) deals with right or wrong. Moral belief changes over time within a given society. In shaping the moral ideas, social environment is important. Cultures have very diverse moral value systems. Moral values along with traditions, laws, behaviour patterns and beliefs are the defining features of a cultures. They were effective in small communities even before laws were formalised.

The purpose of morality is to respect persons by being fair and just with them, respecting their rights, keeping our promises, avoiding unnecessary actions. It implies caring for others, willing to help them at time of their need, showing our gratitude for kindness extended and sympathizing with their suffering. Moral reasons expect us to encourage engineers and others to seek work where their contributions will result in wiser uses of world's depleting resources. The purpose of morality are to promote human flourishing, to keep society intact, to assign blame reward, punishment, to resolve conflicts etc.

Etiquette determines what is a polite behaviour rather than what is the right behaviour. Ethics, religion, law and etiquette can be distinguished as follows :

| *Etiquette* | *Law* | *Ethics* | *Religion* |
|---|---|---|---|
| Proper & Improper | Legal and Illegal | Right, wrong or permissible | Right or wrong |

Moral reasons, involve respecting persons by being fair with them, respecting their rights, keeping promises, avoiding unnecessary offence and pain to them, and not resorting to cheating or practicing dishonesty. They also concern with caring for others and willing to help showing gratitude for favour, and emphasizing personal interactions or the minimal requirements of work-a-day moral duty.

There are four ethical theories that will be considered, each differing according to what is held to be the most important moral concept.

1. **Utilitarianism** seeks to produce the most utility, defined as a balance between good & bad consequences of an action taking into account the consequences for everyone affected.
2. **Duty ethics** contends that there are duties that should be performed regardless of whether these acts lead to the most good of the actions is not taken in to account.
3. **Virtue ethics** regard those actions as right that manifests good traits and bad that display bad traits.
4. In **right ethics** people have fundamental rights that others have duties to protect it.

What makes an action morally good ? For many of us, what counts are consequences. The right action is the one that produces good consequences. Morality has absorbed from the surroundings right from the childhood like what is wrong or right. Morality is a term which prevails in and many times considered a synonymous of ethics. Moral standards are first grasped by the child from family, friends, media, temples, mosque, churches etc. As the person grow, factors like experience & learnings give impression on him. Characteristic of moral judgements is that they express our values. Not all expressions of values are also moral judgments but all moral judgments do express something about what we value. There are three principle types of values which humans can have –

(*i*) Preferential values

(*ii*) Instrumental values

(*iii*) Intrinsic values.

Though each of these plays an important role in human life, but they do not play equal role.

***(i)* Preferential values:** The expression of preference is the expression of some value a person holds. It means that a person what prefers over the other work. For example, it is not necessary that every body like to work, some one can prefer to enjoy by any means rather then doing the work. Systems based on such though argue that those activities/situations which make us happiest are infact, the ones we should morally choose.

***(ii)* Instrumental values:** Instrumental value plays an important role in teleological moral systems. Moral choices are those which lead to the best possible consequences such as human happiness. When some thing is valued instrumentally that implies– a person only values a particular thing as a means to achieve some more important.

***(iii)* Intrinsic value:** In moral philosophy not all agree that such intrinsic values actually exist. If intrinsic values do exist, then how is it that they occur? Are they like colour or mass, a characteristic which can be detected so long as the right tools are used.

## PROFESSIONAL ETHICS

Professional ethics concerns the moral issues that arise because of the specialist knowledge that professional attains, and how the use of this knowledge should be governed when providing a service to the public. Hence it is necessary to deal with professional responsibilities & codes of practice.

## PROFESSIONAL RESPONSIBILITY

The professional carries additional moral responsibilities to those held by the population in general. This is because professionals are capable of making and acting on an informed decision in situations that the general public cannot, because they have not received the relevant training. For example, a layman member of the public could not be held responsible for failing to act to save a car crash victim because they could not give an emergency tracheotomy. This is because they do not have the relevant knowledge. In contrast, a fully trained doctor (with the correct equipment) would be capable of making the correct diagnosis and carrying out the procedure and we would think it wrong if they stood by and failed to help in this situation. You cannot be held accountable for failing to do something that you do not have the ability to do. Ethics are rules and values used in a professional setting (Changing Minds, n.d.). In the workplace managers and supervisors should set the standard for using ethics by showing respect, being honest, and promoting trust. If the management team uses unethical forms of communication, the team and business can fail. Ethics are used world-wide in large companies and small businesses. Promoting ethics in the workplace gives employees a sense of worth and trust that can help the business and employees succeed (Joseph, Joshua, 2000). This additional knowledge also comes with authority and power. The client places trust in the professional on the basis that the service provided will be of benefit to them. It would be quite possible for the professional to use his authority to exploit the client. An obvious example is that of the dentist who carries out unneeded dental work on his patients in order to gain more money. It is likely that the patient will not have sufficient knowledge to question what is being done, and so will undergo and pay for the treatments.

## CODES OF PRACTICE

Questions arise as to the **ethical** limits of the professional's responsibility and how power and authority should be used in service to the client and society. Most professions have internally enforced codes of practice that members of the profession must follow, to prevent exploitation of the client and preserve the integrity of the profession. This is not only to the benefit of the client but to the benefit of those belonging to the profession. For example, an American business may approach an engineer to certify the safety of a project which is not safe. Whilst one engineer may refuse to certify the project on moral grounds, the business may find a less scrupulous engineer who will be prepared to certify the project for a bribe, thus saving the business the expense of redesigning. Disciplinary codes allow the profession to draw a standard of conduct and ensure that individual practitioners meet this standard, by disciplining them from the professional body if they do not practice accordingly. This allows those professionals who act with conscience to practice in the knowledge that they will not be undermined commercially by those who have less ethical qualms. It also maintains the public's trust in the profession, meaning that the public will continue to seek their services.

## PROBLEMS WITH INTERNAL REGULATION

There are questions surrounding the validity of professional codes of ethics. On a practical level it is very difficult for those independent of the profession to monitor practice, leaving the possibility that a code of practice may be self serving. This is because the nature of professions is that they have almost a complete monopoly on a particular area of knowledge. For example, until recently, the English courts deferred to the professional consensus on matters relating to their practice that lay outside case law and legislation.

## SEPARATISM

On a theoretical level, there is debate as to whether an ethical code for a profession should be consistent with the requirements of morality governing the public. Separatists argue that professions should be allowed to go beyond such confines when they judge it necessary. This is because they are trained to produce certain outcomes which may take moral precedence over other functions of society. For example, it could be argued that a doctor may lie to a patient about the severity of their condition, if there is reason to think that telling the patient could cause them so much distress that it would be detrimental to their health. This would be a disrespect of the patient's **autonomy**, as it denies them information on something that could have a great impact on their life. This would generally be seen as morally wrong. However, if the end of improving and maintaining health is given a moral priority in society, then it may be justifiable to contravene other moral demands in order to meet this goal. Separatism is based on a relativist conception of morality that there can be different, equally valid moral codes that apply to different sections of society and differences in codes between societies (see *moral relativism*). If *moral universalism* is ascribed to, then this would be inconsistent with the view that professions can have a different moral code, as the universalist holds that there is only one valid moral code for all.

## ENVIRONMENTAL ETHICS

**Environmental ethics** is the part of *environmental philosophy* which considers extending the traditional boundaries of ethics from solely including humans to including the non-human world. It exterts influence on a large range of disciplines including *law, sociology, theology, economics ecology* and *geography.*

There are many ethical decisions that human beings make with respect to the environment. For example:

- Should we continue to *clear cut* forests for the sake of human consumption ?
- Should we continue to propagate ?
- Should we continue to make *gasoline powered vehicles*?
- What environmental obligations do we need to keep for future generations?

- Is it right for humans to knowingly cause the **extinction** of a species for the convenience of humanity ?

The academic field of environmental ethics grew up in response to the work of scientists such as *Rachel Carson* and events such as the first *Earth Day* in 1970, when environmentalists started urging philosophers to consider the philosophical aspects of environmental problems. Two papers published in Science had a crucial impact: **Lynn White's** "The Historical Roots of our Ecologic Crisis" and [Garett Hardi Also influential was Garett Hardin's later essay called "Exploring New Ethics for Survival", as well as a essay by **Aldo Leopold** in his *A Sand County Almanac*, called "The Land Ethic," in which Leopold explicitly claimed that the roots of the ecological crisis were philosophical.

The first international academic journals in this field emerged from North America in the late 1970s and early 1980s – the US-based journal *Environmental Ethics* in 1979 and the Canadian based journal *The Trumpeter: Journal of Ecosophy* in 1983. The first British based journal of this kind, *Environmental Values*, was launched in 1992.

## MARSHALL'S CATEGORIES OF ENVIRONMENTAL ETHICS

There have been a number of scholars who've tried to categorise the various ways the natural environment is valued. Alan Marshall and **Michael Smith** are two recent examples of this, as cited by **Peter Vardy** in "The Puzzle of Ethics". For Marshall, three general ethical approaches have emerged over the last 40 years. Marshall uses the following terms to describe them: Libertarian Extension, the Ecologic Extension and Conservation Ethics.

(For more on Marshall's environmental ethics, see also: A. Marshall, 2002, The Unity of Nature, Imperial College Press: London, Alan Marshall is currently at the Univerzita Pavol Jozef Safarik, Slovakia).

## LIBERTARIAN EXTENSION

Marshall's Libertarian extension echoes a civil liberty approach (i.e. a commitment to extend equal rights to all members of a community). In environmentalism, though, the community is generally thought to consist of non-humans as well as humans.

Andrew Brennan was an advocate of ecologic humanism (eco-humanism), the argument that all ontological entities, animate and in-animate, can be given ethical worth purely on the basis that they exist. The work of **Arne Næss** and his collaborator Sessions also falls under the libertarian extension, although they preferred the term "*deep ecology.*" Deep ecology is the argument for the intrinsic value or inherent worth of the environment – the view that it is valuable in itself. Their argument, incidentally, falls under both the libertarian extension and the ecologic extension.

**Peter Singer's** work can be categorized under Marshall's 'libertarian extension'. He reasoned that the "expanding circle of moral worth" should be redrawn to include the rights of non-human animals, and to not do so would be guilty of *speciesism*. Singer found it difficult to accept the argument from intrinsic worth of a-biotic or "non-sentient" (non-conscious) entities, and concluded in his first edition of "Practical Ethics" that they should not be included in the expanding circle of moral worth. This approach is essentially then, bio-centric. However, in a later edition of "Practical Ethics" after the work of Næss and Sessions, Singer admits that, although unconvinced by deep ecology, the argument from intrinsic value of non-sentient entities is plausible, but at the best problematic. We shall see later that Singer actually advocated a humanist ethic.

## ECOLOGIC EXTENSION

Alan Marshall's category of ecologic extension places emphasis not on human rights but on the recognition of the fundamental interdependence of all biological (and some abiological) entities and their essential diversity. Where as Libertarian Extension can be thought of as flowing from a political reflection of the natural world, Ecologic Extension is the best thought of as a scientific reflection of the natural world. Ecological Extension is roughly the same classification of Smith's eco-holism, and it argues for the intrinsic value inherent in collective ecological entitites like ecosystems or the global environment as a whole entity. **Holmes Rolston,** among others, has taken this approach.

This category includes James Lovelock's *Gaia hypothesis*; the theory that the planet earth alters its geo-physiological structure over time in order to ensure the continuation of an equilibrium of evolving organic and inorganic matter. The planet is characterized as a unified, *holistic* entity with ethical worth of which the human race is of no particular significance in the long run.

## CONSERVATION ETHICS

Marshall's category of 'conservation ethics' is an extension of use-value into the non-human biological world. It focuses only on the worth of the environment in terms of its utility or usefulness to humans. It contrasts the intrinsic value ideas of 'deep ecology', hence is often referred to as 'shallow ecology', and generally argues for the preservation of the environment on the basis that it has extrinsic value – instrumental to the welfare of human beings. Conservation is therefore a means to an end and purely concerned with mankind and intergenerational considerations. It could be argued that it is this ethic that formed the underlying arguments proposed by Governments at the **Kyoto summit** in 1997 and three agreements reached in Rio in 1992.

## HUMANIST THEORIES

Following the bio-centric and eco-holist theory distinctions, Michael Smith further classifies Humanist theories as those that require a set of criteria for moral status and ethical worth, such as sentience. This applies to the work of Peter Singer who advocated a hierarchy of value similar to the one devised by **Aristotle** which relies on the ability to reason. This was Singer's solution to the problem that arises when attempting to determine the interests of a non-sentient entity such as a garden weed.

Singer also advocated the preservation of "world heritage sites", unspoilt parts of the world that acquire a "scarcity value" as they diminish over time. Their preservation is a bequest for future generations as they have been inherited from our ancestors and should be passed down to future generations so they can have the opportunity to decide whether to enjoy unspoilt countryside or an entirely urban landscape. A good example of a world heritage site would be the tropical rainforest, a very specialist ecosystem or climatic climax vegetation that has taken centuries to evolve. Clearing the rainforest for farmland often fails due to soil conditions, and once disturbed, can take thousands of years to regenerate.

## STATUS OF THE FIELD

Environmental ethics became a subject of sustained academic philosophic reflection in the 1970s. Throughout the 1980s it remained marginalized within the discipline of philosophy, attracting the attention of a fairly small group of thinkers spread across the world.

Only after 1990 did the field gain institutional recognition at programs such as Colorado State, the University of Montana, Bowling Green State, and the University of North Texas. In 1991, Schumacher College of Dartington, England, was founded and now provides an MSc in Holistic Science.

**These programs began to offer a masters degree with a speciality in environmental ethics/** philosophy. Beginning in 2005 the Dept of Philosophy and Religion Studies at the University of North Texas offered a PhD program with a concentration in environmental ethics/philosophy.

## CURRENT ETHICAL ISSUES

Efforts to promote ethical practice continue. In addition to the professional societies and chartering organizations efforts with their members, the Canadian Iron Ring and American Order of the Engineer trace their roots to the 1907 Quebee Bridge collapse. Both require members to swear an oath to uphold ethical practice and wear a symbolic ring as a reminder.

Currently, bribery and political corruption is being addressed very directly by several professional societies and business groups around the world. However, new issues have arisen, such as offshoring, sustainable development, and environmental protection, that the profession is having to consider and address.

## CASE STUDIES AND KEY INDIVIDUALS

Petroski notes that most engineering failures are much more involved than simple technical mis-calculations and involve the failure of the design process or management culture. However, not all engineering failures involve ethical issues. The infamous collapse of the first Tacoma Narrows Bridge, and the losses of the Mars Polar Lander and Mars Climate Orbiter were technical and design process failure.

These episodes of engineering failure include ethical as well as technical issues.

- Space Shuttle Columbia disaster (2003)
- Space Shuttle Challenger disaster (1986)
- Therac-25 accidents (1985 to 1987)
- Chernobyl disaster (1986)
- Bhopal disaster (1984)
- Kansas City Hyatt Regency walkway collapse (1981)
- Love Canal (1980), Lois Gibbs
- Three Mile Island accident (1979)
- Citigroup Center (1978), William LeMessurier
- Ford Pinto safety problems (1970s)
- Minamata disease (1908-1973)
- Chevrolet Corvair safety problems (1960s), Ralph Nader, and Unsafe at Any Speed
- Boston molasses disaster (1919)
- Quebec Bridge collapse (1907), Theodore Cooper
- Johnstown Flood (1889), South Fork Fishing and Hunting Club
- Tay Bridge Disaster (1879), Thomas Bouch, William Henry Barlow, and William Yolland
- Ashtabula River Railroad Disaster (1876), Amasa Stone

## ETHICAL SITUATIONS

As we know ethics is the science of character or habit. It evaluate human habits, character and voluntary determinations. It pronounced ethical decisions upon conduct. It studies what is right & what is wrong in character. It also studies the ideals of human life. Which slows what ought to be done and what ought not to be done.

Ethics is the science of right. It searches for ethical laws which unrelaxed ideals and man is bound to follow them. All the aims and activities of human life are directed to the ideal good. It **is not possible to decide about right & wrong without knowing ideal good. Ethics discuss the** good & bad qualities of human behaviour by comparing it to these ideals. It informs man of the differences between good & bad, right & wrong, evil and virtue. In these conditions a number of situations occur that require consideration whether it should be done or not, it should be right or wrong etc. These situations may be social, political, confidential, related to accounts, organizational, related to media, sexual, assessment, issues, conduct of staff, publishing issues, related to research, Academic, relation etc.

When word "should" is used for any action/conduct then the situation or ethical situation come into play. Under these circumstances it is difficult to decide what should be done or what should not be done. What ethical law says or what is right or wrong is another part of these situations. To come across with these difficulties/conditions/situation one must take an example. So some examples of ethical situations are given for readers reference.

(*i*) There is involvement of a minister in IPL indirectly but he is denying for same. Later it proved indirectly. Will minister resigne from the post.

(*ii*) UP chief minister was garlanded by her fans. Garland was made up of 1000 & 500 Rs. notes. Would any department of Central Govt. will take any action against it.

(*iii*) Suppose you are inquiring the role of an person who is guilty but he is your friend. What you will do in giving the statement.

(*iv*) You are head of a organization. In installation of a lab you purchased the valuable items from a concern. In return if he gives you any kind, will you accept.

(*v*) Suppose you have achieved a important post by cheating. Will you open this mistry? Or you will withdraw the post.

**(*vi*) Some times it happened in exercising our power to do the best for your company. But on the same time it is harmful for employees. Will you do in company's favour or employees**

(*vii*) In the sequence of the price rise, sugar prices hiked after the cheap statement given by agriculture minister. Public opinions is against the minister but prime minister did not saked him. Will minister take responsibility for it, when the sugar mill owners are with him.

(*viii*) One of my classmate in university exam. was passed. When he saw his marks sheet, the sum of Ist papers & IInd paper of one of the subject was differ from the written. On the basis of that he was declared passed. But actually he was failed. Did he should have disclosed this thing to university authorities?

There are so many examples in our daily life, which are said to be the ethical situations. In this cases these are many tools for decision making, but few guides to indicate when situations might have an ethical implication. This awareness is a first step before decisions are made. Therefore, it is important that we first recognise the moral context of a situation and then make an attempt to resolve it, if needed. To act upon ethical situations, there is no any absolute rule or guide. Secondly, there is no universal guide for every situations. They often conflict with each other in practice. One should simply follow one's intuition and rely on the "inner voice".

## MORAL SITUATION

Human life is not as simple as natural phenomena. Man has both intelligence and sentience. Often, while his feelings, senses, desires and tendencies pull him one way, his intellect pulls him in another direction. He is both selfish and selfless. Feeling is an almost daily occurrence. Then a highly intellectual person is cnfronted will another problem. He goes so deep in each subject and he has so many alternatives before him that he finds it difficult to take up any of the causes of action open to him. Almost everyone has experience of similar situation. In this situation the person becomes paralysed. His duties do not suggest themselves to him. Such a one is a moral situation.

## AN EXAMPLE OF MORAL SITUATION

In the Mahabharata, Arjuna too, was, faced with a similar moral situation. When the armies of the Kauravas and the Pandavas stood face to face at Kurukshetra, Arjuna requested Lord Krishna to drive his chariot between them. When the chariot reached the heart of the field, Arjuna cast a glance at the warriors of the two armies. On the side of the Kauravas he saw his uncle, teacher, cousins and other kinsmen armed to the tooth and ready for action. Seeing them Arjuna was moved, the Gandiva bow slipped down from his hands. His mind was captivated. He saw no fun in wading through the blood of these people to be crowned king. He said to Lord Krishna that he did not see any use in killing them and gaining the throne. "I would not like to kill them even for the kingdom of the three worlds. How will we dwell in happiness, having killed any one of those near and dear to us? Killing them would ruin our family. This, in its own turn, will lead to a degeneration in social order. With the ruin of the social order, sin will increase and the women will be corrupted, leading to an increase in impure stock, consequently leading to degeneration of ancestors". Speaking in his sorrowful vein Arjuna said :

न चैतद्विद्मः कतरन्नो गरीयो यद्वा जयेम यदि वा नो जयेयुः।
यानेव हत्वा न जिजीविषाम स्तेऽवस्थिताः प्रमुखे धार्तराष्ट्राः॥२॥६
कार्पण्यदोषोपहत स्वभावः पृच्छामि त्वां धर्मसंममूढ़चेताः।
यच्छ्रेयः स्यान्निश्चितम् ब्रूहि तन्मे शिष्यस्तेऽहं शाधि मां त्वां प्रपन्नम्॥२॥७॥

Meaning, that "we do not know the course of action which is the best for us and we do not ever know whether we will defeat them or they will beat us down and those sons of Dhratrashtra whom we do not want to kill and succeed are arrayed against us. With my very being tainted by the vice of faint heartedness and my mind puzzled with regard to duty, I ask you to tell me a decidedly satisfactory means because I am your disciple and quite under your protection. Give me guidance".

## ROLE MORALITY & CONFLICTS

Actually role morality is morality which a individual may adopt in a definite role. For different roles, different morality is there. Hence, a person acting in a role he is behaving differently but where he opt the another role there he has to give up right of that. That is why the importance of role clarity *i.e.* right person at right place is utmost essential for any organisation in the modern business.

Conflicts start in any organization, company, party, institution, groups, family etc. where dis-satisfaction in different morality, conduct or behaviour starts. It is significant when dealing with professional ethics / morality where they find themselves at adds with their best moral judgement The creation of mis-alignment results in dis-satisfaction of both employee & employer ultimately reflects in lowering the productivity of organisation.

Parties involved in moral conflict also tend to have great difficulty in imagining a win-win resolution of the conflict.

Over the course of conflict, the original issues often become irrelevant and new causes for conflict are generated by actions within the conflict itself. This is because in moral conflict, when groups try to act consistently with what they believe is morally good and just, they "prove" to the other side that they are fools or villains. Thus, the means by which the parties seek resolution often just provoke further conflict. As the conflict continues, substantive issues are largely forgotten and "the other side's means of dealing with the conflict is itself the force that drives the interactions among the various conflicted parties." Thus, moral conflicts are self-sustaining.

Each party may try to force the other side in to compliance. The conflict is likely to escalate and become more protracted as a result. In professional life the conflicts of the interest in certain conditions are often at the heat of role morality. Conflicts of interest are well understood in ethics. They involve conflicts between a person's obligations to others and their own self interest. Keeping self interest away, one should use their competence primarily for the benefit of society. On the otherhand, ethically irresponsible management of the conflicts of interests occur when self interest becomes primary without ethical justification.

Those involved in moral conflict may regard perpetuation of the conflict as virtuous or necessary. Moral conflicts often stem from desire to safeguard basic human needs such as security and social recognition of identity. Some times the continuation of the conflict may seem preferable to what would have to be given up if the other party were accommodated. Professional codes of practice and ethics ensure that the individual professionals meet the standards of the codes by disciplining them from the professional body if they do not practice accordingly. The professionals who act with conscience to practice with ethical knowledge will be allowed not be undermined commercially by those professionals who have less ethical qualms.

Dealing with moral conflicts there are some approaches to avoid/respond to ethical conflicts in any profession.

One of the best approach is to keep conflicts from arising because when we will not allow to happen then there is no any problem/effects. The next way to deal with conflict isallow it to happen and then respond to them. There are two aspects of this approach—we can respond according the nature of the conflict. So some energy can be saved. Secondly the professionals can be treated accordingly. Because adverse impact on an organisation's name and culture is. During the ethical conflict organisations integrity & culture is on state.

Some times the people communicated with different ways to change the conflict's pattern. One of the way is to form a different story. Their form of communication & pattern of interaction is differernt. Reproming the conflict as a problem can some times help to get the parties to focus on something they can acieve together rather their non-negotiable differences. Therefore, reparing by any body else is also one of the way to deal with conflicts.

A very important and popular way of dealing with conflicts is Dialogue. Now a days it into practice and successful irrespective of politics, professionals organizations and in domestic life. Very big conflicts are resolved by dialogues. There are many forms and contexts of dialogues like respectful communication, empathic listening, improved understanding, respect etc.

## NON-PROFESSIONAL, QUASI-PROFESSIONAL AND HARD-PROFESSIONAL

As we have seen earlier, the term "professional" was used for the persons who professed adherence to monastic vows of a religious order. But in modern society a professional is one who is "duly qualified" in a specific field. He should have definite characteristics. To carry on his/her life, every body is engaged in doing some thing, it may be occupation or employment. Different types of jobs/employment require diferent types of persons having different types of skills. It may be low or high skill level.

Self employment is practiced by non professionals, quasi professionals or hard professionals. One who is not a professional is generally called "Non-professional". A non-professional has no that type of skill as the professionals have. They may be termed as ameteur, dabbler, dilett-ante, smatterer, uninitiate. Non professional engaged in an activity superficially or non serious intent. They never earn in those activities or specified occupation but accepts money under restrictions – specified by a regulatory body. There are certain persons, they never do work for money or as a profession but do for their interest or as hobby. They are also called the non-professionals for e.g. athlete, musicians, sculptor, artist, environmentalists, charitable workers etc. On the other hand certain persons engaged in such activities are also working as professional. For example if an athlete is paid for his/her appearances is referred as professional otherwise not.

The quasi word is seems to be historical. Generally it was used for situations in judicial process. This word was derived from the Latin word (Kway-zeye, Kwahjee) means "as if" is a substituted identity. When an action, decision, order or proclamations etc are taken to judicial process, then such situations are called as "Quasi". The meaning of quasi is also given as "to a certain extent, seemingly, alike in some sense but not is actually. Some of the academician define Quasi as the semi-(a part of). For example quasi corporation means a body that has some but not all, quasi argument means it is used as, quasi historical means apparently historical. In the same perspective "Quasi professional" means a person is apparently professional. Such professionals have some or part of but not all of the peculiar attributes of a professional. As we discussed the word sem, *i.e.* this quasi professional are semi professionals having the knowledge a part of that. Like doctor's in villages doing practice like a doctor but they do not have knowledge/degree like a recognised doctor. Similarly there are so many examples like Engineers, Advocates, Electricians, etc. which are doing respective work as professionals, chemists also give a "pseudo" word for quasi. Some times it is also used as pseudo chemical reactions, pseudo environmentalists, pseudo workers in different field. Because they do respective work and behave like that but they are not so or they don't have their knowledge.

Contrary to that another word is "Hard-professional" is also in working. It means those persons which have high degree of knowledge/skill in their respective field, are generally called as hard professionals. It is well understand by taking an example of doctor. Normal degree is a doctor, if he did post graduate degree is a good doctor, when he attains the superspecial degree in his field of medication, he will called as doctor specialist in this field. It is hard professional. Similarly hard professional in the field of journalist, teacher, pharmocist, sports can also be understand.

## VALUES, POLICIES AND ORGANIZATION CULTURE

### Values

Value is that which has importance, that for whose possession the individuals and society endeavour, that for which they live and for which they can make sacrifice. Economics makes a quantitative interpretation of value. For it, value is that which has practical utility and can be exchanged. Objects satisfying desires are valuable. Ethics interprets values qualitatively.

Values assist is the fulfilment of man's desire. As Urban puts it, "Value is that which satisfies human desire. Value is of assistance in the preservation and development of life. It directs one towards self realisation and self-development.

## PLEASURE AND VALUE

In Ward's opinion, value is objective, desire itself having no value. Value is implicit in the object which satisfies desire. Fulfilment of desire gives pleasure, thus the feeling of value is in the sense of pleasure. In the words of Mackenzie, "Pleasure may fairly be described as sense of value." Pleasurability is the subjective sign of value. Put pleasure cannot be credited as the sole subject matter of intrinsic value. The reasons for this follow –

**1.** *Pleasure is merely affective, value satisfies every aspect* : Pleasure is related to the affective aspect of man. But man also has an intellectual aspect and its satisfaction is at least as important as that of the affective aspect. Thus pleasurable object cannot be said to be of value to the entire self. Total welfare should be one which can render satisfaction to both affective and rational aspects.

**2.** *Pleasure is subjective and value is objective* : As Ward has maintained quite correctly, values are objective. On the other hand pleasure is subjective. Thus pleasure cannot be said to be of intrinsic value. The supreme value is both subjective and objective.

**3.** *Difference of Standard* : The standard of pleasure is quantitative and that of value qualitative. Thus, values cannot be distinguished on the basis of pleasure, because it cannot pronounce the difference between superior and inferior values. The identification of pleasure with value is incorrect.

**4.** *Difference in nature* : The nature of pleasure and value differs. Pleasure is momentary and transient. It is individual and cannot therefore assimilate egoism and altruism. Thus, pleasure can not be made the standard of value. The supreme value should be permanently self satisfying, intrinsic and an end in itself. It must include the satisfaction of conative aspects besides the effective aspect. It organises all the parts of character in the form of a balanced unit. It creates an assimilation or synthesis between selfishness and selflessness. Thus pleasure cannot be conceded as the solitary subject matter of intrinsic value.

## INTRINSIC AND EXTRINSIC VALUES

Some objects are ends in themselves whilst others are mere means to the attainment of other objects. Extrinsic values are temporary and instrumental values. Intrinsic values are permanent and ultimate values. In the words of Wright, "An intrinsic value is of worth on its own account; an instrumental value because of its consequence". Generally truth, beauty and good or virtue are accredited intrinsic values. They are not the means to an attainment of other objects. Individual goods are subordinate to them. In spite of being attained by individuals they are independent of individuals. In some form or the other they give spiritual satisfaction to the soul. Truth, beauty and good render satisfaction, respectively, to the cognitive, affective, and conative aspects. Like these three aspects of the self, these three absolute values too are inseparably mutually conjoined although this relation of theirs cannot be rationally explained with exactitude.

**Forms of values :** Value has two forms-intrinsic and extrinsic, absolute and instrumental, permanent and transient.

Absolute values and intrinsic values are permanent and ends in themselves. Instrumental values and extrinsic values are transient and merely means. Morality is concerned with the former class of values. Absolute good is the absolute value. That which is the means to some good, is of instrumental value. All pleasant objects possess practical value. The moral value of objects depends upon their utility in the achievement of the absolute value.

**Absolute value :** Moral absolute good is also the absolute value. According to KANT good will as the sole intrinsic value will is the mirror of the character.

In human character, moral will is the absolute good and the absolute and instrinsic value. It is the end and it renders permanent satisfaction to man. According to the advocates of the theory of value, the only thing intrinsically valuable is the one which gives permanent satisfaction and pleasure to man. Moral values are both individual and universal. The absolute good, though achieved by man is still a universal value. Absolute good gives pleasure, though pleasure cannot be said to be the absolute good. Value is that to which man gives importance, which he expriences and to which man concurs in action. In this way, value rules over conduct. Thus, it is essential to know the nature of value.

## VALUE IS DYNAMIC

Value is changeable with time and place. It is dynamic. It is always undergoing an evolutionary process. Every person treats objects as valuable according to his own level of consciousness. It transpires from a comparative study of values that all values are commensurable or every value can be measured and its position in the grades of values be determined. But, because moral evolution is not quantitative, moral values are unique, and not commensurable with other values. Their knowledge is the origin of dutiful reason and moral obligation. Conjoined to them are moral feelings. Moral values can be accepted as commensurable only in one meaning. In self-perfection, knowledge and beauty have an undeniable place along with will. Truth, beauty and good can together grant complete satiation to the self. All three are valuable by virtue of being capable of satisfying the soul in one way or another. Thus they can be considered commensurable in the form of different constituents of the absolute good.

## POLICIES AND ORGANIZATION CULTURE

Organization culture has a significant impact on its policies, programmes and strategies. In democratic country peoples have different castes, creed and culture, where every society of people, organization have different culture. Because culture reflects the personality & working of concerns/society. The organization culture is characterised by the attitudes, experiences, beliefs and values. For more effective and cultural diversity, organizations examine their policies and practices to ensure that not only they are fair but the systems and values actually contribute to the organization culture policies or principles ideas, thoughts all these contribute the organization culture. Different peoples are defining culture in different ways.

- Culture is the integrated pattern of human behaviour that includes thought, speech, action, artefacts and depends on mains capacity for learning and transmitting knowledge to succeeding generations. **(Webster's New Collegiate dictionary)**
- Culture is the set of habitual and traditional ways of thinking, feeling and reacting that are characteristics of the ways a particular society meets its problems at a particular time.

  **(Clyde Kluchohn)**
- A company's culture is defined by its core purpose or core value and companies that successfully attract, hire and retain top talent align their recruitment and retention, practices with their core culture. **(Jim Harris and Joan Brannick-1999)**

The workers in this field have started to use the term culture in corporate sector. Several other theories are given for culture, but the last is more relevant in the present scenario. Every culture has its own values & systems therefore it is necessary to include transcultural human values in education.

In the context of these definitions & policies, organization can develop, implement & established good human resource policies, programmes and strategies. So as to support and strengthen its purpose & values. Thus relationship between the policies and organization culture is important.

Every organization or company is converting into multinational where people are working in a group. In these situations the knowledge of one culture to another is important. Policies have to be made as per the culture of society. Therefore, the appointment of personal in the area of marketing field must be focussed.

There are varieties of cultural attributes need to be addressed and assessed. There are certain attributes or factors which defines the organisation culture are :

(*i*) Values, traditions and celebrations.

(*ii*) Diversity

(*iii*) Interaction of staff members

(*iv*) Orientation

(*v*) Staff titles

(*vi*) Mission & vision of organization

(*vii*) Management. (Leadership & working)

(*viii*) Levels of hierarchy (High, moderate & low)

(*ix*) Work environment.

(*x*) Communication style

(*xi*) Life style of staff

(*xii*) Problems and their method of solution

(*xiii*) Recruitment of staff

(*xiv*) Organizational structures.

Lastly, culture is a combination of current leadership, past dealings and organizational structures. So far as leadership is concerned, staff prefer the experienced leadership. Experience is always prefered or over shadowed the effects of previous management and making substantial changes. By changing the structures, senior management can drive appropriate behaviour and keep an organization aligned with a rapidly changing business climate.

## DIFFERENT ETHICAL VALUES CRITERIA

According to Rashdall, Morality consists in the promotion of true human good, but a good of which pleasure is only an element. It is right to promote one's and other's good, and every body should do it except in the process where moral qualities are destroyed. Moral values become the ultimate aim in Ideal utilitarianism. Pleasure is secondary to moral values.

Thus, in Ideal Utilitarianism greater stress is laid on value. Rashdall holds that virtue is the ultimate value. All other values are subordinate to it. On the subject of virtue, Rashdall writes, "That value is the greatest of all values." Accordingly, knowledge, beauty culture, etc., are subordinate to virtue while the ultimate value is an organic whole. In it, all the values have an appropriate place. All these are its parts. A moral life includes all these and all these are synthesized. In this way moral life is not only an ideal life, it also has pleasure, knowledge and beauty. Here Rashdall confirms to Eudiamonism. But the fact that knowledge, beauty and pleasure have a place in a moral life does not mean that their values are equal. Knowledge and beauty have a high value in the heirarchy of values because they pertain to and satisfy the intellectual and sentient aspects of man whereas pleasure satisfies only the sensual aspect.

As we know generally people understand the ethics as function/actions/behaviour. But Ethics provide us moral principles and rules. Moral principles focus primarily on people's actions and doings. People apply them by asking what they require of us as professionals. There are defferent types of Ethics personal, professional, business, political, Environmental, Institutional etc. All these have their own ethical values. Thus we have different definitions of ethical value. They are:

- Ethical value is people's elementary perspective of value criterion for scaling different things.
- Ethical value is the system of people's faiths, beliefs and ideals of basic values.
- Ethical value, which is the domination of all our actions, is the summation of people's notions, thoughts and attitudes as to what to be good or bad, what to yearn, what to pursue, what to abnegate, what to support, what to oppose and so on.
- Ethical value is people's subjective judgement of the importance of things, the essential impetus to people's actions and the base on which to gain knowledge, master skills and shape attitude, inclination and worldview.

## UTILITARIAN

The doctrine of utalitarian ethics theory enunciated that what is useful is good and the ethical value of conduct is determined by the utility of the results. The utalitarian may be defined by various ways :

The philosophy that something's value may be measured by its usefulness.

The theory that action should be directed toward achieving the "greatest happiness for the greatest number of people".

Judges behaviour in terms of the costs and benefits to society: it suggests choosing actions that result in the greatest net benefit or the lowest net cost. (the distribution of costs and benefits is not a paramount consideration under the utilitarian perspective)

The original theory was based on pleasure and pain, so Peter Singer extended its bounds to non-human animals who were sentient (capable of experience) and could therefore feel pleasure and pain. From 18$^{th}$ century utilitarianism is the doctrine that an action is moral if it tends to promote the sum of human happiness.

Utilitarianism allows for a flexible approach, carefully weighing up the consequences in each situation. For some people in the developing world, development will save many lives.

**Utilitarian ethics was formulated first by Jeremy Bentham in 1781,** and later championed and elaborated by the philosopher John Stuart Mill. This ethic states that the rightness of an action entirely depends on the value of its consequences, and that the usefulness can be rationally estimated.

## RUSHDALL'S UTILITARIANISM

Rashdall's moral theory is a kind of Utilitarianism, it being propounded by the latter that nothing is good of itself but believed good according to its utility. According to Rashdall's thinking, moral judgment is the judgment of values but these judgements correspond to the utility of the act. The greater the promotion, afforded by an act, in human walfare or universal welfare, the greater is its value. In this way, Rashdall does not differ from other utilitarians in looking upon common welfare as good. This general welfare includes moral qualities, knowledge, beauty and is assisted by them.

## RIGHT AND WRONG

In this way, that which helps in human welfare is right and that which hinders it is wrong. Thus, according to Rashdall, the propriety of an action depends upon its results. Right assists good. It would be wrong if it were to hinder it. For example, all pleasures cannot be said to be right. Pleasures enhancing human welfare are right while those hindering it are wrong.

## REASON IS THE DETERMINANT OF GOOD AND BAD

But who will decide whether a pleasure or activity is assisting human welfare or hindering it ? As with Kant, so with Rashdall, practical reason is the determinant of right and wrong. The aim of man is pleasure. Reason knows intuitively what is going to enhance and which is going to detract from happiness. The ultimate good is known to it by intuition.

## EVOLUTION OF IDEAL UTILITARIANISM

It transpires from the short account of Ideal Utilitarianism that it is merely kind of Perfectionism or Eudiamonism. It supports, in a way, the theory of Self realisation. Its main peculiarity is that it has imagined an ultimate good which may satisfy all the aspects of human nature. Knowledge, beauty and pleasure satisfy respectively the cognitive, affective and conative aspects of man. Its second specially lies in the fact that its ultimate good strikes a compromise between individual and social welfare. Besides perfectionism, Ideal Utilitarianism has also an element of Rationalism because reason is taken to be the determinant of right and wrong. Ideal Utilitarianism has also a part of Intuitionism, because it has been granted that reason recognises good intuitively. Rashdall entitles his theory Utilitarianism but according to Mackenzie, "It is doubtful whether it ought to be described as utilitarianism, since it recognises that moral goodness is part of the supreme end of life, and not a mere means to an end other than itself".

## BENTHAM'S UTILITARIANISM

According to Bentham man always wants to attain pleasure and abstain from pain and this constitutes his ethical duty too. He said man is naturally selfish. Bentham said, "To obtain the greatest portion of happiness for himself is the object of every rational being". Behind every act of selfless service there is some selfish aim. Man helps another only if he stands to profit by it.

Bentham is a utilitarian and it is on the basis of this theory that he supports selfless hedonism. The principle of utilitarianism implies that theory which accepts or rejects every action according to that tendency which gives the impression of increasing or decreasing the pleasure or pain of the people whose interests are conjoined with it. Thus, it is utility which is the standard of morality. It is also the real ultimate motivation of actions. According to Bentham the question of quality is meaningless. Being equal in quantity, all pleasures are similar. In the words of Bentham, "Quantity of pleasure being equal, pushpin is as good as poetry."

## HEDONISTIC CALCULUS

The question arising here is about the mode of measuring the quantity of pleasures. It is Bentham's belief, that the pleasures are capable of being measured. He says, "Weigh pleasures and weigh pains, and as the balance stands will stand the question of right and wrong". Bentham wanted a standard which was hard and solid, and also aways from personal thoughts and feelings. In this discussion he was much impressed by the principle of mathematical calculus. It was his understanding, that pleasure can be measured as can be the length and breadth of a room.

Preceding Bentham, Paley and other hedonists had conceived of two qualities for the measurement of quality—duration and intensity. Bentham concieved of five other dimensions. He holds, accordingly, that it is essential to understand these seven dimensions if quantitative measurement is to be understood. These seven dimensions are— intensity, duration, nearness, certainty; purity or lack of pain, fruitfulness or that which produces other pleasures, and extent or the number of people to profit by it. These seven dimensions are well known as Bentham's Hedonistic Calculus or moral arithmetic. By weighing the quantity of any pleasure, according to these seven dimensions, we can compare it with that of any other and the one yielding more pleasure can be accepted as superior. Here the dimension of extent is important because by accepting it Bentham passes from selfishness to selflessness.

## MILL'S REFINED UTILITARIANISM

On the one hand J. S. Mill popularised the Utilitarianism of his father James Mill and his friend Bentham, and on the other hand, he continued his enquiry into truth. Consequently, his Utilitarianism is full of inconsistencies. Mill has defined Utilitarianism as that theory which treats of the principles of utility or benevolent by nature. Secondly, social tendencies are beyond egoism and altruism and third, moral sense pevails upon and balances the mutually contradictory egoistic and altruistic tendencies. Man and human society are not merely conglomerations of separate parts but are rather organised constructions. The internal sense or conscience creates an adjustment between these different parts. It is the most superior element among the four elements of human nature. According to Butler, these elements are—(1) Specific emotion, (2) passion and tendencies, (3) benevolence and (4) self love and conscience. In their own right each element is good but when they violate their limits they become bad. Inner sense is dependent upon other tendencies because it cannot itself create virtue but inspite of it, it supersedes all others.

### INNER SENSE

In this way, inner sense is superior to both benevolence and self love. It is superior to all faculties. It is implicit in every one and is their moral directive. It is normal because it concurs with the nature of human constitution. It is the real self and reason.

## VIRTUE

Virtue is the habit of doing moral good. It is also the habit of abstraining from doing moral evil. In this way virtue is an indication of good character. According to Aristotle, "Virtue is a permanent state of mind, formed with the concurrence of the will and based upon an idea of what is best in actual life an ideal fixed by reason." In this way, virtue is an acquired quality. It is an acquired disposition, the quality of obeying moral law. The nature of virtue will become even more evident from an analysis of the relations it bears to duty, character, habit and value.

### VIRTUE AND DUTY

Virtue is created by the habit of fulfilling duty. A person given to uninterrupted fulfilment of duty has a sterling character and the virtue of duty fulfilment. Thus the fulfilment of duty is intimately related to virtue but the difference between duty and virtue should not be forgotton on this account. Duty implies external activity and is a part of conduct. Virtue implies internal excellence, being a quality of character. In brief, duty is the external manifestation of virtue, the latter being the internal aspect.

## VIRTUE AND CHARACTER

It is evident from the above exposition that virtue and character are intimately related. Virtue indicates good character as vice is a sign of bad character. Virtue is the superiority of character, vice its defect. A virtuous person endeavours for the development and manifestation of the complete self and controls the baser passions. A person in vice is the slave of base passions and he does not distinguish between moral and immoral, Thus, virtue is as much a quality of character as vice is its defect.

## VIRTUE AND HABIT

Moral activities come from virtue but if an individual has done some good activity he cannot be called virtuous on the strength of it. Actually virtue is the habit of doing good work. In the words of Aristotle, virtue is "a state opt to exercise deliberate choice being in the relative mean. determined by reason and as the man of practical wisdom would determine it." The morality of an action depends upon the goodness of its will. The habit of such good volition is virtue.

## VIRTUE IS THE ADOPTION OF THE GOLDEN MEAN

Aristotle treats virtue as the habit of choosing the mean or middle path. According to him, a prudent act is moral. Excess is invariable bad. Intelligent people always adopt the golden mean in respect of any and everything. This habit in choice is virtue. In this way, the virtuous person adopts the middle path in personal as well as social conduct. For example he is neither a coward nor a rash man who invites dangers, he is instead brave and enterprising. He neither runs from difficulties nor wilfully creates them but faces them adamently. In this way a virtuous person is neither extravagent nor stringent. He is liberal and he also conducts himself in a manner befitting his nature. This mean is right for everyone but mean or everyone will be different proportionate to his capability. For example, bravery on the part of soldier may be mischief for an ordinary citizen.

In a general way, Aristotle's theory is a supremely practical theory but it cannot be applied in all circumstances. Muirhead has said quite correctly, "Moderation in all things may be as much of vice as immoderation in one and all."

## VIRTUE IS KNOWLEDGE

Virtue is knowledge, at least for Socrates it is so. This statement means that most immoral activities are the result of ignorance. If some person possesses the knowledge of good he will not stoop to evil doings. Thus it is as natural for a good work to take shape when there is knowledge as it is for it to take shape when there is virtue. Thus knowledge is actually virtue. Ignorance is vice. No person consciously commits crimes. At the base of every bad work there is some ignorance. Plato, too has supported the opinion of Socrates. And many Indian thinkers have presented similar opinions.

## VIRTUE IS ALSO A HABIT

But, in actual fact a man cannot be virtuous by mere knowledge as long as he is not in the habit of performing good activities, and habit does not necessarily reult from mere knowledge. Here, some people can say that if some person indulges in immoral deed even while understanding good then it means that some part of his personality is still ignorant. But the use of such language is an indication of the ignorance of human psychology. Besides the power of comprehension man has affective and volitional aspects too. This volitional aspect is often beyond the control of man's intellectual aspect. One has to inculcate the habit of controlling it

This habit is virtue. Knowledge is not action without which there can be no question of morality. Thus, it is necessary for virtue that knowledge be converted into action. Accordingly, Mackenzie was correct in asserting that, "Virtue is a kind of knowledge as well as a kind of habit". Actually ethics implies that a man is capable of both good and bad, and is responsible for both. To believe that bad activities originate in ignorance is to relieve man of his responsibility in respect of them. Thus Socratic opinion is partially true.

## VIRTUE AND HAPPINESS

Some ethical phiosophers have treated happiness as invariably conjoined to virtue. Happiness is a state of harmony of reason and sentience. Integral self realisation produces happiness. In this way, a virtuous person gets happiness inevitably. He finds happiness in doing good work and abstains from bad work. Thus, a life of virtue is a life of happiness. Aristotle has unflinchingly supported this opinion. According to him happiness is the result of a proper use of reason. Thus, a rational person is always happy. It should here be noted that virtue itself is not happiness nor happiness itself a virtue. Happiness is an indication of virtue.

No one can doubt the intimate relation between happiness and virtue but virtue alone is not enough for happiness. May be in a lesser degree, but happiness does depend upon good results besides good volitions. As Aristotle rightly maintains happiness necessitates too the external instruments along with virtue. Actually, happiness is an indication of self realisation. The higher the self realisation the more will be the happiness. But self realisation is in as much need of external successes and good results as it is of virtue.

## VIRTUE AND VALUE

Virtue is intrinsic value. Virtues are needed for the achievement of values. In this way, virtue and value are intimately related but this is not reason for forgetting the differences between value and virtue. Value certainly requires virtue. But values are instrumental too. Virtues are subjective but values are objective also.

## VIRTUE AND SOCIETY

## VIRTUE IS RELATIVE TO TIME AND PLACE

Virtues keep on changing according to traditions and customs. Thus virtues are relative to time and space. For example, in medieval Rajasthan emphasis was placed upon virile virtues. Today more emphasis is directed to non violence, sacrifices, love, etc., or the sublime virtues.

## VIRTUES ARE RELATIVE TO A PERSON'S STATUS AND ROLE

Different people in society occupy different status and they have different roles to concur to their respective status. Virtues are relative not only to time and place but they also change with the status and roles of person. In this way, in society different people, men, women, young, old, poor, etc., exhibit some difference in qualities, however slight. As with duties of an individual which are determined by his status and role in society, so with virtue which also are determined by his status and role in society.

## VIRTUES IN INDIAN ETHICS

The above classification of virtues specially presents the western viewpoint. From the comparative viewpoint it would be appropriate to introduce here the thoughts on virtue expressed in Indian ethics. In this respect, the thoughts of Buddha, Jaina, Gita, Gandhi, have been expressed in the foregoing description of their moral theories and consequently only the Yoga opinion shall be presented here. In Yoga philosophy, virtues are classified in the notions of yama and niyama and this notion of virtues has generally been accepted in Indian circles.

1. *Yama* (यम)—It is the control of body, mind and will Yamas are five in number—

(*a*) *non-violence* (अहिंसा)— meaning abstaining from injuring any living being. It includes the control of mind, speech and action.

(*b*) *Truth* (सत्य)— meaning being truthful in mind and speech, to retain in the mind just as seen, heard, inferred and to speak it in the same way and hehave accordingly.

(*c*) *Non-stealing* (अस्तेय)— or avoiding taking possession of another's wealth and not stealing it nor entertaining any desire to this effect.

(*d*) *Celibacy* (ब्रह्मचर्य)— not to entertain engrossment in sense organs, especially sex organs.

(*e*) *Non-covetousness* (अपरिग्रह)— meaning avoidance of collecting useless things merely due to greed.

2. *Niyama* (नियम)— the second part of yoga is niyama or the obedience of proper conduct. Niyamas, too, are five in number—

(*a*) *Cleanliness* (शौच)— meaning keeping the external body clean by means of bathing and pure food etc., and keeping the internal mind pure by friendliness, sympathy, indifference etc.

(*b*) *Contentment* (संतोष)— means satisfaction with whatever can be attained by proper means. Mind cannot be peaceful without contentment.

(*c*) *Penance* (तप)— means the practice of enduring the extremes of weather and fulfilling difficult duties. The body hereby becomes capable of enduring hardships.

(*d*) *Self study* (स्वाध्याय)— means a regular study of religious texts.

(*e*) *Meditation on God* (ईश्वर पणिधान)— means contemplating upon God and leaving oneself to Him.

## UNITY OF VIRTUES

According to Plato, "The virtues are not isolated, but are one and whole." Actually all virtues are different aspects of one supreme virtue. Man's ultimate end is self realisation. This is the supreme objective and the ultimate duty. And the habit of obeying this duty is the supreme virtue and all virtues are different aspects of this fundamental virtue. Supporting this opinion, ............ "It is true that the end of all virtues is the same."

# RIGHTS AND DUTIES

Right and duty are relative words. A right from one view becomes a duty in another context. There are duties wherever there are rights.

## MORAL RIGHTS

Man possesses the following moral rights :

### 1. RIGHT TO LIVE

The first and main right of man is his right to live. The object of a moral life is improvement of the self and life is a necessary condition to effecting this improvement. The right to life is a fundamental right. It is a sacred right. The right to life also includes the right to work. Without money, earning of the necessities of life cannot be obtained. The life of a person without occupation becomes difficult. To the right of life is conjoined the duty of respect of life. If everyone has the right to live, it is then his duty to help others in a similar endeavour or at least avoid becoming an obstacle to it. The infringement of this duty can lead to the right of life being taken away. It is for this reason that homicide is rewarded with a sentence of death. Thus man should augment his own life and the lives of others.

But it is to be reçollected that this fundamental right of human beings was not recognised always. In ancient times, infanticide was a common country people of a different sect were murdered while at places prisoners of war were got rid of most expeditiously. The history of the human race is replete with tales of mass murder of defenseless people. Even today millions of people are executed openly under the pretext of revolution or war. This is completely immoral. In the civilised society every person should have the right to live with distinctions of creed, colour, race, physique etc.

## 2. RIGHT TO FREEDOM

The right to freedom, succeeding the right to live is also very important. All moral activities are based on man's free volition. It is impossible to conceive of morality or moral work without freedom. Man's freedom of will is one of the fundamental assumptions or postulates of ethics. For the attainment of the ultimate end, self improvement, the right to freedom is essential. Any person cannot be reduced to being the means of another's benefit. Every individual is himself an end. He is free to attain his objective. Nature gave him life as a free person. He is not the slave of anyone.

But as has been repeatedly said in various contexts in this book, man's freedom does not mean absolute indeterminism, but rather control by the self. One cause of this is that the right to freedom is encumbered by the duty of desisting from becoming an obstacle in the freedom of others. In an organised society individual and social interests do not conflict. Selfishness and selflessness are two aspects of the human self seeking. Uncontrolled misconduct results not from freedom but from servitude of the senses. Freedom really means freedom of acting in regard to right actions. No one has the right to freedom of wrong actions.

## 3. RIGHT OF PROPERTY

The right to freedom attaches with itself the right of property. The right of property is also essential for the right of life. Without wealth, it is impossible to fulfil even the conventional necessities of any person and property is only a form of wealth. Tendency towards mastery over wealth is a natural tendency in man. But it is to be remembered that the misuse of the right of property is an indication not of man's nature but of a distortion of it. It is true that complete destruction of personal property involves many is even necessary for the development of personality. But the right of property is attached to the duty of social welfare. Thus, no man has the right to property to the limit of seriously inconveniencing other members of society. Actually, property and social welfare are so intimately related that Gandhi and other thinkers have persisted in treating people of wealth as trustees of their wealth. Thus the right of property will have to be taken in an extremely limited sense. Some thinkers do not even recognise any such right. According to Plato no less, all people should be able to avail of equal means in an ideal republic and there should not be any right to personal property. According to Aristotle, an ideal state should equip everyone with the right to make free use of wealth for the common good. Marx and Engels objected to the right to personal property. Actually the capitalist system of today has become infested with so many defects that the idea of complete negation of personal property does not seem completely absurd. But it faces many practical difficulties. The conception of the person as a trustiee of wealth, the credit for which goes to Gandhi and Sri Aurobindo, presents a deviation which maintains the right of property but at the same time eleminates the defects of personal property. It is required of an individual that he uses his property for the welfare of society. It hardly need be asserted that it is not wrong to deprive one of the right of property if he infringes the duty of using his wealth for social welfare. It must be kept in view that social welfare includes the person's interest because he is a part of society.

### 4. RIGHT OF FULFILMENT OF CONTRACT

The right of fulfilment of contract is conjoined to the right of property. Actually, this right of fulfilment of contract can have meaning only when the person has right over the means to the contract. The major means among these is wealth. Thus, while on the one hand the right of property is essential for the right to fulfilment of contract, on the other the limits of the right of property apply also to the right of fulfilment of contract. Along with the right of contract, it is the duty of the person to come to a just agreement which can be fulfilled. In this way, no one has the right to agree to a scale of his wife and children nor to his own servitude. Thus the right of contract can be exercised in limited circumstances.

### 5. RIGHT OF EDUCATION

Second in importance only to the right of life is the 'right of education because it is only education which makes a man a rational being in the true sense of the term. Education is the most superior means of developing him from the animal level to the level of human beings. From the moral view, it is incumbent upon every person to receive the best education suited to his abilities. Man's moral objective of self improvement would prove elusive without education. By developing his powers, education makes man capable of self realisation. In society every person should get an opportunity to develop his mental and spiritual powers through education.

## MORAL DUTIES

Similar to the above moral rights, man has also some moral duties. Some of the major ones are as follows :

### 1. RESPECT OF LIFE

It is man's prime duty to respect his own life and the lives of others. Both suicide and homicide are equally serious immoral deeds. Every man has the right to protect his own life but at the same time it is incumbent upon him to protect or at least to desist from injuring the lives of others. Non violence is the ultimate duty. No man should harm the health or life of any other person.

### 2. RESPECT OF FREEDOM

The respect of freedom is contiguous to the right of freedom. In the modern age especially is there an urgent need for stressing this duty because everywhere people are raising their voices against any form of oppression. Every person is himself an end. No one is invested with the authority to make use of him as a means. One's behaviour should be conducted in a manner befitting a man as a human being. Thus, making other slave for oneself and exploiting them is an act of extreme immorality.

### 3. RESPECT OF CHARACTER

Another of man's duties is the respect of character. It is the supreme quality of mankind. It is the basis of his morality. Every person should be an individual and respect all others as individuals. Character is an essential part of personality and from the moral view it is character which is the foundation of personality.

## 4. RESPECT OF PROPERTY

Along with the right of property one is confronted by the duty of respect of property. Every individual possesses the right to make a constructive or proper use of his personal property. Thus we should not interfere with others doing so. Theft is a moral crime for this reason. No one has the right to appropriate another's property by wrong methods. Thus to the respect of property is attached the idea of the propriety of ends.

Thus, it is the duty of every person to respect the social order. His doing so will ensure the welfare of both society and the individual. A person's rights can be protected only if some social order is maintained. If the social order is shattered, the rights of everyone will be endangered and the moral development of the person shall be in jeopardy. A proper moral development of an individual can be the outcome only of an organised social structure. Thus social order is a sacred organisation. It should be respected by all.

## 5. RESPECT OF TRUTH

The respect of truth is attached to the right of contract. It is a moral maxim that you will refrain from false statement. A man should be faithful to his promises which he should fulfil. Every person should incorporate thought and promise as well as promise and action. It is for him to say what he thinks and to do what he has said. Thus it is a moral duty to respect truth in thought, speech and action.

## 6. RESPECT OF PROGRESS

Ethics places its faith in the progress of mankind. Every person should evince a faith in progress and respect progress. Work is worship. Society can progress only if man does so. The actual extent of progress in society is a debatable question but it is man's duty and supreme morality to exert himself for progress without getting involved in this question.

# CODES OF ETHICS

Every profession including public sector, organisation, business, Government, private etc. have their principles/norms/codes and these cannot expect any professional to do anything that goes against their code. Codes of ethics are concerned with all including Data privacy, accountability, software piracy, academic honesty etc. They define the ideal behaviour of a professional for the purpose of enhancing the public image. Code of ethics also establish rules of conduct & encourage value – laden decisions for concerns. It also provide a framework for ethical judgement for professionals. These (codes of ethics) also help any body to arive at a proper good ethical choice. It is not a legal document.

Every code is a formalization of experience. Adoption of code is because of it accept the adherence by people. The code of ethics sets the standards in the field and mechanism to maintain autonomy. It is expected that each profession has a code of ethics. People concern must adhere to the code, no matter what their employment context. No two code of ethics are identical, they vary with culture group, profession & discipline. Though medieval guilds codified their rules of conduct, but no carry information about professional code is available. Therefore, their chronological order & development can not be expressed.

The codes are expected to play an important role in following fields —

(*i*) Protection of status

(*ii*) Promotion of business

(*iii*) Difference and discipline

(*iv*) Inspiration & guidance

(*v*) Support

(*vi*) Contribution to the profession

(*vii*) Education & mutual understanding.

## NECESSITY & PURPOSE OF CODES OF ETHICS

Essentially all members of the profession take another to abide by the code, when they join as member. Why? So that they can serve the people as effectively as possible, without any fear & improving their abilities. Code of ethics inform an individual that what its expect from professional. It is true that without cose of established ethics professional organisation can be considered. The reasons for the existence of code of ethics for professionals may be summarised below :

(*i*) Codes should be meaningful and should set up a benchmark of minimal duty and specify clear and enforceable rules.

(*ii*) Codes should denote particular profession.

(*iii*) Codes should indicate the maturity of a profession and provide professional identity as well.

(*iv*) It should provide freedom of speech.

(*v*) Codes should have role in promoting understanding among those affected by the codes, guidance and educational functions.

(*vi*) Codes protect consumers and further the professional stance of the organisations.

(*vii*) Codes protect the profession from Govt. and function autonomously instead of being controlled by legislation.

(*viii*) Codes provides guidlines in some situations/areas that are not covered by law.

(*ix*) There should be tolerance towards criticism of the codes themselves. Codes should help to increase trust with Govt. & public.

(*x*) Codes promote stability, within the profession.

(*xi*) Codes offer the professional a framework of guidelines even in the critical situations.

(*xii*) Ethical codes provide some measures of defence for the professional in case of litigation, who practice in accordance with accepted professional codes.

(*xiii*) Codes also protect practitioners from the public to malpractice suits & protect the public from quakes.

## LIMITATIONS OF CODES OF ETHICS

We have seen that codes of ethics play an important role in professions to make certain decisions and guiding professionals for good practice/working. But there are certain limitations, which are as –

(*i*) Codes should not be contradictory, landing the professional in great dilemma. It should be so clear that in case of conflict, one can take the decision in the interest of public.

(*ii*) Though it is not recipe for making any decision but there should be situations, which can be addressed in the context of the code. They are not there.

(*iii*) In the developmental bodies where changes are possible, some of the codes do not suit at every occassion. Therefore, skilful and creative applications of codes guidlines are necessary to serve the need of an individual or group.

(*iv*) Ethical codes do not address every possible issue or codes generally cover the limitted topics.

(*v*) As we know codes are not the legal document hence we can not hope for good judgement.

(*vi*) The enforcement of ethical codes is not red easy, therefore, public should believe the committee as their peers.

(*vii*) The code items should include public interest, truth, honest, and professional performance.

(*viii*) The code inforce at a given time should not be taken as the final words. It should be open for healthy debate.

(*ix*) There is often difficulty in bringing the interests of all parties involved in an ethical dispute together systematically.

(*x*) Some times conflicts may arise between ethical and legal codes.

(*xi*) Some legal and ethical issues are not covered in codes.

(*xii*) Codes are historical documents hence some times they are acceptable & some times not.

(*xiii*) Ethical codes do not address cross-cultural issues.

(*xiv*) If some of the professionals are not the members of the societies, then why these will abide by the codes. (eg. engineers)

(*xv*) In case of interval conflicts in the codes, but there no method to resolve them.

(*xvi*) A code cannot substitute for good lines of communication between employees and upper management.

Inspite of the above limitations / objections, the codes of ethics are in widespread use to day and are generally thought to serve in useful way. Codes provide only general philosophical guidance for conceptualising and responding to conflicts.

## CODES OF PROFESSIONAL ETHICS

Most of the institution and organisations have their code of professional ethics. Inspite of having some limitations in their codes, they are useful advantageous & purposeful. Some of the organisations & institutions which have their code of professional ethics are as follows :

1. The American Association of Computing Machinery (ACM) Code of Ethics and Professional Conduct.
2. The Institution of Engineers (India) Code of Ethics.
3. The American Medical Association Principles of Medical Ethics.
4. The American Psychological Association's Ethics.
5. Association for Information Systems (AIS) Code of Research Conduct.
6. The Australian Computer Society Code of Ethics.
7. The British Computer Society (BCS) Code of Conduct Code of Practice.
8. Canadian Information Processing Society-Professionalism Computer Society of South Africa.
9. The Hong Kong Computer Society's Code of Ethics.
10. Institution of Electrical and Electronics Engineers (IEEE) Code of Ethics.
11. The Institute for the Management of Information Systems Code of Professional Conduct.

12. Librarian and Information Manager Associations Ethics.
13. Singapore Computer Society Professional Code of Conduct.

## INSTITUTION OF ENGINEERS (INDIA) CODE OF ETHICS

This was effective from 1st March 2004. The important points in code of ethics for Institution of Engineers (India) is as —

(*i*) Engineers serve all members of the community in enhancing their welfare, health and safety by a creative process utilizing the engineers' knowledge, expertise and experience.

(*ii*) Pursuant to the avowed objectives of the Institution of Engineers (India) as enshrined in the presents of the Royal Charter granted to the Institution, the Council of the Institution prescribed a set of "Professional Conduct Rules" in the year 1944 replacing the same with the "Code of Ethics for Corporate Members" in the year 1954 which was revised in the year 1997.

(*iii*) In view of globalisation, concern for the environment and the concept of sustainable development, it has been felt that the prevailing "Code of Ethics for Corporate Members" needs review and revision in letter and spirit. The engineering organisations world over have updated their Code of Ethics.

(*iv*) The Council of the Institution vested with the authority in terms of the Present 2(j) of the Royal Charter adopted at its 656th meeting held on 21.12.2003 at Lucknow the "Code of Ethics for Corporate Members" as provided hereinafter.

The Code of Ethics is based on broad principles of truth, honesty, justice, trustworthiness, respect and safeguard of human life and welfare, competence and accountability which constitute the moral values every Corporate Member of the Institution must recognize, uphold and abide by.

This "Code of Ethics for Corporate Members" shall be in force till the same is revised by decision of the Council of the Institution.

## CODE OF ETHICS FOR CORPORATE MEMBERS

**1.0 Preamble**

1.1 The Corporate Members of the Institution of Engineers (India) are committed to promote and practice the profession of engineering for the common good of the community bearing in mind the following concerns:

1.1.1 Concern for ethical standard;

1.1.2 Concern for social justice, social order and human rights;

1.1.3 Concern for protection of the environment;

1.1.4 Concern for sustainable development;

1.1.5 Public safety and tranquility.

**2.0 The Tenets of the Code of Ethics**

2.1 A Corporate Member shall utilise his knowledge and expertise for the welfare, health and safety of the community without any discrimination for sectional or private interests.

2.2 A Corporate Member shall maintain the honor, integrity and dignity in all his professional actions to be worthy of the trust of the community and the profession.

2.3 A Corporate Member shall act only in the domains of his competence and with diligence, care, sincerity and honesty.

2.4 A Corporate Member shall apply his knowledge and expertise in the interest of his employer or the clients for whom he shall work without compromising with other obligations to these Tenets.

2.5 A Corporate Member shall not falsify or misrepresent his own or his associates' qualifications, experience, etc.

2.6 A Corporate Member, wherever necessary and relevant, shall take all reasonable steps to inform himself, his employer or clients, of the environmental, economic, social and other possible consequences, which may arise out of his actions.

2.7 A Corporate Member shall maintain utmost honesty and fairness in making statements or giving witness and shall do so on the basis of adequate knowledge.

2.8 A Corporate Member shall not directly or indirectly injure the professional reputation of another member.

2.9 A Corporate Member shall reject any kind of offer that may involve unfair practice or may cause avoidable damage to the ecosystem.

2.10 A Corporate Member shall be concerned about and shall act in the best of his abilities for maintenance of sustainability of the process of development.

2.11 A Corporate Member shall not act in any manner which may injure the reputation of the Institution or which may cause damage to the Institution financially or otherwise.

**3.0 General Guidance**

The Tenets of the Code of Ethics are based on the recognition that

3.1 A common tie exists among the humanity and that The Institution of Engineers (India) derives its value from the people, so that the actions of its Corporate Members should indicate the member's highest regard for equality of opportunity, social justice and fairness;

3.2 The Corporate Members of the Institution hold a privileged position in the community so as to make it a necessity for their not using the position for personal and sectional interests.

**4.0 And, as such, a Corporate Member**

4.1 should keep his employer or client fully informed on all matters in respect of his assignment which are likely to lead to a conflict of interest or when, in his judgment, a project will not be viable on the basis of commercial, technical, environmental or any other risks;

4.2 should maintain confidentiality of any information with utmost sincerity unless expressly permitted to disclose such information or unless such permission, if withheld, may adversely affect the welfare, health and safety of the community.

4.3 should neither solicit nor accept financial or other considerations from anyone related to a project or assignment of which he is in the charge;

4.4 should neither pay nor offer direct or indirect inducements to secure work;

4.5 should compete on the basis of merit alone;

4.6 should refrain from inducing a client to breach a contract entered into with another duly appointed engineer;

4.7 should, if asked by the employer or a client, to review the work of another person or organization, discuss the review with the other person or organization to arrive at a balanced opinion;

4.8 should make statements or give evidence beforea tribunal or a court of law in an objective and accurate manner and express any opinion on the basis of adequate knowledge and competence; and

4.9 should reveal the existence of any interest pecuniary or otherwise which may affect the judgment while giving an evidence or making a statement.

5.0 Any decision of the Council as per provisions of the relevant Bye-Laws of the Institution shall be final and binding on all Corporate Members.

## IEEE CODE OF ETHICS

(IEEE means the Institute of Electrical and Electronics Engineers)

We, the members of the IEEE, in recognisation of the importance of our technologies in affecting the quality of life throughout the world, and in accepting a personal obligation to our profession, its members and the communities we serve, do here by commit ourselves to the highest ethical and professional conduct and agree:

1. to accept responsibility in making decisions consistent with the safety, health and welfare of the public, and to disclose promptly factors that might endanger the public or the environment.
2. to avoid real or perceived conflicts of interest when ever possible, and to disclose them to affected parties when they do exist.
3. to be honest and realistic in stating claims or estimates based on available data.
4. to reject bribery in all its forms.
5. to improve the understanding of technology, its appropriate application, and potential consequences.
6. to maintain and improve our technical competence and to undertake technological tasks for others only if qualified by training or experience, or after full disclosure of pertinent limitations.
7. to seek, accept, and offer honest criticism of technical work, to acknowledge and correct errors, and to credit properly the contributions of others.
8. to treat fairly all persons regardless of such factors as race, religion, gender, disability, age, or national origin.
9. to avoid injuring others, their property, reputation, or employment by false or malicious action.
10. to assist colleagues and co-workers in their professional development and to support them in following this code of ethics.

(*Approved by the IEEE Board of Directors February 2006*)

## ACM CODE OF ETHICS AND PROFESSIONAL CONDUCT

The American Association of Computing Machinery adopted the Code of Ethics and Professional Conduct 1972. Twenty years later, in May 1992, the ACM published a proposal for a revised and extended version. **On 16th October 1992**, the proposal, with only minor modifications, was adopted by the executive council of ACM as its new official Code of Ethics and Professional Conduct. Commitment to ethical professional conduct is expected of every member (voting members, associate members, and student members) of the American Association of Computing Machinery (ACM).

ACM Code of ethics & professional conduct contains most of the points/issues/articles. But some of the issues are likely to face by professionals. It consists of twenty four imperatives formulated as statements of personal responsibility which identifies the elements of commitment to ethical professional conduct by its every member. These twenty four imperatives are discussed under four broad imperatives or sections, namely :

(I) General Moral Imperatives;

(II) More Specific Professional Responsibilities;

(III) Organizational Leadership Imperatives; and

(IV) Compliance with the code.

Primarily, these imperatives are intended to serve as a basis for ethical decision making in the conduct of professional work. Secondarily they may serve as a basis for judging the merit of a formal complaint pertaining to violation of professional ethical standards.

## (I) GENERAL MORAL IMPERATIVES

General moral imperatives (Section I) outline the fundamental ethical considerations. They are stated as under :

**1. *Contribute to society and human well-being*** : This principle concerning the quality of life of all people affirms an obligation to protect fundamental human rights and to respect the diversity of all cultures. An essential aim of computing professionals is to minimize negative consequences of computing systems, including threats to health and safety. When designing or implementing systems, computing professional must attempt to ensure that the products of their efforts will be used in socially responsible ways, will meet social needs, and will avoid harmful effects to health and welfare.

In addition to a safe social environment, human well-being includes a safe natural environment. Therefore, computing professionals who design and develop systems must be alert to, and make others aware of, any potential damage to the local or global environment.

**2. *Avoid harm to others*** : "Harm" means injury or negative consequences, such as undesirable loss of information, loss of property, property damage, or unwanted environmental impacts. This principle prohibits use of computing technology in ways that result in harm to any of the following: users, the general public, employees, employers. Harmful actions include intentional destruction or modification of files and programs leading to serious loss of resources or unnecessary expenditure of human resources such as the time and effort required to purge systems of "computer viruses".

Well-intended actions, including those that accomplish assigned duties, may lead to harm unexpectedly. In such an event the responsible person or persons are obligated to undo or mitigate the negative consequences as much as possible. One way to avoid unintentional harm is to carefully consider potential impacts on all those affected by decisions made during design and implementation.

To minimize the possibility of indirectly harming others, computing professionals must minimize malfunctions by following generally accepted standards for system design and testing. Furthermore, it is often necessary to assess the social consequences of systems to project the likelihood of any serious harm to others. If system features are misrepresented to users, coworkers, or supervisors, the individual computing professional is responsible for any resulting injury.

In the work environment the computing professional has the additional obligation to report any signs of system dangers that might result in serious personal or social damage. If one's superiors do not act to curtail or mitigate such dangers, it may be necessary to "blow the whistle" to help correct the problem or reduce the risk. However, capricious or misguided reporting of violations can, itself, be harmful. Before reporting violations, all relevant aspects of the incident must be thoroughly assessed. In particular, the assessment of risk and responsibility must be credible.

**3. *Be honest and trustworthy*** : Honesty is an essential component of trust. Without trust an organization cannot function effectively. The honest computing professional will not make deliberately false or deceptive claims about a system or system design, but will instead provide full disclosure of all pertinent system limitations and problems.

A computer professional has a duty to be honest about his or her own qualifications, and about any circumstances that might lead to conflict of interest.

Membership in volunteer organizations such as ACM may at times place individuals in situations where their statements or actions could be interpreted as carrying the "weight" of a larger group of professionals. An ACM member will exercise care to not misrepresent ACM or positions and policies of ACM or any ACM units.

**4. *Be fair and take action not to discriminate*** : The values of equality, tolerance, respect for others, and the principles of equal justice govern this imperative. Discrimination on the basis of race, sex, religion, age, disability, national origin, or other such factors is an explicit violation of ACM policy and will not be tolerated.

Inequities between different groups of people may result from the use or misuse of information and technology. In a fair society, all individuals would have equal opportunity to participate in, or benefit from, the use of computer resources regardless of race, sex, religion, age, disability, national origin or other such similar factors. However, these ideals do not justify unauthorized use of computer resources nor do they provide an adequate basis for violation of any other ethical imperatives of this code.

**5. *Honour property rights including copyrights and patent*** : Violation of copyrights, patents, trade secrets and the terms of license agreements is prohibited by law in most circumstances. Even when software is not so protected, such violations are contrary to professional behaviour. Copies of software should be made only with proper authorization. Unauthorized duplication of materials must not be condoned.

**6. *Give proper credit for intellectual property*** : Computing professionals are obligated to protect the integrity of intellectual property. Specifically, one must not take credit for other's ideas or work, even in cases where the work has not been explicitly protected by copyright, patent, etc.

**7. *Respect the privacy of others*** : Computing and communication technology enables the collection and exchange of personal information on a scale unprecedented in the history of civilization. Thus there is increased potential for violating the privacy of individuals and groups. It is the responsibility of professionals to maintain the privacy and integrity of data describing individuals. This includes taking precautions to ensure the accuracy of data, as well as protecting it from unauthorized access or accidental disclosure to inappropriate individuals. Furthermore, procedures must be established to allow individuals to review their records nd correct inaccuracies.

**8. *Honour confidentiality*** : The principle of honesty extends to issues of confidentiality of information whenever one has made an explicit promise to honour confidentiality or, implicitly, when private information not directly related to the performance of one's duties becomes available. The ethical concern is to respect all obligations of confidentiality to employers, clients, and users unless discharged from such obligations by requirements of the law or other principles of this Code.

## (II) MORE SPECIFIC PROFESSIONAL RESPONSIBILITIES

1. ***Strive to achieve the highest quality, effectiveness and dignity in both the process and products of professional work*** : Excellence is perhaps the most important obligation of a professional. The computing professional must strive to achieve quality and to be cognizant of the serious negative consequences that may result from poor quality in a system.

2. ***Acquire and maintain professional competence*** : Excellence depends on individuals who take responsibility for acquiring and maintaining professional competence. A professional must participate in setting standards for appropriate levels of competence, and strive to achieve those standards. Upgrading technical knowledge and competence can be achieved in several ways: doing independent study; attending serminars, conferences, or courses; and being involved in professional organizations.

3. ***Know and respect existing laws pertaining to professional work*** : ACM members must obey existing local, state, province, national, and international laws unless there is a compelling ethical basis not to do so. Policies and procedures of the organizations in which one participates must also be obeyed. But compliance must be balanced with the recognition that sometimes existing laws and rules may be immoral or inappropriate and, therefore, must be challenged. Violation of a law or regulation may be ethical when that law or rule has inadequate moral basis or when it conflicts with another law judged to be more important. If one decides to violate a law or rule because it is viewed as unethical, or for any other reason, one must fully accept responsibility for one's actions and for the consequences.

4. ***Accept and provide appropriate professional review*** : Quality professional work, especially in the computing profession, depends on professional reviewing and critiquing. Whenever appropriate, individual members should seek and utilize peer review as well as provide critical review of the work of others.

5. ***Give comprehensive and thorough evaluations of computer systems and their impacts, including analysis of possible risks*** : Computer professionals must strive to be perceptive, thorough, and objective when evaluating, recommending, and presenting system descriptions and alternatives. Computer professionals are in a position of special trust, and therefore have a special responsibility to provide objective, credible evaluations to employers, clients users, and the public.

6. ***Honour contracts, agreements, and assigned responsibilities*** : Honouring one's commitments is a matter of integrity and honesty. For the computer professional this includes ensuring that system elements perform as intended. Also, when one contracts for work with another party, one has an obligation to keep that party properly informed about progress toward completing that work.

A computing professional has a responsibility to request a change in any assignment that he or she feels cannot be completed as defined. Only after serious consideration and with full disclosure of risks and concerns to the employer or client, should one accept the assignment. The major underlying principle here is the obligation to accept personal accountability for professional work. On some occasions other ethical principles may take greater priority.

A judgement that a specific assignment should not be performed may not be accepted. Having clearly identified one's concerns and reasons for that judgment, but failing to procure a change in that assignment, one may yet be obligated, by contract or by law, to proceed as directed. The computing professional's ethical judgment should be the final guide in deciding whether or not to proceed. Regardless of the decision, one must accept the responsibility for the consequences.

However, performing assignments "against one's own judgment" does not relieve the professional of responsibility for any negative consequences.

7. ***Improve public understanding of computing and its consequences*** : Computing professionals have a responsibility to share technical knowledge with the public by encouraging understanding of computing, including the impacts of computer systems and their limitations. This imperative implies an obligation to counter any false views related to computing.

8. ***Access computing and communication resources only when authorized to do so*** : Theft or destruction of tangible and electronic property is prohibited by imperative I(2) - "Avoid harm to others." Trespassing and unauthorized use of a computer or communication system is addressed by this imperative. Trespassing includes accessing communication networks and computer systems, or accounts and/or files associated with those systems, without explicit authorization to do so. Individuals and organizations have the right to restrict access to their systems so long as they do not violate the discrimination principle. No one should enter or use another's computer system, software, or data files without permission. One must always have appropriate approval before using system resources, including communication ports, file space, other system peripherals, and computer time.

## (III) ORGANIZATIONAL LEADERSHIP IMPERATIVES

These imperatives (Section 3) pertain more specifically to individuals who have a leadership role, whether in the workplace or in a volunteer capacity. They are stated as under :

1. ***Articulate social responsibilities of members of an organizational unit and encourage full acceptance of those responsibilities*** : Because organizations of all kinds have impacts on the public, they must accept responsibilities to society. Organizational procedures and attitudes oriented toward quality and the welfare of society will reduce harm to members of the public, thereby serving public interest and fulfilling social responsibility. Therefore, organizational leaders must encourage full participation in meeting social responsibilities as well as quality performance.

2. ***Manage personnel and resources to design build information systems that enhance the quality of working life*** : Organizational leaders are responsible for ensuring that computer systems enhance, not degrade, the quality of working life. When implementing a computer system, organizations must consider the personal and professional development, physical safety, and human dignity of all workers. Appropriate human-computer ergonomic standards should be considered in system design and in the workplace.

3. ***Acknowledge and support proper and authorized uses of an organization's computing and communication resources*** : Because computer systems can become tools to harm as well as to benefit an organization, the leadership has the responsibility to clearly define appropriate and inappropriate uses of organizational computing resources. While the number and scope of such rules should be minimal, they should be fully enforced when established.

4. ***Ensure that users and those who will be affected by a system have their needs clearly articulated during the assessment and design of requirements; later the system must be validated to meet requirements*** : Current system users, potential users and other persons whose lives may be affected by a system must have their needs assessed and incorporated in the statement of requirements. System validation should ensure compliance with those requirements.

5. ***Articulate and support policies that protect the dignity of users and others affected by a computing system*** : Designing or implementing systems that deliberately or inadvertently demean individuals or groups is ethically unacceptable. Computer professionals who are in decision making positions should verify that systems are designed and implemented to protect personal privacy and enhance personal dignity.

**6. *Create opportunities for members of the organization to learn the principles and limitations of computer systems*** : Educational opportunities are essential to facilitate optimal participation of all organizational members. Opportunities must be available to all members to help them improve their knowledge and skills in computing, including courses that familiarize them with the consequences and limitations of particular types of systems. In particular, professionals must be made aware of the dangers of building systems around oversimplified models, the improbability of anticipating and designing for every possible operating condition, and other issues related to the complexity of this profession.

### (IV) COMPLIANCE WITH THE CODE

Principles involving compliance with this Code are given in these imperatives (Section IV). They are stated as under :

**1. *Uphold and promote the principles of this Code*** : The future of the computing profession depends on both technical and ethical excellence. Not only is it important for ACM computing professionals to adhere to the principles expressed in this Code, each member should encourage and support adherence by other members.

**2. *Treat violations of this Code as inconsistent with membership in the ACM*** : Adherence of professionals to a code of ethics is largely a voluntary matter. However, if a member does not follow this code by engaging in gross misconduct, membership in ACM may be terminated.

## QUESTIONS

1. Whats do you mean by "Ethics"?
2. Differentiate between, moral, values & ethics.
3. What are ethical situations ?
4. Define professional & environmental ethics.
5. Is duty not the objective morality ? Comment upon.
6. Write note on utilitarianism.
7. Explain moral values.
8. What is organizational leadership ?
9. What are general principles of ethics ? Discuss.
10. Discuss purpose of codes of ethics.
11. What is quasi - professional ethics ? Explain.
12. Justify that ethics is essential for man making & character building ?
13. What is role morality ? How conflicts come in its way ? Discuss to deal with this.
14. Write an essay on Codes of ethics.
15. Discuss the statement. "Virtues are habits".
16. Write short notes on —
    (*i*) Virtue ethics
    (*ii*) Moral values
    (*iii*) Ethical situations
    (*iv*) Right ethics
    (*v*) Duty ethics.
17. Write note on ACM Code of Ethics & professional conduct.
18. Discuss the attributes of organization culture.
19. Differentiate between non-professional, quasi professional and hard professional.

# APPENDIX – 1

## Rio-Declaration on Environment and Development (Earth Charter)

The United Nations Conference on Environment and Development (UNCED) which is also known as "The Earth Summit" took place between June 3 to June 14, 1992 in Rio de Janeiro, the capital of Brazil. The summit marks the 20th anniversary of the first UN conference on Human Environment which took place from June 5 to June 10, 1972 at Stockholm and the founding of the United Nations Environment Programme (UNEP). The conference had generated a new global awareness and debate on issues concerning evironment and ecology and the inter-relationship between man and nature.

The Earth Summit, Agenda 21 of the United Nations programme of Action from Rio, while reaffirming the Stockholm Declaration of 1972 and recognising the integral and independent nature of our mother Earth, and with the goal of establishing a new and equitable partnership through the creation of new levels of cooperation among States, Key sectors of the Society and people, proclaimed the following 27 principles :

**Principle 1** - Human beings are at the centre of concerns for sustainable development. They are entitled to a healthy and productive life in harmony with nature.

**Principle 2** - States have, in accordance with the chapter of the United Nations and the principles of international law, the sovereign right to exploit their own resources pursuant to their own environment and policies, and the responsibility to ensure that activities within their jurisdiction or control do not cause damage to the environment of other states or areas beyond the limits of national jurisdiction.

**Principle 3** - The right to development must be fulfilled so as to equitably meet development and environmental needs of present and future generations.

**Principle 4** - In order to achieve sustainable development, environmental protection shall constitute in integral part of the development process and can not be considered in isolation from it.

**Principle 5** - All States and all people shall co-operate in the essential task of eradicating poverty as an indispensable requirement for sustainable development.

**Principle 6** - The special situation and needs of developing countries, particularly the least developed and those most environmentally vulnerable, shall be given special priority. International actions in the fields of environment and development should also address the interests and needs of all countries.

**Principle 7** - States shall co-operate in a spirit of global partnership to conserve, protect and restore the health and integrity of the earth's eco-system. In view of the different contributions to global environmental degradation, states have common but differentiated responsibilities. The developed countries acknowledge the responsibility that they bear in the international pursuit of sustainable development in view of the pressure their societies place on the global environment and of the technologies and financial resources they command.

**Principle 8** - To achieve sustainable development and a higher quality of life of all people, states should reduce and eliminate unsustainable patterns of production and consumption and promote appropriate demographic policies.

**Principle 9** - States should co-operate to strengthen indigenous capacity building for sustainable development by improving scientific understanding through exchanges of scientific and technological knowledge, and by enhancing the development, adaption, diffusion and transfer of technologies, including new and innovative technologies.

**Principle 10** - Environmental issues are best handled with the participation of all concerned citizens at the relevant level. At the national level, each individual shall have appropriate access to information concerning the environment that is held by public authorities, including information on hazardous materials and activities in their communities, and the opportunity to participate in decision making processes. States shall facilitate and encourage public awareness and participation by making information widely available. Effective access to judicial and administrative proceedings, including redress and remedy, shall be provided.

**Principle 11** - States shall enact effective environmental legislation. Environmental standards, management objectives and priorities should reflect the environmental and development context to which they apply. Standards applied by some countries may be inappropriate and of unwarranted economics and social cost of other countries in particular developing countries.

**Principle 12** - States should co-operate to promote a supportive and open international economic system that would lead to economic growth and sustainable development in all countries, to better address the problems of environmental degradation. Trade policy measures for environmental purposes should not constitute a means of arbitrary or unjustifiable discrimination or a disguised restriction on international trade. Unilateral actions to deal with environmental challenges outside the jurisdiction of the importing country should be avoided. Environmental measures addressing trans-boundary or global environmental problems, should, as far as possible, be based on an international consensus.

**Principle 13** - States shall develop national law regarding liability and compensation for the victims of pollution and other environmental damage. States shall also co-operate in an expenditious and more determined manner to develop further international law regarding liability and compensation for adverse effects of environmental damage caused by activities within their jurisdiction or control to areas beyond their jurisdiction.

**Principle 14** - The state should effectively co-operate to discourage or pevent the relocation and transfer to other states of any activities and substances that cause severe environmental degradation or are found to be harmful to human health.

**Principle 15** - In order to protect the environment, the precautionary approach shall be widely applied by states according to their capabilities. Where there are threats of serious or irreversible damage, lack of full scientific certainty shall not be used as a reason for postponing cost effective measures to prevent environmental degradation.

**Principle 16** - National authorities should endeavour to promote the internalization of environmental costs and the use of economic instruments, taking into account the approach that the polluter should, in principle, bear the cost of pollution, with due regard to the public interest and without distorting international trade and investment.

**Principle 17** - Environmental impact assessment, as a national instrument, shall be undertaken for proposed activities that are likely to have a significant adverse impact on the environment and are subject to a decision of a competent national authority.

**Principle 18** - States shall immediately notify other states of any natural disaster or other emergencies that are likely to produce sudden harmful effect on the environment of those states. Every effort shall be made by the international community to help states so afflicted.

**Principle 19** - States shall provide prior and timely notification and relevant information to potentially affected states on activities that may have a significant adverse trans-boundary environmental effect and shall consult with those states at an early stage and in good faith.

**Principle 20** - Women have a vital role in environmental management and development. Their full participation, is, therefore, essential to achieve sustainable development.

**Principle 21** - The creativity, ideals, and courage of the youth of the world should be mobilized to forge a global partnership in order to achieve sustainable development and ensure a better future to all.

**Principle 22** - Indigeneous people and their communities, and other local communities, have a vital role in environmental management and development, because of their knowledge and traditional practices, states should recognize and duly support their identity, culture and interests and enable their effective participation in the achievement of sustainable development.

**Principle 23** - The environmental and natural resources of people under oppression, domination and occupation shall be protected.

**Principle 24** - Warfare is inherently destructive of sustainable development. States shall, therefore, respect international law providing protection for the environment in times of armed conflict and co-operate in its further development as necessary.

**Principle 25** - Peace, development and environmental protection are inter dependent and indivisible.

**Principle 26** - States shall resolve all their environmental disputes peacefully and by appropriate means in accordance with the charter of the United Nations.

**Principle 27** - States and people shall co-operate in good faith and in a spirit of partnership in the fulfilment of the principles embodied in this declaration and in the further development of international law in the field of sustainable development.

**Appendix-2 : Drinking Water Standards (IS : 10500 - 1991)**

| S. No. | Substance or Characteristic | Requirement (Described limit) | Undesirable effects outside the desirable limit | Permissible limit in absence of alternate source |
|---|---|---|---|---|
| (1) | (2) | (3) | (4) | (5) |
| **Essential Characteristics** | | | | |
| 1. | Colour, Hazen unit | 5 | Above, consumer acceptance decreases | 25 |
| 2. | Odour | | Unobjectionable | -- |
| 3. | Taste | | Agreeable | -- |
| 4. | Turbidity, NTU | 5 | Above, consumer acceptance decreases | 10 |
| 5. | pH value | 6.5-8.5 | Beyond this range the water will affect the mucous membrane and/or water supply system | No relaxation |
| 6. | Total hardness, mg/L as $CaCO_3$ | 300 | Encrustation on water supply structure and adverse effects on domestic use | 600 |
| 7. | Iron (as Fe), mg/L | 0.3 | Beyond this limit, tste/appearance are affected, has adverse effect on domestic uses and water supply structures, and promotes iron bacteria. | 1.0 |
| 8. | Chlorides (as Cl), mg/l | 250 | Beyond this limit, taste, corrosion and palatability are affected | 1000 |
| 9. | Residual free chlorine, mg/L | 0.2 | -- | -- |
| **Desirable Characteristics** | | | | |
| 10. | Dissolved Solids mg/L | 500 | Beyond this palatability decrease and may cause gastrointestinal irritation | 2000 |
| 11. | Calcium (as Ca), mg/L | 75 | -- | 200 |
| 12. | Copper (as Cu), mg/L | 0.05 | Astringent, taste discoloration of pipes, fitting and utensils will be caused beyond this | 1.5 |

| S. No. | Substance or Characteristic | Requirement (Described limit) | Undesirable effects outside the desirable limit | Permissible limit in absence of alternate source |
|---|---|---|---|---|
| (1) | (2) | (3) | (4) | (5) |
| 13. | Manganese (as Mn) mg/L | 0.1 | Astringent taste, discoloration of pipes, fitting and utensils will be caused beyond this | 0.3 |
| 14. | Sulphates (as $SO_4$), mg/L | 200 | Beyond this causes gastro intestinal irritation when magnesium or sodium are present | 400 |
| 15. | Nitrates (as $NO_3$) mg/L | 45 | Beyond this methaemoglobinemia may be caused in infants | 100 |
| 16. | Fluoride (as F) mg/L | 1.0 | Fluoride may be kept as low as possible High fluoride may cause fluorosis | 1.5 |
| 17. | Phenolic substances, mg/L as ($C_6H_5OH$) | 0.001 | Beyond this, it may cause objectionable taste and odour | 0.002 |
| 18. | Mercury (as Hg), mg/L | 0.001 | Beyond this, the water becomes toxic | No relaxation |
| 19. | Cadmium (as Cd) mg/L | 0.01 | Beyond this, the water becomes toxic | No relaxation |
| 20. | Selenium (as Se) mg/L | 0.01 | Beyond this, the water becomes toxic | No relaxation |
| 21. | Arsenic (As) mg/L | 0.05 | Beyond this, the water becomes toxic | No relaxation |
| 22. | Cyanide (CN), mg/L | 0.05 | Beyond this, the water becomes toxic | No relaxation |
| 23. | Lead (Pb), mg/L | 0.05 | Beyond this, the water becomes toxic | No relaxation |
| 24. | Zinc (as Zn) mg/L | 5 | Beyond this limit it can cause astringent taste and an opalescence in water | 15 |
| 25. | Anionic detergents, mg/L (as MBAS) | 0.2 | Beyond this limit, it can cause a light froth in water | 1.0 |

| S. No. | Substance or Characteristic | Requirement (Described limit) | Undesirable effects outside the desirable limit | Permissible limit in absence of alternate source |
|---|---|---|---|---|
| (1) | (2) | (3) | (4) | (5) |
| 26. | Hexavalent Chromium, mg/L | 0.01 | May be carcinogenic above this limit | 0.05 |
| 27. | Polynuclear aromatic hydrocarbons (as (PAH), mg/L | -- | May be carcinogenic | -- |
| 28. | Mineral oil, mg/L. | 0.01 | Beyond this limit undesirable taste and odour after chlorination takes place. | 0.03 |
| 29. | Pesticides, mg/L | Absent | Toxic | 0.001 |
| 30. | Radioactive materials | | | |
| | (a) Alpha emitters Bq/L | -- | -- | 0.1 |
| | (b) Beta emitters pci/L | -- | -- | 1.0 |
| 31. | Alkalinity (as $CaCO_3$), mg/L | 200 | Beyond this limit taste becomes unpleasant | 600 |
| 32. | Aluminium (as Al), mg/L | 0.03 | Cumulative effect is reported to cause dementia | 0.2 |
| 33. | Boron (as B), mg/L | 1 | -- | 5 |

## Appendix-3 : National Ambient Air Quality Standards (NAAQS) (1994)

| Pollutant | Time weighted average | Concentration in ambient air | | | |
|---|---|---|---|---|---|
| | | Industrial area | Residential Rural & mixed use area | Sensitive area | Method of measurement |
| Sulphur dioxide ($SO_2$) | Annual average | 80 μg/m$^3$ | 60 μg/m$^3$ | 15 μg/m$^3$ | - Improved West & Gaeke method |
| | 24 hours | 120 μg/m$^3$ | 80 μg/m$^3$ | 30 μg/m$^3$ | - Ultraviolet fluorescence |
| Oxides of Nitrogen method (as $NO_2$) | Annual average | 80 μg/m$^3$ | 60 μg/m$^3$ | 15 μg/m$^3$ | - Jacab & Hochheiser (Na-Arsenite) |
| | 24 hours | 120 μg/m$^3$ | 80 μg/m$^3$ | 30 μg/m$^3$ | - Gas phase chemiluminiscence |
| Suspended Particulate Matter (SPM) | Annual average | 360 μg/m$^3$ | 140 μg/m$^3$ | 70 μg/m$^3$ | - High volume sampling (average |
| | 24 hours | 500 μg/m$^3$ | 200 μg/m$^3$ | 100 μg/m$^3$ | flow rate not less than 1.1 m$^3$/min |
| Respirable Particulate Matter (size less than 0 μm) (RPM) | Annual average | 120 μg/m$^3$ | 60 μg/m$^3$ | 50 μg/m$^3$ | - Respirable particulate matter sampler |
| | 24 hours | 150 μg/m$^3$ | 1000 μg/m$^3$ | 75 μg/m$^3$ | |
| Lead (Pb) | Annual average | 1.0 μg/m$^3$ | 0.75 μg/m$^3$ | 0.50 μg/m$^3$ | - AAS method after sampling using |
| | 24 hours | 1.5 μg/m$^3$ | 1.00 μg/m$^3$ | 0.75 μg/m$^3$ | EPM 2000 or equivalent filter paper |
| Carbon Monoxide (CO) | Annual average | 5.0 μg/m$^3$ | 2.0 μg/m$^3$ | 1.00 μg/m$^3$ | - Non-dispersive infrared spectroscopy |
| | 1 hour | 10.0 μg/m$^3$ | 4.0 μg/m$^3$ | 2.00 μg/m$^3$ | |

**Appendix-4 : Standards for noise levels (Notification of MoEF, GOI dated 26.12.1989)**

| Category of Area | Noise Level in Leq dB (A) | |
|---|---|---|
| | Day Time | Night Time |
| Industrial Area | 75 | 70 |
| Commercial Area | 65 | 55 |
| Residential Area | 55 | 45 |
| Silence zone | 50 | 40 |

**Note :**

1. Day time is reckoned in between 6 am and 10 pm.
2. Night time is reckoned in between 10 pm and 6 am.
3. Silence zone is defined as areas up to 100 meters around such premises as hospitals, educational institutions and courts. The silence zones are to be declared by the Competent Authority. Use of vehicular horns, loudspeakers and bursting of crackers shall be banned in these zones.
4. Mixed categories of areas should be declared as one of the four above mentioned categories by the Competent Authority and the corresponding standards shall apply.

**Appendix-5 : Standards for discharge of Industrial and Sewage Effluents**

| Sr. No. | Parameters | Industrial Effluent | | | |
|---|---|---|---|---|---|
| | | Into Inland Surface Water | On land for Irrigation | Into Marine Coastal Area | Into Public Sewers |
| 1. | Colour/Odour | -- | -- | -- | -- |
| 2. | Suspended Solids mg/L | 100 | 200 | 100 (For process waste | 600 |
| 3. | Particle Size suspended solids | Shall pass 850 micron IS sieve | -- | Floatable Solids Max 3 mm Settleable Solids Max 850 microns | -- |
| 4. | Dissolved Solids (Inorganic), mg/L, Max | 2100 | 2100 | -- | 2100 |
| 5. | pH Value | 5.5-9 | 5.5-9 | 5.5-9 | 5.5-9.0 |
| 6. | Temperature °C | Shall not 40 in any section of the stream within 15 mts. downstream from the effluent outlet | -- | 45 at the point of discharge | |
| 7. | Oil & Grease, mg/L, Max | 10 | 10 | 20 | 20 |
| 8. | Total Residual Chlorine, mg/L, Max | 1 | -- | 1 | -- |
| 9. | Ammonical Nitrogen (as N) mg/L, Max | 50 | -- | 50 | 50 |
| 10. | Total Kjeldahl Nitrogen (as N), mg/L, Max | 100 | -- | 100 | -- |
| 11. | Free Ammonia (as $NH_3$) mg/L, Max | 5 | -- | 5 | -- |
| 12. | Biochemical Oxygen Demand (5 Days at 20°C), Max | 30 | 100 | 100 | 350 |

| | | | | | |
|---|---|---|---|---|---|
| 13. | Chemical Oxygen Demand mg/l, Max | 250 | -- | 250 | |
| 14. | Arsenic (as As), mg/l Max. | 0.2 | 0.2 | 0.2 | 0.2 |
| 15. | Mercury (as Hg), mg/l Max. | 0.01 | -- | 0.01 | 0.01 |
| 16. | Lead (as Pb), mg/l Max. | 0.1 | -- | 1.0 | 1.0 |
| 17. | Cadmium (as Cd), mg/l Max. | 2 | -- | 2 | 1 |
| 18. | Hexavalent Chromium mg/l, Max. | 0.1 | -- | 1 | 2 |
| 19. | Total chromium (as Cr) mg/l, Max | 2 | -- | 2 | 2 |
| 20. | Copper (as Cu), mg/l Max. | 3 | -- | 3 | 3 |
| 21. | Zinc (as Zn), mg/l Max. | 5 | -- | 15 | 15 |
| 22. | Selenium (as Se), mg/l Max. | 0.05 | -- | 0.05 | 0.05 |
| 23. | Nickel (as Ni), mg/l Max. | 3 | -- | 5 | 3 |
| 24. | Boron (as B), mg/l Max | 2 | 2 | -- | 2 |
| 25. | Percent Sodium, Max | -- | 60 | 60 | -- |
| 26. | Residual Sodium Carbonate, mg/l, Max | -- | 50 | -- | -- |
| 27. | Cyanide (as CN), mg/l, Max | 0.2 | 0.2 | 0.2 | 0.2 |
| 28. | Chloride (as Cl), mg/l, Max | 1000 | 600 | -- | 1000 |
| 29. | Fluoride (as F), mg/l, Max | 2 | -- | 15 | 15 |
| 30. | Dissolved Phosphate (as P), mg/l, Max | 5 | -- | -- | -- |
| 31. | Sulphate (as $SO_4$), mg/l, Maxs | 1000 | 1000 | -- | 1000 |

| | | | | | |
|---|---|---|---|---|---|
| 32. | Sulphide (as S), mg/l, Max | 2 | -- | 5 | -- |
| 33. | Phenolic Compounds (as $C_6H_5OH$), mg/l, Max | 1 | -- | 5 | 6 |
| 34. | Radioactive materials | | | | |
| | (a) Alpha emitters μc/ml, Max | $10^{-7}$ | $10^{-8}$ | $10^{-7}$ | $10^{-7}$ |
| | (b) Beta emitters μc/ml, Max | $10^{-6}$ | $10^{-7}$ | $10^{-6}$ | $10^{-6}$ |
| 35. | Manganese (as Mn), mg/l | 2 | 2 | -- | 2 |
| 36. | Iron (as Fe), mg/l | 3 | 3 | -- | 3 |
| 37. | Vanadium (as V), mg/l | 0.2 | -- | 0.2 | 0.2 |
| 38. | Nitrate Nitrogen, mg/l | 18 | 20 | -- | 0.2 |

## Appendix-6 : Standards of Maharashtra Pollution Control Board for Treated Wastewater

| Sr.No. | Parameter | Value |
|---|---|---|
| 1. | pH | 5.5 to 9.0 |
| 2. | Suspended Solids | Not to exceed 100 mg/l |
| 3. | BOD 3 days at 27°C | 30 mg/l |
| 4. | COD | 250 mg/l |
| 5. | Oil & Grease | 10 mg/l |
| 6. | Total Chromium | 2 mg/l |
| 7. | Sulphide | 2 mg/l |
| 8. | Phenolic Compounds | 5 mg/l |

# BIBLIOGRAPHY


1. "Environmental Science and Engineering" by J. Glynn Henry and Gary W. Heinke, Prentice Hall, N.J., 1989.
2. "Environmental Science" by Charles E. Kupchella and Margaret C. Hyland, Allyn and Bacon, Inc., Boston, 1986.
3. "Environmental Engineering" By G.N. Pandey and G.C. Carney, Tata McGraw-Hill Publishing Co. Ltd., New Delhi, 1989.
4. "Handbook of Noise and Vibration Control", 4th Edition, Trade and Technical Press Ltd., Surrey, England, 1979.
5. "Fundamentals of Environmental Pollution" By Krishnan Kannan, S. Chand & Co Ltd., New Delhi, 1991.
6. "Environmental Chemistry" by A.K. De, Wiley Eastern Limited, New Delhi, 1989.
7. "Power Plant Technology" BY M.M.El. Wakil, McGraw Hill Book Co., New York, 1989.
8. "Nuclear Chemistry - Principles and Applications by G.R. Choppin and J. Rydberg, Pergamon Press, 1980.
9. "Handbook of Solid Waste Management", New York, 1977.
10. "Air Pollution Control Theory", By M. Crawford, McGraw Hill, New York, 1976.
11. "Air Pollution" by H.C. Perkins, McGraw Hill, New York, 1974.
12. "Air Pollution - Physical and Chemical Fundamentals" by J.H. Seinfeld, McGraw Hill, New York, 1975.
13. "Air Pollution" by A.C. Stern, Academic Press, New York, 1976.
14. "Fundamentals of Air Pollution" by J. Williamson, Addison-Wesley, Reading, Mass, 1973.
15. "Pathways of Pollutants in the Atmosphere" by T.M. Sugden (Ed), The Royal Society, London, 1978.
16. "Air Pollution"-its origin and control" By K. Wark and C. Warner, IEP-Dun-Connelly, New York, 1978.
17. "Environmental Chemistry" By S.E. Manahan, Willard Grant Press, Boston, 1983.
18. "Atmospheric Chemistry" BY J. Heicklen. Academic Press, New York, 1976.
19. "Soil Chemistry" By G.H. Bolt and M.G.M. Bruggenwort (Eds), Elsevier, New York, 1976.
20. "Atmospheric Pollution" BY W. Buch, McGraw Hill, New York, 1972.
21. "Standard Methods for the Examination of Water and Waste Water" by M.C. Rand, A.E. Greenberg and M.J. Taras (Eds), American Public Health Association, Washington DC, USA, 1976.
22. "Environmental Chemical Analysis" By I.L. Marr and M.S. Cresser, International Text Book Co., Glasgow, 1983.
23. "Introduction to Environmental Science and Technology", By Gilbert Masler, Wiley, 1978.
24. "Environmental Chemistry" By J.W. Moore and F.A. Moore, Academic Press, 1976.
25. "Environmental Chemistry" by J.OM. Bokris, Plenum Press.

26. "Air Pollution" By M.N. Rao and H.V.N. Rao, Tata McGraw Hill, 1976.
27. Wastewater Engineering : Pretreatment, Disposal and Reuse" By Metcalf, Tata McGraw Hill, 1989.
28. "Industrial Water Pollution Control" By Ecken Felder, Tata McGraw Hill, 1989.
29. "Water and Wastwater Engineering" By Raju, Tata McGraw Hill, 1989.
30. "Microbiology for Sanitary Engineers" By M.C. Kinney and E. Ross, McGraw Hill, 1962.
31. "The Chemistry of our Environment" By R.A. Hame and A. Wiley. Interscience, 1978.
32. "Managing the Heavy Metals on the land" by G.W. Leeper, Marcel Dekker, Inc., New York, 1978.
33. "Air Pollution Control and Design Handbook", Vol. I & II, By P.N. Cheremisinoff and R.A. Young (Eds), Marcel Dekker, Inc., New York, 1977.
34. "The Challenge of Global Warming" By Dean Edwin Abrahamson (Ed), Island Press, Washington, D.C., 1989.
35. "Water and Wastewater Technology" By Mark J. Hammer, John Wiley & Sons., Inc., New York, 1975.
36. "Water Pollution" By V.P. Kadesia, Pragati Prakashan, Meerut, 1985.
37. "Solid Wastes - Engineering Principles and Management Issues" By G. Techbanoglous, H. Theison and R. Eliassen, McGraw Hill Book Co., New York, 1977.
38. "Solid Waste Management" By D.J. Hagerty, J.L. Pavoni and S.E. Heer, Jr. Van Nostrand Reinhold Co., New York, 1973.
39. "Air Pollution Control Equipment" By H. Brauer and Y.B.G. Varma, Springer-Verlage, New York, 1981.
40. "Hand-book of Solid Waste Disposal" By J.L. Pavoni, J.E. Heer, Jr., and D.J. Hagerty, Van Nostrand Reinhold Co., New York, 1975.
41. "Air Pollution", Vols 1 to 8, By A.C. Stern (Ed), Academic Press, Inc., New York, 1986.
42. "Noise Control in Mechanical Services" By R.I. Woods (Ed), Sound Attenuators Ltd & Sound Research Laboratories Ltd., Essex, 1976.
43. "Unit Operations for Treatment of Hazardous Industrial Wastes" By D.J. De Renzo (Ed), Noyes Data Corporation, N.J. U.S.A., 1978.
44. "Air Pollution Modelling and its Application", Vol. 1 to 4, By C. De Wispelaere (Ed), Plenum Press, N.Y. 1981.
45. "The Nalco Water Handbook", Frank N. Kemmer (Ed)., McGraw Hill Book Co., New York, 1988.
46. "Handbook of Solid Waste Management" By David Gordon Wilson, Van Nostrand Reinhold, N.Y., 1977.
47. Chemtech, Vol. 1, By D. Venkateswarlu (Ed), Manual of Chemical Technology, Chemical Engineering Education Development Centre, I.I.T., Madras, 1975.
48. Air Pollution, Part I, II and III, by W. Straubs, Wiley Interscience, New York, 1971, 1972 and 1978.
49. "Industrial Effluents - Origin, Characteristics, Effects, Analysis and Treatment", By N. Manivasakam, Sakthi Publication, Kovaipudur, Coimbator, 1987.
50. "Industrial Water Pollution, Origin, Characteristics and Treatment", By N.L. Nemerow, Addison-Wesley Publishing Co., 1978.

51. "The Treatment of Industrial Wastes", By B. Besselievre, McGraw Hill Book Co., New York, 1969.
52. "Industrial Pollution Control Hand Book" By H.F. Lund, McGraw Hill Book Co., New York, 1971.
53. "Chemical Process Industries" By R.N. Shreve and J. Brink, McGraw Hill International Book Co., 1978.
54. "Chemistry for Environmental Engineering", By C.N. Sawer and P.L. McCarty, McGraw Hill Book Co., New York, 1978.
55. "Waste Water Treatment", By M.N. Rao and A.K. Datta, Oxford & IBH Publishing Co., Calcutta, 1979.
56. "Noise and Vibration", By R.G. White & J.G. Walker (Eds), Ellis Horwood Publishers, 1986.
57. "Handbook of Noise and Vibration Control", Trade & Technical Press Ltd., Morden, England, 1979.
58. "Solid Waste Management" By D.J. Hagerty, J.L. Pavoim and J.E. Heev, Jr., Van Nostrand Reinhold Co., New York, 1973.
59. "Solid Waste Management in Developing Countries" By A.D. Bhide & B.B. Sundaresan, Insdoc Publication, State of Art Report Series No. 2, 1983.
60. "Toxic Metals and their analysis", By, E. Berman, Heyden, Cook Country Hospital, Chicago, Illinois, 1980.
61. "The Pharmacological Basis of Therapeutics", By L.S. Goodman and A. Gilman (Eds), MacMillan Publishing Co, Inc., New York, 1975.
62. "Trade Elements in Human and Animal Nutrition". 4th Edn. By E.J. Underwood, Academic Press Inc., New York, 1977.
63. "Biological Monitoring of Heavy Metal Pollution, Land and Air, Pollution Monitoring Series", by M.H. Marlin & Coughtrey, P.J., Applied Science Publisheres Ltd., London & New York 1982.
64. "Pollution Threat of Heavy Metals in Aquatic Environments" by Geoffrey Mance, Elsevier Applied Science Publishers Ltd., London & New York, 1987.
65. "Radionuclides in the Environment" By K.S.V. Numbi, Vol. 88, No. 1, IANCAS Bulletin, Bombay, Jan. 1992.
66. "Environmental Science" by Bernard J. Nebel and Richard T. Wright, Prentice Hall, New Jersey, 1993.
67. "Environment and Biotechnology" Edited by Harvinder S. Sohol and Ashok K. Srivastava, Ashish Publishing House, New Delhi, 1994.
68. "A Text Book of Bio-Technology" by R.C. Dubey, S. Chand & Co., 1995.
69. "Introduction to Environmental Engineering and Science" by G.M. Masters, Prentice Hall, New Delhi, 1994.
70. "An Introduction to Environmental Chemistry" by P. Brimblecombe and J.E. Andrews, Blackwell Science, 1996.
71. Environment Regulations and Policy in India by Ram Boojh, Page 277-296, "Human Health and Environment" Vol. II (Chief Editor, Dr. A.K. Sinha), APH Publishing Corporation, New Delhi, 1997.
72. "Environmental Management, Law & Administration" By Paras Diwan and Parag Diwan, Vanity Books International, New Delhi, 1998.

73. "Environmental Impact Assessment (Theory and Practice)" Edited by Peter Wathern, Unwin Hyman, London, 1988.

74. "A procedure for Evaluating Environmental Impact" By Leopold, L.B., Glarke, F.E., Hanshaw, B.B., and Balsey, J.R., U.S. Geological Survey Circular 645, Government Printing Office, Washington D.C., 1971.

75. "Electromagnetic Fields – Health Effect" By stephen F. Cleary, Wiley Encyclopedia series in Environmental Science, Energy, Technology and the Environment, Vol. II, Editors – Attilio Bisio and Sharon Boots, John Wiley & Sons, Inc., New York (1995).

76. "Non-ionising Radiations", Encyclopedia of Environmental Science and Engineering, 3rd Edition (Revised and updated), Edited by James R. Pfafflin and Edward N. Ziegler, Vol. 2, Gordon & Breach Science Publishers, USA (1992).

77. "Electric and Magnetic Fields in the Environment" By Marcus Zahn, Wiley Encyclopedia series in Environmental Science, Energy, Technology and the Environment, Vol. II, Editors – Attilio Bisio and Sharon Boots, John Wiley & Sons, Inc., New York (1995).

78. "Process Analytical Chemistry", Edited by F. McLennan and B.R. Kowalski, Blackie Academic and Professional, An Imprint of Chapman Hall, London (1995).

79. "Environment – Problems and Solutions" By D.K. Asthana and Meera Asthana, S. Chand & Co. Ltd., New Delhi (1998).

80. "Environmental Science" By Andrew R.W. Jackson and Julie M. Jackson, Longman, England.

81. "Environmental Chemistry" by Nigel J. Bunce, Wuerz Publishing Ltd., Canada, 1991.

82. "Introduction to Environmental Engineering and Science" by Gilbert M. Masters, Prentice Hall of India Pvt. Ltd., New Delhi, 1994.

83. "Environmental Technology and Economics – Sustainable Development in Industry" by Claire M. Soares, P.E., Butterworth/Heinemann, 1999.

84. "Chemistry of the Environment" by Thomas G. Spiro and William M. Stigliani, Prentice Hall of India Pvt. Ltd., 2002.

85. " Handbook of Pollution Control and Waste Minimization" edited by Abbas Ghassemi, Marcel Dekker, Inc., 2002.

86. "Green Engineering" by David T. Allenand David R. Shonnard, Prentice Hall PTR, NJ, 2002.

87. "Bioprocess Engineering – Basic Concepts" by Michael L. Shuler and Fikret Kargi, Prentice Hall of India Pvt. Ltd., 2002.

88. Green Engineering – Environmentally Conscious Design of Chemical Processes by David T. Allen and David R. Shonnart, Prentice Hall, 2003.

89. Environmental Engineering – A Design Approach by A.P. Sincero and G.A. Sincero, Prentice Hall, 1999.

# INDEX

## D



## E

## P



## R

## T



## U



## V

W



Z